T0186047

CAMBRIDGE LIBRARY COLLECTION

Books of enduring scholarly value

Mathematical Sciences

From its pre-historic roots in simple counting to the algorithms powering modern desktop computers, from the genius of Archimedes to the genius of Einstein, advances in mathematical understanding and numerical techniques have been directly responsible for creating the modern world as we know it. This series will provide a library of the most influential publications and writers on mathematics in its broadest sense. As such, it will show not only the deep roots from which modern science and technology have grown, but also the astonishing breadth of application of mathematical techniques in the humanities and social sciences, and in everyday life.

Scientific Papers

Lord Rayleigh (1842-1919) won the Nobel Prize for Physics in 1904. His early research was in optics and acoustics but his first published paper, from 1869, was an explanation of Maxwell's electromagnetic theory. In 1871, he related the degree of light scattering to wavelength (part of the explanation for why the sky is blue), and in 1872 he wrote his classic Theory of Sound (not included here). He became a Fellow of the Royal Society and inherited his father's peerage in 1873. Rayleigh nevertheless continued ground-breaking research, including the first description of Moiré interference (1874). In 1881, while president of the London Mathematical Society (1878–1880) and successor to Maxwell as Cavendish Professor of Experimental Physics at Cambridge (1879–1884), Rayleigh published a paper on diffraction gratings which led to improvements in the spectroscope and future developments in high-resolution spectroscopy. This volume contains papers published between 1869 and 1881.

Scientific Papers

VOLUME 1: 1869-1881

BARON JOHN WILLIAM STRUTT RAYLEIGH

CAMBRIDGE
UNIVERSITY PRESS

CAMBRIDGE UNIVERSITY PRESS

Cambridge New York Melbourne Madrid Cape Town Singapore São Paolo Delhi

Published in the United States of America by Cambridge University Press, New York

www.cambridge.org
Information on this title: www.cambridge.org/9781108005425

© in this compilation Cambridge University Press 2009

This edition first published 1899
This digitally printed version 2009

ISBN 978-1-108-00542-5

SCIENTIFIC PAPERS

SCIENTIFIC PAPERS

BY

JOHN WILLIAM STRUTT,

BARON RAYLEIGH,

D.Sc., F.R.S.,

HONORARY FELLOW OF TRINITY COLLEGE, CAMBRIDGE,
PROFESSOR OF NATURAL PHILOSOPHY IN THE ROYAL INSTITUTION.

VOL. I.

1869—1881.

CAMBRIDGE:
AT THE UNIVERSITY PRESS.
1899

PREFACE.

THE papers of the present Collection are reprinted very nearly as they originally appeared, and with a few partial exceptions in order of date. Obvious misprints have been corrected, in several cases with the aid of the original manuscript. Other alterations of the slightest significance are indicated by the use of square brackets [ ], while additional matter is introduced with the proper date in the form of footnotes or at the end of a memoir. In a few cases, where it has not been thought worth while to reproduce a paper in full, a brief statement of the principal results is given.

Some short papers of a rather slender character have been included. These may serve to mitigate the general severity. In consulting similar collections I have usually felt even more grateful for the reproduction of short and often rather inaccessible notes than for the larger and better known memoirs.

TERLING PLACE, WITHAM,
October 1899.

The works of the Lord are great,
Sought out of all them that have pleasure therein.

CONTENTS.

ART.   PAGE

1. On some Electromagnetic Phenomena considered in connexion with the Dynamical Theory 1
   [*Phil. Mag.* XXXVIII. pp. 1—15, 1869.]

2. On an Electromagnetic Experiment 14
   [*Phil. Mag.* XXXIX. pp. 428—435, 1870.]

3. On the values of the Integral $\int_0^1 Q_n Q_{n'} d\mu$, Q_n, $Q_{n'}$ being Laplace's Coefficients of the orders n, n', with an application to the Theory of Radiation 21
   [*Phil. Trans.* CLX. pp. 579—590; read June 1870.]

4. Remarks on a paper by Dr Sondhauss 26
   Postscript 32
   [*Phil. Mag.* XL. pp. 211—217, 1870.]

5. On the Theory of Resonance 33
   Introduction 33
   Part I. 37
      Several Openings 39
      Double Resonance 41
      Open Organ-pipes 45
      Long Tube in connexion with a Reservoir . . . 48
      Lateral Openings 50
   Part II. 51
      Long Tubes 51
      Simple Apertures 52
      Cylindrical Necks 53
      Potential on itself of a uniform Circular Disk . . . 55

ART.   PAGE

5. On the Theory of Resonance—*continued*.
    Nearly Cylindrical Tubes of Revolution 62
    Upper Limit 62
    Application to straight Tube of Revolution whose end lies on
    two infinite Planes 64
    Tubes nearly Straight and Cylindrical but not necessarily
    of Revolution 64
    Tubes not nearly Straight 66
        Part III. 67
        Experimental 67
        [*Phil. Trans.* CLXI. pp. 77—118 ; Read Nov. 1870.]

6. Note on the explanation of Coronas, as given in Verdet's *Leçons
    d'Optique Physique*, and other works 76
        [*London Math. Soc. Proc.* III. pp. 267—269, 1871.]

7. Some Experiments on Colour 79
    Yellow 85
        [*Nature,* III. pp. 234—237, 264, 265, 1871.]

8. On the Light from the Sky, its Polarization and Colour . 87
    Appendix 96
        [*Phil. Mag.* XLI. pp. 107—120, 274—279, 1871.]

9. On the Scattering of Light by small Particles . . . 104
        [*Phil. Mag.* XLI. pp. 447—454, 1871.]

10. On Double Refraction 111
        [*Phil. Mag.* XLI. pp. 519—528, 1871.]

11. On the Reflection of Light from Transparent Matter . . 120
        [*Phil. Mag.* XLII. pp. 81—97, 1871.]

12. On a Correction sometimes required in Curves professing to
    represent the connexion between two Physical Magnitudes . 135
        [*Phil. Mag.* XLII. pp. 441—444, 1871.]

13. On the Vibrations of a Gas contained within a Rigid Spherical
    Envelope 138
        [*London Math. Soc. Proc.* IV. pp. 93—103, 1872.]

14. Investigation of the Disturbance produced by a Spherical Obstacle
    on the Waves of Sound 139
        [*London Math. Soc. Proc.* IV. pp. 253—283, 1872.]

ART.   PAGE

15. Notes on Bessel's Functions 140
        [*Phil. Mag.* XLIV. pp. 328—344, 1872.]

16. On the Reflection and Refraction of Light by Intensely Opaque
    Matter 141
        [*Phil. Mag.* XLIII. pp. 321—338, 1872.]

17. Preliminary note on the Reproduction of Diffraction-Gratings by
    means of Photography 157
        [*Proceedings of the Royal Society,* XX. pp. 414—417, 1872.]

18. On the Application of Photography to copy Diffraction-Gratings 160
        [*British Association Report,* 1872, p. 39.]

19. On the Diffraction of Object-Glasses 163
        [*Astron. Soc. Month. Not.* XXXIII. pp. 59—63, 1872.]

20. An Experiment to illustrate the Induction on itself of an Electric
    Current 167
        [*Nature,* VI. p. 64, 1872.]

21. Some General Theorems relating to Vibrations . . . 170
        Section I.
            The natural periods of a conservative system, vibrating
            freely about a configuration of stable equilibrium, fulfil
            the stationary condition 170
        Section II.
            The Dissipation Function 176
        Section III. 179
        [*London Math. Soc. Proc.* IV. pp. 357—368, 1873.]

22. On the Nodal Lines of a Square Plate 182
        [*Phil. Mag.* XLVI. pp. 166—171, 246, 247, 1873.]

23. Note on a Natural Limit to the Sharpness of Spectral Lines 183
        [*Nature,* VIII. pp. 474, 475, 1873.]

24. On the Vibrations of Approximately Simple Systems . . 185
        [*Phil. Mag.* XLVI. pp. 357—361, 1873; XLVIII. pp. 258—262, 1874.]

25. On the Fundamental Modes of a Vibrating System . . 186
        [*Phil. Mag.* XLVI. pp. 434—439, 1873.]

ART.   PAGE

26. Vibrations of Membranes 187
[*London Math. Soc. Proc.* v. pp. 9, 10, 1873.]

27. Harmonic Echoes 188
[*Nature*, VIII. pp. 319, 320, 1873.]

28. Note on the Numerical Calculation of the Roots of Fluctuating
Functions 190
[*London Math. Soc. Proc.* v. pp. 119—124, 1874.]

29. A History of the Mathematical Theories of Attraction and the
Figure of the Earth from the time of Newton to that of
Laplace. By I. Todhunter, M.A., F.R.S. Two Volumes.
(London, Macmillan & Co., 1873.) 196
[*The Academy*, v. pp. 176, 177, 1874.]

30. On the Manufacture and Theory of Diffraction-Gratings . . 199
[*Phil. Mag.* XLVII. pp. 81—93, 193—205, 1874.]

31. Insects and the Colours of Flowers 222
[*Nature*, XI. p. 6, 1874.]

32. A Statical Theorem 223
[*Phil. Mag.* XLVIII. pp. 452—456, 1874 ; XLIX. pp. 183—185, 1875.]

33. Mr Hamilton's String Organ 230
[*Nature*, XI. pp. 308, 309, 1875.]

34. General Theorems relating to Equilibrium and Initial and Steady
Motions 232
[*Phil. Mag.* XLIX. pp. 218—224, 1875.]

35. On the Dissipation of Energy 238
[*Roy. Instit. Proc.* VII. pp. 386—389, 1875 ; *Nature*, XI. pp. 454, 455, 1875.]

36. On the Work that may be gained during the Mixing of Gases 242
[*Phil. Mag.* XLIX. pp. 311—319, 1875.]

37. Vibrations of a Liquid in a Cylindrical Vessel . . . 250
[*Nature*, XII. p. 251, 1875.]

ART.   PAGE

38. On Waves 251
    The Solitary Wave 256
    Periodic Waves in Deep Water 261
    Oscillations in Cylindrical Vessels 265

[*Phil. Mag.* I. pp. 257—279, 1876.]

39. On the Approximate Solution of Certain Problems relating to
    the Potential 272

[*London Math. Soc. Proc.* VII. pp. 70—75, 1876.]

40. Our Perception of the Direction of a Source of Sound . . 277

[*Nature,* XIV. pp. 32, 33, 1876.]

41. Questions from Mathematical Tripos Examination for 1876 . 280
    January 6. 9—12 280
    January 6. 1½—4 281
    January 17. 9—12 282
    January 19. 1½—4 283
    January 20. 1½—4 284
    January 21. 9—12 285
    January 21. 1½—4 286

[*Cambridge University Calendar,* 1876.]

42. On the Resistance of Fluids 287

[*Phil. Mag.* II. pp. 430—441, 1876.]

43. Notes on Hydrodynamics 297
    The Contracted Vein 297
    Meeting Streams 302

[*Phil. Mag.* II. pp. 441—447, 1876.]

44. On the Application of the Principle of Reciprocity to Acoustics 305

[*Proceedings of the Royal Society,* XXV. pp. 118—122, 1876.]

45. On a Permanent Deflection of the Galvanometer-Needle under
    the influence of a rapid series of equal and opposite Induced
    Currents 310

[*Phil. Mag.* III. pp. 43—46, 1877.]

ART.		PAGE

46. Acoustical Observations. I. 314

    Perception of the Direction of a Source of Sound . . 314

    The Head as an Obstacle to Sound 315

    Reflection of Sound 316

    Audibility of Consonants 317

    Interference of Sounds from two unisonant Tuning-forks . 317

    Symmetrical Bell. 317

    Octave from Tuning-forks 318

    Influence of a Flange on the Correction for the Open End

      of a Pipe 319

    The Pitch of Organ-pipes 320

      [*Phil. Mag.* III. pp. 456—464, 1877.]

47. On Progressive Waves 322

      [*London Math. Soc. Proc.* IX. pp. 21—26, 1877.]

48. On the Amplitude of Sound-Waves 328

      [*Proceedings of the Royal Society*, XXVI. pp. 248, 249, 1877.]

49. Absolute Pitch 331

      [*Nature*, XVII. pp. 12—14, 1877.]

50. On Mr Venn's Explanation of a Gambling Paradox . 336

      [*Mind*, II. pp. 409, 410, 1877.]

51. On the Relation between the Functions of Laplace and Bessel 338

      [*London Math. Soc. Proc.* IX. pp. 61—64, 1878.]

52. Note on Acoustic Repulsion 342

      [*Phil. Mag.* VI. pp. 270, 271, 1878.]

53. On the Irregular Flight of a Tennis-Ball . . . 344

      [*Messenger of Mathematics*, VII. pp. 14—16, 1877.]

54. A simple Proof of a Theorem relating to the Potential . 347

      [*Messenger of Mathematics*, VII. p. 69, 1878.]

55. The Explanation of certain Acoustical Phenomena . . 348

      [*Roy. Inst. Proc.* VIII. pp. 536—542, 1878 ; *Nature*, XVIII. pp. 319—321, 1878.]

56. Uniformity of Rotation 355

      [*Nature*, XVIII. p. 111, 1878.]

ART. PAGE

57. On the Determination of Absolute Pitch by the Common Harmonium 357
[*Nature*, XIX. pp. 275, 276, 1879.]

58. On the Instability of Jets 361
[*London Math. Soc. Proc.* X. pp. 4—13, 1879.]

59. The Influence of Electricity on Colliding Water Drops . . 372
[*Proceedings of the Royal Society*, XXVIII. pp. 406—409, 1879.]

60. On the Capillary Phenomena of Jets 377
Appendix I.. 396
Appendix II. 400
[*Proceedings of the Royal Society*, XXIX. pp. 71—97, 1879.]

61. Acoustical Observations. II. 402
Pure Tones from Sounding Flames 402
Points of Silence near a Wall from which a Pure Tone is reflected 403
Sensitive Flames 406
Aerial Vibrations of very Low Pitch maintained by Flames 407
Rijke's Notes on a large scale 408
Mutual Influence of Organ-Pipes nearly in Unison . . 409
Kettledrums 411
The Æolian Harp 413
[*Phil. Mag.* VII. pp. 149—162, 1879.]

62. Investigations in Optics, with special reference to the Spectroscope 415
§ 1. Resolving, or Separating, Power of Optical Instruments 415
§ 2. Rectangular Sections 418
§ 3. Optical Power of Spectroscopes 423
§ 4. Influence of Aberration 428
§ 5. On the Accuracy required in Optical Surfaces . . 436
§ 6. The Aberration of Oblique Pencils 440
§ 7. Aberration of Lenses and Prisms 444
§ 8. The Design of Spectroscopes 453
[*Phil. Mag.* VIII. pp. 261—274, 403—411, 477—486, 1879; IX. pp. 40—55, 1880.]

63. On Reflection of Vibrations at the Confines of two Media between which the Transition is Gradual 460
[*London Math. Soc. Proc.* XI. pp. 51—56, 1880.]

ART.   PAGE

64. On the Minimum Aberration of a Single Lens for Parallel Rays   466
    [*Cambridge Phil. Soc. Proc.* III. pp. 373—375, 1880.]

65. Acoustical Observations. III.   468
    Intermittent Sounds   468
    A New Form of Siren   471
    The Acoustical Shadow of a Circular Disk . . .   472
    [*Phil. Mag.* IX. pp. 278—283, 1880.]

66. On the Stability, or Instability, of certain Fluid Motions .   474
    [*London Math. Soc. Proc.* XI. pp. 57—70, 1880.]

67. On the Resolving-Power of Telescopes   488
    [*Phil. Mag.* X. pp. 116—119, 1880.]

68. On the Resultant of a large number of Vibrations of the same
    Pitch and of arbitrary Phase   491
    [*Phil. Mag.* X. pp. 73—78, 1880.]

69. Note on the Theory of the Induction Balance . . .   497
    [*British Association Report*, Swansea, pp. 472, 473, 1880.]

70. On a New Arrangement for Sensitive Flames   500
    [*Cambridge Phil. Soc. Proc.* IV. pp. 17, 18, 1880.]

71. The Photophone   501
    [*Nature*, XXIII. pp. 274, 275, 1881.]

72. On Copying Diffraction-Gratings, and on some Phenomena con-
    nected therewith   504
    [*Phil. Mag.* XI. pp. 196—205, 1881.]

73. On Images formed without Reflection or Refraction . .   513
    [*Phil. Mag.* XI. pp. 214—218, 1881.]

74. On the Electromagnetic Theory of Light   518
    [*Phil. Mag.* XII. pp. 81—101, 1881.]

75. On the Velocity of Light   537
    [*Nature*, XXIV. pp. 382, 383 ; XXV. p. 52, 1881.]

ART. PAGE

76. On a Question in the Theory of Lighting 541

[*British Association Report*, 1881, p. 526.]

77. Experiments on Colour 542

[*Nature*, xxv. pp. 64—66, 1881.]

78. On the Infinitesimal Bending of Surfaces of Revolution . . 551

[*London Math. Soc. Proc.* xiii. pp. 4—16, 1881.]

1.

ON SOME ELECTROMAGNETIC PHENOMENA CONSIDERED IN CONNEXION WITH THE DYNAMICAL THEORY.

[*Phil. Mag.* XXXVIII. pp. 1—15, 1869.]

It is now some time since general equations applicable to the conditions of most electrical problems have been given, and attempts, more or less complete, have been made to establish an analogy between electrical phenomena and those of ordinary mechanics. In particular, Maxwell has given a general dynamical theory of the electromagnetic field*, according to which he shows the mutual interdependence of the various branches of the science, and lays down equations sufficient for the theoretical solution of any electrical problem. He has also in scattered papers illustrated the solution of special problems by reference to those which correspond with them (at least in their mathematical conditions) in ordinary mechanics. There can be no doubt, I think, of the value of such illustrations, both as helping the mind to a more vivid conception of what takes place, and to a rough quantitative result which is often of more value from a physical point of view, than the most elaborate mathematical analysis. It is because the dynamical theory seems to be far less generally understood than its importance requires that I have thought that some more examples of electrical problems illustrated by a comparison with their mechanical analogues might not be superfluous.

As a simple case, let us consider an experiment first made by De la Rive, in which a battery (such as a single Daniell cell) whose electromotive force is insufficient to decompose water, becomes competent to do so by the intervention of a coil or electromagnet. Thus, let the primary wire of a Ruhmkorff coil be connected in the usual manner with the battery, and the electrodes of the voltameter (which may consist of a test-tube containing dilute sulphuric acid into which dip platinum wires) with the points where

* *Philosophical Transactions* for 1865.

in the ordinary use of the instrument the contact is made and broken. There will thus be always a complete conducting circuit through the voltameter; but when the contact is made the voltameter will be *shunted*, and the poles of the battery joined by metal. Now when the shunt is open the battery is unable to send a steady current through the voltameter, because, as has been shown by Thomson, the mechanical value of the chemical action in the battery corresponding to the passage of any quantity of electricity is less than that required for the decomposition of the water in the voltameter. When, however, the shunt is closed, a current establishes itself gradually in the coil, where there is no permanent opposing electro-motive force, and after the lapse of a fraction of a second reaches its full value as given by Ohm's law. If the contact be now broken, there is a momentary current through the voltameter, which causes bubbles of gas to appear on the electrodes, and which is often (but not, I think, well) called the extra current. Allowing the rheotome to act freely we get a steady evolution of gas.

To this electrical apparatus Montgolfier's hydraulic ram is closely analogous. The latter, it will be remembered, is a machine in which the power of a considerable quantity of water falling a small height is used to raise a portion of the water to a height twenty or thirty times as great. The body of water from the reservoir flows down a closed channel to the place of discharge, which can be suddenly closed with a valve. When this takes place, the moving mass by its momentum is able for a time to overcome a pressure many times greater than that to which it owes its own motion, and so to force a portion of itself to a considerable height through a suitably placed pipe. Just as the electromotive force of the battery is unable directly to overcome the opposing polarization in the voltameter, so of course the small pressure due to the fall cannot lift a valve pressed down by a greater. But when an independent passage is opened, the water (or electricity) begins to flow with a motion which continues to accelerate until the moving force is balanced by friction (resistance), and then remains steady. At the moment the discharge-valve is closed (or, in the electrical problem, the shunt-contact is broken), the water, by its inertia, tends to continue moving, and thus the pressure instantly rises to the value required to overcome the weight of the great column of water. The second valve is accordingly opened, and a portion of the water is forced up. Now the electrical current, in virtue of self-induction, can no more be suddenly stopped than the current of water; and so in the above experiment the polarization of the voltameter is instantly overcome, and a quantity of electricity passes.

If no second means of escape were provided for the water in the hydraulic ram, the pipe would in all probability be unable to withstand the shock, and in any case could only do so by yielding within the limits of its elasticity, so as gradually, though of course very quickly, to stop the flow of water. The bursting of the pipe may properly be compared to the passage of a spark at the place where a conductor carrying an electric current is opened. Just as the natural elasticity of the pipe or the compressibility of the air in a purposely connected air-vessel greatly diminishes the strain, so the electrical spark may be stopped by connecting the breaking-points with the plates of a condenser, as was done by Fizeau in the induction-coil. Contrary to what might at first sight have been expected, the fall of the primary current is thus rendered more sudden, and the power of the instrument for many purposes increased. Of course the spark is equally prevented when the breaking-points are connected by a short conducting circuit, as in our experiment by the voltameter. In fact the energy of the actual motion which exists the moment before contact is broken is in the one case transformed into that of the sound and heat of the spark, and in the other has its equivalent partly in the potential energy of the decomposed water, partly in the heat generated by the passage of the momentary current in the voltameter branch.

The experiment will be varied in an instructive manner if we replace the voltameter by a coil (with or without soft iron), according to the resistance and self-induction of the latter. In order to know the result, we must examine closely what takes place at the moment when contact is broken. The original current, on account of its self-induction or inertia, tends to continue. At the same time the inertia in the branch circuit tends to prevent the sudden rise of a current there. A force is thus produced at the breaking-points exactly analogous to the pressure between two bodies, which we will suppose inelastic, one of which impinges on the other at rest. The pressure or electrical tension continues to vary until the velocities or currents become equal. All this time the motion of each body or current is opposed by a force of the nature of friction proportional to the velocity or current. Whether this resistance will affect the common value of the currents (or velocities) at the moment they become equal, will depend on its magnitude as compared with the other data of the problem.

There is for every conducting circuit a certain time-constant which determines the rapidity of the rise or fall of currents, and which is proportional to the self-induction and conductivity of the circuit. Thus, to use Maxwell's notation, if L and R be respectively the coefficient of self-induction and the resistance, the time-constant is $L/R = \tau$. If the current c exist at any moment in the circuit and fall undisturbed by external electromotive force, the value at any time t afterwards is given by $x = c \cdot e^{-t/\tau}$. Any action which takes place in a time much smaller than τ will be sensibly unaffected by resistance.

We see, then, that we may neglect the effects of resistance during the time of equalization of the currents, provided that the operation is completed in a time much smaller than the time-constants of either circuit. And this I shall suppose to be the case. The value of the common current or velocity at the moment the impact is over will of course be given by the condition that the momentum, electromagnetic or ordinary, is unchanged. If L and N be the coefficients of self-induction for the main and branch circuits respectively, x and X the original and required currents, the analytical expression of the above condition is

$$(L + N)\, X = Lx, \qquad \text{or} \qquad X = \frac{L}{L + N}\, x.$$

It is here supposed that there is no sensible mutual induction between the two circuits.

The spark is the result of the excess of the one current over the other, and lasts until its cause is removed. Its mechanical value is the difference between that of the original current in the main circuit and that of the initial current in the combined circuit, and is expressed by

$$\tfrac{1}{2} L x^2 - \tfrac{1}{2} (L + N)\, X^2;$$

or if the value of X be substituted,

$$\tfrac{1}{2} \frac{L N x^2}{L + N}. *$$

Exactly the same expression holds good for the heat produced during the collision of the inelastic bodies, which is necessarily equal to the loss of ordinary actual energy, at least if the permanent change of their molecular state may be neglected. From the value X the current gradually increases or diminishes to that determined according to Ohm's law, by the resistance of the combined circuit. It may be seen from the expression just found that the resistance of the branch may be varied without affecting the spark, provided always that it is not so great in relation to the self-induction as to make the time-constant comparable in magnitude with the duration of the spark. The spark depends only on the comparative self-induction of the branch circuit, being small when this is small, and when this is great approximating to its full value $\tfrac{1}{2} L x^2$.

These results are easily illustrated experimentally. I have two coils of thick wire belonging to an electromagnet, which for convenience I will call A and B. Each consists of two wires of equal length, which are coiled together. These may be called $A_1\, A_2$, $B_1\, B_2$. When $A_1\, A_2$ are joined consecutively, so that the direction of the current is the same in the two wires, we have a circuit whose self-induction is four times that of either

* [1898—An erratum is here corrected.]

wire taken singly. But if, on the contrary, the current flows opposite ways in the two wires, the self-induction of the circuit becomes quite insensible.

The main circuit may be composed of the wire A_1 (A_2 remaining open) into which the current from a single Daniell cell is led, and which can be opened or closed at a mercury cup. One end of the branch circuit dips into the mercury while the other communicates with the wire whose entrance or withdrawal from the cup closes or opens the main circuit. In this way the coils of the branch may be said to be *thrown in* at the break.

If the branch is open, we obtain at break the full spark, whose value is $\frac{1}{2}Lx^2$. If the wire B_1 be thrown in, the spark is still considerable, having approximately the value $\frac{1}{4}Lx^2$, for $N = L$. And if B_1 B_2 are thrown in, so that the currents are parallel, the spark is still greater and is measured by $\frac{1}{2}Lx^2 \times \frac{4}{9}$. But if the currents are opposed, the spark disappears, because now $N = 0$; so that the addition of the wire B_2, whereby the resistance of the branch is doubled, diminishes the spark. It is true that to this last case our calculation is not properly applicable, inasmuch as the time-constant of the branch is so exceedingly small. But it is not difficult to see that in such a case (where the self-induction of the branch may be neglected) the tension at the breaking-points, or more accurately the difference of potential between them, cannot exceed that of the battery more than in the proportion of the resistances of the branch and main circuits, so that it could not here give rise to any sensible spark. Soft iron wires may be introduced into the coils in order to exalt the effects; but solid iron cores would allow induced currents to circulate which might interfere with the result.

In this form of the experiment there was no sensible mutual induction between the coils A and B. Should there be such, the result may be considerably modified. For instance, let the wire A_2 be thrown at the break into the circuit of A_1 and the battery. This may happen in two ways. If the connexions are so made that the currents are parallel in A_1 A_2, there will be no sensible spark; but if the directions of the currents are opposed, the spark appears equal to the full spark $\frac{1}{2}Lx^2$.

And this is in accordance with theory. The current X is given by the same condition as before, which leads to the equation

$$Lx + Mx = (L + 2M + N)\, X,$$

M being the coefficient of mutual induction between the two circuits. The spark is therefore

$$\tfrac{1}{2}Lx^2 - \tfrac{1}{2}(L + 2M + N)\, X^2 = \frac{x^2}{2}\frac{L - M}{2}, \quad \text{as } N = L.$$

Now in the first-mentioned connexion $M = L$ very nearly, and in the second $M = -L$; so that the observed sparks are just what theory requires.

With regard to those electrical phenomena which depend on the mutual induction of two circuits, it may be remarked that it is not easy to find exact analogues in ordinary mechanics which are sufficiently familiar to be of much use as aids to conception. A rough idea of the reaction of neighbouring currents may be had from the consideration of the motion of a heavy bar to whose ends forces may be applied. If, when the bar is at rest one end is suddenly pushed forwards in a transverse direction, the inertia of the material gives the centre of gravity in some degree the properties of a fulcrum, and so the other end begins to move backwards. This corresponds to the inverse wave induced by the rise of a current in a neighbouring wire. If the motion be supposed infinitely small, so that the body never turns through a sensible angle, the kinetic energy is proportional to

$$\tfrac{1}{2}(a^2 + k^2)\,x^2 + \tfrac{1}{2}(b^2 + k^2)\,y^2 + (ab - k^2)\,xy,$$

where a and b are the distances of the driving-points (whose velocities are x and y) from the centre of gravity, k^2 the radius of gyration about the latter point. This corresponds to the expression for the energy of the electromagnetic field due to two currents,

$$\tfrac{1}{2}Lx^2 + Mxy + \tfrac{1}{2}Ny^2;$$

and if we imagine the motion of the driving-points to be resisted by a frictional force proportional to the velocity, we get a very tolerable representation of the electrical conditions.

Or we may take an illustration, which is in many respects to be preferred, from the disturbance of a perfect fluid, by the motion of solid bodies in its interior. Thus if in an infinite fluid two spheres move parallel to each other and perpendicularly to the line joining them, and with such small velocities that their relative position does not sensibly change, the kinetic energy may as usual be expressed by

$$\tfrac{1}{2}Lx^2 + Mxy + \tfrac{1}{2}Ny^2,$$

x, y denoting the velocities of the two spheres, and L, M, N being approximately constants*. When the spheres move in the same direction, the reaction of the fluid tends to press them together; but if the motions are opposed, the force changes to a repulsion. We see here the analogues of the phenomena of attraction and repulsion discovered by Ampère. If when all is at rest a given velocity is impulsively impressed on one sphere, the other immediately starts backwards, and, as Thomson† has shown, with such velocity that the energy of the whole motion is the least possible under the given condition.

This theorem is general, and leads directly to the solution of a large class of electrical problems connected with induction; for whenever a current is

---

* Thomson and Tait's *Natural Philosophy*, §§ 331, 332.

† Thomson and Tait, § 317.

suddenly generated in one of the circuits of a system, the initial currents in all the others are to be determined so as to make the energy of the field a minimum. These initial currents are formed unmodified by resistance whenever the electromotive impulses to which they owe their existence last only for a time which may be regarded as vanishingly small compared with the time-constants of the circuits. The sudden fall of a current when a circuit is opened generates the same currents, except as to sign, in neighbouring circuits as those due to a rise of the first current, and the condition as to sufficient suddenness is more generally fulfilled; at the same time it is more convenient in explaining the theory to take the case of the establishment of the primary current.

Suppose, then, that in the wire A_1 of our coil a current x is suddenly generated, while the ends of A_2 are joined by a short wire. The condition of minimum energy is obviously fulfilled if there arise in A_2 a current represented by $-x$; for then the energy of the field is approximately zero. But if the self-induction of the wire joining the ends of A_2 be sensible, the annihilation of the energy can no longer be perfect. Thus, let the circuit of A_2 be completed by $B_1 B_2$, then the general expression for the energy of two currents becomes in this case

$$\tfrac{1}{2}Lx^2 + Lxy + \tfrac{1}{2}Ly^2 \times (5 \text{ or } 1),$$

according to the connexions; and the value of y for which this is a minimum is $-x\,(1 \text{ or } \tfrac{1}{5})$. In the first case, the exterior part of the induced circuit having no sensible self-induction, takes away nothing from the initial current; but in the second there is a reduction to one-fifth. On the other hand, it makes no difference to the total current $(-xM/S)^*$, as measured by the deflection of the galvanometer-needle, which way the connexion is made; for the smaller initial current, in virtue of its greater inertia, sustains itself proportionally longer against the damping action of resistance, which is the same in the two cases. The heating-power and the effect on the electro-dynamometer, which depend on the integral of the square of the current while it lasts $(\tfrac{1}{2}x^2M^2/NS)$, will be different; but the easiest proof of the diversity of the currents is to be had by comparing their powers of magnetizing steel.

Thus, if we include in the induced circuit a magnetizing spiral in which is placed a new sewing-needle, we shall find an immense difference in the magnetization produced by a break-induced current, according as its direction is the same or otherwise in the wires $B_1 B_2$. In the actual experiment the diluted current was unable, even after several repetitions, to give the needle any considerable magnetization (the vibrations were only about three per minute), while after one condensed current the needle gave sixteen, raised

---

* R, S are the resistances of the primary and secondary circuits respectively.

by repetition to nineteen*. A new needle submitted to the action of several condensed currents also gave nineteen per minute. The magnetic moments, which are as the squares of these numbers, show a still greater disproportion.

The truth seems to be that the time required for the permanent magnetization of steel is so small as compared even with the duration of our induced currents, that the amount of acquired magnetism depends essentially on the initial or maximum current without regard to the time for which it lasts.

The increased heating-effect when the two parts of the current in B are opposed in direction is, of course, at the expense of the spark in the mercury-cup. The mechanical value of the spark is the difference between the values of the currents which exist at the moments before and after the breaking of the contact, and

$$= \tfrac{1}{2}Lx^2 - \tfrac{1}{2}Ny^2 = \tfrac{1}{2}x^2\left(L - \frac{M^2}{N}\right) = \tfrac{1}{2}x^2\left(L - \frac{L^2}{N}\right) \text{ nearly.}$$

Now, according to the connexions, $N = L$ or $5L$; and so in the first case the spark disappears, while in the second it falls short of the full spark by only one-fifth.

While considering the dynamics of the field of two currents, I noticed that the initial induced current due to a sudden fall of a given current in the primary wire is theoretically greater the smaller the number of terms of which the secondary consists; for in calculating the energy of the field, it makes no difference whether we have a current of any magnitude in a doubled circuit, or twice that current in a single circuit. The same conclusion may be arrived at by the consideration of the analytical expression for the initial induced current

$$y_0 = -\frac{M}{N}x;$$

for if the secondary circuit consists essentially of a single coil of n terms, we have, *cæteris paribus*, $M \propto n$, while $N \propto n^2$, so that $y_0 \propto 1/n$. The whole induced current $\int y\,dt \propto M \propto n$. Intermediate to these is the heating-effect $\int y^2 dt$, which $\propto M^2/N$, and is therefore independent of n. Thus it was evident that neither the galvanometer nor electrodynamometer was available for the verification of this rather paradoxical deduction from theory, at least without commutators capable of separating one part of the induced current from the rest. On the other hand, it appeared probable that the smaller total current, in virtue of its greater maximum, might be the most powerful in its magnetizing action on steel.

---

* These were *complete* vibrations.

With the view of putting this idea to the test of experiment, I bound three wires of ·001 inch diameter, and about 20 feet long, together into a coil whose opening was sufficient to allow it to pass over the coil A. The ends of the wires were free, so that they could be joined up in any order into one circuit, which was also to contain the magnetizing spiral. It is evident that if the currents are parallel in the three wires (an arrangement which I will call a), then

$$M = 3M_0, \quad N = 9N_0,$$

$M_0 N_0$ being the values of the induction-coefficients for *one* wire; while if in the two wires the current flows one way round and in the third the opposite (b), we shall have $M = M_0$, $N = N_0$. Inasmuch as the self-induction of the magnetizing spiral was relatively very small, these may be regarded as the induction-coefficients for the secondary circuit as a whole. This arrangement was adopted in order that there might be no change in the resistance in passing from one case to the other. The primary current was excited by a Daniell cell in the two wires of A arranged collaterally, and was interrupted at a mercury-cup. The needle was submitted to the *break* induction-currents only—although the make currents had no perceptible magnetizing-power, on account of the relatively large time-constant of the primary circuit, and the consequent slow rise of its current to the maximum.

On actually submitting a new needle to the current a, I obtained after one discharge 12 vibrations (complete) per minute, a number raised after several discharges to 15. On the other hand, a new needle after one discharge b gave only 5 per minute, and was not much affected by repetition. The last needle being now submitted to discharge a gave $8\frac{1}{2}$, and after several 12. Other trials having confirmed these results, there seemed to be no doubt that the current a was the most efficient magnetizer. There remained, however, some uncertainty as to whether the time-constant, especially in b, was sufficiently large relatively to the time for which the spark at the mercury-cup lasted to allow of the initial current being formed undiminished by resistance. In order to make the fall of the primary current more sudden, I connected the breaking-points with the plates of a condenser belonging to a Ruhmkorff coil, and now found but little difference between the magnetizing-powers of a and b. Seeing that the theoretical condition had not been properly fulfilled, I prepared another triple coil of much thicker wire, and, for greater convenience, arranged a mercury-cup commutator, by means of which it was possible to pass at once from the one mode of connexion to the other. The magnetizing spiral was still of fine wire coiled, without any tube, closely over the needle, and its ends were soldered to the thicker wire of the triple coil.

The experiment was now completely successful. Out of the large number of results obtained, the following are selected as an example. A new needle was submitted to the break discharge of arrangement *b*, and gave,

After 1 discharge, 19 per minute.
  „   3   „   23   „
  „   6   „   24   „

Another needle was now taken and magnetized by discharge *a*. It gave,

After 1 discharge, 11 per minute.
  „   3   „   12   „
  „   10   „   12½   „

On submitting this needle, which had received all the magnetism that *a* could give it, to current *b*, I obtained,

After 1 discharge, 21 per minute.
  „   3   „   24   „

In fact it was the general result of the experiments that more magnetism is always given to the needle by arrangement *b* than by *a*. In order, however, that the difference may be striking, it is advisable not to approach too nearly the point of magnetic saturation. The numbers quoted were obtained with the condenser, which was still necessary, in order to make the break sufficiently sudden. I have no doubt, however, that it might have been dispensed with had the triple coil consisted of a larger number of turns.

The circumstances of this experiment are in some degree represented by supposing, in the hydrodynamical analogue, one of the balls to vary in size. When a given motion is suddenly impressed on the other ball, the corresponding velocity generated in the first would vary inversely with its magnitude; for the larger the ball the greater hold, as it were, would it have on the fluid.

It is interesting also to examine the influence of neighbouring soft iron on the character of the induced current. This influence is of two sorts; but I refer here to the modifications produced by the magnetic character of iron. The circulation of induced currents in its mass may generally be prevented from exercising any injurious influence on the result by using only wires, or fragments of small size. The proximity of soft iron always increases the coefficient of self-induction N, while M may be either increased or diminished. The latter statement is true also for the initial current y_0, which is proportional to M/N. For the two wires of the coil A, however, it is easy to see that M and N are approximately equal, whether there be soft iron in their neighbourhood or not. Thus, if A_1 be connected with a Daniell cell while the circuit of A_2 is completed by the magnetizing spiral, the magnetism acquired by the needle, after a break-induced current, is not much altered, even if a considerable number of iron wires are placed in the coil. The total current is increased fifteen times or more; but this is because the current lasts longer, the maximum or initial value being no greater than before. This experiment strikingly illustrates the comparative independence of the magnetizing effect of a current on its duration. It seems probable *à priori*, and is partly confirmed by some of my experiments, that this is more especially true if we take the limiting magnetism which an induced current can produce, after repetition, as the measure of its magnetizing power.

The same kind of reasoning may be applied to more complicated problems. As an example, we may recur to a former combination, in which the primary current is excited in the wire A_1, while the secondary circuit includes A_2, B_1, and the magnetizing spiral. The initial current y_0, on which, as we have seen, the magnetizing power mainly depends, will be greatly increased if the ends of the wire B_2 are joined so as to make a tertiary circuit; for a current in B_2 is developed, which, being equal and contrary to that in B_1, neutralizes its action on the magnetic field, and so allows the energy, immediately after the sudden rise of the current x in A_1, to be vanishingly small, exactly as when the secondary circuit consisted of A_2 alone. The effect of closing B_2 is therefore to increase the current y_0 from $-\frac{1}{2}x$ to $-x$, and at the same time to produce a new current denoted by $+x$ in B_2 itself. The following were some of the experimental results:—

B_2 open $\left\{ \begin{array}{l} \text{A new needle,} \\ \quad \text{After 1 break-discharge, gave } 7\frac{1}{2} \text{ per minute.} \\ \quad \text{,,} \quad 8 \quad \text{,,} \quad \text{,,} \quad 9 \quad \text{,,} \end{array} \right.$

On closing B_2 we had, with the same needle,

        After 1 discharge, 15 per minute.

        ,,   8   ,,   17   ,,

A *new* needle gave,

        After 1 discharge, 17 per minute.

        ,,   8   ,,   19   ,,

Another new needle in the tertiary circuit gave,

        After 1 discharge, 16 per minute.

        ,,   4   ,,   19   ,,

        ,,   8   ,, $19\frac{1}{2}$,,

The magnetizing spiral was here removed from the secondary to the tertiary circuit; and although its resistance was by no means relatively small, the results are none the less comparable; for in this experiment resistances (within limits) are of no account, and the *self-induction* of the spiral was quite insensible.

Had there been twenty coils A B C D......similar to A B, with the wires $B_2 C_1$, $C_2 D_1$, &c. connected, as in the experiment just described, the magnetizing power of the current in the last would not, I imagine, be much less than in the first; for the condition of minimum energy would still be fulfilled by currents in the series of coils all equal in numerical value, and alternately opposite in algebraic sign. On this subject much confusion seems to have prevailed, as shown by the numerous inquiries into the *direction* of the induced currents of high orders. The currents, as a whole, at least after the first, cannot properly be said to have any direction at all, as they involve, when complete, no transfer of electricity in any direction. Nevertheless the positive and negative parts are not similar; and if they were, one must necessarily precede the other; so that in this way directional effects may be produced. The magnetizing power, for instance, depends essentially on the initial maximum magnitude of the induced current, and is probably but little affected by the character of the diluted but comparatively long-continued remaining parts. This being understood, the alternately opposite magnetizations observed by Henry in a series of induced currents of high order, is an immediate consequence of the dynamical theory.

The circuits being denoted by the numbers $1, 2, 3, \ldots$, let the coefficient of mutual induction between 2 and 3 be denoted by (2 3), and of self-induction of 2 by (2 2), and so on. The result is only generally true when there is no mutual induction except between immediate neighbours in the series; and it will therefore be supposed that

$$(1\ 3),\quad (1\ 4),\quad (1\ 5)\ldots(2\ 4)\ldots$$

vanish, as indeed they practically would in the ordinary arrangement of the experiment. The energy of the field is given by

$$E = \tfrac{1}{2}(1\ 1)\,x_1{}^2 + \tfrac{1}{2}(2\ 2)\,x_2{}^2 + \tfrac{1}{2}(3\ 3)\,x_3{}^2 + \ldots$$
$$+ (1\ 2)\,x_1 x_2 + (2\ 3)\,x_2 x_3 + (3\ 4)\,x_3 x_4 + \ldots$$

Here x_1 is the given current in the first circuit, and x_2, $x_3, \ldots$ are to be determined so as to make E a minimum. Now, E being homogeneous in $x_1, x_2, \ldots$, we have identically

$$2E = x_1 \frac{dE}{dx_1} + x_2 \frac{dE}{dx_2} + \ldots$$

And since, when E is a minimum,

$$dE/dx_2,\quad dE/dx_3, \ldots \text{ all vanish,}$$

we see that

$$2E\,(\text{min.}) = x_1 \frac{dE}{dx_1} = (1\ 1)\,x_1{}^2 + (1\ 2)\,x_1 x_2.$$

But if x_2, $x_3, \ldots$ had been all zero, $2E$ would have been equal to $(1\ 1)\,x_1{}^2$. It is clear therefore that $(1\ 2)\,x_1 x_2$ is negative; or, as $(1\ 2)$ is taken positive, the sign of x_2 is the opposite of that of x_1.

Again, supposing x_1, x_2 both given, we must have, when E is a minimum,

$$dE/dx_3, \quad dE/dx_4, \ldots = 0,$$

and thus

$$2E \text{ (min.)} = x_1 [(1\ 1)\, x_1 + (1\ 2)\, x_2]$$
$$+ x_2 [(1\ 2)\, x_1 + (2\ 2)\, x_2 + (2\ 3)\, x_3]$$
$$= (1\ 1)\, x_1^2 + 2\, (1\ 2)\, x_1 x_2 + (2\ 2)\, x_2^2 + (2\ 3)\, x_2 x_3.$$

As before, $2E$ might have been

$$(1\ 1)\, x_1^2 + 2\, (1\ 2)\, x_1 x_2 + (2\ 2)\, x_2^2 ;$$

and therefore the minimum value is necessarily less than this, and accordingly the signs of x_2 and x_3 are opposite. This process may be continued, and shows that, however long the series, the initial induced currents are alternately opposite in sign. In any definite example, the actual values of the initial currents are to be found from the solution of the linear equations

$$dE/dx_2 = 0, \quad dE/dx_3 = 0, \ldots\ldots ;$$

but the *sign* of the result does not appear at once from the form of the expression so obtained. In order to exhibit it, it is necessary to introduce a number of relations which exist between the induction-coefficients, and which are the analytical expression of the fact that the energy is always positive, whatever may be the values of x_2, x_3,...

It has been assumed throughout that the time of rise or fall of the current in the primary wire is very small as compared with the time-constants of the other circuits. In the case of coils, such as are generally used in induction-experiments, and which are not clogged by great external resistances, this condition is abundantly fulfilled at the break of the voltaic current*. The time of rise depends more on the nature of the circuit, but may be made as small as we please by sufficiently increasing the resistance in proportion to the self-induction; of course, in order to get an equally strong current, a higher electromotive force must be employed. In this way the rise may be made sufficiently sudden to fulfil the condition. Indeed, with a battery intense enough the rise of the current at *make* may become more sudden than the fall when contact is broken. In some of Henry's experiments this seems actually to have occurred. Thus, with a single cell as electromotor, he found the shock at make barely perceptible; but when the battery was increased to thirty cells, the shock became more powerful at make than at break.

---

* A rough measurement by Maxwell's method (*Phil. Trans.* 1865) gave for the time-constant of the circuit composed of the two wires of coil *A* ·0023″. The time-constant is the same whether the wires are collateral or consecutive, the greater self-induction of the latter arrangement being balanced by its greater resistance. For *one* wire only, the time-constant would be *half* the above. [1898—But see next paper.]

2.

ON AN ELECTROMAGNETIC EXPERIMENT.

[*Phil. Mag.* xxxix. pp. 428—435, 1870.]

THE experiment referred to is one described in the *Philosophical Magazine* for July, 1869, p. 9 [Art. 1], where it was shown that, within certain limits, the magnetizing effect of a break-induced current on steel needles is greater the smaller the number of turns of which the secondary circuit consists, the opposite, of course, being true of the effect on a galvanometer. The ground of the distinction is that the galvanometer takes account of the induced transient current as a whole; while the magnetizing-power depends mainly on the magnitude of the current at the first moment of its formation, without regard to the time which it takes to subside.

But even with this explanation, few, I imagine, would be prepared for the result who had not been accustomed to look at electrical phenomena in the light of some dynamical theory. It was for this reason that I considered the matter worthy of experimental investigation, the fruits of which were given in the paper referred to. One point, however, still required a little clearing up; and it is this which I now propose to deal with. I mean the mode of action of the condenser, which was employed, as in the inductorium, for the purpose of rendering the break more sudden, and which I had found necessary for the success of the experiment as then arranged. At this necessity I was not surprised; for, according to the indications of theory, the effect was only to be expected when the fall of the primary current is sudden compared to that of the secondary. Now the duration of free transient currents in a circuit varies, *cœteris paribus*, as the self-induction; so that when the number of turns in the secondary is too much reduced, there is danger of the condition not being fulfilled. If it be objected that as much would be gained by improved conductivity as lost by diminished self-induction, I answer this is not the fact, the resistance varying as the number of turns simply, while the self-induction varies as the square of the same number. Besides, I had reasons for keeping the resistance in all cases invariable.

Wishing, however, to obtain the effect without the aid of a condenser, I prepared a quadruple coil by bending into the form of a compact ring a bundle consisting of four No. 16 copper wires, each 70 feet long. Into one of these the current from a Daniell cell was permitted to flow, subject to interruption at a mercury-cup. The secondary circuit consisted of the other three wires arranged consecutively, and of the magnetizing spiral, which contained the needle destined to measure the effect of the momentary current. The three wires could be joined so that the current circulated the same way round them all (a), or so that in one of them the direction was the opposite to that in the two others (b). It will be seen that the resistance was always the same, the only change being in the coefficient of mutual induction (M), and of self-induction of induced circuit (N). In the former paper it was shown that the *initial* induced current, being proportional to M/N, is three times as great in (b) as in (a) (a sufficient suddenness of break being assumed), while the total currents are in the reciprocal proportion. In carrying out the experiments, I submitted the needle (a *new* one in each measurement) to the action of six break-induced currents, always removing it from the spiral when the battery contact was made. In this way a more constant result is obtained than from one discharge only, which is liable to vary from slight differences in the character of the break. The needle was then swung by a silk fibre and the number of complete vibrations in a minute observed. The numbers were :—

Arrangement (a) $4\frac{1}{2}$,   6,   6 :   mean   5.

„   (b)   21, 19, 19 :   „   20.

The superior efficiency of (b) is very conspicuous.

There is another way in which the subject may be investigated. If the secondary current containing a galvanometer be broken so quickly after the primary that the induced current has not time during the interval sensibly to diminish, the deflection of the needle may be considered to measure the initial value of the induced current. To carry out this experiment properly would require rather elaborate apparatus, on account of the necessity of a *constant* interval of time between the breaks. The contrivance that I used was of home manufacture and very rough, and acted by the almost simultaneous withdrawal of wires from two mercury-cups. The secondary circuit in case (a) consisted of the two wires of a large coil, A, joined consecutively, and of a short wire galvanometer. Iron wires were inserted in A in order to increase the duration of the induced current. In (b) only one wire of A was used, the resistance being made up by the substitution of another wire, whose self-induction might relatively be neglected. The total currents in the two cases would be as 2 : 1, and the initial currents as 1 : 2. The deflections of the galvanometer-needle were rather irregular; but the sum of ten throws in case (*a*) was 317°, while in (*b*) it rose to 480°, so that there could be no doubt as to the reality of the phenomenon.

Returning to the experiments with steel needles, I thought it desirable to compare the permanent magnetisms developed in two cases where the initial currents were equal. With this object the primary current (originating in a Grove cell) was passed through two wires, Q_1, Q_2, of the quadruple coil combined for self-induction. The induced circuit included Q_3, Q_4 and the magnetizing spiral. The arrangements in other respects being as before, I obtained—

13, 14, 15, 16 : mean 14½ vibrations per minute*.

Q_2 was now removed from the primary circuit (the resistance being made up to former value) and Q_4 from the induced circuit. The numbers now were—

11, 10, 12, 12 : mean 11½.

Q_2 being next replaced, but not Q_4, there resulted,—

35, 35, 33 : mean 34.

In the first two arrangements the initial currents would be equal, while in the third they would be doubled.

On the whole, I think these experiments confirm the view that the acquired magnetism depends principally on the initial current. The exact laws regulating the connexion between the current and the magnetism produced by it are doubtless complicated, and not of much interest. The facts here detailed should, however, be borne in mind by any one who wishes to pursue this subject, and they do something towards explaining the discordant results of previous experimenters.

And now as to the effect of a condenser. Considering that, in consequence of the length of wire in the quadruple coil, the duration of a current in it, even under arrangement (*b*), must be much greater than the time occupied by the break, or, which is the same thing, the duration of the spark, I did not anticipate that a condenser (whose plates were connected with the breaking-points) would have any influence. But to my surprise I found that although, of course, the superiority of (*b*) was not disturbed, the magnetic effects were all increased. The explanation is, I believe, to be found by an examination of what takes place in the two circuits after the primary current is thrown on the condenser by the removal of the wire from the mercury-cup.

The charge (or discharge) of a condenser through a circuit endowed with sensible self-induction was first investigated by Sir W. Thomson (see *Phil. Mag.*, vol. v. p. 393, or Wiedemann's *Galvanismus*, vol. ii. p. 1007). There

---

* The experiments were not made exactly in the order here adopted for convenience, but were broken up.

are two cases according to the comparative magnitudes of the three elements of the problem, which are (1) the capacity of the condenser, (2) the self-induction of the circuit, (3) the resistance of the same. If these, reckoned in absolute measure, be denoted by S, L, R respectively, the motion of electricity is of an oscillatory character if $S < 4L/R^2$; otherwise the charge is completed and equilibrium established without a retrograde motion. The motion of a pendulum in a viscous liquid is exactly analogous and may serve as an illustration. If the viscosity of the liquid exceed a certain limit, the pendulum withdrawn from the vertical and then let go will subside gradually back again into its position of equilibrium without ever passing it; but if the viscosity be small, rest is only attained after a number (theoretically indefinite) of oscillations of continually decreasing amplitude. To our case of currents mutually influencing one another, Thomson's calculations are not immediately applicable; indeed the exact solution would be rather complicated*. However, we are concerned principally with the first part of the electrical motion, the manner in which the currents wear down under the action of resistance being of subordinate importance. Now it appears that, if the motion be decidedly of the oscillatory type, the first few oscillations will take place almost uninfluenced by resistance; and on this supposition the calculation becomes remarkably simple.

L, M, N being the induction-coefficients, as before, let the total flow of electricity in the two circuits from the moment of the break be x, y, so that the currents at any moment are dx/dt, dy/dt. Then the equations to the currents are

$$L\frac{d^2x}{dt^2} + M\frac{d^2y}{dt^2} + \frac{x}{S} = 0 \dots\dots\dots\dots\dots\dots\dots(1),$$

$$M\frac{d^2x}{dt^2} + N\frac{d^2y}{dt^2} \qquad = 0 \dots\dots\dots\dots\dots\dots\dots(2),$$

where S is the capacity of the condenser.

Eliminating y, we get

$$\left(L - \frac{M^2}{N}\right)\frac{d^2x}{dt^2} + \frac{x}{S} = 0.$$

The oscillation in the primary wire is accordingly the same as if the secondary were open and the self-induction changed from L to $L - M^2/N$. (2) gives immediately the connexion between x and y,

$$M\frac{dx}{dt} + N\frac{dy}{dt} = \text{const.},$$

which shows that the currents in the two circuits oscillate synchronously, the maximum of one coinciding in time with the minimum of the other. Since

---

* It would depend upon a cubic equation.

$dy/dt = 0$ at the moment of break, the constant of integration must be equal to $M\,dx/dt_0$; dx/dt_0 denoting the value of the primary current at or before the break. Accordingly

$$\frac{dy}{dt} = \frac{M}{N}\left(\frac{dx}{dt_0} - \frac{dx}{dt}\right);$$

so that when after half an oscillation $dx/dt = -\,dx/dt_0$,

$$\frac{dy}{dt} = 2\,\frac{M}{N}\,\frac{dx}{dt_0}.$$

This is its maximum value, and is *double* of that which would be generated by a simple stoppage of the primary current, however sudden. In this way I am inclined to explain the increase of effect produced by the condenser. It is true that, having reached its maximum value, the secondary current rapidly declines and then changes sign; but from what is known of the behaviour of permanent magnets when submitted afresh to the action of magnetizing force, it does not seem likely that much disturbance would arise from this cause*.

A more plausible objection may be founded on the exceeding rapidity of the oscillations, for *some* time must be necessary for the magnetization of steel. Indeed in our case the period of oscillation is unusually short, on account of the smallness of $L - M^2/N$. When the two circuits are composed mainly of wires coiled side by side, L, M, N are approximately equal, and therefore $L - M^2/N$ very small compared with either of them. The current is then transferred, with almost indefinite rapidity, from one wire to the other and back again.

I made some experiments to examine this point, which will also serve as examples of the general increase of magnetic effect produced by a condenser. The primary current from a Grove's cell was passed through the two wires of coil A joined consecutively, and then through one wire of a similar coil B. According as the wires of A are joined, $L = 5$ or 1. The induced circuit included the other wire of B and the magnetizing spiral, so that $N = 1$ and $M = 1$ approximately. In the first case, therefore, $L - M^2/N = 4$, while in the second it is very much smaller. The experiments, conducted in other respects as before, gave the following results :—

$L = 5$, without condenser   16,   13,   13,   14 ;   mean 14.

$L = 5$, with   condenser $24\frac{1}{2}$, 28,   31,   28 ;   „   28.

$L = 1$, without condenser   14,   15, $13\frac{1}{2}$, 14 ;   „   14.

$L = 1$, with   condenser   17,   18,   19,   21 ;   „   19.

* [1898—This seems to need qualification.]

It will be seen that, while without the condenser it made little difference whether $L = 1$ or $L = 5$, the increase produced by the condenser is much greater in the latter case. This is so far in agreement with the explanation just given; but I confess I should have been better satisfied had the influence of the condenser been less marked when $L = 1$. In order that the reader may better judge of the correctness of the view here taken, I subjoin an estimate of the period of oscillation in the actual arrangement of the experiment.

The time of a complete oscillation of a current in simple connexion with a condenser $= 2\pi \sqrt{(LS)}$*, where L and S are measured in absolute measure. If τ be the time-constant of the circuit, $L = R\tau$, R being the resistance in absolute measure. Now the capacity called a [micro-]Farad is [10^{-15} C. G. S.], and the B.A. unit is [10^9] on the same system.

If therefore we take as practical units the B.A. unit of resistance, and the [micro-]Farad as unit of capacity,

$$t = \frac{\pi}{500} \sqrt{(RS\tau)}.$$

t and τ are here measured in seconds. The condenser employed (made by Elliott Brothers) had a capacity of half a [micro-]Farad, so that $S = \frac{1}{2}$. For one wire of coil A or B,

$$R = \tfrac{1}{4}, \quad \tau = \cdot004\dagger.$$

For one of these wires in simple connexion with the condenser the time of oscillation would be

$$\frac{\pi}{500} \sqrt{\frac{1}{4} \times \frac{1}{2} \times \frac{4}{1000}} = \text{about } \cdot00014^s.$$

Comparing this with the value of τ, we see that the first ten or so oscillations would be comparatively unaffected by resistance. By what has been proved, the time, when $L = 5$ (in different unit of course), must be just double of this, or $\cdot00028^s$.

The action of the condenser in the inductorium is very imperfectly explained in the text-books, and is no doubt in many cases rather complicated. From the reasoning of this paper, it appears that it is by no means a complete account of the matter to say that the advantage derived from the use of the condenser depends only on the increased suddenness with which the primary current is stopped. In a complete investigation (which I do not mean to enter on here) a distinction would probably have to be made, according as the secondary circuit when open allows the passage

* The effect of resistance being neglected.
† The value of τ previously given for coil A is erroneous.

of a spark or not, or, as a third case, is completely closed.   I would, however, remark that a good deal of misapprehension arises in this and similar cases from forgetting that a condenser* is powerless to make away with electrical energy.   Such energy may be disposed of in the form of a spark, or it may be converted into heat by the operation of electrical resistance; but the absorption in this way cannot take place instantaneously, requiring as it does a time comparable with the time-constants of the circuits concerned.   So far, indeed, is a condenser from itself absorbing electrical energy, that in many cases it actually prolongs the duration of motion; for an oscillatory current, in consequence of its smaller mean square, sustains itself twice as long against the damping action of resistance as a comparatively steady current of the same maximum value.

* [1898—This must be understood to refer to an ideal condenser.]

ON THE VALUES OF THE INTEGRAL $\int_0^1 Q_n Q_{n'}\, d\mu$, Q_n, Q_n BEING LAPLACE'S COEFFICIENTS OF THE ORDERS n, n', WITH AN APPLICATION TO THE THEORY OF RADIATION.

[*Phil. Trans.* CLX. pp. 579—590; read June 1870.]

IN the course of an investigation concerning the potential function which is subject to conditions at the surface of a sphere which vary discontinuously in passing from one hemisphere to the other, it became necessary to know the values of the integral

$$\int_0^1 Q_n Q_{n'}\, d\mu,$$

Q_n, $Q_{n'}$ being Laplace's coefficients of the orders n, n' respectively. The expression for Q_n in terms of μ is

$$Q_n = \frac{1.3.5 \ldots (2n-1)}{1.2.3 \ldots n}\left\{\mu^n - \frac{n(n-1)}{2.(2n-1)}\mu^{n-2} + \frac{n(n-1)(n-2)(n-3)}{2.4(2n-1)(2n-3)}\mu^{n-4} - \ldots\right\};$$

but the multiplication of two such series together and subsequent integration with respect to μ would be very laborious even for moderate values of n and n'.

By the following method the values of the integrals in question may be obtained without much trouble. According to the definition of the functions Q,

$$\frac{1}{\sqrt{1 + e^2 - 2e\mu}} = 1 + Q_1 e + Q_2 e^2 + \ldots + Q_n e^n + \ldots$$

so that

$$\int_0^1 \frac{d\mu}{\sqrt{1 + e^2 - 2e\mu}\,\sqrt{1 + e'^2 - 2e'\mu}} = \sum_{n=0}^{n=\infty} \sum_{n'=0}^{n'=\infty} \int_0^1 Q_n Q_{n'} d\mu \cdot e^n e'^{n'},$$

which shows that $\int_0^1 Q_n Q_{n'} d\mu$ is the coefficient of $e^n e'^{n'}$ in the expansion of the integral on the left in powers of e and e'.

[1898—By effecting the integration and expansion of the result certain properties were established, of which the most interesting is perhaps the equation

$$\int_0^1 Q_{2n} Q_{2n-1} d\mu = \int_0^1 Q_{2n} Q_{2n+1} d\mu.$$

The method however is cumbrous and is superseded by that of Todhunter (*Proc. Roy. Soc.* vol. XXIII. p. 300, 1875), who remarked that general results are readily obtained from the differential equation satisfied by Legendre's functions on integration by parts. From the well-known results applicable to the complete sphere it is easily seen that the integral vanishes when n, n' are both odd or both even, exception being made of the case where n' and n are equal. For the remaining case when one suffix is odd and the other even, Todhunter obtains

$$\{2m (2m + 1) - (2n - 1) 2n\} \int_0^1 Q_{2m} Q_{2n-1} d\mu$$

$$= (- 1)^{m+n} \frac{1 \cdot 3 \cdot 5 \ldots (2m - 1)}{2 \cdot 4 \cdot 6 \ldots 2m} \frac{1 \cdot 3 \cdot 5 \ldots (2n - 1)}{2 \cdot 4 \cdot 6 \ldots (2n - 2)}.$$

Substituting m, $m + 1$ in succession for n, we find

$$\int_0^1 Q_{2m} Q_{2m-1} d\mu = \int_0^1 Q_{2n} Q_{2m+1} d\mu = \tfrac{1}{2} \cdot \frac{1^2 \cdot 3^2 \cdot 5^2 \ldots (2m - 1)^2}{2^2 \cdot 4^2 \cdot 6^2 \ldots (2m)^2}.]$$

As an application of some of the results of this investigation I will take the following physical problem. A spherical ball of uniform material is exposed to the radiation from infinitely distant surrounding objects. It is required to find the *stationary* condition. For the sake of simplicity, the surface of the sphere will be supposed to be perfectly black, that is, to absorb all the radiant heat that falls upon it, and Newton's law of cooling will be employed, at least provisionally.

If V denote the temperature, it is to be determined by the equations

$$\left(\frac{d^2}{dx^2} + \frac{d^2}{dy^2} + \frac{d^2}{dz^2}\right) V = 0, \quad \ldots\ldots\ldots\ldots\ldots\ldots(A)$$

$$k \frac{dV}{dr} = F(E) - hV, \quad \ldots \ldots\ldots\ldots\ldots\ldots(B)$$

where $F(E)$ is a function of the position of the point E on the surface, and denotes the heat received per unit area at that point, k is the conductivity, and h the coefficient of radiation. Equation (A) is to be satisfied throughout the interior and (B) over the surface of the sphere.

If V be expanded in Laplace's series,

$$V = S_0 + S_1 \frac{r}{a} + S_2 \frac{r}{a^2} + \dots ; \quad \frac{dV}{dr} = a = \frac{1}{a}(S_1 + 2S_2 + 3S_3 + \dots);$$

and if

$$F = F_0 + F_1 + F_2 + \dots$$

be the expansion of F in a similar series of surface harmonics, we obtain, on substituting in (B) and equating to zero the terms of any order,

$$S_0 = \frac{F_0}{h}, \quad S_1 = \frac{F_1}{h + k/a}, \quad \dots\dots S_n = \frac{F_n}{h + nk/a} \dots\dots\dots(C)$$

The mean temperature S_0 is seen to be independent of the conductivity and of the size of the sphere.

The case where the heat which falls on the sphere proceeds from a single radiant-point is not only important in itself, but may be made the foundation of the general solution in virtue of the principle of superposition. Taking the axis in the direction of the radiant-point, we have

$$F(E) = \mu$$

over the positive hemisphere, that is, from $\theta = 0$ to $\theta = \frac{1}{2}\pi$, while over the negative hemisphere $F(E) = 0$. It is required to expand F in a series of spherical harmonics.

Let $F = \frac{1}{2}\mu + f$, then f is a function of μ, which is equal to $\frac{1}{2}\mu$ over the positive hemisphere and to $-\frac{1}{2}u$ over the negative. The problem therefore reduces itself to the expression of $\frac{1}{2}\mu$ over the positive hemisphere in a series of functions Q *of even order*. The same series will then give $-\frac{1}{2}\mu$ over the negative hemisphere.

Assume

$$\tfrac{1}{2}\mu = A_0 + A_2 Q_2 + A_4 Q_4 + \dots$$

Multiplying by Q_{2n}, and integrating with respect to μ from $\mu = 0$ to $\mu = 1$,

$$\tfrac{1}{2}\int_0^1 Q_1 Q_{2n}\, d\mu = A_{2n} \int_0^1 (Q_{2n})^2\, d\mu,$$

all the other terms on the right vanishing.

Now $\displaystyle\int_0^1 (Q_{2n})^2\, d\mu = \frac{1}{4n+1}$.

$$\int_0^1 Q_1 Q_{2n}\, d\mu = \text{coefficient of } e^{2n} \text{ in the expansion of } \tfrac{1}{3}\frac{(1+e^2)^{\frac{3}{2}} - 1}{e^2}$$

$$= -(-1)^n \frac{1 . 1 . 3 . 5 \dots (2n-3)}{2 . 4 . 6 \dots (2n+2)},$$

or

$$A_{2n} = -(-1)^n \frac{4n+1}{2} \frac{1 . 1 . 3 \dots (2n-3)}{2 . 4 . 6 \dots (2n+2)}.$$

Accordingly

$$F(E) = \tfrac{1}{4} + \tfrac{1}{2}Q_1 + \tfrac{5}{16}Q_2 - \tfrac{3}{32}Q_4 + \ldots - (-1)^n \frac{4n+1}{2} \frac{1.1.3.5\ldots(2n-3)}{2.4.6\ldots(2n+2)} Q_{2n} + \ldots$$

When n is great the coefficient of Q_{2n} approximates to $\dfrac{1}{2\sqrt{\pi}.n^{\frac{3}{2}}}$.

This completes the solution for a sphere exposed to the radiation from an infinitely distant source of heat situated over the point $\mu = 1$.

If its coordinates are μ', ϕ', it is only necessary to replace μ in Q_{2n} by

$$\cos\theta \cos\theta' + \sin\theta \sin\theta' \cos(\phi - \phi').$$

Hence if H denote the intensity of the radiation which comes in *direction* μ', ϕ', the general value of S_n is

$$S_{2n} = -(-1)^n \frac{4n+1}{2\left(h + \dfrac{nk}{a}\right)} \frac{1.1.3.5\ldots(2n-3)}{2.4.6\ldots(2n+2)}$$

$$\times \iint H Q_{2n} (\cos\theta \cos\theta' + \sin\theta \sin\theta' \cos(\phi - \phi'))\, d\mu'd\phi',$$

the integration going *all* round the sphere.

Now $(4n+1) \iint H Q_{2n}\, d\mu'\, d\phi'$ is the same as $4\pi H_{2n}$, where H_{2n} is the harmonic element of H of order $2n$; so that

$$S_{2n} = -(-1)^n \frac{1.1.3\ldots(2n-3)}{2.4.6\ldots(2n+2)} \frac{2\pi H_{2n}}{h + 2nk/a},$$

$$S_0 = \frac{\pi}{h} H_0, \quad S_1 = \frac{2\pi}{3\left(h + \dfrac{k}{a}\right)} H_1, \quad S_2 = \frac{\pi}{4\left(h + \dfrac{2k}{a}\right)} H_2, \quad S_4 = -\frac{\pi}{24\left(h + \dfrac{4k}{a}\right)} H_4.$$

It is remarkable that the odd terms in H (except H_1) are altogether without influence. The reason is simply that they do not affect the total heat falling on any point of the surface.

For this is expressed by

$$\int_0^1 \int_0^{2\pi} \mu H_n\, d\mu d\phi,$$

the point considered being taken as pole of μ, which involves no loss of generality.

Now (Thomson and Tait, Ch. I., Appendix B)

$$H_n = \sum_{s=0}^{s=n} (A_s \cos s\phi + B_s \sin s\phi)\, \Theta_n{}^s,$$

where $\Theta_n{}^s$ is a function of μ not containing ϕ.

When the integration with respect to ϕ is effected, all the terms will vanish except that whose coefficient is A_0. For this purpose, therefore, we may take

$$H_n = A_0 \Theta_n{}^0 \text{ or } A_0'Q_n,$$

and we know that $\int_0^1 \mu\, Q_n d\mu$ vanishes if n be odd and different from unity [*].

The same thing is true for an ellipsoid or body of any figure which lies altogether on one side of every tangent plane, namely, that the terms of odd order in H (except one) are wholly without influence on it, and for the same reason.

We saw that in the case of a sphere the mean temperature was independent of the conductivity, and also of the size of the sphere; but this depends on Newton's law of cooling. A comparison, however, may be made which shall hold good whatever be the law of variation of radiation with temperature; for if the conducting-power of any uniform body (which need not be oval) be increased in the same proportion as its linear dimensions, a corresponding distribution of temperature will satisfy all the conditions. Conclusions of interest from a physical point of view may be deduced from the foregoing considerations, but I refrain from pursuing the subject at present, as the physical problem was only brought forward in illustration of the mathematical results developed in this paper.

---

[*] The proof given is sufficient for the object in view; but it may be well to notice that the essential thing is that the two surface harmonics which are multiplied together are either both odd or both even. A harmonic of even order has identical values at opposite points of the sphere, and one of odd order has contrary values. The product of two harmonics which are either both even or both odd has therefore the same value when integrated over any portion of the sphere, or over what may be called the opposite portion, or as a particular case over two opposite hemispheres. The last two integrals are the halves of the integral over the whole sphere, which vanishes by a well-known property of these functions.

4.

REMARKS ON A PAPER BY DR SONDHAUSS.

[*Phil. Mag.* XL. pp. 211—217, 1870.]

IN Nos. 5 and 6 of Poggendorff's *Annalen* for this year there is a paper by Dr Sondhauss "On the Tones of Heated Tubes and Aërial Vibrations in Pipes of various forms," in which are given formulæ of considerable generality embodying the results of original and other experiments. Many years ago* Dr Sondhauss had investigated the influence of the size and form of flask- or bottle-shaped vessels on the pitch of the sounds produced when a stream of air is blown across their mouth, and had obtained as an empirical formula for flasks with rather long cylindrical necks,

$$n = C \sqrt{\frac{\sigma}{SL}}, \quad \dots\dots\dots\dots\dots\dots\dots(\text{I.})$$

where n is the number of vibrations per second, σ the area of the section of the neck whose length is L, and S the volume of the body of the flask. C is a constant determined by the experiments. On the other hand, when L is very small compared with the diameter of the neck, which then becomes a mere hole,

$$n = C \frac{\sigma^{\frac{1}{4}}}{S^{\frac{1}{2}}}. \quad \dots\dots\dots\dots\dots\dots\dots(\text{II.})$$

In the paper now under discussion it is sought to fill up the gap, as it were, and the following formula is arrived at as applicable for all proportions of L and $\sigma^{\frac{1}{4}}$,

$$n = \frac{a}{4} \sqrt{\frac{\sigma}{(Sc + L\sigma)(L + \sqrt{\sigma})}}, \quad \dots\dots\dots\dots(\text{VII.})$$

or, as I prefer to write it,

$$n = \frac{a}{4\sqrt{c}} \sqrt{\frac{\sigma}{(S + L\sigma/c)(L + \sqrt{\sigma})}},$$

* Pogg. *Ann.* vol. lxxxi.

in which $a =$ velocity of sound. c is a constant, of which Dr Sondhauss says that it relates to the change in the velocity of sound in closed spaces from which the sound-waves have only a restricted exit; and its value, as found from the experiments, is approximately 2·3247.

In (VII.), if σ be small,

$$n = \frac{a}{4\sqrt{c}}\sqrt{\frac{\sigma}{LS}}. \quad\dots\dots\dots\dots\dots\text{(VIII.)}$$

If, on the contrary, L be very small,

$$n = \frac{a}{4\sqrt{c}}\frac{\sigma^{\frac{1}{4}}}{S^{\frac{3}{4}}} \quad\dots\dots\dots\dots\dots\dots\text{(IX.)}$$

If in (VII.) we further put $S = 0$,

$$n = \frac{a}{4}\sqrt{\frac{1}{L(L + \sqrt{\sigma})}}, \quad\dots\dots\dots\dots\dots\text{(X.)}$$

a result which Dr Sondhauss applies to cylindrical tubes closed at one end. This being admitted, it readily follows that for a pipe open at both ends,

$$n = \frac{a}{2}\sqrt{\frac{1}{L(L + 2\sqrt{\sigma})}}. \quad\dots\dots\dots\dots\text{(XI.)}$$

An extension is next made to the case of more than one neck, but it will not be necessary for my purpose to repeat the formulæ. A few days before I saw Dr Sondhauss's work I had myself completed a paper on a similar subject, which has since been sent to the Royal Society*. The formulæ there given were in the first instance obtained theoretically, though some of them were afterwards verified by a rather laborious series of experiments. But on the present occasion I shall leave the theory on one side, and wish only to discuss some differences between the results of Dr Sondhauss and my own, regarded from an experimental point of view. The rational formula corresponding to (VII.) is

$$n = \frac{a}{2\pi}\sqrt{\frac{\sigma}{S(L + \frac{1}{2}\sqrt{\pi}\sqrt{\sigma})}}; \quad\dots\dots\dots\dots\text{(A)}$$

where, however, S has not quite the same meaning as with Dr Sondhauss, but includes the volume of about half the neck, and is therefore nearly identical with the $(S + L\sigma/c)$ of (VII.). On this understanding (VII.) may be written

$$n = \frac{a}{6\cdot0988}\frac{\sigma^{\frac{1}{4}}}{S^{\frac{1}{2}}\sqrt{L + \sigma^{\frac{1}{2}}}}, \quad\dots\dots\dots\dots\text{(VII.)}$$

[* *Phil. Trans.* for 1871; Art. 5 of this collection.]

while (A) expressed in numbers is

$$n = \frac{a}{6\cdot2832} \frac{\sigma^{\frac{1}{2}}}{S^{\frac{1}{2}} \sqrt{L + \cdot8863 \sqrt{\sigma}}} . \quad\dots\dots\dots\text{(A)}$$

If $\sigma^{\frac{1}{2}}$ be very small against L,

$$n = \frac{a}{6\cdot0988} \frac{\sigma^{\frac{1}{2}}}{S^{\frac{1}{2}} L^{\frac{1}{2}}}, \quad\dots\dots\dots\text{(VIII.)}$$

$$n = \frac{a}{6\cdot2832} \frac{\sigma^{\frac{1}{2}}}{S^{\frac{1}{2}} L^{\frac{1}{2}}} . \quad\dots\dots\dots\text{(B)}$$

But if L be very small,

$$n = \frac{a}{6\cdot0988} \frac{\sigma^{\frac{1}{2}}}{S^{\frac{1}{2}}}, \quad\dots\dots\dots\text{(IX.)}$$

$$n = \frac{a}{5\cdot9149} \frac{\sigma^{\frac{1}{2}}}{S^{\frac{1}{2}}} . \quad\dots\dots\dots\text{(C)}$$

The rational formula (C) was first given by Helmholtz in his admirable paper in Crelle, on Vibrations in Open Pipes; it is only strictly applicable to openings of circular form. The difference between (A) and (VII.) is never very great, being on one side when L is small, and on the other when L is large, and accordingly vanishing for some intermediate value. The greatest difference is shown in (VIII.) and (B) when L is very large. I therefore consider Dr Sondhauss's opinion and anticipation to be in the main justified by my investigation, when he says, "I remark that I regard the formula (VII.)...not merely as an empirical formula useful for interpolation, but am convinced that it forms the theoretical expression of a natural law. From the zeal with which the field of mathematical physics is now cultivated, we may expect that the laws which I have discovered experimentally will soon be proved by analysis." But I must observe that (A) is only true subject to a series of limitations, which Dr Sondhauss seems scarcely, if at all, to have contemplated. All the dimensions of the vessel (with a partial exception of the length of the neck) must be small compared with the quarter wave-length, and the diameter of the neck must be small against the linear dimension of the body of the vessel. The latter condition excludes the case of S small or nothing, to which Dr Sondhauss pushes the application of his formula. But there *is* a rational formula proper for closed cylindrical tubes, as has been proved by Helmholtz in his paper on open pipes, to which Dr Sondhauss refers, but apparently without availing himself of the results. It runs,

$$n = \frac{a}{4(L + \frac{1}{4}\sqrt{\pi} . \sigma^{\frac{1}{2}})}, \quad\dots\dots\dots\text{(D)}$$

but is only strictly true when $\sigma^{\frac{1}{2}}$ is small against L. Although I am of opinion that (A) and (D) and the transition between them cannot be comprehended in the same theoretical investigation, yet it is easy to adjust (A) so as algebraically to include (D).   Thus

$$n = \frac{a}{2\pi} \cdot \frac{\sigma^{\frac{1}{2}}}{\sqrt{\left(S + \frac{4}{\pi^2}\sigma L\right)\left(L + \frac{\sqrt{\pi}}{2}\sigma^{\frac{1}{2}}\right)}} \quad \ldots\ldots\ldots\ldots(A)$$

becomes, when S (which now refers to the volume of the body only) is put equal to zero,

$$n = \frac{a}{4} \cdot \frac{1}{\sqrt{L\left(L + \frac{1}{2}\sqrt{\pi}\cdot\sigma^{\frac{1}{2}}\right)}} = \frac{a}{4\left(L + \frac{1}{4}\sqrt{\pi}\cdot\sigma^{\frac{1}{2}}\right)} \text{ approx.}, \ldots\ldots(D)$$

supposing the condition fulfilled as to the relative magnitudes of L and $\sigma^{\frac{1}{2}}$. Now this form of (A) is perfectly legitimate, the value of $4/\pi^2$ being ·405. In the formula (VII.) ·405 is replaced by $1/c$ or ·430.

But Dr Sondhauss will naturally point to the comparison of (X.) with the experiments of Wertheim, which he justly regards as very satisfactory. I have examined the series of twenty-two experiments with cylindrical tubes of circular form closed at one end, and have calculated for comparison the results of (D). It will be seen from the annexed Table that, good as is the agreement of the observations with (X.), it is *still better* (on the whole) with (D). Indeed I must confess that the differences in some cases, where $\sigma^{\frac{1}{2}}$ is by no means small compared with L, are much less than I should have expected.

$\left[\frac{1}{100}a \text{ in mm./sec.}\right]$	L [in mm.].	Diameter [in mm.].	n, observed by Wertheim.	n, calculated by Sondhauss from (X.).	n, calculated by me from (D).
3371	785	80	103·4	102·9	103·2
3395	500	......	158·0	158·8	159·7
3413	355	......	219·2	219·4	220·8
3371	280	......	272·4	268·8	270·6
3418	157	......	460·4	451·8	453·5
3413	1000	60	83·3	83·1	83·3
3418	640	......	128	128·3	128·8
......	340	......	237	233·7	235
3413	1000	48	84	83·6	83·8
3427	350	......	234·4	231·2	232·3
......	......	25	238·3	237·4	238·1
......	......	20	239·7	238·8	239·3
......	......	11	241·5	241·5	241·8
......	......	5	243·8	243·3	243·4
3390	206	68	372·1	361·8	364·2
......	116	46·5	640	627·6	631·0
......	208	43·5	376·5	374·2	376·4
3471	487	200	154·2	152·6	153·5
......	243·5	......	274·1	271·1	269·5
......	121·7	......	416·9	455·0	433·4
3390	305	25	270·1	268·3	269·2
......	500	20	166·6	166·6	166·8

The foregoing Table shows that although (X.) represents Wertheim's observations with considerable accuracy, yet Helmholtz's rational formula (D) is on all grounds to be preferred.

Dr Sondhauss expresses himself strongly as to the difficulty which exists in determining accurately the pitch of the very uncertain sound produced by tubes whose diameter is not small compared with their length, an opinion which I entirely share. It is indeed difficult to understand how Wertheim obtained results of such precision. But I cannot agree with Dr Sondhauss when he goes on to say that resonance is not a sure guide in determining accurately the pitch of a pipe; for it was by this method exclusively that the determinations recorded in my paper were made. I have there given at length my reasons for adopting it, and for doubting the results of the method of blowing, although such experiments as those of Wertheim go to show à *posteriori* that in his hands at least it was not unworthy of dependence.

Other experiments of Wertheim are calculated from formula (IX.) and show a tolerable agreement. The difference between (IX.) and Helmholtz's theoretical formula (C) relates only to a constant multiplier, and corresponds to a difference of pitch of about a quarter of a semitone. The discordances are attributed (no doubt correctly) to the unsuitable form of some of the vessels, and consequent imperfect fulfilment of the theoretical condition to which (C) is subject.

We come next to vessels in the form of flasks with a cylindrical neck of sensible length. Dr Sondhauss gives a Table containing the results of a comparison of (VII.) with some experiments of his own. The average discordance amounts to about a semitone. Although it was evident before-hand that in most cases the limitations on formula (A) were grossly violated, I thought it worth while to calculate in accordance with (A) the theoretical pitch, and have given the results in the form of a Table :—

No. of exp.	a.	Shape of vessel.	S, in cubic centims.	L, in millims.	Diam., in millims.	n, observed.	n, calculated from (VII.).	n, calculated from (A).
1.	341260	Sphere.	17·1	60	5·5	241·6	246·9	251·3
2.	....	Cylinder.	60·9	19	12·5	430·5	454·5	453·6
3.	....	......	10·7	15	10	966·5	959	970
4.	....	......	97·7	13	9	287·3	311·2	309·2
5.	....	......	66·2	175	11·6	143·7	155·2	158·8
6.	....	......	117·8	183	18	170·9	170·4	178·5
7.	....	Octagon.	654·5	193	26·5	114	106·7	107·6
8.	....	Sphere.	76·3	118	20·5	256·0	285·6	306·1
9.	....	Cylinder.	117·8	15	25·3	574·7	589·5	600·2
10.	....	......	132·4	44	27	362	429·2	441·4
11.	342740	Sphere.	923	205	25	85·4	83·2*	83·7
12.	344210	Cylinder.	8920	30	36	76·1	76·6	76·6
13.	341260	Sphere.	178	160	18·5	152·2	155·8	159·1
14.	344210	......	1·09	11	2	812·7	842·6	825·4
15.	....	......	·80	2·2	2·2	1933	1926·2	1902

* The result of the formula (VII.) ought evidently here to be greater than that of (A). On a recalculation I find 85·8 instead of 83·2.

All the columns except the last are copied from Dr Sondhauss's paper. It will be found that the observations are better represented by (VII.) than by (A); but it must be remembered that (VII.) contains an arbitrary constant, c, which acts nearly as a constant multiplier, although, if I understand Dr Sondhauss aright, its value was not determined from this series of experiments. However this may be, it is certain that nearly all the values of n calculated from (A) are too great. The fact is that (A) is scarcely applicable to the experiments at all. In only five cases is the ratio of the diameter of the neck to the dimension of the vessel even tolerably small. These are 1, 4, 11, 12, 14; but in 1, on account of the extremely small diameter of the neck and its considerable length, the influence of friction is probably sensible; and its effect would be to lower the pitch. The body in 4 is cylindrical, and perhaps too long in proportion to the quarter wave-length. In 11, 12, and 14 the agreement is sufficiently good. I consider accordingly that there is no evidence in the Table unfavourable to formula (A), supposed to be stated with the proper restrictions. In my own experiments, made by the method of resonance, I found a very good agreement between the directly observed and the calculated pitch, the average error being under a quarter of a semitone. Even with formula (VII.) as the basis of calculation there would be a fair agreement, certainly better than is the case with Dr Sondhauss's own experiments. The difference between (VII.) and (A) is, as I have already remarked, comparatively small, and could only be certainly distinguished under favourable circumstances. Not finding the necessary data in Dr Sondhauss's paper, I venture to quote some experiments from the paper on Resonance. There are seven observations in which the necks were sufficiently long to bring out the difference between the formulæ, being more than four times the diameter. It will be seen that the alteration is in every case for the worse if the formula (VII.) is substituted for (A).

n, observed.	n, calculated from (A).	n, calculated approx. from (VII.).
126	127·7	131
108·7	107·7	110
180	179·7	184
228	233·7	239
204	201·9	207
182	186·3	190
384	391·6	400

These experiments seem to decide the question; but it would be interesting to see if Dr Sondhauss obtained a similar result by the method of blowing. The difference, amounting in (VIII.) and (B) to half a semitone, is far greater than any error to be feared in the measurement of pitch or of the dimensions of the vessel, and ought therefore to give a sufficient handle to decide between the formulæ; if proper attention is given to the choice of a suitable resonator. In the foregoing remarks I have naturally dwelt most on my differences with Dr Sondhauss; but I should be sorry to have it supposed that I write in a hostile spirit, or do not recognize the claims of one to whom the science of acoustics is so largely indebted.

Postscript.

I have since calculated the results of the experiments of Wertheim on pipes open at both ends, and find that in this case also the rational formula $\left(n = \dfrac{a}{4\left(\frac{1}{2}L + \frac{1}{8}\pi D\right)}\right)$ agrees best with the observations :—

a.	L.	D.	n, observed by Wertheim.	n, calculated by Sondhauss.	n, calculated by myself.
3371	785	80	202·0	197·6	[198·8]
3395	500	...	306·9	299.6	301·5
3413	355	...	418·3	406·3	408·4
3371	280	...	512·0	490·4	491·7
3418	157	...	836·6	789·2	777·3
3413	1000	60	164·3	162·2	163·0
3418	640	...	251·0	247·3	248·8
......	340	...	450·7	438·8	441·5
3413	1000	48	165·6	163·8	164·5
3427	350	...	452·3	439·1	442·0
3413	1000	25	167·5	167·0	167·4
......	......	20	168·2	167·7	168·0
......	......	11	169·1	169·0	169·2
......	......	5	170·0	169·3	169·9
3390	906	30	182·5	181·8	182·3
......	307	21	526·7	521·2	524·0
......	541·5	20	306·2	303·2	304·3
......	392	...	419·6	413·1	415·7
......	676	38	240·6	239·1	240·1
......	343	...	458·8	451·8	454·6

5.

ON THE THEORY OF RESONANCE.

[*Phil. Trans.* CLXI. pp. 77—118; Read Nov. 1870.*]

Introduction.

ALTHOUGH the theory of aërial vibrations has been treated by more than one generation of mathematicians and experimenters, comparatively little has been done towards obtaining a clear view of what goes on in any but the more simple cases. The extreme difficulty of anything like a general deductive investigation of the question is no doubt one reason. On the other hand, experimenters on this, as on other subjects, have too often observed and measured blindly without taking sufficient care to simplify the conditions of their experiments, so as to attack as few difficulties as possible at a time. The result has been vast accumulations of isolated facts and measurements which lie as a sort of dead weight on the scientific stomach, and which must remain undigested until theory supplies a more powerful solvent than any now at our command. The motion of the air in cylindrical organ-pipes was successfully investigated by Bernoulli and Euler, at least in its main features; but their treatment of the question of the open pipe was incomplete, or even erroneous, on account of the assumption that at the open end the air remains of invariable density during the vibration. Although attacked by many others, this difficulty was not finally overcome until Helmholtz†, in a paper which I shall have repeated occasion to refer to, gave a solution of the problem under certain restrictions, free from any arbitrary assumptions as to what takes place at the open end. Poisson and Stokes ‡ have solved the problem of the

---

* Additions made since the paper was first sent to the Royal Society are inclosed in square brackets [ ].

† Theorie der Luftschwingungen in Röhren mit offenen Enden. *Crelle*, 1860.

‡ *Phil. Trans.* 1868, or *Phil. Mag.* Dec. 1868.

vibrations communicated to an infinite mass of air from the surface of a sphere or circular cylinder. The solution for the sphere is very instructive, because the vibrations outside any imaginary sphere enclosing vibrating bodies of any kind may be supposed to take their rise in the surface of the sphere itself.

More important in its relation to the subject of the present paper is an investigation by Helmholtz of the air-vibrations in cavernous spaces (*Hohlräume*), whose three dimensions are very small compared to the wave-length, and which communicate with the external atmosphere by small holes in their surfaces. If the opening be circular of area σ, and if S denote the volume, n the number of vibrations per second in the fundamental note, and a the velocity of sound,

$$n = \frac{a\sigma^{\frac{1}{4}}}{2^{\frac{1}{2}}\pi^{\frac{5}{4}}S^{\frac{1}{2}}}.$$

Helmholtz's theory is also applicable when there are more openings than one in the side of the vessel.

In the present paper I have attempted to give the theory of vibrations of this sort in a more general form. The extension to the case where the communication with the external air is no longer by a mere hole in the side, but by a neck of greater or less length, is important, not only because resonators with necks are frequently used in practice, but also by reason of the fact that the theory itself is applicable within wider limits. The mathematical reasoning is very different from that of Helmholtz, at least in form, and will I hope be found easier. In order to assist those who may wish only for clear general ideas on the subject, I have broken up the investigation as much as possible into distinct problems, the results of which may in many cases be taken for granted without the rest becoming un-intelligible. In Part I. my object has been to put what may be called the dynamical part of the subject in a clear light, deferring as much as possible special mathematical calculations. In the first place, I have con-sidered the general theory of resonance for air-spaces confined nearly all round by rigid walls, and communicating with the external air by any number of passages which may be of the nature of necks or merely holes, under the limitation that both the length of the necks and the dimensions of the vessel are very small compared to the wave-length. To prevent misapprehension, I ought to say that the theory applies only to the funda-mental note of the resonators, for the vibrations corresponding to the overtones are of an altogether different character. There are, however, cases of multiple resonance to which our theory is applicable. These occur when two or more vessels communicate with each other and with the external air by necks or otherwise; and are easily treated by Lagrange's general dynamical method, subject to a restriction as to the relative magnitudes of the wave-lengths and the dimensions of the system corresponding to that stated above for a single vessel. I am not aware whether this kind of resonance has been investigated before, either mathematically or experimentally. Lastly, I have sketched a solution of the problem of the open organ-pipe on the same general plan, which may be acceptable to those who are not acquainted with Helmholtz's most valuable paper. The method here adopted, though it leads to results essentially the same as his, is I think more calculated to give an insight into the real nature of the question, and at the same time presents fewer mathematical difficulties. For a discussion of the solution, however, I must refer to Helmholtz.

In Part II. the calculation of a certain quantity depending on the form of the necks of common resonators, and involved in the results of Part I., is entered upon. This quantity, denoted by *c*, is of the nature of a length, and is identical with what would be called in the theory of electricity the *electric conductivity* of the passage, supposed to be occupied by uniformly conducting matter. The question is accordingly similar to that of determining the electrical resistance of variously shaped conductors—an analogy of which I have not hesitated to avail myself freely both in investigation and statement. Much circumlocution is in this way avoided on account of the greater completeness of electrical phraseology. Passing over the case of mere holes, which has been already considered by Helmholtz, and need not be dwelt upon here, we come to the value of the resistance for necks in the form of circular cylinders. For the sake of simplicity each end is supposed to be in an infinite plane. In this form the mathematical problem is definite, but has not been solved rigorously. Two limits, however (a higher and a lower), are investigated, between which it is proved that the true resistance must lie. The lower corresponds to a correction to the length of the tube equal to $\frac{1}{4}\pi \times$ (radius) for each end. It is a remarkable coincidence that Helmholtz also finds the same quantity as an approximate correction to the length of an organ-pipe, although the two methods are entirely different and neither of them rigorous. His consists of an exact solution of the problem for an approximate cylinder, and mine of an approximate solution for a true cylinder; while both indicate on which side the truth must lie. The final result for a cylinder infinitely long is that the correction lies between $\cdot785\,R$ and $\cdot828\,R$. When the cylinder is finite, the upper limit is rather smaller. In a somewhat similar manner I have investigated limits for the resistance of a tube of revolution, which is shown to lie between

$$\int \frac{dx}{\pi y^2} \quad \text{and} \quad \int \frac{dx}{\pi y^2}\left\{1 + \tfrac{1}{2}\left(\frac{dy}{dx}\right)^2\right\},$$

where y denotes the radius of the tube at any point x along the axis. These formulæ apply whatever may be in other respects the form of the tube, but are especially valuable when it is so nearly cylindrical that dy/dx is everywhere small. The two limits are then very near each other, and either of them gives very approximately the true value. The resistance of tubes, which are either not of revolution or are not nearly straight, is afterwards approximately determined. The only experimental results bearing on the subject of this paper, and available for comparison with theory, that I have met with are some arrived at by Sondhauss[*] and Wertheim[†]. Besides those quoted by Helmholtz, I have only to mention a series of observations by Sondhauss[‡] on the pitch of flasks with long necks which led him to the empirical formula

$$n = 46705 \, \frac{\sigma^{\frac{1}{2}}}{L^{\frac{1}{2}} S^{\frac{1}{2}}},$$

σ, L being the area and length of the neck, and S the volume of the flask. The corresponding equation derived from the theory of the present paper is

$$n = 54470 \, \frac{\sigma^{\frac{1}{2}}}{L^{\frac{1}{2}} S^{\frac{1}{2}}},$$

which is only applicable, however, when the necks are so long that the corrections at the ends may be neglected—a condition not likely to be fulfilled. This consideration sufficiently explains the discordance. Being anxious to give the formulæ of Parts I. and II. a fair trial, I investigated experimentally the resonance of a considerable number of vessels which were of such a form that the theoretical pitch could be calculated with tolerable accuracy. The result of the comparison is detailed in Part III., and appears on the whole very satisfactory; but it is not necessary that I should describe it more minutely here. I will only mention, as perhaps a novelty, that the experimental determination of the pitch was not made by causing the resonators to speak by a stream of air blown over their mouths. The grounds of my dissatisfaction with this method are explained in the proper place.

[Since this paper was written there has appeared another memoir by Dr Sondhauss[§] on the subject of resonance. An empirical formula is obtained bearing resemblance to the results of Parts I. and II., and agreeing fairly well with observation. No attempt is made to connect it with the fundamental principles of mechanics. In the *Philosophical Magazine* for September 1870 [Art. IV. above], I have discussed the differences between Dr Sondhauss's formula and my own from the experimental side, and shall not therefore go any further into the matter on the present occasion.]

* Pogg. *Ann.* vol. LXXXI.
† *Annales de Chimie*, vol. XXXI.
‡ Pogg. *Ann.* vol. LXXIX.
§ Pogg. *Ann.* 1870.

PART I.

The class of resonators to which attention will chiefly be given in this paper are those where a mass of air confined almost all round by rigid walls communicates with the external atmosphere by one or more narrow passages. For the present it may be supposed that the boundary of the principal mass of air is part of an oval surface, nowhere contracted into anything like a narrow neck, although some cases not coming under this description will be considered later. In its general character the fundamental vibration of such an air-space is sufficiently simple, consisting of a periodical rush of air through the narrow channel (if there is only one) into and out of the confined space, which acts the part of a reservoir. The channel spoken of may be either a mere hole of any shape in the side of the vessel, or may consist of a more or less elongated tube-like passage.

If the linear dimension of the reservoir be small as compared to the wave-length of the vibration considered, or, as perhaps it ought rather to be said, the quarter wave-length, the motion is remarkably amenable to deductive treatment. Vibration in general may be considered as a periodic transformation of energy from the potential to the kinetic, and from the kinetic to the potential forms. In our case the kinetic energy is that of the air in the neighbourhood of the opening as it rushes backwards or forwards. It may be easily seen that relatively to this the energy of the motion inside the reservoir is, under the restriction specified, very small. A formal proof would require the assistance of the general equations to the motion of an elastic fluid, whose use I wish to avoid in this paper. Moreover the motion in the passage and its neighbourhood will not differ sensibly from that of an incompressible fluid, and its energy will depend only on the rate of total flow through the opening. A quarter of a period later this energy of motion will be completely converted into the potential energy of the compressed or rarefied air inside the reservoir. So soon as the mathematical expressions for the potential and kinetic energies are known, the determination of the period of vibration or resonant note of the air-space presents no difficulty.

The motion of an incompressible frictionless fluid which has been once at rest is subject to the same formal laws as those which regulate the flow of heat or electricity through uniform conductors, and depends on the properties of the potential function, to which so much attention has of late years been given. In consequence of this analogy many of the results obtained in this paper are of as much interest in the theory of electricity as in acoustics, while, on the other hand, known modes of expression in the former subject will save circumlocution in stating some of the results of the present problem.

Let h_0 be the density, and ϕ the velocity-potential of the fluid motion through an opening. The kinetic energy or *vis viva*

$$= \tfrac{1}{2} h_0 \iiint \left[\left(\frac{d\phi}{dx} \right)^2 + \left(\frac{d\phi}{dy} \right)^2 + \left(\frac{d\phi}{dz} \right)^2 \right] dx\, dy\, dz,$$

the integration extending over the volume of the fluid considered

$$= \tfrac{1}{2} h_0 \iint \phi \frac{d\phi}{dn}\, dS, \quad \text{by Green's theorem.}$$

Over the rigid boundary of the opening or passage, $d\phi/dn = 0$, so that if the portion of fluid considered be bounded by two equipotential surfaces, ϕ_1 and ϕ_2, one on each side of the opening,

$$\textit{vis viva} = \tfrac{1}{2} h_0 (\phi_1 - \phi_2) \iint \frac{d\phi}{dn}\, dS = \tfrac{1}{2} h_0 (\phi_1 - \phi_2)\, \dot{X},$$

if $\dot{X}$ denote the rate of total flow through the opening.

At a sufficient distance on either side ϕ becomes constant, and the rate of total flow is proportional to the difference of its values on the two sides. We may therefore put

$$\phi_1 - \phi_2 = \frac{1}{c} \iint \frac{d\phi}{dn}\, dS = \frac{\dot{X}}{c},$$

where c is a linear quantity depending on the size and shape of the opening, and representing in the electrical interpretation the reciprocal of the *resistance* to the passage of electricity through the space in question, the specific resistance of the conducting matter being taken for unity. The same thing may be otherwise expressed by saying that c is the side of a cube, whose resistance between opposite faces is the same as that of the opening.

The expression for the *vis viva* in terms of the rate of total flow is accordingly

$$\textit{vis viva} = \frac{h_0}{2} \frac{\dot{X}^2}{c} . \quad \ldots\ldots\ldots\ldots\ldots\ldots\ldots(1)$$

If S be the capacity of the reservoir, the condensation at any time inside it is given by X/S, of which the mechanical value is

$$\tfrac{1}{2} h_0 a^2 \frac{X^2}{S}, \quad \ldots\ldots\ldots\ldots\ldots\ldots\ldots\ldots\ldots(2)$$

a denoting, as throughout the paper, the velocity of sound.

The whole energy at any time, both actual and potential, is therefore

$$\frac{h_0}{2} \frac{\dot{X}^2}{c} + \frac{h_0}{2} a^2 \frac{X^2}{S}, \quad \ldots\ldots\ldots\ldots\ldots\ldots(3)$$

and is constant. Differentiating with respect to time, we arrive at

$$\ddot{X} + \frac{a^2 c}{S} X = 0 \quad \ldots\ldots\ldots\ldots\ldots\ldots\ldots\ldots(4)$$

as the equation to the motion, which indicates simple oscillations performed in a time

$$2\pi \div \sqrt{\left(\frac{a^2 c}{S}\right)}.$$

Hence if n denote the number of vibrations per second in the resonant note,

$$n = \frac{a}{2\pi} \sqrt{\left(\frac{c}{S}\right)}. \quad\quad\quad\quad\quad\quad\quad\dots(5)$$

The wave-length λ, which is the quantity most immediately connected with the dimensions of the resonant space, is given by

$$\lambda = \frac{a}{n} = 2\pi \sqrt{\left(\frac{S}{c}\right)}. \quad\quad\quad\quad\quad\dots(6)$$

A law of Savart, not nearly so well known as it ought to be, is in agreement with equations (5) and (6). It is an immediate consequence of the principle of dynamical similarity, of extreme generality, to the effect that *similar* vibrating bodies, whether they be gaseous, such as the air in organ-pipes or in the resonators here considered, or solid, such as tuning-forks, vibrate in a time which is directly as their linear dimensions. Of course the material must be the same in two cases that are to be compared, and the geometrical similarity must be complete, extending to the shape of the opening as well as to the other parts of the resonant vessel. Although the wave-length λ is a function of the size and shape of the resonator only, n or the position of the note in the musical scale depends on the nature of the gas with which the resonator is filled. And it is important to notice that it is on the nature of the gas in and near the opening that the note depends, and *not* on the gas in the interior of the reservoir, whose inertia does not come into play during vibrations corresponding to the fundamental note. In fact we may say that the mass to be moved is the air in the neighbourhood of the opening, and that the air in the interior acts merely as a spring in virtue of its resistance to compression. Of course this is only true under the limitation specified, that the diameter of the reservoir is small compared to the quarter wave-length. Whether this condition is fulfilled in the case of any particular resonator is easily seen, *à posteriori*, by calculating the value of λ from (6), or by determining it experimentally.

Several Openings.

When there are two or more passages connecting the interior of the resonator with the external air, we may proceed in much the same way, except that the equation of energy by itself is no longer sufficient. For simplicity of expression the case of two passages will be convenient, but the same method is applicable to any number. Let X_1, X_2 be the total flow through the two necks, c_1, c_2 constants depending on the form of the necks corresponding to the constant c in formula (6); then T, the *vis viva*, is given by

$$T = \frac{h_0}{2}\left(\frac{\dot{X_1}^2}{c_1} + \frac{\dot{X_2}^2}{c_2}\right),$$

the necks being supposed to be sufficiently far removed from one another not to *interfere* (in a sense that will be obvious). Further,

$$V = \text{Potential Energy} = \tfrac{1}{2}h_0 a^2 \frac{(X_1 + X_2)^2}{S}.$$

Applying Lagrange's general dynamical equation,

$$\frac{d}{dt}\left(\frac{dT}{d\dot{\psi}}\right) - \frac{dT}{d\psi} = -\frac{dV}{d\psi},$$

we obtain

$$\frac{\ddot{X}_1}{c_1} + \frac{a^2}{S}(X_1 + X_2) = 0, \qquad \frac{\ddot{X}_2}{c_2} + \frac{a^2}{S}(X_1 + X_2) = 0 \dots\dots\dots(7)$$

as the equations to the motion.

By subtraction,

$$\ddot{X}_1/c_1 - \ddot{X}_2/c_2 = 0,$$

or, on integration,

$$\frac{X_1}{c_1} = \frac{X_2}{c_2} \dots\dots\dots\dots\dots\dots\dots\dots\dots(8)$$

Equation (8) shows that the motions of the air in the two necks have the same period and are at any moment in the same phase of vibration. Indeed there is no essential distinction between the case of one neck and that of several, as the passage from one to the other may be made continuously without the failure of the investigation. When, however, the separate passages are sufficiently far apart, the constant c for the system, considered as a single communication between the interior of the resonator and the external air, is the simple sum of the values belonging to them when taken separately, which would not otherwise be the case. This is a point to which we shall return later, but in the mean time, by addition of equations (7), we find

$$\ddot{X}_1 + \ddot{X}_2 + \frac{a^2}{S}(c_1 + c_2)(X_1 + X_2) = 0,$$

so that

$$n = \frac{a}{2\pi}\sqrt{\left(\frac{c_1 + c_2}{S}\right)}. \dots\dots\dots\dots\dots(9)$$

If there be any number of necks for which the values of c are c_1, c_2, $c_3 \dots$, and no two of which are near enough to interfere, the same method is applicable, and gives

$$n = \frac{a}{2\pi}\sqrt{\frac{c_1 + c_2 + c_3 + \dots}{S}}. \dots\dots\dots\dots(9')$$

When there are two similar necks $c_2 = c_1$, and

$$n = \sqrt{2} \times \frac{a}{2\pi} \sqrt{\left(\frac{c}{S}\right)}.$$

The note is accordingly higher than if there were only one neck in the ratio of $\sqrt{2} : 1$, a fact observed by Sondhauss and proved theoretically by Helmholtz for the case of openings which are mere holes in the sides of the reservoir.

Double Resonance.

Suppose that there are two reservoirs, S, S', communicating with each other and with the external air by narrow passages or necks. If we were to consider SS' as a single reservoir and to apply equation (9), we should be led to an erroneous result; for the reasoning on which (9) is founded proceeds on the assumption that, within the reservoir, the inertia of the

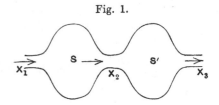

Fig. 1.

air may be left out of account, whereas it is evident that the *vis viva* of the motion through the connecting passage may be as great as through the two others. However, an investigation on the same general plan as before meets the case perfectly. Denoting by X_1, X_2, X_3 the total flows through the three necks, we have for the *vis viva* the expression

$$T = \tfrac{1}{2} h_0 \left\{ \frac{\dot{X}_1{}^2}{c_1} + \frac{\dot{X}_2{}^2}{c_2} + \frac{\dot{X}_3{}^2}{c_3} \right\},$$

and for the potential energy

$$V = \tfrac{1}{2} h_0 a^2 \left\{ \frac{(X_2 - X_1)^2}{S} + \frac{(X_3 - X_2)^2}{S'} \right\}.$$

An application of Lagrange's method gives as the differential equations to the motion,

$$\left. \begin{aligned} &\frac{\ddot{X}_1}{c_1} + a^2 \frac{X_1 - X_2}{S} = 0, \\ &\frac{\ddot{X}_2}{c_2} + a^2 \left\{ \frac{X_2 - X_1}{S} + \frac{X_2 - X_3}{S'} \right\} = 0, \\ &\frac{\ddot{X}_3}{c_3} + a^2 \frac{X_3 - X_2}{S'} = 0. \end{aligned} \right\} \quad \dots\dots\dots\dots(10)$$

By addition and integration $X_1/c_1 + X_2/c_2 + X_3/c_3 = 0$. Hence, on elimination of X_2,

$$\left. \begin{aligned} &\ddot{X}_1 + \frac{a^2}{S} \left\{ (c_1 + c_2) X_1 + \frac{c_1 c_2}{c_3} X_3 \right\} = 0, \\ &\ddot{X}_3 + \frac{a^2}{S'} \left\{ (c_3 + c_2) X_3 + \frac{c_3 c_2}{c_1} X_1 \right\} = 0. \end{aligned} \right\}$$

Assuming $X_1 = A \epsilon^{pt}$, $X_3 = B \epsilon^{pt}$, we obtain, on substitution and elimination of $A : B$,

$$p^4 + p^2 a^2 \left\{ \frac{c_1 + c_2}{S} + \frac{c_3 + c_2}{S'} \right\} + \frac{a^4}{SS'} \left\{ c_1 c_3 + c_2 (c_1 + c_3) \right\} = 0 \quad \ldots\ldots(11)$$

as the equation to determine the resonant notes. If n be the number of vibrations per second, $n^2 = -p^2/4\pi^2$, the values of p^2 given by (11) being of course both real and negative. The formula simplifies considerably if $c_3 = c_1$, $S' = S$; but it will be more instructive to work this case from the beginning. Let $c_1 = c_3 = mc_2 = mc$.

The differential equations take the form

$$\left. \begin{array}{l} \ddot{X}_1 + \dfrac{a^2 c}{S} \left\{ (1 + m) X_1 + X_3 \right\} = 0, \\[2ex] \ddot{X}_3 + \dfrac{a^2 c}{S} \left\{ (1 + m) X_3 + X_1 \right\} = 0, \end{array} \right\} \quad \text{while } X_2 = -\frac{X_1 + X_3}{m}.$$

Hence

$$\left. \begin{array}{l} (X_1 + X_3)^{\cdot\cdot} + \dfrac{a^2 c}{S} (m + 2)(X_1 + X_3) = 0, \\[2ex] (X_1 - X_3)^{\cdot\cdot} + \dfrac{a^2 c}{S} m (X_1 - X_3) = 0. \end{array} \right\}$$

The whole motion may be regarded as made up of two parts, for the first

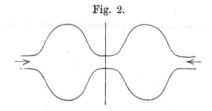

Fig. 2.

of which $X_1 + X_3 = 0$; which requires $X_2 = 0$. This motion is therefore the same as might take place were the communication between S and S' cut off, and has its period given by

$$n^2 = \frac{a^2 c_1}{4\pi^2 S} = \frac{a^2 mc}{4\pi^2 S}.$$

For the other component part, $X_1 - X_3 = 0$, so that

$$X_2 = -\frac{2X_1}{m}, \qquad n'^2 = \frac{a^2 (m + 2) c}{4\pi^2 S} \quad \ldots\ldots\ldots\ldots\ldots(12)$$

Thus $\dfrac{n'^2}{n^2} = \dfrac{m + 2}{m}$, which shows that the second note is the higher. It

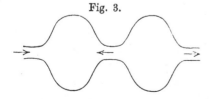

Fig. 3.

consists of vibrations in the two reservoirs opposed in phase and modified by the connecting passage, which acts in part as a second opening to both, and so raises the pitch. If the passage is small, so also is the difference of pitch between the two notes. A particular case worth notice is obtained by putting in the general equation $c_3 = 0$, which amounts to suppressing one of the communications with the external air. We thus obtain

$$p^4 + a^2 p^2 \left(\frac{c_1 + c_2}{S} + \frac{c_2}{S'} \right) + \frac{a^4}{SS'} c_1 c_2 = 0 \, ;$$

or if $S = S'$, $c_1 = mc_2 = mc$,

$$p^4 + a^2 p^2 \frac{c}{S} (m + 2) + \frac{a^4 c^2}{S^2} m = 0,$$

$$n^2 = \frac{a^2 c}{8 \pi^2 S} \{ m + 2 \pm \sqrt{m^2 + 4} \}.$$

If we further suppose $m = 1$ or $c_2 = c_1$,

$$n^2 = \frac{a^2 c}{8 \pi^2 S} (3 \pm \sqrt{5}).$$

If N be the number of vibrations for a simple resonator (S, c),

$$N^2 = \frac{a^2 c}{4 \pi^2 S} \, ;$$

so that

$$n_1^2 \div N^2 = \frac{3 + \sqrt{5}}{2} = 2 \cdot 618, \qquad N^2 \div n_2^2 = \frac{2}{3 - \sqrt{5}} = 2 \cdot 618.$$

It appears therefore that the interval from n_1 to N is the same as from N to n_2, namely, $\sqrt{(2 \cdot 618)} = 1 \cdot 618$, or rather more than a fifth. It will be found that whatever the value of m may be, the interval between the resonant notes cannot be less than $2 \cdot 414$, which is about an octave and a minor third. The corresponding value of m is 2.

A similar method is applicable to any combination of reservoirs and connecting passages, no matter how complicated, under the single restriction as to the comparative magnitudes of the reservoirs and wave-lengths; but the example just given is sufficient to illustrate the theory of multiple resonance. In Part III. a resonator of this sort will be described, which was constructed for the sake of a comparison between the theory and experiment. In applying the formulæ (6) or (12) to an actual measurement, the question will arise whether the volume of the necks, especially when they are rather large, is to be included or not in S. At the moment of rest the air in the neck is compressed or rarefied as well as that inside the reservoir, though not to the same degree; in fact the condensation must vary continuously between the interior of the resonator and the external air. This consideration shows that, at least in the case of necks which are tolerably symmetrical, about half the volume of the neck should be included in S.

[In consequence of a suggestion made by Mr Clerk Maxwell, who reported on this paper, I have been led to examine what kind of effect would be produced by a deficient rigidity in the envelope which contains the alternately compressed and rarefied air. Taking for simplicity the case of a sphere, let us suppose that the radius, instead of remaining constant at its normal value R, assumes the variable magnitude $R + \rho$. We have

$$\text{kinetic energy} = \frac{h_0 \dot{X}^2}{2c} + \frac{m}{2} \dot{\rho}^2,$$

$$\text{potential energy} = \frac{h_0 a^2}{2S} \{X + 4\pi R^2 \rho\}^2 + \tfrac{1}{2}\beta\rho^2,$$

where m and β are constants expressing the inertia and rigidity of the spherical shell. Hence, by Lagrange's method,

$$\left. \begin{array}{l} \ddot{X} + \dfrac{ca^2}{S}(X + 4\pi R^2 \rho) = 0, \\[3mm] m\ddot{\rho} + 4\pi R^2 \dfrac{h_0 a^2}{S}(X + 4\pi R^2 \rho) + \beta\rho = 0, \end{array} \right\}$$

equations determining the periods of the two vibrations of which the system is capable. It might be imagined at first sight that a yielding of the sides of the vessel would necessarily lower the pitch of the resonant note; but this depends on a tacit assumption that the capacity of the vessel is largest when the air inside is most compressed. But it may just as well happen that the opposite is true. Everything depends on the relative magnitudes of the periods of the two vibrations supposed for the moment independent of one another. If the note of the shell be very high compared to that of the air, the inertia of the shell may be neglected, and this part of the question treated statically. Putting in the equations $m = 0$, we see that the phases of X and ρ are opposed, and then X goes through its changes more slowly than before. On the other hand, if it be the note of the air-vibration, which is much the higher, we must put $\beta = 0$, which leads to

$$4\pi R^2 h_0 \ddot{X} - cm\ddot{\rho} = 0,$$

showing that the phases of X and ρ agree. Here the period of X is diminished by the yielding of the sides of the vessel, which indeed acts just in the same way as a second aperture would do. A determination of the actual note in any case of a spherical shell of given dimensions and material would probably be best obtained deductively.

But in order to see what probability there might be that the results of Part III. on glass flasks were sensibly modified by a want of rigidity, I thought it best to make a direct experiment. To the neck of a flask was fitted a glass tube of rather small bore, and the whole filled with water so as make a kind of water-thermometer. On removing by means of an air-pump the pressure of the atmosphere on the outside of the bulb, the liquid fell in the tube, but only to an extent which indicated an increase in the capacity of the flask of about a ten-thousandth part. This corresponds in the ordinary arrangement to a doubled density of the contained air. It is clear that so small a yielding could produce no sensible effect on the pitch of the air-vibration.]

Open Organ-pipes.

Although the problem of open organ-pipes, whose diameter is very small compared to their length and to the wave-length, has been fully considered by Helmholtz, it may not be superfluous to show how the question may be attacked from the point of view of the present paper, more especially as some important results may be obtained by a comparatively simple analysis. The principal difficulty consists in finding the connexion between the spherical waves which diverge from the open end of the tube into free space, and the waves in the tube itself, which at a distance from the mouth, amounting to several diameters, are approximately plane. The transition occupies a space which is large compared to the diameter, and in order that the present treatment may be applicable must be small compared to the wave-length. This condition being fulfilled, the compressibility of the air in the space mentioned may be left out of account and the difficulty is turned. Imagine a piston (of infinitely small thickness) in the tube at the place where the waves cease to be plane. The motion of the air on the free side is entirely determined by the motion of the piston, and the *vis viva* within the space considered may be expressed by $\frac{1}{2}h_0 \dot{X}^2/c$, where $\dot{X}$ denotes the rate of total flow at the place of the piston, and c is, as before, a linear quantity depending on the form of the mouth. If Q is the section of the tube and ψ the velocity potential, $\dot{X} = Q \, d\psi/dx$. The most general expression for the velocity-potential of plane waves is

$$\psi = \left(\frac{A}{k} \sin kx + B \cos kx\right) \cos 2\pi nt + \beta \cos kx \sin 2\pi nt, \dots (13)$$

$$\frac{d\psi}{dx} = (A \cos kx - Bk \sin kx) \cos 2\pi nt - \beta k \sin kx \sin 2\pi nt,$$

where $k = 2\pi/\lambda = 2\pi n/a$. When $x = 0$,

$$\psi = B \cos 2\pi nt + \beta \sin 2\pi nt, \qquad \frac{d\psi}{dx} = A \cos 2\pi nt.$$

The variable part of the pressure on the tube side of the piston $= - h_0 \, d\psi/dt$. The equation to the motion of the air in the mouth is therefore

$$\frac{Q}{c} \frac{d}{dt} \frac{d\psi}{dx} + \frac{d\psi}{dt} = 0,$$

or, on integration,

$$\frac{Q}{c} \frac{d\psi}{dx} + \psi = 0. \quad \dots\dots\dots\dots\dots\dots\dots(14)$$

This is the condition to be satisfied when $x = 0$.

Substituting the values of ψ and $d\psi/dx$, we obtain

$$\cos 2\pi nt \left(A \frac{Q}{c} + B \right) + \beta \sin 2\pi nt = 0,$$

which requires

$$A \frac{Q}{c} + B = 0, \qquad \beta = 0.$$

If there is a node at $x = -l$, $A \cos kl + Bk \sin kl = 0$; so that

$$k \tan kl = -\frac{A}{B} = -\frac{c}{Q}. \qquad \dots\dots\dots\dots\dots(15)$$

This equation gives the fundamental note of the tube closed at $x = -l$; but it must be observed that l is not the length of the tube, because the origin $x = 0$ is not in the mouth. There is, however, nothing indeterminate in the equation, although the origin is to a certain extent arbitrary; for the values of c and l will change together so as to make the result for k approximately constant. This will appear more clearly when we come, in Part II., to calculate the actual value of c for different kinds of mouths. In the formation of (14) the pressure of the air on the positive side at a distance from the origin small against λ has been taken absolutely constant. Across such a loop surface no energy could be transmitted. In reality, of course, the pressure is variable on account of the spherical waves, and energy continually escapes from the tube and its vicinity. Although the pitch of the resonant note is not affected, it may be worth while to see what correction this involves.

We must, as before, consider the space in which the transition from plane to spherical waves is effected as small compared with λ. The potential in free space may be taken

$$\psi = \frac{A'}{r} \cos (kr + g - 2\pi nt), \qquad \dots\dots\dots\dots\dots(16)$$

expressing spherical waves diverging from the mouth of the pipe, which is the origin of r. The origin of x is still supposed to lie in the region of plane waves.

$* \, 4\pi r^2 \dfrac{d\psi}{dr}$ = rate of total flow across the surface of the sphere whose radius is r

$$= - 4\pi A' [\cos 2\pi nt \, \{\cos (kr + g) + kr \sin (kr + g)\}$$
$$+ \sin 2\pi nt \, \{\sin (kr + g) - kr \cos (kr + g)\}].$$

---

* Throughout Helmholtz's paper the mouth of the pipe is supposed to lie in an infinite plane, so that the diverging waves are hemispherical. The calculation of the value of c is thereby simplified. Except for this reason it seems better to consider the diverging waves completely spherical as a nearer approximation to the actual circumstances of organ-pipes, although the sphere could never be quite complete.

If the compression in the neighbourhood of the mouth is neglected, this must be the same as

$$Q \frac{d\psi}{dx} {}_{=0} = QA \cos 2\pi nt.$$

Accordingly

$$AQ = -4\pi A' \left\{ \cos (kr + g) + kr \sin (kr + g) \right\},$$

$$0 = \sin (kr + g) - kr (\cos kr + g).$$

These equations express the connexion between the plane and spherical waves. From the second, $\tan (kr + g) = kr$, which shows that g is a small quantity of the order $(kr)^2$. From the first

$$A' = -\frac{AQ}{4\pi},$$

so that

$$\psi_r = -\frac{AQ}{4\pi r} \cos 2\pi nt - \frac{AQk}{4\pi} \sin 2\pi nt,$$

the terms of higher order being omitted.

Now within the space under consideration the air moves according to the same laws as electricity, and so

$$\frac{Q}{c} \frac{d\psi}{dx} {}_{=0} = -\psi_{x=0} + \psi_r,$$

$$\frac{d\psi}{dx} {}_{=0} = A \cos 2\pi nt,$$

$$\psi_{x=0} = B \cos 2\pi nt + \beta \sin 2\pi nt.$$

Therefore on substitution and equation of the coefficients of $\sin 2\pi nt$, $\cos 2\pi nt$, we obtain

$$AQ \left(\frac{1}{c} + \frac{1}{4\pi r} \right) = -B, \qquad \beta = -\frac{AQk}{4\pi}.$$

When the mouth is not much contracted c is of the order of the radius of the mouth, and when there is contraction it is smaller still. In all cases therefore the term $1/4\pi r$ is very small compared to $1/c$; and we may put

$$\frac{AQ}{c} = -B, \qquad \beta = -\frac{AQk}{4\pi}, \quad \dots\dots\dots\dots\dots(17)$$

which agree nearly with the results of Helmholtz. In his notation a quantity α is used defined by the equation $-A/Bk = \cot k\alpha$, so that $\cot k\alpha = \tan kl$ by (15), or $k(l + \alpha) = \frac{1}{2}(2m + 1)\pi$; α may accordingly be considered as the correction to the length of the tube (measured, however, in our method only on the negative side of the origin), and will be given by $\cot k\alpha = -c/kQ$.

The value of c will be investigated in **Part II.**

The original theory of open pipes makes the pressure absolutely constant at the mouth, which amounts to neglecting the inertia of the air outside. Thus, if the tube itself were full of air, and the external space of hydrogen, the correction to the length of the pipe might be neglected. The first investigation, in which no escape of energy is admitted, would apply if the pipe and a space round its mouth, large compared to the diameter, but small compared to the wave-length, were occupied by air in an atmosphere otherwise composed of incomparably lighter gas. These remarks are made by way of explanation, but for a complete discussion of the motion as determined by (13) and 17, I must refer to the paper of Helmholtz.

Long Tube in connexion with a Reservoir.

It may sometimes happen that the length of a neck is too large compared to the quarter wave-length to allow the neglect of the compressibility of the air inside. A cylindrical neck may then be treated in the same way as the organ-pipe. The potential of plane waves inside the neck may, by what has been proved, be put into the form

$$\psi = A' \sin k (x - \alpha) \cos 2\pi nt,$$

if we neglect the escape of energy, which will not affect the pitch of the resonant note.

$$d\psi/dt = -2\pi n A' \sin k (x - \alpha) \sin 2\pi nt,$$

$$d\psi/dx = kA' \cos k (x - \alpha) \cos 2\pi nt,$$

where α is the correction for the outside end.

The rate of flow out of $S = Q \, d\psi/dx.$

$$\text{Total flow} = Q \int \frac{d\psi}{dx} \, dt = kA'Q \cos kL \, \frac{\sin 2\pi nt}{2\pi n},$$

the reduced length of the tube, including the corrections for both ends, being denoted by L. Thus rarefaction in S

$$= k \, \frac{A'Q \cos kL}{S} \, \frac{\sin 2\pi nt}{2\pi n} = \frac{1}{a^2} \frac{d\psi}{dt} = \frac{2\pi n A' \sin kL}{a^2} \sin 2\pi nt.$$

This is the condition to be satisfied at the inner end. It gives

$$\tan kL = \frac{a^2}{4\pi^2 n^2} \frac{kQ}{S} = \frac{Q}{kS}. \quad \dots\dots\dots\dots\dots(18)$$

When kL is small,

$$\tan kL = kL + \tfrac{1}{3}(kL)^2 = \frac{Q}{kS};$$

so that

$$k^2 = \frac{Q}{LS}\left(1 - \tfrac{1}{3}\frac{LQ}{S}\right),$$

$$n = \frac{a}{2\pi}\sqrt{\frac{Q}{LS}}\left(1 - \tfrac{1}{6}\frac{LQ}{S}\right) = \frac{a}{2\pi}\sqrt{\frac{Q}{L\left(S + \tfrac{1}{3}LQ\right)}}. \quad \ldots\ldots(19)$$

In comparing this with (5), it is necessary to introduce the value of c, which is Q/L. (5) will accordingly give the same result as (19) if *one-third* of the contents of the neck be included in S. The first overtone, which is often produced by blowing in preference to the fundamental note, corresponds approximately to the length L of a tube open at both ends, modified to an extent which may be inferred from (18) by the finiteness of S.

The number of vibrations is given by

$$n = \frac{a}{2}\left(\frac{1}{L} + \frac{Q}{\pi^2 S}\right). \quad \ldots\ldots\ldots\ldots\ldots\ldots\ldots\ldots(20)$$

[The application of (20) is rather limited, because, in order that the condensation within S may be uniform as has been supposed, the linear dimension of S must be considerably less than the quarter wave-length; while, on the other hand, the method of approximation by which (20) is obtained from (18) requires that S should be large in comparison with QL.

A slight modification of (18) is useful in finding the pitch of pipes which are cylindrical through most of their length, but at the closed end expand into a bulb S of no great capacity. The only change required is to understand by L the length of the pipe down to the place where the enlargement begins, with a correction for the *outer* end. Or if L denote the length of the tube simply, we have

$$\tan k\,(L + \alpha) = \frac{Q}{kS}, \quad \ldots\ldots\ldots\ldots\ldots\ldots(20\,a)$$

and $\alpha = \tfrac{1}{4}\pi R$ approximately.

If S be very small we may derive from $(20\,a)$

$$n = \frac{a}{4\,(L + \alpha + S/Q)}. \quad \ldots\ldots\ldots\ldots\ldots\ldots(20\,b)$$

In this form the interpretation is very simple, namely, that at the closed end the shape is of no consequence, and only the volume need be attended to. The air in this part of the pipe acts merely as a spring, its inertia not coming into play. A few measurements of this kind will be given in Part III.

The overtones of resonators which have not long necks are usually very high. Within the body of the reservoir a nodal surface must be formed, and the air on the further side vibrates as if it was contained in a completely closed vessel. We may form an idea of the character of these vibrations from the case of a sphere, which may be easily worked out from the equations given by Professor Stokes in his paper " On the Communication of Motion from a vibrating Sphere to a Gas"*. The most important vibration within a sphere is that which is expressed by the term of the first order in Laplace's series, and consists of a swaying of the air from side to side like that which takes place in a doubly closed pipe. I find that for this vibration

$$\text{radius : wave-length} = \cdot 3313,$$

so that the note is higher than that belonging to a doubly closed (or open) pipe of the length of the diameter of the sphere by about a musical fourth. We might realize this vibration experimentally by attaching to the sphere a neck of such length that it would by itself, when closed at one end, have the same resonant note as the sphere.

Lateral Openings.

In most wind instruments the gradations of pitch are attained by means of lateral openings, which may be closed at pleasure by the fingers or otherwise. The common crude theory supposes that a hole in the side of, say, a flute establishes so complete a communication between the interior and the surrounding atmosphere that a loop or point of no condensation is produced immediately under it. It has long been known that this theory is inadequate, for it stands on the same level as the first approximation to the motion in an open pipe in which the inertia of the air outside the mouth is virtually neglected. Without going at length into this question, I will merely indicate how an improvement in the treatment of it may be made.

Let ψ_1, ψ_2 denote the velocity-potentials of the systems of plane waves on the two sides of the aperture, which we may suppose to be situated at the point $x = 0$. Then with our previous notation the conditions evidently are that when $x = 0$,

$$\psi_1 = \psi_2, \qquad \frac{Q}{c}\left(\frac{d\psi_1}{dx} - \frac{d\psi_2}{dx}\right) + \psi = 0, \ldots\ldots\ldots(20\,c)$$

the escape of energy from the tube being neglected. These equations determine the connexion between the two systems of waves in any case that may arise, and the working out is simple. The results are of no particular interest, unless it be for a comparison with experimental measurements, which, so far as I am aware, have not hitherto been made.]

----

* Professor Stokes informs me that he had himself done this at the request of the Astronomer Royal. [1899. See *Theory of Sound,* §§ 330, 331.]

PART II.

In order to complete the theory of resonators, it is necessary to determine the value of c, which occurs in all the results of Part I., for different forms of mouths. This we now proceed to do. Frequent use will be made of a principle which might be called that of minimum *vis viva*, and which it may be well to state clearly at the outset.

Imagine a portion of incompressible fluid at rest within a closed surface to be suddenly set in motion by an arbitrary normal velocity impressed on the surface, then the actual motion assumed by the fluid will have less *vis viva* than any other motion consistent with continuity and with the boundary conditions*.

If u, v, w be the component velocities, and ρ the density at any point,

$$vis\ viva = \tfrac{1}{2} \iiint \rho\,(u^2 + v^2 + w^2)\,dx\,dy\,dz,$$

the integration extending over the volume considered. The minimum *vis viva* corresponding to prescribed boundary conditions depends of course on ρ; but if in any specified case we conceive the value of ρ in some places diminished and nowhere increased, we may assert that the minimum *vis viva* is *less* than before; for there will be a decrease if u, v, w remain unaltered, and therefore, *à fortiori*, when they have their actual values as determined by the minimum property. Conversely, an increase in ρ will necessarily raise the value of the minimum *vis viva*. The introduction of a rigid obstacle into a stream will always cause an increase of *vis viva*; for the new motion is one that might have existed before consistently with continuity, the fluid displaced by the obstacle remaining at rest. Any kind of obstruction in the air-passages of a musical instrument will therefore be accompanied by a fall of the note in the musical scale.

Long Tubes.

The simplest case that can be considered consists of an opening in the form of a cylindrical tube, so long in proportion to its diameter that the corrections for the ends may be neglected. If the length be L and area of section σ, the electrical resistance is L/σ, and

$$c = \frac{\sigma}{L}. \quad \dots\dots\dots\dots\dots\dots\dots\dots\dots\dots\dots(21)$$

For a circular cylinder of radius R

$$c = \frac{\pi R^2}{L}. \quad \dots\dots\dots\dots\dots\dots\dots\dots\dots(22)$$

---

* Thomson and Tait's *Natural Philosophy*, § 317.

Simple Apertures.

The next in order of simplicity is probably the case treated by Helmholtz, where the opening consists of a simple hole in the side of the reservoir, considered as indefinitely thin and approximately plane in the neighbourhood of the opening. The motion of the fluid in the plane of the opening is by the symmetry normal, and therefore the velocity-potential is constant over the opening itself. Over the remainder of the plane in which the opening lies the normal velocity is of course zero, so that ϕ may be regarded as the potential of matter distributed over the opening only. If the there constant value of the potential be called ϕ_1, the electrical resistance for *one side only* is

$$\phi_1 \div \iint \frac{d\phi}{dn}\, d\sigma,$$

the integration going over the area of the opening.

Now

$$\iint \frac{d\phi}{dn}\, d\sigma = 2\pi \times \text{ the whole quantity of matter;}$$

so that if we call M the quantity necessary to produce the unit potential, the resistance for one side $= 1/2\pi M$.

Accordingly

$$c = \pi M. \quad \dots\dots\dots\dots\dots\dots\dots\dots(23)$$

In electrical language M is the *capacity* of a conducting lamina of the shape of the hole when situated in an open space.

For a circular hole $M = 2R/\pi$, and therefore

$$c = 2R. \quad \dots\dots\dots\dots\dots\dots\dots\dots(24)$$

When the hole is an ellipse of eccentricity e and semimajor axis R,

$$c = \frac{\pi R}{F(e)}, \quad \dots\dots\dots\dots\dots\dots\dots(25)$$

where F is the symbol of the complete elliptic function of the first order. Results equivalent to (23), (24), and (25) are given by Helmholtz.

When the eccentricity is but small, the value of c depends sensibly on the area (σ) of the orifice only. As far as the square of e,

$$F(e) = \frac{\pi}{2}\left(1 + \tfrac{1}{4}e^2\right),$$

$$\sigma = \pi R^2 \sqrt{1 - e^2} = \pi R^2 \left(1 - \tfrac{1}{2}e^2\right), \qquad R = \sqrt{\frac{\sigma}{\pi}}\left(1 + \tfrac{1}{4}e^2\right);$$

that

$$c = \pi \sqrt{\frac{\sigma}{\pi}} \div \frac{\pi}{2} = 2 \sqrt{\frac{\sigma}{\pi}}, \quad \dots\dots\dots\dots\dots (26)$$

the fourth power of e being neglected—a formula which may be applied without sensible error to any orifice of an approximately circular form. In fact for a given area the circle is the figure which gives a minimum value to c, and in the neighbourhood of the minimum the variation is slow.

Next, consider the case of two circular orifices. If sufficiently far apart they act independently of each other, and the value of c for the pair is the simple sum of the separate values, as may be seen either from the law of multiple arcs by considering c as the electric *conductivity* between the outside and inside of the reservoir, or from the interpretation of M in (23). The first method applies to any kind of openings with or without necks. As the two circles (which for precision of statement we may suppose equal) approach one another, the value of c diminishes steadily until they touch. The change in the character of the motion may be best followed by considering the plane of symmetry which bisects at right angles the line joining the two centres, and which may be regarded as a rigid plane precluding normal motion. Fixing our attention on half the motion only, we recognize the plane as an obstacle continually advancing, and at each step more and more obstructing the passage of fluid through the circular opening. After the circles come into contact this process cannot be carried further; but we may infer that, as they amalgamate and shape themselves into a single circle (the total area remaining all the while constant), the value of c still continues to diminish till it approaches its minimum value, which is less than at the commencement in the ratio of $\sqrt{2} : 2$ or $1 : \sqrt{2}$. There are very few forms of opening indeed for which the exact calculation of $\dot{M}$ or c can be effected. We must for the present be content with the formula (26) as applying to nearly circular openings, and with the knowledge that the more elongated or broken up the opening, the greater is c compared to σ. In the case of similar orifices or systems of orifices c varies as the linear dimension.

Cylindrical Necks.

Most resonators used in practice have necks of greater or less length, and even where there is nothing that would be called a neck, the thickness of the side of the reservoir could not always be neglected. For simplicity we shall take the case of circular cylinders whose inner ends lie on an approximately plane part of the side of the vessel, and whose outer ends are also supposed to lie in an infinite plane, or at least a plane whose dimensions are considerable compared to the diameter of the cylinder. Even under this form the problem does not seem capable of exact solution; but we shall be able to fix two slightly differing quantities between which the true value of c must lie, and which

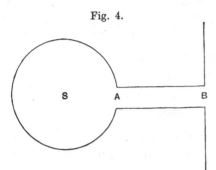

Fig. 4.

determine it with an accuracy more than sufficient for acoustical purposes. The object is to find the *vis viva* in terms of the rate of flow. Now, according to the principle stated at the beginning of Part II., we shall obtain too small a *vis viva* if at the ends A and B of the tube we imagine infinitely thin laminæ of fluid of infinitely small density. We may be led still more distinctly perhaps to the same result by supposing, in the electrical analogue, thin disks of perfectly conducting matter at the ends of the tube, whereby the effective resistance must plainly be lessened. The action of the disks is to produce uniform potential over the ends, and the solution of the modified problem is obvious. Outside the tube the question is the same as for a simple circular hole in an infinite plane, and inside the tube the same as if the tube were indefinitely long.

Accordingly

$$\text{resistance} = \frac{L}{\pi R^2} + \frac{1}{2R} = \frac{1}{\pi R^2}\left(L + \frac{\pi}{2}R\right). \quad \ldots\ldots\ldots(27)$$

The correction to the length is therefore $\frac{1}{2}\pi R$, that is, $\frac{1}{4}\pi R$ for each end,

and

$$c = \frac{\pi R^2}{L + \frac{1}{2}\pi R}. \quad \ldots\ldots\ldots\ldots\ldots\ldots\ldots\ldots\ldots(28)$$

Helmholtz, in considering the case of an organ-pipe, arrives at a similar conclusion,—that the correction to the length (α) is approximately $\frac{1}{4}\pi R$. His method is very different from the above, and much less simple. He begins by investigating certain forms of mouths for which the exact solution is possible, and then, by assigning suitable values to arbitrary constants, identifies one of them with a true cylinder, the agreement being shown to be everywhere very close. Since the curve substituted for the generating line of the cylinder lies entirely outside it, Helmholtz infers that the correction to the length thus obtained is too small.

If, at the ends of the tube, instead of layers of matter of no density, we imagine rigid pistons of no sensible thickness, we shall obtain a motion whose *vis viva* is necessarily *greater* than that of the real motion; for the motion with the pistons might take place without them consistently with continuity. Inside the tube the character of the motion is the same as before, but for the outside we require the solution of a fresh problem :— To determine the motion of an infinite fluid bounded by an infinite plane, the normal velocity over a circular area of the plane being a given constant, and over the rest of the plane zero. The potential may still be regarded as due to matter confined to the circle, but is no longer constant over its area; but the density of matter at any point, being proportional to $d\phi/dn$ or to the normal velocity, is constant.

The *vis viva* of the motion $= \frac{1}{2}\iint \phi \frac{d\phi}{dn} d\sigma = \frac{1}{2}\frac{d\phi}{dn}\iint \phi d\sigma$, the integration going over the area of the circle.

The rate of total flow through the plane $= \iint \frac{d\phi}{dn} d\sigma = \pi R^2 \frac{d\phi}{dn}$; so that

$$\frac{2 \text{ vis viva}}{(\text{rate of flow})^2} = \frac{\iint \phi d\sigma}{\pi^2 R^4 d\phi/dn}. \quad \dots\dots\dots\dots\dots(29)$$

We proceed to investigate the value of $\iint \phi d\sigma$, which is the *potential on itself* of a circular disk of unit density.

Potential on itself of a uniform circular disk.

r denoting the distance between any two points on the disk, the quantity to be evaluated is expressed by

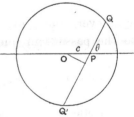

Fig. 5.

$$\iint d\sigma \iint \frac{d\sigma'}{r}.$$

The first step is to find the potential at any point P, or $\iint \frac{d\sigma'}{r}$. Taking this point as an origin of polar coordinates, we have

$$\text{potential} = \iint \frac{d\sigma'}{r} = \iint \frac{r dr d\theta}{r} = \int r d\theta = \int (PQ + PQ') d\theta.$$

Now from the figure

$$\tfrac{1}{4}(QQ')^2 = R^2 - c^2 \sin^2 \theta,$$

where c is the distance of the point P from the centre of the circle whose radius is R. Thus potential at P

$$= 2R \int_0^\pi \sqrt{1 - \frac{c^2}{R^2} \sin^2 \theta}\, d\theta = 4R \int_0^{\frac{\pi}{2}} \sqrt{1 - \frac{c^2}{R^2} \sin^2 \theta}\, d\theta. \quad \dots\dots(30)$$

Hence potential of disk on itself

$$= 4\pi R^3 \int_0^1 dx \int_0^{\frac{1}{2}\pi} \sqrt{1 - x \sin^2 \theta}\, d\theta,$$

if for the sake of brevity we put $c^2/R^2 = x$.

In performing first the integration with respect to θ we come upon elliptic functions, but they may be avoided by changing the order of integration.

$$\int_0^1 dx \sqrt{1 - x \sin^2 \theta} = \left\{ -\frac{2}{3 \sin^2 \theta} (1 - x \sin^2 \theta)^{\frac{3}{2}} \right\}_0^1$$

$$= \frac{2}{3 \sin^2 \theta} (1 - \cos^3 \theta) = \tfrac{2}{3} \frac{1}{1 + \cos \theta} + \tfrac{2}{3} \cos \theta;$$

so that potential on itself

$$= \tfrac{8}{3} \pi R^3 \int_0^{\frac{\pi}{2}} d\theta \left\{ \frac{1}{1 + \cos \theta} + \cos \theta \right\} = \tfrac{8}{3} \pi R^3 \{1 + 1\} = \tfrac{16}{3} \pi R^3. \ldots (31)$$

This, therefore, is the value of $\iint \phi \, d\sigma$ when the density is supposed equal to unity. The corresponding value of $d\phi/dn = 2\pi$, and so from (29)

$$\frac{2 \; vis \; viva}{(\text{rate of flow})^2} = \frac{8}{3\pi^2 R} \cdot \quad \ldots\ldots\ldots\ldots\ldots\ldots\ldots(32)$$

This is for the space outside one end. For the whole tube and both ends

$$\frac{2 \; vis \; viva}{(\text{rate of flow})^2} = \frac{L}{\pi R^2} + \frac{16}{3\pi^2 R} \cdot \quad \ldots\ldots\ldots\ldots (33)$$

Whatever, then, may be the ratio of $L : R$, the electrical resistance to the passage in question, or $1/c$, is limited by

$$\frac{L}{\pi R^2} + \frac{16}{3\pi^2 R} > \frac{1}{c} > \frac{L}{\pi R^2} + \frac{1}{2R} \cdot \quad \ldots\ldots\ldots\ldots (34)$$

In practical application it is sometimes convenient to use the quantity α or correction to the length. In terms of α (34) becomes

$$\frac{16}{3\pi} R > \alpha > \frac{\pi}{2} R,$$

or in decimals

$$\left.\begin{array}{l} \alpha > (1{\cdot}571 R = 2 \times {\cdot}785 R) \\ \alpha < (1{\cdot}697 R = 2 \times {\cdot}849 R) \end{array}\right\} \quad \ldots\ldots\ldots\ldots\ldots(34')$$

The corrections for *both* ends is the thing here denoted by α. Of course for one end it is only necessary to take the half*.

---

* Though not immediately connected with our present subject, it may be worth notice that if at the centre of the tube, or anywhere else, the velocity be constrained (by a piston) to be constant across the section, as it would approximately be if the tube were very long without a piston, the limiting inequalities (34) still hold good. For large values of L the two cases do not sensibly differ, but for small values of L compared to R the true solution of the original problem tends to coincide with the lower limit, and of the modified (central piston) problem with the higher.

I do not suppose that any experiments hitherto made with organ-pipes could discriminate with certainty between the two values of α in (34'). If we adopt the mean provisionally, we may be sure that we are not wrong by so much as $\cdot032R$ for each end.

Our upper limit to the value of α expressed in (34') was found by considering the hypothetical case of a uniform velocity over the section of the mouth, and we fully determined the non-rotational motion both for the inside and for the outside of the tube. Of course the velocity is not really uniform at the mouth; it is, indeed, infinite at the edge. If we could solve the problem for the inside and outside when the velocity (normal) at the mouth is of the form $a + br^2$, we should with a suitable value of $b : a$ get a much better approximation to the true *vis viva*. The problem for the outside may be solved, but for the inside it seems far from easy. It is possible, however, that we may be able to find some motion for the inside satisfying the boundary conditions and the equation of continuity, which, though of a rotational character, shall yet make the whole *vis viva* for the inside and outside together less than that previously obtained. At the same time this *vis viva* is by Thomson's law necessarily greater than the one we seek.

Motion in a finite cylindrical tube, the axial velocity at the plane ends $(x = 0$ and $x = l)$ *being*

$$u = u_0 + \chi(r), \quad\text{...........................(35)}$$

where

$$\int_0^1 r\chi(r)\,dr = 0, \quad\text{........................(36)}$$

r being the transverse coordinate, and the radius of the cylinder being put equal to 1.

If u, v be the component velocities, the continuity equation is

$$\frac{du}{dx} + \frac{1}{r}\frac{d(rv)}{dr} = 0, \quad\text{........................(37)}$$

whence

$$ru = d\psi/dr, \qquad rv = -d\psi/dx, \quad\text{................(37')}$$

where ψ is arbitrary so far as (37) is concerned.

Take

$$\psi = \tfrac{1}{2}u_0 r^2 + \phi(x)\int_0^r r\chi(r)\,dr,$$

so that

$$u = u_0 + \phi(x)\chi(r), \qquad v = -\phi'(x)\frac{1}{r}\int_0^r r\chi(r)\,dr. \quad\text{.........(38)}$$

It is clear from (38) that if

$$\phi(0) = \phi(l) = 1, \quad\text{........................(39)}$$

$$u_{x=0} = u_0 + \chi(r) = u_{x=l}, \qquad v_{r=1} = 0 \quad\text{for all values of } x.$$

Thus (38) satisfies the boundary conditions including (35), and ϕ is still arbitrary, except in so far as it is limited by (39).

In order to obtain an expression for the *vis viva*, we must integrate $u^2 + v^2$ over the volume of the cylinder.

$$\text{Twice } \textit{vis viva} = u_0{}^2 \pi l + 2u_0 \int_0^l \phi(x)\, dx \int_0^1 \chi(r)\, 2\pi r\, dr$$

$$+ \int_0^l \{\phi(x)\}^2\, dx . \int_0^1 \{\chi(r)\}^2\, 2\pi r\, dr + \int_0^l \{\phi'(x)\}^2\, dx . \int_0^1 2\pi r^{-1}\, dr \left\{\int_0^r r\chi(r)\, dr\right\}^2.$$

$$\dots\dots\dots (40)$$

The second term vanishes in virtue of (36), and we may write

$$\text{Twice } \textit{vis viva} = u_0{}^2 \pi l + \int_0^l (Ay^2 + By'^2)\, dx, \dots\dots\dots (40')$$

where A and B are known quantities depending on χ, and $y = \phi(x)$ is so far an arbitrary function, which we shall determine so as to make the *vis viva* a minimum.

By the method of variations

$$y = Ce^{-x\sqrt{(A/B)}} + C'e^{+x\sqrt{(A/B)}} ; \quad \dots\dots\dots\dots (41)$$

and in order to satisfy (39),

$$1 = C + C', \qquad 1 = Ce^{-l\sqrt{(A/B)}} + C'e^{+l\sqrt{(A/B)}} ; \quad \dots\dots\dots (42)$$

(41) and (42) completely determine y as a function of x, and when this value of y is used in (40) the *vis viva* is less than with any other form of y. On substitution in (40'),

$$\text{Twice } \textit{vis viva} = u_0{}^2 \pi l + 2\sqrt{(AB)} \frac{1 - e^{-2l\sqrt{(A/B)}}}{1 + e^{-2l\sqrt{(A/B)}}} . \quad \dots\dots\dots\dots (43)$$

The *vis viva* expressed in (43) is less than any other which can be derived from the equation (38); but it is not the least possible, as may be seen by substituting the value of ψ in the stream-line equation

$$\frac{d^2\psi}{dx^2} - \frac{1}{r}\frac{d\psi}{dr} + \frac{d^2\psi}{dr^2} = 0,$$

which will be found to be *not* satisfied.

The next step is to introduce special forms of χ. Thus let $u_{x=0} = 1 + \mu r^2$.

Then $\qquad\qquad\qquad u_0 = 1 + \tfrac{1}{2}\mu, \qquad \chi = \mu(-\tfrac{1}{2} + r^2).$

Accordingly

$$A = \frac{\pi\mu^2}{12}, \qquad B = \tfrac{1}{16} . \frac{\pi\mu^2}{12}, \qquad \sqrt{(AB)} = \frac{\pi\mu^2}{48}, \qquad \sqrt{(A/B)} = 4 ;$$

and (43) becomes

$$2 \ vis \ viva = \pi l \left(1 + \tfrac{1}{2}\mu\right)^2 + \frac{\pi \mu^2}{24} \frac{1 - e^{-sl}}{1 + e^{-sl}} \cdot \quad \dots \dots \dots (44)$$

We have in (44) the *vis viva* of a motion within a circular cylinder which satisfies the continuity equation, and which makes over the plane ends $u = 1 + \mu r^2$. If $\mu = 0$ we fall back on the simple case considered before; and this is the value of μ for which the *vis viva* in (44) is a minimum compared to the rate of flow $(1 + \tfrac{1}{2}\mu)$. But for the part outside the cylinder the *vis viva* is, as we may anticipate, least when μ has some finite value; so that when we consider the motion as a whole it will be a finite value of μ that gives the least *vis viva*.

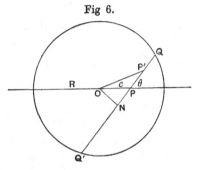

Fig 6.

The *vis viva* of the motion outside the ends is to be found by the same method as before, the first step being to determine the potential at any point of a circular disk whose density $= \mu r^2$;

$$\text{potential at } P = \iint \frac{\rho \, d\rho \, d\theta}{\rho} \, \mu OP'^2,$$

where

$$OP'^2 = c^2 + \rho^2 + 2c\rho \cos \theta ;$$

so that

$$\text{potential at } P = \mu \int d\theta \left\{ c^2 \rho + \tfrac{1}{3} \rho^3 + c\rho^2 \cos \theta \right\} ;$$

or if previously to integration with respect to θ we add together the elements from Q to Q',

$$= \mu \int_0^\pi d\theta \, (PQ + PQ') \left\{ c^2 + \frac{PQ^2 + PQ'^2 - PQ \cdot PQ'}{3} + c \cos \theta \, (PQ - PQ') \right\}.$$

Now

$$PQ + PQ' = 2\sqrt{R^2 - c^2 \sin^2 \theta}, \quad PQ - PQ' = -2c \cos \theta, \quad PQ \cdot PQ' = R^2 - c^2.$$

Thus potential at $P = \tfrac{4}{3}\mu R^3 \int_0^{\frac{1}{2}\pi} \sqrt{(1 - c^2 \sin^2 \theta)} \cdot (1 + 2c^2 \sin^2 \theta) \, d\theta$, c being written for c/R. To this must be added the potential for a uniform disk found previously, and the result must be multiplied by the compound density and integrated again over the area, the order of integration being changed as before so as to take first the integration with respect to c. In this way elliptic functions are avoided; but the process is too long to be given here, particularly as it presents no difficulty. The result is that the potential on itself of a disk whose density $= 1 + \mu r^2/R^2$ is expressed by

$$\frac{16\pi R^3}{3} (1 + \tfrac{14}{15}\mu + \tfrac{5}{21}\mu^2)^*. \quad\dots\dots\dots\dots\dots (45)$$

Thus if for brevity we put $R = 1$, we may express the *vis viva* of the whole motion (both extremities included) by

$$2 \text{ vis viva} = \pi l (1 + \tfrac{1}{2}\mu)^2 + \frac{\pi\mu^2}{24}\frac{1 - e^{-8l}}{1 + e^{-8l}} + \tfrac{16}{3}(1 + \tfrac{14}{15}\mu + \tfrac{5}{21}\mu^2),$$

which corresponds to the rate of flow $\pi u_0 = \pi(1 + \tfrac{1}{2}\mu)$.

Thus

$$\frac{2 \text{ vis viva}}{(\text{rate of flow})^2} = \frac{l}{\pi} + \frac{1}{3\pi} \frac{\dfrac{\Lambda}{8}\mu^2 + \dfrac{16}{\pi}(1 + \tfrac{14}{15}\mu + \tfrac{5}{21}\mu^2)}{(1 + \tfrac{1}{2}\mu)^2}, \quad\dots(46)$$

where

$$\Lambda = \frac{1 - e^{-8l}}{1 + e^{-8l}}.$$

The second fraction on the right of (46) is next to be made a minimum by variation of μ. Putting it equal to z and multiplying up, we get the following quadratic in μ:—

$$\mu^2 \left\{\frac{\Lambda}{8} + \frac{5 \cdot 16}{21\pi} - \frac{z}{4}\right\} + 2\mu \left\{\frac{16 \cdot 7}{15\pi} - \frac{z}{2}\right\} + \frac{16}{\pi} - z = 0.$$

---

* [Mr Clerk Maxwell has pointed out a process by which this result may be obtained much more simply. Begin by finding the potential at the *edge* of the disk whose density is $1 + \mu r^2$. Taking polar coordinates (ρ, θ), the pole being at the edge, we have

$$r^2 = \rho^2 + a^2 - 2a\rho \cos \theta$$

and

$$V = \iint \{1 + \mu (\rho^2 + a^2 - 2a\rho \cos \theta)\} \, d\theta \, d\rho,$$

the limits of ρ being 0 and $2a \cos \theta$, and those of θ being $-\tfrac{1}{2}\pi$ and $+\tfrac{1}{2}\pi$. We get at once

$$V = 4a + \tfrac{20}{9}\mu a^3.$$

Now let us cut off a strip of breadth da from the edge of the disk, whose mass is accordingly

$$2\pi a (1 + \mu a^2) \, da.$$

The work done in carrying this strip off to infinity is

$$2\pi a \, da \, (1 + \mu a^2) (4a + \tfrac{20}{9}\mu a^3).$$

If we gradually pare the disk down to nothing and carry all the parings to infinity, we find for the total work by integrating with respect to a from 0 to R,

$$\frac{8\pi R^3}{3} (1 + \tfrac{14}{15}\mu + \tfrac{5}{21}\mu^2),$$

μ being written for μR^2. This is, as it should be, the half of the expression in the text.]

The smallest value of z consistent with a real value of μ is therefore given by

$$\left(\frac{16\cdot7}{15\pi}-\frac{z}{2}\right)^2-\left(\frac{16}{\pi}-z\right)\left(\frac{\Lambda}{8}+\frac{5\cdot16}{21\pi}-\frac{z}{4}\right)=0,$$

$$z=\frac{2\Lambda+\dfrac{8192}{1575\pi}}{\dfrac{\Lambda\pi}{8}+\frac{12}{35}}=\frac{2\Lambda+1\cdot6556}{\cdot3927\Lambda+\cdot3429}=\frac{3\cdot6556-\cdot3444\,e^{-8l}}{\cdot7356-\cdot0498\,e^{-8l}}.$$

Thus

$$\frac{2\ vis\ viva}{(\text{rate of flow})^2}=\frac{l}{\pi}+\frac{1}{3\pi}\frac{3\cdot6556-\cdot3444\,e^{-8l}}{\cdot7356-\cdot0498\,e^{-8l}}.\quad\ldots\ldots(47)$$

This gives an upper limit to $1/c$. In terms of α (including both ends)

$$\alpha<2\cdot305R\,\frac{10\cdot615-e^{-8l/R}}{14\cdot771-e^{-8l/R}}.\quad\ldots\ldots\ldots\ldots\ldots(47')$$

From $(47')$ we see that the limit for α is smallest when $l=0$, and gradually increases with l.

When $l=\infty$, it becomes

$$1\cdot6565R=2\times\cdot8282R.$$

Thus the correction for one end of an infinite tube is limited by

$$\cdot8282R>\alpha>\cdot785R.\quad\ldots\ldots\ldots\ldots\ldots(48)$$

When l is not infinitely great the upper limit may be calculated from $(47')$, the lower limit remaining as before; but it is only for quite small values of l that the exponential terms in $(47')$ are sensible. It is to be remarked that the *real* value of α is least when $l=0$, and gradually increases to its limit when $l=\infty$. For consider the actual motion for any finite value of l. The *vis viva* of the motion going on in any middle piece of the tube is greater than corresponds merely to the length. If the piece therefore be removed and the ends brought together, the same motion may be supposed to continue without violation of continuity, and the *vis viva* will be more diminished than corresponds to the length of the piece cut out. A *fortiori* will this be true of the real motion which would exist in the shortened tube. Thus α steadily decreases as the tube is shortened until when $l=0$ it coincides with the lower limit $\frac14\pi R$.

In practice the outer end of a rather long tube-like neck cannot be said generally to end in an infinite plane, as is supposed in the above calculation. On the contrary, there could ordinarily be a certain flow back round the edge of the tube, the effect of which must be sensibly to diminish α. It would be interesting to know the exact value of α for an infinite tube projecting into unlimited space free from obstructing bodies, the thickness of the cylindrical tube being regarded as vanishingly small. Helmholtz has solved what may be called the corresponding problem in two dimensions; but the difficulty in the two cases seems to be of quite a different kind. Fortunately our ignorance on this point is not of much consequence for acoustical purposes, because when the necks are short the hypothesis of the infinite plane agrees nearly with the fact, and when the necks are long the correction to the length is itself of subordinate importance.

Nearly Cylindrical Tubes of Revolution.

The non-rotational flow of a liquid in a tube of revolution or of electricity in a similar solid conductor can only in a few cases be exactly determined. It may therefore be of service to obtain formulæ fixing certain limits between which the *vis viva* or resistance must lie. First, considering the case of electricity (for greater simplicity of expression), let us conceive an indefinite number of infinitely thin but at the same time perfectly conducting planes to be introduced perpendicular to the axis. Along these the potential is necessarily constant, and it is clear that their presence must *lower* the resistance of the conductor in question. Now at the point x (axial coordinate) let the radius of the conductor be y, so that its section is πy^2. The resistance between two of the above-mentioned planes which are close to one another and to the point x will be in the limit $dx \div \pi y^2$, if dx be the distance between the planes, the resistance of the unit cube being unity. Thus

$$\text{resistance} = \int \frac{dx}{\pi y^2}. \quad\dots\dots\dots\dots\dots\dots(49)^*$$

Upper Limit.

Secondly, we know that in the case of a liquid the true *vis viva* is less than that of any other motion which satisfies the boundary conditions and the equation of continuity. Now u, v being the axial and transverse velocities, it will always be possible so to determine v as to satisfy the conditions if we assume $u = $ constant over the section, and therefore

$$u = \frac{u_0}{\pi y^2}. \quad\dots\dots\dots\dots\dots\dots\dots(50)$$

* [It is easy to show formally that no error can arise from neglecting the effect of the curved rim. Imagine the planes at x and $x + dx$ extended, and the curves in which they cut the surface of the conductor projected by lines parallel to the axis. In this way a cylinder is formed which contains the whole surface between x and $x + dx$, and another cylinder which is entirely contained by the surface. The small cylinder may be obtained by supposing part of the matter not to conduct, and therefore gives too great a resistance. On the other hand, the real solid may be obtained from the large cylinder by the same process. The resistance of the slice lies accordingly between those of the two cylinders which are themselves equal in the limit. Hence, on the whole, the parts neglected vanish compared to those retained.]

This may be seen by imagining rigid pistons introduced perpendicular to the axis. To determine v it is convenient to use the function ψ, which is related to u and v according to the equations (37),

$$ru = d\psi/dr, \qquad rv = -d\psi/dx.$$

These forms for u and v secure the fulfilment of the continuity equation. Since

$$u = u_0/\pi y^2, \qquad \frac{d\psi}{dr} = ru_0/\pi y^2,$$

$$\psi = \psi_0(x) + \frac{u_0}{2\pi y^2} r^2,$$

and therefore

$$v = -\frac{\psi'_0(x)}{r} - \frac{u_0}{2\pi} r \frac{d}{dx}\left(\frac{1}{y^2}\right).$$

But since v cannot be infinite on the axis, but must, on the contrary, be zero, $\psi'_0(x) = 0$, and we have

$$u = \frac{u_0}{\pi y^2}, \qquad v = -\frac{u_0}{2\pi} r \frac{d}{dx}\left(\frac{1}{y^2}\right). \quad \ldots\ldots\ldots\ldots(51)$$

From the manner in which these were obtained, they must satisfy the condition of giving no normal motion at the surface of the tube. That this is actually the case may be easily verified à posteriori, but it is scarcely necessary for our purpose to do so. To find the vis viva,

$$\int_0^y u^2\, 2\pi r\, dr = \frac{u_0^2}{\pi y^2}, \qquad \iint u^2\, 2\pi r\, dr\, dx = \frac{u_0^2}{\pi}\int \frac{1}{y^2}\, dx,$$

$$\int v^2\, 2\pi r\, dr = \frac{u_0^2 y^4}{8\pi}\left[\frac{d}{dx}\left(\frac{1}{y^2}\right)\right]^2, \qquad \iint v^2\, 2\pi r\, dr\, dx = \frac{u_0^2}{2\pi}\int \frac{1}{y^2}\left(\frac{dy}{dx}\right)^2 dx.$$

Thus

$$vis\ viva = \frac{u_0^2}{2\pi}\int \frac{1}{y^2}\left\{1 + \tfrac{1}{2}\left(\frac{dy}{dx}\right)^2\right\} dx.$$

The total flow across any section is $\pi y^2 u = u_0$.

Therefore

$$\frac{2\ vis\ viva}{(\text{rate of flow})^2} = \frac{1}{\pi}\int \frac{1}{y^2}\left\{1 + \tfrac{1}{2}\left(\frac{dy}{dx}\right)^2\right\} dx. \quad \ldots\ldots\ldots\ldots(52)$$

This is the quantity which gives an upper limit to the resistance. The first term, which corresponds to the component u of the velocity, is the same as that previously obtained for the lower limit, as might have been foreseen. The difference between the two, which gives the utmost error involved in taking either of them as the true value, is

$$\frac{1}{2\pi}\int \frac{1}{y^2}\left(\frac{dy}{dx}\right)^2 dx.$$

In a nearly cylindrical tube dy/dx is a small quantity, and so the result found by this method is closely approximate. It is not necessary that the section of the tube should be nearly constant, but only that it should vary slowly. The success of the approximation in this and similar cases depends in great measure on the fact that the quantity to be estimated is a minimum. Any reasonable approximation to the real motion will give a *vis viva* very near the minimum, according to the principles of the differential calculus.

Application to straight tube of revolution whose end lies on two infinite planes.

For the lower limit to the resistance we have

$$\frac{1}{\pi} \int \frac{dx}{y^2} + \frac{1}{4R_1} + \frac{1}{4R_2},$$

R_1, R_2 being the radii at the ends; and for the higher limit

$$\frac{1}{\pi} \int \frac{1}{y^2} \left\{ 1 + \tfrac{1}{2} \left(\frac{dy}{dx}\right)^2 \right\} dx + \frac{8}{3\pi^2} \left(\frac{1}{R_1} + \frac{1}{R_2}\right).$$

The first expression is obtained by supposing infinitely thin but perfectly conducting planes perpendicular to the axis to be introduced from the ends of the tube inwards, while in the second the conducting planes in the electrical interpretation are replaced by pistons in the hydrodynamical analogue. For example, let the tube be part of a cone of semivertical angle θ.

The lower limit is

$$\frac{1}{\pi \tan \theta} \left(\frac{1}{R_1} \sim \frac{1}{R_2}\right) + \tfrac{1}{4}\left(\frac{1}{R_1} + \frac{1}{R_2}\right),$$

and the higher

$$\frac{1 + \tan^2 \theta}{\pi \tan \theta} \left(\frac{1}{R_1} \sim \frac{1}{R_2}\right) + \frac{8}{3\pi^2}\left(\frac{1}{R_1} + \frac{1}{R_2}\right).$$

Tubes nearly straight and cylindrical but not necessarily of revolution.

Taking the axis of x in the direction of the length, we readily obtain by the same process as before a *lower* limit to the resistance

$$\int \frac{dx}{\sigma}, \quad \dots\dots\dots\dots\dots\dots\dots\dots(53)$$

where σ denotes the section of the tube by a plane perpendicular to the axis at the point x, an expression which has long been known and is sometimes given as rigorous. The conductor (for I am now referring to the electrical interpretation) is conceived to be divided into elementary slices by planes perpendicular to the axis, and the resistance of any slice is calculated as if its faces were at constant potentials, which is of course not the case. In fact it is meaningless to talk of the resistance of a limited solid at all, unless with the understanding that certain parts of its surface are at constant potentials, while other parts are bounded by non-conductors. Thus, when the resistance of a cube is spoken of, it is tacitly assumed that two of the opposite faces are at constant potentials, and that the other four faces permit no escape of electricity across them. In some cases of unlimited conductors, for instance one we have already contemplated—an infinite solid almost divided into two separate parts by an infinite insulating plane with a hole in it—it is allowable to speak of the resistance without specifying what particular surfaces are regarded as equipotential; for at a sufficient distance from the opening on either side the potential is constant, and any surface no part of which approaches the opening is approximately equipotential. After this explanation of the exact significance of (53), we may advantageously modify it into a form convenient for practical use.

The section of the tube at n different points of its length l is obtained by observing the length λ of a mercury thread which is caused to traverse the tube. Replacing the integration by a summation denoted by the symbol Σ, we arrive at the formula

$$\text{resistance} = \frac{l^2}{n^2 V}\Sigma\lambda \cdot \Sigma\lambda^{-1}, \quad \ldots\ldots\ldots\ldots\ldots(54)$$

which was used by Dr Matthiessen in his investigation of the mercury unit of electrical resistance, and was the subject of some controversy[*]. It is perfectly correct in the sense that when the number of observations is increased without limit it coincides with (53), *itself, however, only an*

---

[*] See Sabine's *Electric Telegraph*, p. 329. To prove (54), we have

$$\text{resistance} = \frac{l}{n}\Sigma\sigma^{-1},$$

and $\sigma\lambda = \text{constant} = \kappa$ (say), so that $\Sigma\sigma^{-1} = \kappa^{-1}\Sigma\lambda$.

But $V = \text{volume} = \frac{l}{n}\Sigma\sigma = \frac{l}{n}\kappa\Sigma\lambda^{-1}$, so that $\frac{1}{\kappa} = \frac{l}{nV}\Sigma\lambda^{-1}$,

and $$\Sigma\sigma^{-1} = \frac{l}{nV}\Sigma\lambda \cdot \Sigma\lambda^{-1}.$$

The correction for the ends of the tube employed by Siemens is erroneous, being calculated on the supposition that the divergence of the current takes place from the curved surface of a hemisphere of radius equal to that of the tube. This is tantamount to assuming a constant potential over the solid hemisphere conceived as of infinite conductivity, and gives of course a result too small—R for both ends together. The proper correction, which probably is not of much importance, would depend somewhat upon the mode of connexion of the tube with the terminal cups, but cannot differ much from $\frac{1}{2}\pi R$ (for both ends), as we have seen. (I have since found that Siemens was aware of the small error in this correction.)

R. I.

approximation to the magnitude sought. The extension of our second method (for the higher limit) to tube not of revolution would require the general solution of the potential problem in two dimensions. It may be inferred that the difference between the two limits is of the order of the square of the inclination of the tangent plane to the axis, and is therefore very small when the section of the tube alters but slowly.

Tubes not nearly straight.

In applying (53) to such cases, we are at liberty to take any straight line we please as axis; but if the tube is much bent, even though its cross section remain nearly constant, the approximation will cease to be good. This is evident, because the planes of constant potential must soon become very oblique, and the section σ used in the formula much greater than the really effective section of the tube. To meet this difficulty a modification in the formula is necessary. Instead of taking the artificial planes of equal potential all perpendicular to a straight line, we will now take them normal to a curve which may have double curvature, and which should run, as it were, along the middle of the tube. Consecutive planes intersect in a straight line passing through the centre of curvature of the "axis" and perpendicular to its plane.

The resistance between two neighbouring equipotential planes is in the limit

$$\delta\theta \div \iint \frac{d\sigma}{r},$$

where $\delta\theta$ is the angle between the planes, and r is the distance of any element $d\sigma$ of the section from the line of intersection of the planes. Now $\delta\theta = ds \div \rho$, if ds be the intercept on the axis between the normal planes, and ρ the radius of curvature at the point in question. The lower limit to the resistance is thus expressed by

$$\int \frac{ds}{\iint r^{-1} \rho\, d\sigma}. \quad\quad\quad\quad\dots\dots\dots\dots\dots\dots(55)$$

In the particular case of a tube of revolution (such as an anchor-ring) $\iint r^{-1} \rho\, d\sigma$ is a constant, and the limit which now coincides with the true resistance varies as the length of the axis, and is evidently independent of its position. In general the value of the integral will depend on the axis used, but it is in every case less than the true value of the resistance. In choosing the axis, the object is to make the artificial planes of constant potential agree as nearly as possible with the true equipotential surfaces.

A still further generalization is possible by taking for the artificial equipotential surfaces those represented by the equation $F = \text{const.}$

For all systems of surfaces, with one exception, the resistance found on this assumption will be too small. The exception is of course when the surfaces $F = \text{const.}$ coincide with the undisturbed equipotential surfaces. The element of resistance between the surfaces F and $F + dF$ is

$$\frac{1}{\iint \frac{1}{dn}\, d\sigma},$$

where dn is the distance between the surfaces at the element $d\sigma$, and the integration goes over the surface F as far as the edge of the tube. Now

$$dn = dF \div \sqrt{\left(\frac{dF}{dx}\right)^2 + \left(\frac{dF}{dy}\right)^2 + \left(\frac{dF}{dz}\right)^2};$$

and limit to resistance

$$= \int \frac{dF}{\iint d\sigma \sqrt{\left(\frac{dF}{dx}\right)^2 + \left(\frac{dF}{dy}\right)^2 + \left(\frac{dF}{dz}\right)^2}}, \quad \ldots\ldots\ldots\ldots(56)$$

an expression whose form remains unchanged when $f(F)$ is written for F. If $F = r$, so that the surfaces are spheres,

$$\Sigma\, (dF/dx)^2 = (dF/dr)^2 = 1\ ;$$

and

$$\text{limit} = \int \frac{dr}{\iint d\sigma} = \int \frac{dr}{\sigma}.$$

This form would be suitable for approximately conical tubes, the vertex of the cone being taken as origin of r.

The last formulæ, (55) and (56), are perhaps more elaborate than is required in the present state of acoustical science, and it is rather in the theory of electricity that their interest would lie; but they present themselves so readily as generalizations of previous results that I hope that they are not altogether out of place in the present paper. In all these cases we have the advantage that the quantity sought is determined by a minimum property, and is therefore subject to a much smaller error than exists in the conditions which determine it.

PART III.

Experimental.

The object of this Part is to detail some experiments on resonators instituted with a view of comparing some of the formulæ of Parts I. and II. with observation. Helmholtz in his paper on organ-pipes has compared his own theory with the experiments of Sondhauss and Wertheim for the case of resonators whose communication with the external atmosphere is by simple holes in their sides. The theoretical result is embodied in (5) and (23), or for circular holes (24), and runs,

$$n = \frac{a}{2\pi} \sqrt{\frac{2R}{S}} = \frac{a}{\pi} \sqrt{\frac{R}{2S}}; \dots\dots\dots\dots(57)$$

or when the area of the opening is approximately circular and of magnitude σ,

$$n = \frac{a\sigma^{\frac{1}{4}}}{2^{\frac{1}{2}}\pi^{\frac{5}{4}}S^{\frac{1}{2}}} . \dots\dots\dots\dots\dots\dots(58)$$

On calculation Helmholtz finds[*] $n = 56174\,\sigma^{\frac{1}{4}}S^{-\frac{1}{2}}$, the unit of length being metrical.

The empirical formula found by Sondhauss is $n = 52400\,\sigma^{\frac{1}{4}}S^{-\frac{1}{2}}$, which agrees completely with theory as regards its form, but not so well in the value it assigns to the constant multiplier. The difference corresponds to more than a semitone, and is in the direction that the observed notes are all too low. I can only think of two explanations for the discordance, neither of which seems completely satisfactory. In the first place, Sondhauss determined his resonant notes by the pitch of the sound produced when he blew obliquely across the opening through a piece of pipe with a flattened end. It is possible that the proximity of the pipe to the opening was such as to cause an obstruction in the air-passage which might sensibly lower the pitch. Secondly, no account is taken of the thickness of the side of the vessel, the effect of which must be to make the calculated value of n too great. On the other hand, two sources of error must be mentioned which would act in the opposite direction. The air in the vicinity of the opening must have been sensibly warmer than the external atmosphere, and we saw in Part I. how sensitive resonators of this sort must be to small changes in the physical properties of the gas which occupy the air-passages. Indeed Savart long ago remarked on the instability of the pitch of short pipes, comparing them with ordinary organ-pipes. The second source of disturbance is of a more recondite character, but not, I think, less real. It is proved in works on hydrodynamics that in the steady motion of fluids, whether compressible or not, an increased velocity is always accompanied by a diminished pressure. In the case of a gas the diminished pressure entails a diminished density. There seems therefore every reason to expect a diminution of density in the stream of air which plays over the orifice of the resonator, which must cause a rise in the resonant note. But independently of these difficulties, the theory of pipes or other resonators made to speak by a

---

[*] The velocity of sound is taken at the freezing-point; otherwise the discordance would be greater.

stream of air directed against a sharp edge is not sufficiently understood to make this method of investigation satisfactory. For this reason I have entirely abandoned the method of causing the resonators to speak in my experiments, and have relied on other indications to fix the pitch. The only other experiments that I have met with on the subject of the present paper are also by Sondhauss, who has been very successful in unravelling the complications of these phenomena without much help from theory[*]. For flasks with long necks he found the formula

$$n = 46705 \frac{\sigma^{\frac{1}{2}}}{L^{\frac{1}{2}} S^{\frac{1}{2}}}$$

as applicable when the necks are cylindrical and not too short, corresponding to the theoretical

$$n = \frac{a}{2\pi} \frac{\sigma^{\frac{1}{2}}}{L^{\frac{1}{2}} S^{\frac{1}{2}}}, \quad \dots\dots\dots\dots\dots\dots\dots\dots(59)$$

obtained by combining (5) and (21), or, in numbers with metrical units,

$$n = 54470 \frac{\sigma^{\frac{1}{2}}}{L^{\frac{1}{2}} S^{\frac{1}{2}}}.$$

The discrepancy is no doubt to be attributed (at least in great measure) to the omission of the correction to the length of the neck.

In the experiments about to be described the pitch of the resonator was determined in various ways. Some of the larger ones had short tubes fitted to them which could be inserted in the ear. By trial on the piano or organ the note of maximum resonance could be fixed without difficulty, probably to a quarter of a semitone. In most of the experiments a grand piano was used, whose middle c was in almost exact unison with a fork of 256 vibrations per second. Whenever practicable the harmonic under-tones were also used as a check on any slight difference which might be possible in the quality of consecutive notes. Indeed the determination was generally easier by means of the first undertone (the octave), or even the second (the twelfth), than when the actual note of the resonator was used. The explanation is, I believe, not so much that the overtones belonging to any note on the piano surpass in strength the fundamental tone, although that is quite possible[†], as that the ear (or rather the attention) is more sensitive to an increase in the strength of an overtone than of the funda-mental. However this may be, there is no doubt that a little practice greatly exalts the power of observation, many persons on the first trial being apparently incapable of noticing the loudest resonance. Another plan very convenient, though not to be used in measurements without caution, is

---

[*] Pogg. Ann. t. lxxxi.
[†] In this respect pianos, even by the same maker, differ greatly.

to connect one end of a piece of india-rubber tubing * with the ear, while the other end is passed into the interior of the vessel. In this way the resonance of any wide-mouthed bottle, jar, lamp-globe, &c. may be approximately determined in a few seconds; but it must not be forgotten that the tube in passing through the air-passage acts as an obstruction, and so lowers the pitch. In many cases, however, the effect is insignificant, and can be roughly allowed for without difficulty. For large resonators this method is satisfactory, but in other cases is no longer available. I have, however, found it possible to determine with considerable precision the pitch of small flasks with long necks by simply holding them rather close to the wires of the piano while the chromatic scale is sounded. The resonant note announces itself by a quivering of the body of the flask, easily perceptible by the fingers. Since it is not so easy by this method to divide the interval between consecutive notes, I rejected those flasks whose pitch neither exactly agreed with any note on the piano nor exactly halved the interval. In some cases it is advantageous to sing into the mouth, taking care not to obstruct the passage; the resonant note is recognized partly by the tremor of the flask, and partly by a peculiar sensation in the throat or ear, hard to localize or describe.

The precision obtainable in any of these ways may seem inferior to that reached by several experimenters who have used the method of causing the resonators or pipes to speak by a stream of air. That the apparent precision in the last case is greater I of course fully admit; for any one by means of a monochord could estimate the pitch of a continuous sound within a smaller limit of error than a quarter of a semitone. But the question arises, *what* is it that is estimated? Is it the natural note of the resonator? I have already given my reasons for doubting the affirmative answer; and if the doubt is well grounded, the greater precision is only apparent and of no use theoretically. I may add, too, that many of the flasks that I used could not easily have been made to speak by blowing. If they sounded at all it was more likely to be the first overtone, which is the note rather of the neck than the flask: see equation (20). In carrying out the measurements of the quantities involved in the formula, the volume of the flask or reservoir was estimated by filling it with water halfway up the neck, which was then measured, or in some cases weighed. The measurements of the neck were made in two ways according to the length. Unless very short their capacity was measured by water, and the expression for the resistance (54) in a simplified form was used. The formula for n then runs,

$$n = \frac{a}{2\pi L} \sqrt{\frac{\text{vol. of neck}}{\text{vol. of flask} \times (1 + \frac{1}{2}\pi R/L)}}. \quad \cdots\cdots\cdots (60)$$

* The black French tubing, about $\frac{1}{2}$ inch in external diameter, is the pleasantest to use.

When, on the other hand, the necks were short, or simply holes of sensible thickness, the following formula was used,

$$n = \tfrac{1}{4}a\,(r_1 + r_2)\{\pi LS\,[1 + \pi\,(r_1 + r_2)/4L]\}^{-\frac{1}{2}}, \quad \ldots\ldots\ldots\ldots (61)$$

r_1, r_2 being the radii, or halves of the diameters, as measured at each end. It is scarcely necessary to say that the estimation of pitch was made in ignorance of the theoretical result; otherwise it is almost impossible to avoid a certain bias in dividing the interval between the consecutive notes.

TABLE I.

No. of observation.	S, in cub. centims.	V, in cub. centims.	L, in inches.	R, in inches.	n, by calculation.	n, by observation.	Difference, in mean semitones.
1	805	68	$4\frac{1}{2}$	$\frac{1}{2}$	127·7	126	+ ·23
2	1350	126·7	$5\frac{5}{8}$	$\frac{5}{8}$	107·7	108·7	− ·16
3	7100	430	$1\frac{3}{4}$	$2\frac{3}{16}$, $1\frac{5}{8}$	122·3	120	+ ·33
4	405	49·9	$3\frac{1}{16}$	$\frac{1}{2}$	179·7	180	− ·03
5	180	21·26	$2\frac{3}{8}$	$\frac{3}{4}$	233·7	228	+ ·42
6	785	36·84	$\frac{3}{4}$	$\frac{17}{32}$, $1\frac{1}{8}$	174·3	176	− ·16
7	210	32·5	$3\frac{7}{8}$	$1\frac{3}{4}$	201·9	204	− ·18
8	312	29·32	$3\frac{1}{4}$	$\frac{3}{8}$	186·3	182	+ ·41
10	6300	270	$3\frac{3}{8}$	$\frac{3}{8}$, $\frac{1}{4}$	104·2	102·4	+ ·29
12	54·69	11·6	$2\frac{5}{16}$	not recorded	391·6	384*	+ ·34

In Table I. the first column gives the number of the experiment, the second the volume of the reservoirs, including half the necks, the third the volume of the necks themselves, the fourth their lengths, and the fifth their radii measured, when necessary, at both ends. In the sixth column is given the number of vibrations per second calculated from (60), the velocity of sound being taken at 1123 feet per second, corresponding to 60° F., about the temperature of the room in which the pitch was determined. Column 7 contains the values of n estimated by means of the pianoforte, while in 8 is given for convenience the discrepancy between the observed and calculated values expressed in parts of a mean semitone.

1, 2, 4, 5, 6, 7, 8, 12 were glass flasks with well-defined nearly cylindrical necks, the body of the flask being approximately spherical. Of these 1 and 2 had small tubes cemented into them, which were inserted in the ear; the pitch of the rest was estimated mainly by their quivering to the resonant note. 3 and 10 were globes intended for burning phosphorus in oxygen gas, and their pitch was fixed principally by the help of the india-rubber tube passed through the neck. A good ear would find no difficulty in identifying the note produced when the body of the globe is struck with the soft part of the hand. The agreement is I think very satisfactory, and is certainly better

* 12 was originally estimated an octave too low, so that the number in the Table is the double of what was put down as the result of observation.

than I expected, having regard to the difficulties in the measurements of pitch and of the dimensions of the flasks. The average error in Table I. is about a quarter of a semitone, and the maximum error less than half a semitone. It should be remembered that there is no arbitrary constant to be fixed as best suits the observations, but that the calculated value of n is entirely determined by the dimensions of the resonator and the velocity of sound. If a lower value of the latter than 1123 were admissible, the agreement would be considerably improved.

TABLE II.

No. of experiment.	S, in cub. centims.	L, in inches.	d, in inches.	n, by calculation.	n, by observation.	Difference, in mean semitones.
9	1245	$2\frac{11}{16}$	$1\frac{1}{8}$, $1\frac{5}{16}$	107·3	108	− ·11
11	216·6	$\frac{1}{6}$	1, $1\frac{5}{16}$	526	538	− ·39
13	1245	1	1	163·2	170	− ·71
14	1245	$\frac{1}{4}$	1, $1\frac{5}{16}$	219·4	213	+ ·51
15	3090	$\frac{3}{32}$	$2\frac{1}{16}$	218·1	227·5	− ·73
16	3240	$1\frac{1}{4}$	$1\frac{3}{4}$	131·3	142	− 1·36
17	3240	$\frac{1}{8}$	$1\frac{1}{8}$, $1\frac{1}{16}$	149·1	153·5	− ·50
18	3240	$\frac{5}{16}$	$1\frac{5}{16}$	153·2	153·5	− ·03
19	3240	$\frac{5}{16}$	1	129·1	132	− ·38
20	3040	neglected	$\frac{3}{8}$	128·6	128	+ ·08
21	3240	$1\frac{1}{8}$	$1\frac{1}{16}$, $1\frac{1}{32}$	101·5	103·5	− ·34
22	3240	$\frac{11}{16}$	$2\frac{7}{32}$	216	229	− 1·01

Table II. contains the results of the comparison between theory and observation for a number of resonators whose necks were too short for the convenient measurement of the volume. The length and diameter were measured with care and used in formula (61). In 9, 13, 14 the reservoir consisted of the body of a flask whose neck had been cut off close, and which was fitted with a small tube for insertion in the ear. In 9 and 13 there was a short glass or tin tube fitted into the opening[*], while in 14 the mouth was covered (air-tight) with a piece of sheet gutta-percha pierced by a cork borer; 11 was a small globe treated in the same way. 15 to 22 were all experiments with a globe of a moderator-lamp, which also had a tube for the ear, one opening being closed by a piece of plate glass cemented over it. Sometimes a little water was poured in for greater convenience in determining the pitch, whence the slightly differing values of S. In 15 the opening was clear, and in 16 fitted with a brass tube; in 17 it was covered with a gutta-percha face, in 18, 19, 21 with a wooden face bored by a centre-bit, and in 20 with a piece of tin plate carrying a circular hole; 22 contains the result when the other opening of the globe was used clear.

On inspection of Table II. it appears that the discrepancy between theory and observation is decidedly greater than in Table I., in fact about double,

---

[*] Gutta-percha softened in hot water is very useful for temporary fittings of this sort.

whether we consider the maximum or the mean error. The cause of some of the large errors may, I think, be traced. 13 and 16 had necks of just the length for which the correction $\frac{1}{4}\pi R$ may not be quite applicable. A decided flow back round the edge of the outer end must take place with the effect of diminishing the value of α. In order to test this explanation, a piece of millboard was placed over the outer end of the tube in 16 to represent the infinite plane. A new estimation of n, as honest as possible, gave $n = 137$, which would considerably diminish the error. I fancied that I could detect a decided difference in the resonance according as the millboard was in position or not; but when the theoretical result is known, the difficulty is great of making an independent observation. In 15 and 22, where the apertures of the globe were used clear, the error is, I believe, due to an insufficient fulfilment of the condition laid down at the commencement of this paper. Thus in 15 the wave-length $= 1123 \div 227 = 4\cdot9$ feet; or $\frac{1}{4}\lambda = 1\cdot2$ feet, which is not large enough compared to the diameter of the globe (6 inches). The addition of a neck lowers the note, and then the theory becomes more certainly applicable.

It may perhaps be thought that the observations on resonance in Tables I. and II. do not extend over a sufficient range of pitch to give a satisfactory verification of a general formula. It is true that they are for the most part confined within the limits of an octave, but it must be remembered that if the theory is true for any resonant air-space, it may be extended to include all *similar* air-spaces in virtue of Savart's law alone—a law which has its foundations so deep that it hardly requires experimental confirmation. If this be admitted, the range of comparison will be seen to be really very wide, including all *proportions* of L and R. When the pitch is much higher or much lower than in the Tables, the experimental difficulties are increased. For much lower tones the ear is not sufficiently sensitive, while in the case of the higher tones some of the indications relied on to fix the pitch are no longer available.

Some experiments were next made with the moderator globe and two openings. The theoretical formulæ are

$$n = \frac{a}{2\pi} \sqrt{\frac{c_1 + c_2}{S}}, \qquad c = \frac{\pi R^2}{L + \frac{1}{2}\pi R}.$$

TABLE III.

No. of experiment.	S, in cub. centims.	c_1, in inches.	c_2, in inches.	$c_1 + c_2$, in inches.	n, by calculation.	n, by observation.	Difference, in mean semitones.
23	3240	2·00	1·95	3·95	303·4	320	·92
24	3240	2·00	1·01	3·01	264·8	282	1·09
25	3240	·715	1·01	1·725	200·4	204	·31

In 23 both holes of the globe were clear, and in 24, 25 they were covered with wooden faces carrying holes of various diameters. The error in 23, 24 is to be ascribed to the same cause as in 15 and 22 above.

The last experiment that I shall describe was made in order to test the theory of double resonance, but is not quite satisfactory, for the same reason as 15, 22, 23, 24. Two moderator globes were cemented together so as to form two chambers communicating with each other and with the external air. The natural openings were used clear, and the resonance (which was not very good) was estimated by means of a tube connecting the ear with one of them. The observations gave for the values of n,

$$\text{High note} = 384, \qquad \text{Low note} = 213.$$

The result of calculation from the dimensions of the globes and openings by means of the formulæ of Parts I. and II. was

$$\text{High note} = 360, \qquad \text{Low note} = 212.$$

The error in the high note is about a semitone.

[The two moderator globes were fitted up again as a double resonator, only with bored wooden disks over the holes, so as to lower the note and render the theory more strictly applicable. The pitch was much better defined than before, and gave

$$\text{Low note} = 152\cdot5, \qquad \text{High note} = 240.$$

$$c_1 = c_3 = 1\cdot008; \quad c_2 = \cdot7152. \quad a = 1133 \,(70^\circ \text{ F.}).$$

Thus

$$n_1 = \frac{1133 \times 12}{6\cdot28} \sqrt{\frac{16\cdot39 \times 1\cdot008}{3200}} = 155\cdot6,$$

$$n_2 = \frac{1133 \times 12}{6\cdot28} \sqrt{\frac{16\cdot39 \times 2\cdot438}{3200}} = 241\cdot9.$$

The agreement is now very good.

One of the outer holes was stopped with a plate of glass. The resonance of the high note was feeble though well defined; that of the low was rather loud but badly defined.

The high note was put at 225, low note at 90.

$$S = 3150, \quad S' = 3250,$$

$$c_2 = \cdot7152, \quad c_1 = 1\cdot008, \quad c_1 + c_2 = 1\cdot7232.$$

Calculating from these data, we get

$$n_1 = 225\cdot2, \quad n_2 = 90\cdot5.$$

The agreement is here much better than was expected, and must be in part fortuitous.

I will now detail two experiments made to verify the formula marked (20a). A moderator [lamp] chimney was plugged at the lower end with gutta-percha, through which passed a small tube for application to the ear. The bulb was here represented by the enlargement where the chimney fits on to the lamp. On measurement,

$$S/\sigma = 4\!\cdot\!16 \text{ inches}, \quad L = 5\!\cdot\!367 \text{ inches}, \quad \alpha = \tfrac{1}{4}\pi R = \cdot471.$$

Thus

$$\tan k \times 9\!\cdot\!611 = \frac{1}{k \times 4\!\cdot\!161};$$

from this the value of k was calculated by the trigonometrical tables. Finally,

$$n = 251\!\cdot\!4.$$

As the result of observation n had been estimated at 252.

In another case,

$$L = 5\!\cdot\!767, \quad \alpha = \cdot537, \quad \text{or } S/\sigma = 3\!\cdot\!737, \quad n \text{ by observation} = 351.$$

The result of calculation is $n = 350\!\cdot\!3$. These are the only two instances in which I have tried the formula (20a). It is somewhat troublesome in use, but appears to represent the facts very closely; though I do not pretend that the above would be average samples of a large series. There is no necessity for the irregularity at the lower end taking the form of an enlargement. For example, the formula might be applied to a truly cylindrical pipe with a ball of solid material resting at the bottom.]

I had intended to have made these experiments more complete, particularly on multiple resonance, but have not hitherto had time. However, the results obtained seem quite sufficient to establish a substantial agreement between theory and fact. It should be understood that those here presented are not favourable specimens selected out of a large number, but include, with one exception, all the measurements attempted. There are many kinds of bottles and jars, and among them some of the best resonators, which do not satisfy the fundamental condition on which our theory rests. The deductive treatment of the problem in such cases presents great difficulties of a different kind from any encountered in this paper. Until they are surmounted the class of resonators referred to are of no use for an exact comparison between theory and observation, though they may be of great service as aids to investigation in other directions.

6.

NOTE ON THE EXPLANATION OF CORONAS, AS GIVEN IN VERDET'S *LEÇONS D'OPTIQUE PHYSIQUE*, AND OTHER WORKS.

[*Proceedings of the London Mathematical Society*, Vol. III. pp. 267—269, 1871.]

CORONAS are formed whenever, between the eye and the source of light, a large number of small opaque bodies, regular in size but irregular in distribution, are to be found. By a principle known as Babinet's, it is permitted to substitute for the layer of obstructing globules an opaque screen having apertures, which admit the light precisely where before it was stopped. From each aperture diverges a secondary wave, whose phase is arbitrarily related to those of the other waves on account of the irregularity in the distribution of the centres from which they emanate. In the theory of coronas, the intensities of the partial waves are supposed to be equal.

Any vibration may be represented by a straight line drawn from a given point, whose length is proportional to the amplitude, and angular position depends on the phase. When several vibrations affect the same point, the resultant may be found from the representative lines, according to the ordinary rules for compounding directed magnitudes. We have, therefore, to consider the resultant of a great number of lines radiating from an origin whose lengths are all equal and directions thoroughly irregular. From this statement, it appears at once that the question is perfectly indeterminate, and that no conclusion whatever can be drawn as to the probable value of the phase of the resultant, while all that could be inferred as to the amplitude relates merely to the order of its magnitude. Nevertheless, Verdet (vol. i., p. 297) comes to the conclusion that, the number of partial vibrations being n, the resultant is definitely $\sqrt{n}$ times greater than each component, provided, of course, that n is a large number. As to its phase, he is silent; and, if asked the question, would, I think, have found it difficult to explain how a resultant vibration could have a definite amplitude in conjunction with an indeterminate phase. I propose to point out what I conceive to be the error which vitiates his reasoning, and afterwards to show that, after all, his theory is substantially correct.

The component vibrations may be expressed by

$$\sin 2\pi \left(\frac{t}{\tau}\right), \quad \sin 2\pi \left(\frac{t}{\tau} + \frac{\delta_1}{\lambda}\right), \quad \sin 2\pi \left(\frac{t}{\tau} + \frac{\delta_2}{\lambda}\right), \quad \ldots\ldots$$

The intensity of the light corresponding to the sum of these is

$$\left(1 + \cos 2\pi \frac{\delta_1}{\lambda} + \cos 2\pi \frac{\delta_2}{\lambda} + \ldots\ldots\right)^2 + \left(\sin 2\pi \frac{\delta_1}{\lambda} + \sin 2\pi \frac{\delta_2}{\lambda} + \ldots\ldots\right)^2,$$

or

$$n + 2 \cos 2\pi \frac{\delta_1}{\lambda} + 2 \cos 2\pi \frac{\delta_2}{\lambda} + \ldots\ldots$$

$$\ldots\ldots + 2 \cos 2\pi \frac{\delta_1 - \delta_2}{\lambda} + 2 \cos 2\pi \frac{\delta_1 - \delta_3}{\lambda} + \ldots\ldots + 2 \cos 2\pi \frac{\delta_2 - \delta_3}{\lambda} + \ldots\ldots$$

According to Verdet, this reduces to n when n is great enough, in consequence of the tendency of the series of cosines to zero. The inference is, I believe, unwarranted. All that we are entitled to say is, that when n is great, the sum of the cosines of the arbitrary angles tends to vanish, *in comparison with the number of terms of the series*, which is of the order n^2. There is no reason for supposing that the series becomes small in comparison with n, as the argument requires. The analysis, as it here stands, is taken from Billet (*Traité d'Optique physique*, tom. i., p. 235). In Verdet's book there is, in addition to the above fallacy, a mistake in the work, which I need not further refer to.

I conclude, therefore, that the effect of the diffraction of a plane wave coming through a great number of small holes in an opaque screen is not correctly given. Yet the error disappears in the application to the theory of coronas as they occur in Nature. For Verdet's result comes only to this, that the sources of light are to be treated as independent, and the illuminations due to them, and not the vibrations, are to be compounded. As it is commonly though not very correctly expressed, there is to be no interference. For this the sensible apparent magnitude of the sun, or other source of light, is an amply sufficient cause.

At page 106 of his excellent lectures, Verdet shows that the disturbance from the sun cannot be considered as a system of plane waves over a space greater than a circle of $\frac{1}{20}$ of a millimetre diameter. Between the vibrations at two points whose mutual distance is much greater than this, there is no permanent relation of phase. This shows that the vibrations corresponding to two holes in the imaginary screen cannot interfere regularly, but behave as if they were due to thoroughly independent sources of light.

[1899. One or two further remarks may be made. The assumption that the phases of the components due to the various apertures are distributed at random can be justified only when two restrictions are admitted. The first is that the direction in which the effect is sought must be much more oblique than corresponds to any important part of the diffraction pattern due to the entire aperture. If we consider a direction corresponding to an extreme relative retardation of only a wave length or two, we see that a central diffraction pattern is formed upon the same scale as if the aperture were complete. This shows that the screening entails no loss of resolving-power, as regards the direct image of the source of light. If the number of apertures of small area α each be n, and if the whole aperture be denoted by unity, the amplitude at any point of the central diffraction pattern is diminished by the screening in the ratio $n\alpha : 1$, or the brightness in the ratio $n^2\alpha^2 : 1$.

The second restriction is that $n\alpha$ must be small. Otherwise apertures situated at random will sensibly overlap. The calculation according to which the mean intensity in a sufficiently oblique direction is n times that of a single aperture then fails, since each aperture is supposed to produce its full effect, whereas in practice the common part can act only once. If we suppose that overlapping is precluded, the calculation still fails, since the phases are no longer distributed at random. It is easy to see that the calculated result then becomes an overestimate.

The subject of random vibrations is further considered in a subsequent paper (*Phil. Mag.* Vol. x., p. 73, 1880).]

7.

SOME EXPERIMENTS ON COLOUR.

[*Nature*, III. pp. 234—237, 264, 265; 1871.]

THE theory of colour perception, although in England it has not yet made its way into the text-books, still less into the popular works on science, is fully established with regard to many important points. It is known that our perception of colour is threefold, that is, that any colour may be regarded as made up of definite quantities of three primary colours, the exact nature of which is, however, still uncertain. More strictly stated, the fundamental fact in the doctrine of colour is that, between any four colours whatever given as well in quantity as in quality, there exists what mathematicians call a linear relation, that is, that either a mixture of two of them (in proper proportions) can be found identical, so far as the eye is able to judge, with a mixture of the other two, or else that one of them can be matched by a mixture of the other three. There are various optical contrivances by which the mixture spoken of may be effected. In the year 1857, Mr Maxwell published an account of some experiments with the colour top undertaken to test the theory. From six coloured papers, black, white, red, green, yellow, and blue, discs of two sizes were prepared, which were then slit along a radius so as to admit of being slipped one over the other. Any five out of the six being taken, a match or colour equation between them is possible. For instance, if yellow be excluded, the other five must be arranged so that a mixture of red, green, and blue is matched with a mixture of black and white. The large discs of the three colours are taken and slipped on to each other, and similarly the small discs of black and white. When the small discs are placed over the others and the whole made to rotate rapidly on any kind of spinning machine, the colours are blended, those of the large discs and those of the small, each into a uniform tint.

By adjustment of the discs an arrangement may be found after repeated trials, such that the colour of the inner circle is exactly the same both in tint and luminosity with that of the outer rim. The quantities of each colour exposed may then be read off on a graduated circle, and the result recorded. For instance (the circle being divided into 192 parts), eighty-two parts red mixed with fifty-six green and fifty-four blue, match thirty-seven parts white mixed with 155 black. In this way Maxwell observed the colour equations between each set of five, in all six sets, formed by leaving out in turn each of the six colours. Moreover, for greater accuracy each set was observed six times, and the mean taken. But according to the theory these six final equations are not all independent of each other, but if any two of them are supposed known, the others can be found by a simple calculation. Accordingly, the comparison of the calculated and observed equations furnishes a test of the theory; but in practice, in order to ensure greater accuracy, instead of founding the calculations on two of the actually observed equations chosen arbitrarily, it is preferable to combine all the observations into two equations, which may then be made the basis of calculation. In this way, a system of equations is found necessarily consistent with itself, and agreeing as nearly as possible with the actually observed equations. A comparison of the two sets gives evidence as to the truth of the theory according to which the calculations are made, or if this be considered beyond doubt, tests the accuracy of the observations. In Maxwell's experiments the average difference between the calculated and observed systems amounted to ·77 divisions of which the circle carried 100. So good an agreement is regarded by him as a confirmation of the whole theory; but it seems to me, I confess, that only a very limited part of it is concerned. The axioms, in virtue of which it is permitted to combine the colour equations in the manner required for the calculations, are only such as the following:—If colours which match are mixed with colours which match, the results will match. It is difficult to imagine any theory of colour which will not include them. What proves the threefold character of colour—the most important part of the doctrine—is simply the fact that with any five coloured papers *whatever* a match can be made, while with less than five it cannot (except in certain particular cases). In regard to this point the value of the quantitative experiments is rather that they show of what sort of accuracy the eye is capable in this kind of observation. Those to whom the subject is new may think at first that if colour be threefold a match ought to be possible between any *four* colours. And so it is possible if there is no other limitation; but in experiments with revolving discs we are subject to a limitation, being obliged to fill up the whole circumference somehow. The difficulty will clear itself up, when it is remembered that one of the five colours may be black, so that with any *four* colours and *black* a match can be made with revolving discs.

It was rather for my own satisfaction than with the hope of adding anything new to a subject already so fully and ably treated by Maxwell, that I commenced a repetition of his experiments. The colours used were, roughly speaking, the same as his, as was also the general plan of the observations. The agreement of the calculated and directly observed equations was very good, the average error being only ·24 divisions, of which the complete circle contained ninety-six, or one-third of the corresponding average error in Mr Maxwell's Table. A second set of observations and calculations made after a year's interval with a different set of colours gave about the same result. I am inclined to attribute the considerably greater accuracy of my observations rather to an excellent perception of minute differences of colour (to which I have always found my eyes very sensitive) than to greater care in conducting the experiments. One precaution, however, I have found so important as to be worth mentioning. Unless the small discs are very accurately cut and centred, a coloured rim appears on rotation between the two uniform tints to be compared and adjusted to identity, which is exceedingly distracting to the eye, and interferes much with the accuracy of the comparison. One set of observations made with the same care, and apparently as satisfactory as any of the others, puzzled me for some time on account of the great discrepancies with the others which it exhibited. I have no doubt that the cause lay in the different character of the light on the day in question, which came from the unusually blue sky which some-times accompanies a high wind. On the other days the light came principally from clouds. I have had no opportunity of confirming this opinion by a repetition of the experiment with a sky of the same degree of blueness, but that the disagreement was not the result of unusually large errors of observa-tion, is, I think, to be inferred from the fact that the observations under the blue sky were as consistent among themselves as any of the other sets. As the point is of some interest, I give the figures in full.

July 23, blue sky.

Black.	White.	Red.	Green.	Yellow.	Blue.	
0	+ 30	+ 122	+ 40	− 77	− 115	obs.
0	+ 32·2	+ 120·8	+ 39·1	− 78·8	− 113·2	calcd.
+ 94	0	− 132	− 60	+ 55	+ 43	obs.
+ 91·6	0	− 133·5	− 58·5	+ 54·4	+ 45·9	calcd.
− 138	− 54	0	+ 24	+ 50	+ 118	obs.
− 138·3	− 53·7	0	+ 23·1	+ 49·5	+ 119·5	calcd.
+ 92	+ 50	+ 50	0	− 66	− 126	obs.
+ 94·1	+ 49·5	+ 48·5	0	− 65·2	− 126·7	calcd.
− 154	− 38	+ 86	+ 52	0	+ 54	obs.
− 154·6	− 37·5	+ 84·6	+ 53·1	0	+ 54·3	calcd.
+ 139	+ 18	− 128	− 64	+ 35	0	obs.
+ 138·5	+ 19·7	− 127·5	− 64·5	+ 33·9	0	calcd.

The numbers read off for the big discs are written with the sign + prefixed, and those corresponding to the little discs with −. Thus the first line may be read:—30 parts white together with 122 red and 40 green, match 77 yellow and 115 blue. The upper line of each pair represents the actual observation, and the second is the theoretical equation calculated from two in the manner described. The average difference between the two sets of numbers which may be taken as a measure of the inaccuracy of the observations amounts to 1·1. A similar table, formed from the observations of July 20 (cloudy), and which agreed very well with the results of other days, is as follows*:—

Black.	White.	Red.	Green.	Yellow.	Blue.
0	+ 30	+ 117	+ 45	− 79	− 113
0	31·1	116·2	44·8	79·9	112·2
+ 90	0	− 128	− 64	+ 56	+ 46
85·9	0	128·4	63·5	57·0	49·0
− 136	− 56	0	+ 22	+ 52	+ 118
137	55	0	22·3	50	119·6
+ 100	+ 50	+ 42	0	− 64	− 128
99·2	51	41·9	0	65	127·1
+ 135	+ 21	− 123	− 69	+ 36	0
135·7	21·5	122·7	69·3	34·8	0
− 152	− 40	+ 80	+ 56	0	+ 56
152·6	39·5	81	56	0	55

The average error is here ·95, showing only a trifling better agreement than the former set, so that the blue sky observations are nearly as self-consistent as those made with cloud-light. Moreover, the agreement is itself very good, being decidedly better than Maxwell's, though his calculations refer to a *mean* of six sets of observations.

While therefore there is no reason to distrust the results of July 23 any more than of July 20, the differences between them are much greater than can be ascribed to errors of observation. It will be found that they relate principally to the quantities of red, the numbers under that head being considerably greater for the case of the blue light from the sky. I am not aware whether the difference of sky and cloud light has ever been made the subject of direct investigation, but it would seem a fair inference that it must consist mainly in a relative deficiency of the red rays. If this be so, as I have other grounds for suspecting, the light of the sky would be similar in composition to that of dilute solutions of copper, which acquire their light

---

* These calculations were made by means of Prof. Everett's Proportion table, which seems admirably adapted to work of this sort.

blue tint by a partial suppression of the extreme red*. There is no doubt that the colour equations are dependent on the character of the light, as may easily be proved by taking an observation looking all the time through a layer of coloured liquid. It is not, however, the most brilliantly coloured solutions that cause the most disturbance, for anything like a complete stoppage of all the rays which are capable of exciting one of the primary colour sensations would affect both the mixtures to be compared in nearly the same manner, putting the observer in fact very much into the position of a colour-blind person. Those liquids will be most efficient which have a different action on parts of the spectrum allied in colour. For instance, an aqueous infusion of litmus has a strongly marked action on the yellow ray, stopping it with great energy, even in rather dilute solutions. It is easy to trace the effect of looking through this on most of the colour equations. Consider, for example, the fifth equation of July 20 (that from which the blue is absent) wherein red and green are matched against black, white, and yellow. The red and green will for the most part escape absorption, but the white and yellow will be shorn of a part of their yellow rays. The match supposed to have been adjusted without the litmus must evidently be spoiled; the red and green mixture becoming strongly yellow in comparison with the other. In order to restore equivalence the yellow must be considerably increased. On trial I found that 124 black + 19 white + 49 yellow matched 121 red + 71 green.

It is only the impurity of the colours on the discs that prevents the effect being still more strongly marked, for with the pure colours of the spectrum the most violent alterations are possible. When a match is made between the simple yellow and that compounded of pure red and green, almost any coloured liquid acts unequally on the two parts and destroys the balance. The simple yellow, of course, retains its colour under any absorbing influence, and can only be changed in luminosity. Chloride of copper extinguishes the red component of the compound yellow, which accordingly becomes green. Litmus would leave the compound colour nearly unchanged, while it extinguishes the simple yellow. It is needless to multiply instances.

Before leaving the compound yellow, of whose very existence many are incredulous, I will mention an easy way of obtaining it, which is the more desirable as the use of the pure spectral colours is not very convenient. In

---

* Direct observations, made since the above was written, show that there is no *peculiar* deficiency at the red end of the spectrum, but a general falling off as the refrangibility diminishes from one end to the other. If lights from sky and cloud are of equal intensity at the line *C* in the red, the first will be somewhere about twice as bright as the other at *b* in the green. This is for a well-developed blue light taken from the zenith; but, even with a large allowance, enough difference remains to account for the discrepancies in the two sets of colour disc observations. I have lately found from theory that the power of very small particles to scatter the rays belonging to different parts of the spectrum varies as the inverse fourth power of the wavelength.

order to isolate the red and green rays of the spectrum by means of absorption, the first thing is to find a liquid capable of removing the intermediate yellow and orange. With this object we may fall back on the alkaline solution of litmus, whose opacity to the yellow, and particularly to the orange, rays is so marked. The next step is to remove the blue and bluish green, for which nothing is 'more convenient than the chromate of potash. A mixture of these two liquids in proper proportions, easily found by trial, isolates the green and extreme red rays with considerable perfection, and exhibits in a high degree the phenomenon of Dichromatism. According to the thickness traversed by the light the red or the green predominates, and there is no difficulty with a given thickness in arranging the strength of the solution so as to give a full compound yellow. It is worth notice in confirmation of the opinion expressed as to the character of the sky-blue, that when a cloud seen through the liquid appears a full yellow, or even orange, the former, if at all intense, acquires a decided green colour. A window backed by well-lighted clouds, when looked at across a room through the liquid and a prism, has a very splendid appearance, the red being isolated on one side, and the green on the other; while the intermediate space, where the two overlap, shows the compound yellow in great perfection. Another liquid, in some respects preferable, which answers the same purpose, may be made by mixing chloride of chromium and bichromate of potash. Through either of them the sodium flame is invisible, though they may easily be made to correspond with it in colour very closely. I tried to obtain a liquid capable of isolating the pure yellow ray, but only with partial success. The best was a mixture of bichromate and permanganate of potash with a salt of copper (sulphate or chloride). The first removes the blue and violet, the second the green, and the third the red, and thus the yellow is isolated in considerable purity. This liquid is very unstable. The comparison of the simple and compound yellow (which nearly matched) was interesting. One was transparent to the sodium flame, the other completely opaque to it. When the two are brought together so that the light has to traverse both, almost complete darkness results, even when the brightest clouds are used. I should mention that it is only when the light is strong that any of these liquids give yellow in full perfection; otherwise the colour is more nearly described as brown, which is, in fact, identical with a dark yellow or orange. The best natural yellows, such as chrome, are partly simple and partly compound, returning all the light which falls upon them except the blue and violet. It is clear that neither a purely simple nor a purely compound yellow can rival them in brilliancy.

Impartial observers, unprejudiced by the results of mixing pigments, or, on the other hand, by experiments on the spectrum, see, so far as I can make out, no connection between the four principal colours—red, yellow, green, and blue. It seems to them quite as absurd that yellow should be compounded of red and green, as it most unquestionably is, as that green should be a compound of blue and yellow, though many have accepted the latter alternative on the authority of painters, and some have even worked themselves into the belief that it is only necessary to look at the colours in order to recognise the compound nature of green. My own prejudice would be on the other side, the result of experiments on the compound yellow, which is seen so easily to pass into green on the one side or red on the other. The most impartial opinion that I can form is that there is no real *resemblance* between any of the four, and if this be so it is certainly a most remarkable, if not unaccountable, fact. The difficulty is not so much that we are unable to analyse the compound sensation, as to explain why our inability is limited to yellow (and white). For everyone, I imagine, sees in purple a resemblance to its components red and blue, and can trace the primary colours in a mixture of green and blue. Sir John Herschel even thinks that our inability to resolve yellow leaves it doubtful whether our vision is trichromic or tetra-chromic, but this seems to me to be going much too far. Surely the fact that the most saturated yellow can be compounded of red and green, deprives it of any right to stand in the same rank with them as primary colours, however little resemblance it may bear to them and blue. Besides, if yellow is to be considered primary, why not also white, which is quite as distinct a sensation as any of the others? Undoubtedly there is much that is still obscure in the mutual relations of the colours—why, for instance, as mentioned by Sir John Herschel, a dark yellow or orange suggests its character so little as to be called by a new name (brown), while a dark blue is blue still. But difficulties such as these should make us all the more determined to build our theories of colour on the solid ground that normal vision is threefold, and that the three primary elements of colour correspond nearly with red, green, and blue.

Yellow.

It was not from any experiments of my own, but on the authority of Helmholtz, that I asserted [supra, p. 84] the identity of brown with a dark yellow or orange. He found that the pure red and yellow of the spectrum gave the various shades of brown when seen by the side of more brilliantly lighted white surfaces. (*Physiologische Optik*, p. 281.) There is therefore nothing in the nature of the colour to exclude complete saturation, although it may well happen that most of the browns we ordinarily see fall somewhat short of it.

In *Nature* of Jan. 26, Mr Munro calls attention to the great brilliancy and saturation of many natural yellows as accounting for the difficulty of resolving them into their components. It is, no doubt, quite true that a full yellow could not be compounded of such reds and greens as we come across in daily life, but it is equally certain that a drab or dilute yellow could be; and yet no one recognises the fact by his unaided senses, or thinks it anything but strange and unlikely when told of it. And after all can it properly be said that natural yellows are more saturated than other colours? That they approach more nearly the corresponding tints in the spectrum is admitted; but is that test a fair one? It seems to me that the homogeneous yellow itself must be considered as dilute when brought into comparison with the nearly primary red and green.

I have another difficulty in accepting Mr Munro's explanation. A suitable mixture of any red, green or blue will give a neutral grey. All four come within our every day experience; but such a result seemed to Goethe, soon after Newton proved it, a paradox of paradoxes; and I believe to unsophisticated minds it seems so still.

Mr Munro has ingeniously shown from the colour equations that there is no more primary blue in my blue disc than about $2\frac{1}{2}$ as much as in the red *plus* $1\frac{1}{3}$ as much as in the green—a conclusion which seems somewhat startling. In choosing the coloured papers and cards for the discs, I had great difficulty in finding a green that was even tolerably good, and the one that I finally used reflected large quantities of blue light. I had some thought of trying a green silk disc, which was of a much better colour, but feared errors depending on the different character of the surface.

It is not hard to see a reason for the comparative scarcity of good greens. To obtain a good red orange or yellow by means of absorption, all that is necessary is to cut away the spectrum above a certain point; for a good blue, the rays standing below a given one in refrangibility must be got rid of; but in order to isolate a green in anything like purity, the absorbing agent must hit off *two* points of the spectrum, removing all below one point and all above the other. The result is, that while nearly saturated yellows and reds abound—the scarlet of the geranium is almost perfect—hardly a good green is to be met with. The best I know is a mixture, prepared by adding bichromate of potash to a strong solution of sulphate of copper. The addition of a little chloride of chromium to remove the yellow more effectually is perhaps an improvement.

[1899. Further experiments upon the subject of this paper are described in *Nature*, vol. XXV. pp. 64—66, 1882.]

8.

ON THE LIGHT FROM THE SKY, ITS POLARIZATION AND COLOUR.

[*Phil. Mag.* XLI. pp. 107—120, 274—279; 1871.]

IT is now, I believe, generally admitted that the light which we receive from the clear sky is due in one way or another to small suspended particles which divert the light from its regular course. On this point the experiments of Tyndall with precipitated clouds seem quite decisive. Whenever the particles of the foreign matter are sufficiently fine, the light emitted laterally is blue in colour, and, in a direction perpendicular to that of the incident beam, is *completely polarized.*

About the colour there is no *prima facie* difficulty; for as soon as the question is raised, it is seen that the standard of linear dimension, with reference to which the particles are called small, is the wave-length of light, and that a given set of particles would (on any conceivable view as to their mode of action) produce a continually increasing disturbance as we pass along the spectrum towards the more refrangible end; and there seems no reason why the colour of the compound light thus scattered laterally should not agree with that of the sky.

On the other hand, the direction of polarization (perpendicular to the path of the primary light) seems to have been felt as a difficulty. Tyndall says, "......the polarization of the beam by the incipient cloud has thus far proved itself to be *absolutely independent of the polarizing-angle.* The law of Brewster does not apply to matter in this condition; and it rests with the undulatory theory to explain why. Whenever the precipitated particles are sufficiently fine, no matter what the substance forming the particles may be, the direction of maximum polarization is at right angles to the illuminating beam, the polarizing angle for matter in this condition being invariably 45°. This I consider to be a point of capital importance with reference to the present question"*. As to the importance there will not be two opinions; but I venture to think that the difficulty is imaginary, and is caused mainly by misuse of the word reflection. Of course there is nothing in the etymology of reflection or refraction to forbid their application in this sense; but the words have acquired technical meanings, and become associated with certain well-known laws called after them. Now a moment's consideration of the principles according to which reflection and refraction are explained in the wave theory is sufficient to show that they have no application unless the surface of the disturbing body is larger than many square wave-lengths; whereas the particles to which the sky is supposed to owe its illumination must be *smaller* than the wave-length, or else the explanation of the colour breaks down. The idea of polarization by reflection is therefore out of place; and that "the law of Brewster does not apply to matter in this condition" (of extreme fineness) is only what might have been inferred from the principles of the wave theory.

Nor is there any difficulty in foreseeing what, according to the wave theory, the direction of polarization ought to be. Conceive a beam of plane-polarized light to move among a number of particles, all small compared with any of the wave-lengths. The foreign matter, if optically denser than air, may be supposed to *load* the æther so as to increase its *inertia* without altering its resistance to distortion, provided that we agree to neglect effects analogous to chromatic dispersion. If the particles were away, the wave would pass on unbroken and no light would be emitted laterally. Even with the particles retarding the motion of the æther, the same will be true if, to counterbalance the increased inertia, suitable forces are caused to act on the æther at all points where the inertia is altered. These forces have the same period and *direction* as the undisturbed luminous vibrations themselves. The light actually emitted laterally is thus the same as would be caused by forces exactly the opposite of these acting on the medium otherwise free from disturbance; and it only remains to see what the effect of such forces would be.

On account of the smallness of the particles, the forces acting throughout the volume of any one are all of the same intensity and direction, and may be considered as a whole. The determination of the motion in the æther, due to the action of a periodic force at a given point, requires, of course, the aid of mathematical analysis; but very simple considerations will lead us to a conclusion on the particular point now under discussion. In the first place there is a complete symmetry round the direction of the force. The disturbance, consisting of transverse vibrations, is propagated outwards in all directions from the centre; and in consequence of the symmetry, the

* *Phil. Mag.* S. 4, vol. xxxvii. p. 388.

direction of vibration in any ray lies in the plane containing the ray and the axis; that is to say, the direction of vibration in the scattered or diffracted ray makes with the direction of vibration in the incident or primary ray the least possible angle. The symmetry also requires that the intensity of the scattered light should vanish for the ray which would be propagated along the axis; for there is nothing to distinguish one direction transverse to the ray from another. We have now got what we want. Suppose, for distinctness of statement, that the primary ray is vertical, and that the plane of vibration is that of the meridian. The intensity of the light scattered by a small particle is constant, and a maximum for rays which lie in the vertical plane running east and west, *while there is no scattered ray along the north and south line.* If the primary ray is unpolarized, the light scattered north and south is entirely due to that component which vibrates east and west, and is therefore *perfectly polarized*, the direction of its vibration being also east and west. Similarly any other ray scattered horizontally is perfectly polarized, and the vibration is performed in the horizontal plane. In other directions the polarization becomes less and less complete as we approach the vertical, and in the vertical direction itself altogether disappears.

So far, then, as disturbance by very small particles is concerned, theory appears to be in complete accordance with the experiments of Tyndall and others. At the same time, if the above reasoning be valid, the question as to the direction of the vibrations in polarized light is decided in accordance with the view of Fresnel. Indeed the observation on the plane of polarization of the scattered light is virtually only another form of Professor Stokes's original test with the diffraction-grating. In its present shape, however, it is free from certain difficulties both of theory and experiment, which have led different physicists who have used the other method to contradictory conclusions. I confess I cannot see any room for doubt as to the result it leads to*.

The argument used is apparently open to a serious objection, which I ought to notice. It seems to prove too much. For if one disturbing particle is unable to send out a scattered ray in the direction of original vibration, it would appear that no combination of them (such as a small body may be supposed to be) could do so, at least at such a distance that the

* I only mean that *if* light, as is generally supposed, consists of transversal vibrations similar to those which take place in an elastic solid, the vibration must be normal to the plane of polarization. There is unquestionably a formal analogy between the two sets of phenomena extending over a very wide range; but it is another thing to assert that the vibrations of light are really and truly to-and-fro motions of a medium having mechanical properties (with reference to small vibrations) like those of ordinary solids. The fact that the theory of elastic solids led Green to Fresnel's formulæ for the reflection and refraction of polarized light seems amply sufficient to warrant its employment here, while the question whether the analogy is more than formal is still left open.

body subtends only a small solid angle.   Now we know that when light vibrating in the plane of incidence falls on a reflecting surface at an angle of 45°, light *is* sent out according to the law of ordinary reflection, whose direction of vibration is perpendicular to that in the incident ray.   And not only is this so in experiment, but it has been proved by Green* to be a consequence of the very same view as to the nature of the difference between media of various refrangibilities as has been adopted in this paper.   The apparent contradiction, however, is easily explained.   It is true that the disturbance due to a foreign body of any size is the same as would be caused by forces acting through the space it fills in a direction parallel to that in which the primary light vibrates; *but these forces must be supposed to act on the medium as it actually is—that is, with the variable density.*   Only on the supposition of complete uniformity would it follow that no ray could be emitted parallel to the line in which the forces act.   When, however, the sphere of disturbance is small compared with the wave-length, the want of uniformity is of little account, and cannot alter the law regulating the intensity of the vibration propagated in different directions.

Having disposed of the polarization, let us now consider how the intensity of the scattered light varies from one part of the spectrum to another, still supposing that all the particles are many times smaller than the wave-length even of violet light.   The whole question admits of analytical treatment; but before entering upon that, it may be worth while to show how the principal result may be anticipated from a consideration of the *dimensions* of the quantities concerned.

The object is to compare the intensities of the incident and scattered rays; for these will clearly be proportional.   The number (i) expressing the ratio of the two amplitudes is a function of the following quantities:—T, the volume of the disturbing particle; r, the distance of the point under consideration from it; λ, the wave-length; b, the velocity of propagation of light; D and D', the original and altered densities: of which the first three depend only on space, the fourth on space and time, while the fifth and sixth introduce the consideration of mass.   Other elements of the problem there are none, except mere numbers and angles, which do not depend on the fundamental measurements of space, time, and mass.   Since the ratio i, whose expression we seek, is of no dimensions in mass, it follows at once that D and D' only occur under the form $D : D'$, which is a simple number and may therefore be omitted.   It remains to find how i varies with T, r, λ, b.

Now, of these quantities, b is the only one depending on time; and therefore, as i is of no dimensions in time, b cannot occur in its expression. We are left, then, with T, r, and λ; and from what we know of the dynamics

* *Camb. Phil. Trans.* vol. VII. 1837.

of the question, we may be sure that i varies directly as T and inversely as r, and must therefore be proportional to $T \div \lambda^2 r$, T being of three dimensions in space. In passing from one part of the spectrum to another λ is the only quantity which varies, and we have the important law :—

When light is scattered by particles which are very small compared with any of the wave-lengths, the ratio of the amplitudes of the vibrations of the scattered and incident light varies inversely as the square of the wave-length, and the intensity of the lights themselves as the inverse fourth power.

I will now investigate the mathematical expression for the disturbance propagated in any direction from a small particle which a beam of light strikes.

Let the vibration corresponding to the incident light be expressed by $A \cos (2\pi bt/\lambda)$. The acceleration is

$$- A \left(\frac{2\pi}{\lambda} b\right)^2 \cos \frac{2\pi}{\lambda} bt \, ;$$

so that the force which would have to be applied to the parts where the density is D', in order that the wave might pass on undisturbed, is, per unit of volume,

$$- (D' - D) A \left(\frac{2\pi b}{\lambda}\right)^2 \cos \frac{2\pi}{\lambda} bt.$$

To obtain the total force which must be supposed to act over the space occupied by the particle, the factor T must be introduced. The opposite of this conceived to act at O (the position of the particle) gives the same disturbance in the medium as is actually caused by the presence of the particle. Suppose, now, that the ray is incident along OY, and that the direction of vibration makes an angle α with the axis of x, which is the line of the scattered ray under consideration—a supposition which involves no loss of generality, because of the symmetry which we have shown to exist round the line of action of the force. The question is now entirely reduced to the discovery of the disturbance produced in the æther by a given periodic force acting at a fixed point in it. In his valuable paper " On the Dynamical Theory of Diffraction "[*], Professor Stokes has given a complete investigation of this problem; and I might assume the result at once. The method there used is, however, for this particular purpose very indirect, and accordingly I have thought it advisable to give a comparatively short cut to the result, which will be found at the end of the present paper. It is proved that if the total force acting at O in the manner supposed be $F \cos (2\pi bt/\lambda)$, the resulting disturbance in the ray propagated along OX is

$$\zeta = \frac{F \sin \alpha}{4\pi b^2 D r} \cos \frac{2\pi}{\lambda} (bt - r).$$

[*] *Camb. Phil. Trans.* vol. IX. p. 1, 1849.

Substituting for F its value, we have

$$\zeta = A \frac{D' - D}{D} \frac{\pi T}{r\lambda^2} \sin \alpha \, \cos \frac{2\pi}{\lambda} (bt - r)*,$$

an equation which includes all our previous results and more.

One reservation, however, must not be omitted. Since we have supposed the medium uniform throughout, whereas it really has a different density at the place where the force acts, our investigation does not absolutely correspond to the actual circumstances of the case. As before remarked, no error is on that account to be feared in the law determining the intensity of the vibration in different directions; but it is probable that the coefficient, so far as it depends on $D : D'$, may be changed†, and there may be a change in the phase comparable with $(2\pi/\lambda) \times$ the linear dimension of the particle, which is of importance when the scattered and primary waves have to be compounded.

So much for a single particle. In actual experiments, as, for instance, with Professor Tyndall's " clouds," we have to deal with an immense number of such particles; and the question now is to deduce what their effect must be from the results already obtained. Were the particles absolutely motionless, the partial waves sent out in any direction from them would have permanent relations as to phase, and the total disturbance would have to be found by compounding the *vibrations* due to all the particles. Such a supposition, however, would be very wide of the mark; for, in consequence of the extreme smallness of λ, the slightest motion of any particle will cause an alteration of phase passing through many periods in a less time than the eye could appreciate. Our particles are, then, to be treated as so many *unconnected* sources of light; and instead of adding the *vibrations*, we must take the *intensities* represented by their squares. Only in one direction is a different treatment necessary, namely along the course of the primary light. I mention this because it would not otherwise appear how the reduction in the intensity of the transmitted light is effected; but we do not require to follow the details of the process, because, when once we know the intensity of the light emitted laterally, the principle of energy will tell us what the primary wave has lost.

The intensity of the light scattered from a cloud is thus equal to

$$A^2 \frac{(D' - D)^2}{D^2} \sin^2 \alpha \, \frac{\pi^2 . \Sigma T^2}{\lambda^4 r^2},$$

---

* [1898. The factor π was omitted in the original paper.]

† I find that no alteration of any kind is needed.—Jan. 20. [1899. See *Phil. Mag.* XLI. p. 452, 1871 ; This Collection Art. 9 below.]

where ΣT^2 is the sum of all the squares of T. If T^2 be understood to denote the mean square of T (*not* the square of the mean value of T), and m be the number of particles,

$$\Sigma T^2 = m \cdot T^2.$$

If the primary light be unpolarized, the intensity in a direction making an angle β with its course becomes

$$A^2 \frac{(D'-D)^2}{D^2} (1 + \cos^2 \beta) \frac{m\pi^2 T^2}{\lambda^4 r^2}.$$

Backwards from the cloud the light is thus twice as bright as normally. To the light scattered nearly in the direction of the primary ray our expression does not apply.

Fig. 1.

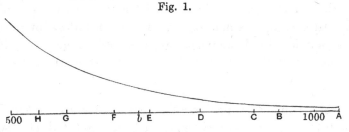

Fig. 1 shows the curve representing the intensity of the scattered light for each part of the spectrum, referred to the intensity in the primary light as a standard. The abscissa being proportional to λ, the base line represents the diffraction-spectrum with the principal fixed lines. Over the brighter portion of the spectrum from B to G the curve differs but little from a straight line, while the small curvature is turned downwards, indicating a deficiency in the green and yellow.

Before making out the theory, I had endeavoured to ascertain by observation the actual prismatic composition of the blue of the sky, and had obtained preliminary results. The experimental method (the description of which I must reserve for another opportunity) was fully adequate to the comparison of two given lights; but the difficulty was to find something to compare the blue light with. In the only complete set of observations that I have hitherto been able to make, the blue of the sky (apparently a very good one) taken from the neighbourhood of the zenith was compared with sunlight diffused through white paper. About thirty consistent comparisons were made, ranging over the spectrum from C to beyond F, and a curve drawn on the plan of fig. 1. I do not give the complete curve, because I hope before long to complete and confirm the observations; but the following numbers will give an idea of the results:—

$C.$	$D.$	$b_3.$	$F.$	
25	40	63	80	from fig. 1.
25	41	71	90	observed.

The upper line gives the theoretical intensities for the fixed lines C, D, b_3, F, while the lower gives the observed ratios between the lights (sky and diffused sunlight), the two sets of numbers being made to agree at C. Considering the difficulties and uncertainties of the case, the two curves agree very well; and it should be noticed that the sky compared with diffused light was even bluer than theory makes it, on the supposition that the diffused light through the paper may be taken as similar to that whose scattering illuminates the sky. It is possible that the paper was slightly yellow; or the cause may lie in the yellowness of sunlight as it reaches us compared with the colour it possesses in the upper regions of the atmosphere. It would be a mistake to lay any great stress on the observations in their present incomplete form; but at any rate they show that a colour more or less like that of the sky would result from taking the elements of white light in quantities proportional to λ^{-4}. I do not know how it may strike others; but individually I was not prepared for so great a difference as the observations show, the ratio for F being more than three times as great as for C.

There is one point in which our calculations do not exactly meet the case of the sky. In the experiments with precipitated clouds the total quantity of light scattered is quite insignificant compared with the incident beam; but it is by no means so clear that the same is the case with the sky. Each particle is thus struck, not only by the direct light of the sun, but also by that scattered from others. It does not seem that the chromatic effects would be much affected by this consideration; but it is worth notice that the conclusion as to complete polarization perpendicular to the incident ray would have to be modified. To see this, imagine, as before, the light (unpolarized) incident along OY upon a particle O; we have seen that the ray diffracted along OX contains no vibration parallel to OY. By the aid, however, of another particle P in the xy plane such a vibration may be communicated to it; for in the ray diffracted from P to O there is a component vibration in the xy plane perpendicular to PO, which, when again diffracted along OX, will give a component parallel to OY. This is perhaps the explanation of the incomplete polarization of sky-light at right angles to the solar beams; but it must be remembered that an insufficient fineness in some of the particles of foreign matter would have a like result.

By many physicists, from Newton downwards, the light of the sky has been supposed to be reflected from thin plates, and the colour to be the blue of the first order in Newton's scale. Such a view is fundamentally different from that adopted in this paper, though it might not at first seem so. In support of this assertion, it may be sufficient to notice that the two theories are at variance as to the law connecting the intensity with wave-length. By an argument from dimensions similar to that already used, it is easy to find how the intensity of the light reflected from a thin plate (thin, that is, compared with *any* of the wave-lengths) varies with λ. Instead of our former quantities, T, r, λ, we now have merely λ, and δ the thickness of the plate. Since the reflected vibration necessarily varies as δ, it must also be proportional to λ^{-1}, and so the *intensity* of the reflected light $\propto \lambda^{-2}$ instead of λ^{-4}. The ordinary analytical expression for the reflected light leads readily to the same conclusion (Airy's *Tracts*, p. 297). There can, I think, be no question that the composition of the light of the sky agrees more nearly with the latter than with the former law.

The principle of energy makes it clear that the light emitted laterally is not a new creation, but only diverted from the main stream. If I represent the intensity of the primary light after traversing a thickness x of the turbid medium, we have

$$dI = -kI\lambda^{-4}\,dx,$$

where k is a constant independent of λ. On integration,

$$I = I_0 \epsilon^{-k\lambda^{-4}x},$$

if I_0 correspond to $x = 0$,—a law altogether similar to that of absorption, and showing how the light tends to become yellow and finally red as the thickness of the medium increases. Fig. 2 shows a series of curves representing the composition of the originally white light after passing through thicknesses in the ratio of 1, 2, 4, 8, 16, 32. The reader will observe how little of the violet light remains when the red is still in nearly its original force. I cannot but think that this rapid diversion of the rays of short

Fig. 2.

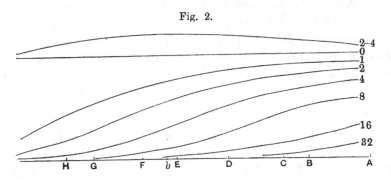

wave-length has a good deal to do with the absence of light of the highest refrangibility from the direct rays of the sun. For the line A at the extreme red and R near the upper limit of the photographic spectrum the wave-lengths are 7617 and 3108. The ratio of the fourth powers is about 36 : 1 ; so that, whatever the fraction representing the transmission of A may be, its 36th power will give the transmission of R. To take an instance, if ·9 of the ray A gets through, only ·018 of R would be able to penetrate. For the rays of still higher refrangibility, which Professor Stokes found abundant in the electric light but missing in the solar rays, the fraction would be smaller still; but I am not aware of any measurements of smaller wave-length on which to found a calculation.

We have hitherto supposed that the light scattered by the finely divided matter reaches the eye without modification, and we have taken no account of any change in the composition of the primary light before diffraction. If x be the total length of the path of the ray through the turbid medium, we may express the quality of the light in terms of x; for it makes no difference whether the lateral leakage takes place before diffraction or after. In fact

$$I \propto \lambda^{-4} \epsilon^{-k\lambda^{-4}x},$$

an expression which shows that I vanishes for very small as well as for very large values of λ, while for some definite value (say Λ) it rises to a maximum (I_0). Expressing I in terms of I_0 and Λ, we have

$$\frac{I}{I_0} = \frac{\Lambda^4}{\lambda^4} \epsilon^{1 - \Lambda^2/\lambda^2};$$

from which we may fall back on our original law by supposing Λ indefinitely small, and replacing $\Lambda^4 I_0$ by a finite constant. An approximate idea of the character of these lights may be obtained by subtracting the successive curves of Fig. 2. Thus the difference of the curves marked 2 and 4 represents a light having its maximum brightness (of course relatively to the primary light) in the blue-green portion of the spectrum. I find by calculation that, if the maximum intensity be at b and be taken as unity, the intensities at G and C are given by the numbers ·713, ·710 respectively. The colour would be greenish; but whether the green of the sky is to be accounted for in this way I am not able to say. Some, I believe, consider it to be entirely a contrast effect.

APPENDIX.

Within a space T, small in all its dimensions against λ, and situated at the origin of coordinates, let a force parallel to OZ, and, so far as it depends upon the time, expressed by a simple circular function, act on the medium. If ξ, η, ζ denote the displacements parallel to the axes at the point x, y, z, and

$$\delta = \frac{d\xi}{dx} + \frac{d\eta}{dy} + \frac{d\zeta}{dz},$$

$$\left.\begin{aligned}
\frac{d^2\xi}{dt^2} &= b^2\nabla^2\xi + (a^2 - b^2)\frac{d\delta}{dx}, \\[2mm]
\frac{d^2\eta}{dt^2} &= b^2\nabla^2\eta + (a^2 - b^2)\frac{d\delta}{dy}, \\[2mm]
\frac{d^2\zeta}{dt^2} &= b^2\nabla^2\zeta + (a^2 - b^2)\frac{d\delta}{dz} + Z,
\end{aligned}\right\} \quad \ldots\ldots\ldots\ldots(A)^*$$

where ∇^2 stands for $d^2/dx^2 + d^2/dy^2 + d^2/dz^2$; a^2 and b^2 are constants depending on the nature of the medium supposed to be isotropic. For the luminiferous æther, Green has shown that a is to be regarded as indefinitely great†.

To represent the periodic force, write for Z, $Z\epsilon^{int}$. Similar transformations will then apply to ξ, η, ζ, and δ; so that on substitution in (A) and dividing out the common factor ϵ^{int}, there results

$$\left.\begin{aligned}
(b^2\nabla^2 + n^2)\,\xi + (a^2 - b^2)\,d\delta/dx &= 0, \\[1mm]
(b^2\nabla^2 + n^2)\,\eta + (a^2 - b^2)\,d\delta/dy &= 0, \\[1mm]
(b^2\nabla^2 + n^2)\,\zeta + (a^2 - b^2)\,d\delta/dz &= -Z.
\end{aligned}\right\} \quad \ldots\ldots\ldots(B)$$

Writing

$$\frac{d\xi}{dy} - \frac{d\eta}{dx} = \varpi_3 \ \&c.,$$

we obtain from (B) by differentiation and subtraction,

$$\left.\begin{aligned}
(b^2\nabla^2 + n^2)\,\varpi_3 &= 0, \\[1mm]
(b^2\nabla^2 + n^2)\,\varpi_1 &= dZ/dy, \\[1mm]
(b^2\nabla^2 + n^2)\,\varpi_2 &= -dZ/dx.
\end{aligned}\right\} \quad \ldots\ldots\ldots\ldots(C)$$

ϖ_1, ϖ_2, ϖ_3 are the *rotations* of the elements of the medium round axes parallel to those of coordinates.

The disturbance which we are investigating is that caused and maintained by the force Z acting within the space T. Accordingly‡

$$\varpi_3 = 0,$$

$$\varpi_1 = -\frac{1}{4\pi b^2}\iiint\frac{dZ}{dy}\frac{\epsilon^{\pm ikr}}{r}\,dx\,dy\,dz,$$

r being the distance between the element $dx\,dy\,dz$ and the point where ϖ_1 is estimated, and

$$k = 2\pi/\lambda = n/b. \quad \ldots\ldots\ldots\ldots\ldots(D)$$

---

* Thomson and Tait's *Natural Philosophy*, § 698.
† *Camb. Phil. Trans.* vol. VII.
‡ Helmholtz, *Crelle's Journal*, 1860.

Since $\epsilon^{\pm ikr}$ will be finally multiplied by ϵ^{int}, and the disturbance which we are dealing with is propagated *outwards* from T, it is evident that the *lower* sign is to be employed. Now

$$\int \frac{\epsilon^{-ikr}}{r} \frac{dZ}{dy} dy = \left[Z \frac{\epsilon^{-ikr}}{r} \right] - \int Z \frac{d}{dy} \left(\frac{\epsilon^{-ikr}}{r} \right) dy,$$

of which the term within brackets vanishes, because the value of Z is only finite within the space T. Thus

$$\varpi_1 = \frac{1}{4\pi b^2} \iiint Z \frac{d}{dy} \left(\frac{\epsilon^{-ikr}}{r} \right) dx\, dy\, dz.$$

The factor $\dfrac{d}{dy} \left(\dfrac{\epsilon^{-ikr}}{r} \right)$ within the space T is sensibly constant, so that, if Z stand for the mean value of Z over the volume T,

$$\varpi_1 = \frac{TZ}{4\pi b^2} \frac{d}{dy} \left(\frac{\epsilon^{-ikr}}{r} \right), \qquad \varpi_2 = - \frac{TZ}{4\pi b^2} \frac{d}{dx} \left(\frac{\epsilon^{-ikr}}{r} \right),$$

$$x\varpi_1 + y\varpi_2 \propto \left(x \frac{d}{dy} - y \frac{d}{dx} \right) . \frac{\epsilon^{-ikr}}{r} = 0.$$

And if $R = \sqrt{x^2 + y^2}$,

$$\varpi = \frac{x\varpi_2 - y\varpi_1}{R} = - \frac{TZ}{4\pi b^2 R} \left(x \frac{d}{dx} + y \frac{d}{dy} \right) . \frac{\epsilon^{-ikr}}{r}$$

$$= - \frac{TZ}{4\pi b^2} \frac{d}{dR} \cdot \frac{\epsilon^{-ikr}}{r} = \frac{TZ \sin \alpha}{4\pi b^2} \frac{d}{dr} \cdot \frac{\epsilon^{-ikr}}{r},$$

where α denotes the angle between r and z.

The resultant rotation at any point is thus about an axis perpendicular to the plane passing through the point and the axis of Z; and its magnitude is given by ϖ. In differentiating $(r^{-1} e^{-ikr})$ with respect to r, we may neglect the term divided by r^2 as altogether insensible, kr being an exceedingly great quantity at any ordinary distance from the origin of disturbance. Thus

$$\varpi = \frac{- ik . TZ \sin \alpha}{4\pi b^2} \cdot \frac{\epsilon^{-ikr}}{r}, \quad \dots\dots\dots\dots\dots(E)$$

which completely determines the rotation at any point. For a given disturbance it is seen to be everywhere about an axis perpendicular to r and the direction of the force, and in magnitude dependent only on the angle between these two directions (α) and on the distance (r).

The intensity of the light, however, is more usually expressed in terms of the actual displacement in the plane of the wave. In order to find the connexion between the two quantities, it will be more convenient to suppose the scattered ray parallel to x, and that the force F (for Z is no longer appropriate) acts in the plane of zx at an angle α with Ox. ϖ becomes identical with ϖ_2; that is, with $d\zeta/dx$; for ξ as well as η is zero; so that

$$\zeta = \int \varpi \, dr = \frac{TF \sin \alpha}{4\pi b^2} \cdot \frac{\epsilon^{-ikr}}{r}.$$

Restoring the factor ϵ^{int}, we have

$$\zeta = \frac{TF \sin \alpha}{4\pi b^2} \cdot \frac{\epsilon^{i(nt-kr)}}{r};$$

or throwing away the imaginary part,

$$\zeta = \frac{TF \sin \alpha}{4\pi b^2 r} \cdot \cos \frac{2\pi}{\lambda} (bt - r). \quad \dots\dots\dots\dots(F)$$

This corresponds to a total accelerating force equal to $FT \cos (2\pi bt/\lambda)$; a result which agrees with that of Professor Stokes's more complete investigation, with the exception of a slight difference of notation.

In the February Number of the *Philosophical Magazine* I have propounded a theory of the scattering of light by particles which are small in *all* their dimensions compared with the wave-length of light, and have applied the results to explain the phenomena presented by the sky. Another theory has been given by Clausius, who attributes the light of the sky to reflection from water-bubbles, and has developed his views at length in a series of papers in Poggendorff's *Annalen* and Crelle's *Journal**.

Starting from the ordinary laws of reflection and refraction, he has no difficulty in showing that, were the atmosphere charged with globes of water in sufficient quantity to send us the light which we actually receive, a star instead of appearing as a point would be dilated into a disk of considerable magnitude. But the requirements of the case are satisfied if we suppose the spheres hollow, like bubbles; for then, on account of the parallelism of the surfaces, but little effect is produced by refraction on a wave of light. At the same time, if the film be sufficiently thin, the light reflected from it will be the blue of the first order, and so the colour of the sky is apparently accounted for.

Apart from the difficulty of seeing how such bubbles could be formed, there is a formidable objection to this theory, mentioned by Brücke (Pogg. *Ann.* vol. LXXXVIII. p. 363)—that the blue of the sky is a much better colour than the blue of the first order. That it is so appears clearly from the

---

* Pogg. *Ann.* vols. LXXII. LXXVI. LXXXVIII.   *Crelle*, vols. XXXIV. XXXVI.

measurements quoted in the February Number, and from the theoretical composition of the blue of the first order*. Nor can we escape from this difficulty by supposing, with Brücke, that the greater part of the light from the sky has been reflected more than once.

Brücke also brings forward an experiment of great importance when he shows that mastic precipitated from an alcoholic solution scatters light of a blue tint. He remarks that it is impossible to suppose that the particles of mastic are in the form of bubbles.

In his last utterance on this subject†, Clausius replies to the objections urged by Brücke and others against his theory, and shows that, if the illumination of the sky is due to thin plates at all, those thin plates *must* be in the form of bubbles. While admitting that if the particles are very small the ordinary laws of reflection and refraction no longer apply‡, and that therefore this case is not excluded by his argument, he still holds to his original view as to the nature of the reflecting matter in the sky, considering that the polarization of the light indicates that it has undergone *regular* reflection. His concluding paragraph so well sums up the case that I cannot do better than quote it. "Das Resultat der vorstehenden Betrachtungen kann ich hiernach kurz so zusammenfassen. Soweit man die gewöhnlichen Brechungs- und Reflexionsgesetze als gültig anerkennt, glaube ich auch meine früheren Schlüsse festhalten zu müssen, nämlich, dass in der Atmosphäre Dampfbläschen vorhanden seyen, und dass sie die Hauptursache der in ihr stattfindenden Lichtreflexion und ihrer Farben bilden. Nimmt man aber an, die in der Atmosphäre wirksamen Körperchen seyen so klein, dass jene Gesetze auf sie keine Anwendung mehr finden, dann sind auch diese Schlüsse ungültig. Auf diesen Fall ist aber auch die Theorie der Farben dünner Blättchen nicht mehr anwendbar, und er bedarf vielmehr einer neuen Entwickelung, bei welcher noch besonders berücksichtigt werden muss, in wiefern diese Annahme mit der Polarisation des vom Himmel kommenden Lichtes und mit der angenähert bekannten Grösse der in den Wolken vorhandenen Wassertheilchen vereinbar ist."

* I find that I omitted to explain why it is that the light dispersed from small particles is of so much richer a hue than that reflected from very thin films. In the latter case the reflected wave may be regarded as the sum of the disturbances originating in the elementary parts of the film, and these elementary parts may be assimilated to the small particles of the former supposition. The integration is best effected by dividing the surface into the *zones of Huyghens*; and it is proved in works on physical optics that the total effect is just half of that due to the first zone. Now the zones of Huyghens vary as the wave-length; and thus it appears that in the integration the long waves gain an advantage which diminishes the original preponderance of their quicker-timed rivals.

† Pogg. *Ann.* vol. LXXXVIII. p. 543.

‡ In many departments of science a tendency may be observed to extend the field of familiar laws beyond their proper limits. Thus the properties of gross matter are often assumed to hold equally good for molecules. An example more analogous to that which suggests this remark is to be found in the common explanation of the mode of action of the speaking-trumpet.

Clausius does not seem to have followed up the line of research here indicated. My investigation (written, it so happens, before seeing Clausius's papers) shows in the clearest manner the connexion between the smallness of the particles and the polarization of the light scattered from them. Indeed I must remark that in this respect there is an advantage over the theory of thin plates, according to which the direction of complete polarization would be about 76° from the sun. It would be a singular coincidence if the action of secondary causes were to augment this angle to 90°—its observed magnitude. It seems, therefore, not too much to say that, if the illumination of the sky were due to suspended water-bubbles, neither its colour nor its polarization would agree with what is actually observed.

In his celebrated paper on Fluorescence*, Professor Stokes makes the following significant remark :—" Now this result appears to me to have no remote bearing on the question of the direction of the vibrations in polarized light. So long as the suspended particles are large compared with the waves of light, reflection takes place as it would from a portion of the surface of a large solid immersed in the fluid, and no conclusion can be drawn either way. But if the diameter of the particles be small compared with the length of a wave of light, it seems plain that the vibrations in a reflected ray cannot be perpendicular to the vibrations in the incident ray." This is the only passage that I have met with in which the theory of the reflection of light from very small particles is touched upon.

If it be assumed, as in the theories of Green and Cauchy of reflection at plane surfaces, that the effect of dense matter is merely to load the ether, it follows rigorously that the direction of vibration cannot be turned through a right angle when light is scattered from small particles. But all we know in the first instance is that the velocity of propagation of luminous waves is less in ordinary transparent matter than in vacuum; and this may be accounted for as well by a diminished rigidity as by an increased density. In the first case a scattered ray *might* be composed of vibrations perpendicular to those of the incident beam; so that the matter is not quite so clear as it would seem from the argument of Professor Stokes. I believe, however, that good reasons may be given for rejecting the view that the difference between media of varying refrangibility is one of rigidity. The point is an important one, and I propose to recur to it later.

The experiments of Professor Tyndall† with precipitated clouds exhibit more clearly than had been done by Brücke the relation between the size of the particles and the nature of the dispersed light. The observation that the polarization is complete perpendicular to the track of the incident light is in itself sufficient to disprove the theory of bubbles. As the particles increase

* *Phil. Trans.* 1852, p. 526.
† *Phil. Mag.* vol. xxxvii. p. 385. *Phil. Trans.* 1870.

in magnitude, the azure and polarization are gradually lost. During the transition a different and more complicated set of phenomena present themselves, which will furnish a test for the theory when it is extended so as to include the consideration of particles which are no longer very small in comparison with the waves of light.

All who have written on this subject seem to have taken for granted that the foreign matter in the atmosphere is water or ice. Even Tyndall, who expressly says that any particles, if small enough, will do, still believes in the presence of water-particles. But this view is encumbered with considerable difficulty; for even if, in virtue of its transparency to radiant heat, the air in the higher regions of our atmosphere is at a very low temperature, it would still be capable of absorbing the very small quantity of water which is sufficient to explain the blue of the sky. At any rate it is difficult to imagine particles of water smaller than the wave-length endowed with any stability. These difficulties might perhaps be got over if there were any strong argument in favour of the water-particles; but of the existence of such I am not aware. Every one knows that a blue haze evidently akin to the azure of the sky obliterates the details and modifies the colour of a distant mountain; and this, when it occurs on a hot day, cannot possibly be attributed to aqueous particles. On the face of it, there is no reason for supposing that near the earth's surface the foreign matter is of one kind and at a great altitude another. If it were at all probable that the particles are all of one kind, it seems to me that a strong case might be made out for common salt. Be this as it may, the optical phenomena can give us no clue.

The apparatus by means of which the comparison was made between sky light and that of the sun diffused through white paper, was originally arranged for measurements of the absolute absorption of coloured fluids for the various rays of the spectrum, and had been applied rather extensively in experiments having that object. In the shutter of a darkened room were placed two slits in the same vertical line, each about three inches long, and a foot apart. At the other end of the room was an arrangement of prisms and lenses for producing a pure spectrum on a screen in the ordinary way. At first only one prism was used; but I soon introduced another, and the number might probably be further increased with advantage. It is even more important to have a great dispersion in these experiments than in the ordinary spectroscope. Two spectra would thus be thrown on the screen one over the other, but by means of a very obtuse-angled prism situated in front of the dispersion-prisms they are brought together so as exactly to overlap. The double spectrum thus formed passes through a horizontal slit in the screen placed so as to receive it. Close behind is an opaque card carrying a small vertical slit, which can be slid along so as to allow any desired part of the spectrum to pass through. At the beginning and end of a set of experiments the card is removed, and the principal fixed lines are observed through an eyepiece and referred to a scale situated just over the horizontal aperture.

When the experimenter looks through the eye-slit in the direction of the lens, he sees the two parts of the obtuse prism illuminated with light, in each case homogeneous, and, if the adjustments are properly made, belonging to the same part of the spectrum. By varying the breadths of the original slits, the two parts of the field may be made equally bright; and when the match is attained, the breadths are inversely proportional to the richness of the lights behind them in the homogeneous ray under consideration. But if the object be to make a complete comparison between two lights, it is often more convenient to leave the widths of the slits arbitrary, and then, by sliding the card, to seek that part of the spectrum which allows a match. It was in this way that the observations on the light of the sky were made. To give an idea of the degree of accuracy to which the comparisons may be made, I may mention that in my experiments on absorption, the means of six observations were usually correct to about one in 50 or 60. In the less-luminous parts of the spectrum the error might be somewhat greater.

The difficulty, however, of getting a satisfactory result with the blue of the sky does not lie in the inaccuracy of the measurements, but in the arbitrary character of the light with which it is compared. In order to test the theory in a strict manner, the second light ought to be similar in composition to that which lights up the sky. Now the sky is lit not only by the direct rays of the sun, but also by itself and by the bright surface of the earth. It is evident, therefore, that the requirements of the case are very imperfectly met by taking as the second light that of the sun as received by us, even if the translucent material through which we diffuse it effects no change in the quality. A nearer approximation to what we want would probably be found in the diffused light of a thoroughly cloudy day. But here we meet with an experimental difficulty; for the method described is only available to compare two lights both given at once. A suitable artificial light might no doubt be used as a middle term to be afterwards eliminated; but a candle or a lamp would hardly be available, on account of the yellowness of their light. On the other hand, the bluer radiation from burning magnesium would probably be inconvenient, and difficult to keep constant in quality from day to day. I am, however, in hopes that, by a method founded on a different principle, I may be able to compare the blue of one day's clear sky with the white light from the clouds on another.

9.

ON THE SCATTERING OF LIGHT BY SMALL PARTICLES.

[*Phil. Mag.* XLI. pp. 447—454, 1871.]

THE investigation of the diffraction of light by small particles, contained in the February Number of this Magazine [Art. VIII.], proceeds throughout on the assumption that the difference between two media which differ in refractive power is a difference of density and not a difference of rigidity. My object in the present communication is to attack the problem more generally, and to show that the more special hypothesis is in no degree arbitrary, but forced upon us by the phenomena themselves. The words "density," "rigidity" need not be interpreted literally, but are used in a generalized sense analogous to that given to "velocity" and "force" in the higher mechanics.

The first step is to find the equation of motion of an isotropic elastic medium whose density and rigidity may vary from point to point. If D denote the density and n the rigidity, a process similar to that used in Thomson and Tait's *Natural Philosophy*, § 698, leads us to the following:—

$$\frac{d}{dx}(m\delta) + \nabla_n \xi - D\frac{d^2\xi}{dt^2} - \frac{dn}{dx}\frac{d\eta}{dy} - \frac{dn}{dx}\frac{d\zeta}{dz} + \frac{dn}{dy}\frac{d\eta}{dx} + \frac{dn}{dz}\frac{d\zeta}{dx} = 0, \quad \ldots\ldots(1)$$

and two similar equations, where ξ, η, ζ are the displacements parallel to the coordinate axes,

$$\delta \equiv \frac{d\xi}{dx} + \frac{d\eta}{dy} + \frac{d\zeta}{dz},$$

$$\nabla_n \equiv \frac{d}{dx}\left(n\frac{d}{dx}\right) + \frac{d}{dy}\left(n\frac{d}{dy}\right) + \frac{d}{dz}\left(n\frac{d}{dz}\right).$$

If n and D were constant, equations (1) would be satisfied by

$$\xi = \eta = \delta = 0, \qquad \zeta = \zeta_0 = \epsilon^{ik(bt+x)}. \qquad \ldots\ldots\ldots\ldots\ldots(2)$$

In the application that we have to make, n and D may be supposed to be constant except within a small space T at the origin of coordinates, where they assume the values $n + \Delta n$, $D + \Delta D$. In consequence of this variation the equations of motion are no longer satisfied by (2); but we may take as the true values of the displacements, ξ, η, $\zeta_0 + \zeta$, δ, where ξ, η, ζ, δ are small quantities of the order T, which are to be neglected when multiplied by Δn, ΔD*. Substituting in equations (1) and dropping the factor relating to the time, we get

$$
\left.
\begin{aligned}
\frac{d}{dx}(m\delta) + \nabla_n \xi + Dk^2 b^2 \xi + \frac{dn}{dz}\frac{d\zeta_0}{dx} &= 0, \\
\frac{d}{dy}(m\delta) + \nabla_n \eta + Dk^2 b^2 \eta &= 0, \\
\frac{d}{dz}(m\delta) + \nabla_n \zeta + Dk^2 b^2 \zeta + \nabla_n \zeta_0 + Dk^2 b^2 \zeta_0 &= 0,
\end{aligned}
\right\} \quad \cdots\cdots\cdots\cdots(3)
$$

and
$$
k = 2\pi/\lambda, \quad b^2 = n/D,
$$

or, since
$$
\nabla_n \zeta_0 = \nabla_{\Delta n} \zeta_0 + \nabla_n \zeta_0,
$$
$$
D\zeta_0 = (D + \Delta D)\zeta_0 = D\zeta_0 + \Delta D \cdot \zeta_0.
$$

and
$$
\nabla_n \zeta_0 + nk^2 \zeta_0 = 0,
$$
$$
\left.
\begin{aligned}
\frac{1}{n}\frac{d}{dx}(m\delta) + \nabla\xi + k^2\xi + \frac{1}{n}\frac{dn}{dz}\frac{d\zeta_0}{dx} &= 0, \\
\frac{1}{n}\frac{d}{dy}(m\delta) + \nabla\eta + k^2\eta &= 0, \\
\frac{1}{n}\frac{d}{dz}(m\delta) + \nabla\zeta + k^2\zeta + \frac{1}{n}\nabla_{\Delta n}\zeta_0 + \frac{\Delta D}{D}k^2\zeta_0 &= 0.
\end{aligned}
\right\}
$$

Hence if $\varpi_3 = d\xi/dy - d\eta/dx$ &c. be the *rotations* of the medium,

$$
\left.
\begin{aligned}
\nabla\varpi_3 + k^2\varpi_3 + \frac{1}{n}\frac{d}{dy}\left(\frac{dn}{dz}\frac{d\zeta_0}{dx}\right) &= 0, \\
\nabla\varpi_1 + k^2\varpi_1 - \frac{d}{dy}\left\{\frac{1}{n}\frac{d}{dx}\left(\Delta n\frac{d\zeta_0}{dx}\right) + \frac{\Delta D}{D}k^2\zeta_0\right\} &= 0, \\
\nabla\varpi_2 + k^2\varpi_2 + \frac{d}{dx}\left\{\frac{1}{n}\frac{d}{dx}\left(\Delta n\frac{d\zeta_0}{dx}\right) + \frac{\Delta D}{D}k^2\zeta_0\right\} - \frac{1}{n}\frac{d}{dz}\left\{\frac{dn}{dz}\frac{d\zeta_0}{dz}\right\} &= 0.
\end{aligned}
\right\} \quad \cdots (4)
$$

Accordingly

$$
\varpi_3 = \frac{1}{4\pi n}\iiint \frac{\epsilon^{-ikr}}{r}\frac{d\zeta_0}{dx}\frac{d}{dy}\left(\frac{d\Delta n}{dz}\right)dv = -\frac{1}{4\pi n}\iiint \frac{d\Delta n}{dz}\frac{d\zeta_0}{dx}\frac{d}{dy}\left(\frac{\epsilon^{-ikr}}{r}\right)dv
$$

$$
= -\frac{1}{4\pi n}\iiint \Delta n\frac{d\zeta_0}{dx}\frac{d^2}{dy\,dz}\left(\frac{\epsilon^{-ikr}}{r}\right)dv = \frac{\Delta n T}{4\pi n}\frac{d\zeta_0}{dx}\cdot\frac{yz}{r^2}\frac{d^2}{dr^2}\left(\frac{\epsilon^{-ikr}}{r}\right),
$$

---

* The effect of this is subsequently considered.

higher powers of $(1/r)$ being neglected.   By similar reasoning,

$$\varpi_1 = -\frac{T}{4\pi}\frac{\Delta D}{D}k^2\zeta_0\frac{y}{r}\frac{d}{dr}\left(\frac{\epsilon^{-ikr}}{r}\right) - \frac{T}{4\pi}\frac{\Delta n}{n}\frac{d\zeta_0}{dx}\frac{xy}{r^2}\frac{d^2}{dr^2}\left(\frac{\epsilon^{-ikr}}{r}\right),$$

$$\varpi_2 = \frac{T}{4\pi}\frac{\Delta D}{D}k^2\zeta_0\frac{x}{r}\frac{d}{dr}\left(\frac{\epsilon^{-ikr}}{r}\right) + \frac{T}{4\pi}\frac{\Delta n}{n}\frac{d\zeta_0}{dx}\frac{x^2-z^2}{r^2}\frac{d^2}{dr^2}\left(\frac{\epsilon^{-ikr}}{r}\right);$$

or, since $\qquad\qquad \zeta_0 = \epsilon^{ikx}, \quad d\zeta_0/dx = ik\epsilon^{ikx},$

$$\varpi_3 = \frac{ik^3 T}{4\pi}\frac{\epsilon^{-ikr}}{r}\left(-\frac{\Delta n}{n}\frac{yz}{r^2}\right),$$

$$\varpi = \frac{ik^3 T}{4\pi}\frac{\epsilon^{-ikr}}{r}\left(\frac{\Delta D}{D}\frac{y}{r} + \frac{\Delta n}{n}\frac{xy}{r^2}\right), \qquad\qquad \Bigg\} \quad\text{............(5)}$$

$$\varpi_2 = \frac{ik^3 T}{4\pi}\frac{\epsilon^{-ikr}}{r}\left(-\frac{\Delta D}{D}\frac{x}{r} + \frac{\Delta n}{n}\frac{z^2-x^2}{r^2}\right).$$

These are the component rotations. The resultant in the general case would be rather complicated, and is not wanted for our purpose. It is easily seen to be about an axis perpendicular to the scattered ray, inasmuch as

$$x\varpi_1 + y\varpi_2 + z\varpi_3 = 0.$$

Let us consider the particular case of a ray scattered normally to the incident light, so that $x = 0$. Denoting for brevity the common factor by p, we have

$$\varpi_3 = -p\frac{\Delta n}{n}\frac{yz}{r^2}, \qquad \varpi_1 = p\frac{\Delta D}{D}\frac{y}{r}, \qquad \varpi_2 = p\frac{\Delta n}{n}\frac{z^2}{r^2}, \qquad \text{............(6)}$$

whence

$$\varpi^2 = \varpi_1^2 + \varpi_2^2 + \varpi_3^2 = p^2\left(\frac{\Delta n}{n}\right)^2\frac{z^2}{r^2} + p^2\left(\frac{\Delta D}{D}\right)^2\frac{y^2}{r^2}.$$

Here we have reached a result of some importance and one which can be confronted with fact. For from the value of ϖ it appears that there is no direction in the plane perpendicular to an incident ray of polarized light in which the scattered light vanishes, if Δn and ΔD be both finite. Now experiment tells us plainly that there is such a direction, and therefore we may infer with certainty that either Δn or ΔD vanishes. So far we have a choice between two suppositions; either we may assume, as in my former paper, that there is no difference of *rigidity* between one medium and another, and that the vibrations of light are normal to the plane of polarization, or, on the other hand, that there is no difference of *density* between media, and then the vibrations must be supposed to be *in* the plane of polarization. The former view is the one adopted by Green and (virtually) by Cauchy in their theories of reflection; while the latter is that of MacCullagh and Neumann, which I now proceed to show is untenable.

Suppose then that $\Delta D = 0$. Reverting to the general values of ϖ_1, ϖ_2, ϖ_3 in (5), we have

$$\varpi_3 = -p\,\frac{\Delta n}{n}\frac{yz}{r^2}, \quad \varpi_1 = p\,\frac{\Delta n}{n}\frac{xy}{r^2}, \quad \varpi_2 = p\,\frac{\Delta n}{n}\frac{z^2 - x^2}{r^2}; \quad \dots\dots\dots(7)$$

which show that there are in all six directions from O along which there is no scattered ray—two perpendicular to the plane (zx) of original vibration, and four in that plane inclined at angles of 45° to the original ray and its prolongation. No vanishing of the dispersed light in these oblique directions is known from experiment; but before unreservedly discarding the theory which indicates it, we ought to inquire how far our approximation is sufficient to warrant such a step. In neglecting the products of ξ, η, ζ with Δn, we have in reality omitted terms from the result which involve the square and higher powers of Δn, and it may be that the light corresponding to them would not vanish in the specified directions. I have not been able to satisfy myself whether this would be so or not; but I think that, in spite of ignorance on this point, the inference may be safely drawn that the theory is untenable; for the terms in question, depending on the square of the difference of rigidity, are proportional to $\Delta\mu^2/\mu^2$ (where μ is the refractive index), and become of less and less importance as the media approach one another in refrangibility. In the case of particles of mastic suspended in water, the indices are 1·5 and 1·33, and terms depending on the square of Δn must be comparatively small. Yet I could find no indication of a falling off of intensity in the predicted directions in some experiments that I made with precipitated mastic and soap, and accordingly conclude that the hypothesis of a constant density and variable rigidity must be rejected. The only alternative is to suppose, as in the February number, that the æther preserves its statical properties unchanged when associated with matter, whose effect is therefore merely to increase the inertia of the vibrating parts in greater or less degree.

It may be worth notice that, according to the theory here combated, there would be two polarizing-angles, of $22\frac{1}{2}°$ and $67\frac{1}{2}°$ respectively, when light vibrating in the plane of incidence is reflected from the boundary of two media which differ but little in refrangibility, as may be seen from the reasoning of this paper by remembering that the square of Δn may be neglected. I need scarcely say that in such a case the polarizing angle is really 45°, and that the reflected light does not tend to vanish at the two first-mentioned incidences, whichever way the light may be polarized*.

In equations (5), putting $\Delta n = 0$, we have

$$\varpi_3 = 0, \quad \varpi_1 = p\,\frac{D' - D}{D}\frac{y}{r}, \quad \varpi_2 = -p\,\frac{D' - D}{D}\frac{x}{r}, \quad \dots\dots\dots(8)$$

---

* There is a sense in which 45° is the first approximation to the polarizing-angle for all substances. The difference between the true value and 45° may be looked upon as a *correction* depending on the square and higher powers of the difference of optical density.

which correspond to the results already obtained. Since the optical density is proportional to the square of the refractive index,

$$(D' - D)/D = (\mu'^2 - \mu^2)/\mu^2. \quad\ldots\ldots\ldots\ldots\ldots\ldots\ldots(9)$$

In a note to my previous paper I mentioned that no change is required in (8), even though the terms containing the square and higher powers of $(D' - D)/D$ are retained. As I there showed, the density of a medium may always be supposed to be changed, even in the most arbitrary manner, if suitable bodily forces proportional to the variation of density and to the actual acceleration are conceived to act upon it, while the motion remains absolutely the same as before. The waves thrown off from a small particle which lies in the path of a beam of light are those due to a set of forces proportional to $D' - D$ and parallel to the actual vibrations acting through the space T occupied by the particle. In calculating the effect of the forces, the variation of density is to be taken into account, unless we are content to neglect the square of $D' - D$. But by a second application of our principle we see that the density within the space T may be supposed to be D instead of D', provided we introduce a second set of forces proportional to $D' - D$ and to the acceleration at T. Now it may be proved (Thomson and Tait, § 730) that the effect of a bodily force applied through a small space T to an elastic medium diminishes without limit with T *even within the region of application.* Accordingly the acceleration at T caused by our first set of forces is of a higher order of magnitude than the forces themselves, and thus, whether $D' - D$ be small or not, the effect of the second set is to be neglected. The error caused by taking in the calculation of the first set the undisturbed instead of the actual acceleration is evidently smaller still *.

If it were desired to continue the approximation, some further supposition would be necessary as to the *shape* of the disturbing particles. The leading term, we have seen, depends only on the volume ; but the same would not be true for those that follow. However, little exception could be taken to the assumption of a spherical form, and in that case there is no difficulty in proceeding further ; but I have not arrived at any results of interest. Without calculation, we may anticipate that, as the diameter of the particles approaches in magnitude the quarter wave-length, the amplitude of the diffracted vibration will begin to increase less rapidly than T, and that about the time the half wave-length is passed an absolute diminution will set in. Of course, when the incident light is compound, the more refrangible

---

* [1899. The result that the secondary disturbance is proportional to the volume and involves D' and D in the form $(D' - D)/D$, whatever may be the *shape* of the obstacle provided only that it be small enough, is peculiar. In the electromagnetic theory (*Phil. Mag.* XII. p. 81, 1881) the disturbance is proportional to the simple volume and to $(K' - K)/K$ only so long as $(K' - K)$ is small. Otherwise, however small the obstacle may be, its shape enters into the question. In the case of a *sphere* $(K' - K)/K$ is replaced by $3(K' - K)/(K' + 2K)$.]

elements will be the first to show a sensible deviation from the more simple law.

In his interesting experiments with precipitated vapours, Professor Tyndall* found that when the particles of the cloud illuminated by un-polarized light from the electric lamp had attained such a size that the light discharged normally had lost most of its power of affecting the naked eye with the sensation of colour, even then by analyzing the light with a Nicol placed in its position of minimum transmission the azure could be revived in increased splendour. Professor Tyndall calls this the "residual blue." Experimentally it is doubtless more convenient to analyze the light after diffraction from the cloud; but in theoretical explanation and deduction it is simpler, and comes to the same thing in the end, to consider the original beam as polarized before it falls on the cloud. The residual blue is then the light discharged from the cloud in a direction parallel to that in which the incident light swings. The complete explanation of this and other allied phenomena is yet to be made out; but one thing we learn from our theory, if indeed it is at all to be depended on. However large the particles may be, the light scattered or reflected† parallel to the primary vibrations depends on the square and higher powers of $(D' - D)/D$, or, in experimental language, of $(\mu'^2 - \mu^2)/\mu^2$. It is easy to see, too, that the first term in the expression of the amplitude must contain a much higher inverse power of λ than λ^{-2}, and that *if it stood alone* it would correspond to a compound light of a much richer colour than that due to very small particles acting in the ordinary way. Still I cannot honestly say that the residual blue is predicted by theory: before the light discharged in this unfavourable direction could become at all sensible, the particles must have grown to such a size that their diameter would bear no inconsiderable proportion to the waves of light; and then we have no right to suppose that the first term in the expansion proceeding by powers of the diameter may be taken as representing with sufficient approximation the entire series. Indeed the residual blue appears to be rather capricious in its appearance, and to depend on conditions not yet fully known. I may mention that I have not been able to detect any unusually intense coloration in that part of the light from the sky which vibrates in a plane passing through the sun. This is the more remarkable, because it might be supposed that a part at least is light which has twice undergone diffraction, in which case the intensity would vary as λ^{-8} if other-wise undisturbed. But we must not forget that, of the indirect light illumi-nating the higher strata of our atmosphere, a very considerable fraction must come from the earth itself; and this certainly is coloured anything but blue. It would be interesting to observe whether the residual light from parts of

---

* *Phil. Mag.* vol. xxxviii. p. 156. *Phil. Trans.* 1870.

† This may be verified with Fresnel's expression for the intensity of the light regularly reflected when the plane of polarization and plane of incidence include a right angle.

the sky 90° distant from the sun is in any way dependent on the character of the earth's surface—whether, for example, it is the same as usual over water or when the ground is covered with snow. I presume that with the precipitated clouds there is no question of light diffracted more than once.

Theory would lead us to anticipate that the optical density of the particles of foreign matter may have a large influence on the development of the residual blue. If the particles and the medium in which they are suspended have nearly the same refrangibility, the light emitted parallel to the original vibrations may be expected to be very feeble, not only absolutely, but in comparison with that emitted in other directions. Professor Tyndall's method of precipitating organic vapours (some of which may have a high optical density) in air is then more favourable than the suspension of mastic or other moderately dense solids in water, as used by Brücke and other physicists.

I take this opportunity of referring to the observations of Roscoe on the photographic power of skylight, with which I have only lately become acquainted. The comparison of photographic with luminous intensities is well adapted to exhibit differences of quality related in a simple manner to the wave-length. The very small chemical action of the direct solar rays, as compared with what might have been expected from their intense action on the retina, is a striking verification of the theoretical results developed in the February Number of this Magazine [Art. VIII.].

[1899. The electromagnetic theory of the scattering of light by small particles is treated in *Phil. Mag.* vol. XII. p. 81, 1881 ; vol. XLIV. p. 28, 1897. In a recent memoir (*Phil. Mag.* vol. XLVII. p. 375, 1899) I have given reasons for the conclusion that a large part of the light of the sky may be dispersed from the molecules of air themselves.]

10.

ON DOUBLE REFRACTION.

[*Phil. Mag.* XLI. pp. 519—528, 1871.]

IN a former paper* I have shown that, of the various hypotheses which might be made to explain the diminished velocity of light in transparent matter, only one can be reconciled with the observed laws regulating the intensity of polarized light scattered in different directions from an assemblage of particles whose diameters do not exceed a small fraction of the wave-length. We are forced to suppose that the difference between media which is the cause of refraction is a dynamical and not a statical difference, that the rigidity or force with which the æther resists distortion is absolutely invariable. In this view there is nothing novel. Fresnel distinctly adopts it in the investigation of his celebrated formulæ for the intensities of reflected light; and, what is more important, Green's rigorous mechanical theory of reflection† is based on the same assumption. Cauchy also, to whom much of the credit really due to Green has been transferred, starts from the principle of continuity of movement, which asserts that in the passage from one medium to another there is no break in the continuity of the values, either of the displacements *or of their differential coefficients.* I believe that Cauchy has nowhere explained the ground or significance of his principle; but it is easy to see that to assume the continuity of strain is equivalent to asserting a complete continuity of statical properties, so that, as has been pointed out by Haughton‡, Cauchy's theory is essentially the same as Green's.

On the other hand, MacCullagh and Neumann have founded their investigations of reflection on the hypothesis that the difference between media is statical and not dynamical. There is, however, no difficulty in

---

* *Phil. Mag.* for June 1871, p. 447. [Art. IX.]
† *Camb. Phil. Trans.* 1838, or Green's *Math. Papers.*
‡ *Phil. Mag.* S. 4. vol. VI. p. 81.

showing that their hypothesis is as inconsistent with the phenomena of regular reflection as it is with those of diffraction from small particles, as I may perhaps explain in detail on another occasion. But there is one argument urged by them against the rival view which deserves the greatest attention. How, they ask, can double refraction be accounted for if the elastic forces brought into action by a given deformation of the æther are the same in all cases? It is well known that all the theories of double refraction hitherto given by Fresnel and his followers assume expressly that within a doubly refracting medium the elasticity varies in different directions. How is it possible, in investigating the laws of reflection from the surface of isotropic media, to suppose that the statical condition of the æther is invariable, and then, when we come to double refraction, to turn round and say that in them the æther has a rigidity dependent on the direction of displacement? I am not surprised at the importance attached by MacCullagh and Neumann to this objection. Fresnel and Green's investigations of reflection are indeed absolutely inconsistent with the received views as to the cause of double refraction. We find ourselves then in this position: either we must give up Green's theory of reflection, which is the only one hitherto proposed, or easily conceivable, capable of meeting the facts of the case; or else we must abandon the ideas of Fresnel as to the mechanical cause of double refraction.

MacCullagh and Neumann were consistent, though, as I believe, consistently wrong. They rejected the hypothesis of a constant rigidity and variable density as incompatible with the existence of double refraction. How, indeed, conceive a density different in different directions?

Fresnel and Green were inconsistent. The latter has given two rigorous theories of double refraction* which differ from one another in important points, but agree in this, that neither of them can be reconciled with his explanation of reflection; for both assume that the forces which resist displacement within a crystal vary according to the direction of displacement. Precisely the same remark applies to the investigations of Cauchy.

It will readily be anticipated that, having the strongest grounds for believing that the rigidity of the æther is constant whether it be free as in vacuum or entangled with the molecules of matter, I adopt the latter of the two alternatives already mentioned, and look in another direction for the explanation of double refraction. In taking a step which may seem retrograde, I would remark that we are not abandoning a theory in itself very complete or satisfactory. Fresnel's explanation of double refraction will always be considered worthy of his great genius; but it is well known that as a rigorous mechanical theory it will not bear criticism. Nor do the attempts that have been made to improve upon it carry the mark of truth.

* *Camb. Phil. Trans.* 1837. Green's *Math. Papers.*

On this point I refer to the excellent report on double refraction by Professor Stokes in the *British Association's Report* for 1862, and will only say that the analogy between the vibrations of the æther and those which may take place in solids, so striking as long as we confine ourselves to ordinary media, seems to break down when we pass on to consider the case of crystals. For Green has shown that the elasticity of a crystallized medium depends in general on twenty-one constants[*], while the phenomena of double refraction in biaxal crystals involve only six. It is true that, by assumptions more or less arbitrary, the redundant constants may be got rid of, and the result manipulated so as to agree very well with observation; but no one, I suppose, would consider a theory arrived at in such a manner altogether satisfactory. At any rate this is not the opinion of Professor Stokes, who says that in his belief the true theory of double refraction is yet to be found.

We have, then, to consider this question: Can double refraction be explained if the statical properties of the æther are independent of the associated matter? Can we suppose that the density within a crystal is a function of the direction of vibration? I answer, yes. The absurdity is apparent only, and disappears on more attentive examination. As I am conscious the position is one that will need all the light that can be thrown upon it, I think it well to give an illustration of a comparatively simple character which occurred to me at an early stage of this inquiry, and which was of great use in showing me in a general way the possibility of the kind of explanation I was in search of[†].

Let a solid body, such as an ellipsoid, be so supported in space that its centre of inertia is free to move in any direction, but is urged by springs or otherwise towards a certain fixed point with a force symmetrical all round and proportional to the displacement. The arrangement may be supposed to be such that the body always retains its parallelism. Under these circumstances, a vibration may be performed in any direction, and its period is the same in all cases. If the inertia of the body be increased, the only result can be that the motion will become more sluggish and the period longer. Here we have the analogue of singly refracting media. But now suppose that instead of moving in free space the body is immersed in a fluid of sensible density. According to known theorems in hydrodynamics[‡], the inertia of the fluid adds itself to the inertia of the body, and that in a manner dependent on the direction of vibration. An extreme case will make this evident. Suppose that the ellipsoid degenerates into a circular disk of inconsiderable thickness. It is clear that if the vibration be performed in

---

[*] See also Thomson and Tait's *Natural Philosophy*, Appendix C to Ch. VII.

[†] [1899. The suggestion of an inertia different in different directions had been made much earlier by Rankine (*Phil. Mag.* I. p. 441, 1851).]

[‡] *Thomson and Tait*, § 331.

R. I.   8

a line perpendicular to the disk, the fluid, which cannot readily pass from the one side to the other, will greatly impede the motion—that is, increase its period; for there is no question here of a loss of energy from friction or viscosity.

It is equally evident that if the motion be in the plane of the disk, the fluid has no effect and might as well not be there. We see, then, that, to all intents and purposes, the disk has a density, or rather inertia, of variable magnitude dependent on the line of vibration and symmetrical round an axis, and are reminded of a uniaxal crystal. Next suppose that we try to make the disk vibrate in a line oblique to itself. It would at once appear that such a vibration cannot be performed without an additional constraint, which we may suppose applied. The system would then perform pendulous vibrations whose period is a function of the position of the line in which the centre of inertia is made to move. Lagrange's general method leads immediately to a solution of the whole problem :—

$$T = \text{kinetic energy} = \tfrac{1}{2}\{P\dot{x}^2 + Q(\dot{y}^2 + \dot{z}^2)\},$$
$$V = \text{potential energy} = \tfrac{1}{2}\mu(x^2 + y^2 + z^2);$$

whence the equations of vibratory motion,

$$P\ddot{x} + \mu x = 0, \qquad Q\ddot{y} + \mu y = 0, \qquad Q\ddot{z} + \mu z = 0,$$

showing that vibrations along x cannot be performed synchronously with vibrations along y or z. This is on the supposition that the body is free: but if it be constrained to a line ξ making an angle θ with x, we have

$$T = \tfrac{1}{2}\dot{\xi}^2(P\cos^2\theta + Q\sin^2\theta), \qquad V = \tfrac{1}{2}\mu\xi^2,$$

whence

$$(P\cos^2\theta + Q\sin^2\theta)\ddot{\xi} + \mu\xi = 0,$$

so that the period τ is

$$2\pi\sqrt{\frac{P\cos^2\theta + Q\sin^2\theta}{\mu}}.$$

When $\theta = 0$, let $\tau = \tau_1$; when $\theta = 90°$, let $\tau = \tau_2$; then

$$\tau^2 = \tau_1^2\cos^2\theta + \tau_2^2\sin^2\theta.$$

There is, of course, one case which does not bring out the peculiarity for whose sake the illustration is brought forward, I mean when the ellipsoid becomes a sphere. The only effect of the fluid is then to retard the motion, just as if the mass of the sphere itself had been increased.

From the problem generally we may infer that there is nothing absurd in the idea of an inertia varying with the direction of motion, and that the want of symmetry causing double refraction may be attributed with as great probability to the dynamical as to the statical conditions of the question. We know nothing about the real nature of the æther, and, if possible, still less about its relations to ponderable matter; and it is therefore the merest assumption to say that the energy of motion within a crystal is necessarily a symmetrical function of the velocities of displacement. But this has virtually been done in all the theories hitherto given. I would even go further, and ask whether, when we consider the enormous velocity of light and the magnitude of the forces which resist distortion, it is not on the whole more probable that the relatively considerable effect of ponderable matter is due to its action rather on the small quantity (the inertia) than on the great quantity (the rigidity)?

Instead, then, of assuming for the energy of the medium

$$2T = \iiint \rho \left(\dot{\xi}^2 + \dot{\eta}^2 + \dot{\zeta}^2 \right) dv,$$

let us take the most general quadratic function of $\dot{\xi}$, $\dot{\eta}$, $\dot{\zeta}$, containing six constants. Even this form is somewhat restricted; for it may be that the energy cannot be expressed at all as the sum of parts corresponding to the various elements of the æther. Ordinary chromatic dispersion and rotatory polarization, which is a phenomenon of the same nature, show that the mutual influence of the parts is not restricted to a distance which may be regarded as vanishingly small in comparison with the wave-length; and although in Cauchy's theory of dispersion the mutual action is supposed to be of a statical character, yet the fact that there is no dispersion in vacuum, when regarded from the point of view of the present paper, leads rather to the conclusion that the mutual influence is dynamical, by which I mean that it would show itself in the expression of the kinetic rather than of the potential energy. But it will only be following precedents to drop the consideration of dispersion in explaining a theory of double refraction, which may be done consistently by supposing the wave very long.

By a suitable choice of axes the terms involving the products of the velocities may be got rid of, so that

$$2T = \iiint \left(\rho_x \dot{\xi}^2 + \rho_y \dot{\eta}^2 + \rho_z \dot{\zeta}^2 \right) dv,$$

where ρ_x, ρ_y, ρ_z are positive quantities representing the densities corresponding to the three coordinate axes. The expression of the potential energy I suppose to be exactly the same as in vacuum; and thus by Lagrange's general method* we find for the equations of motion,

$$\left. \begin{aligned} \rho_x \frac{d^2\xi}{dt^2} &= a^2 \frac{d\delta}{dx} + b^2 \nabla^2 \xi, \\ \rho_y \frac{d^2\eta}{dt^2} &= a^2 \frac{d\delta}{dy} + b^2 \nabla^2 \eta, \\ \rho_z \frac{d^2\zeta}{dt^2} &= a^2 \frac{d\delta}{dz} + b^2 \nabla^2 \zeta. \end{aligned} \right\} \quad \dots\dots\dots\dots \dots\dots\dots(1)$$

* *Thomson and Tait*, Appendix C. Green, *Camb. Trans.* 1838.

On account of the incompressibility of the æther, δ is very small; but it does not follow that the terms containing it are to be omitted, for a^2 is correspondingly great. We may, however, write p for $a^2\delta$, and p may then be compared to a hydrostatic pressure. The problem of double refraction is solved so soon as the laws are known which regulate the possible directions of vibration and corresponding velocities of propagation for every position of the wave-front.

Let us consider a plane wave whose front is at any time given by

$$lx + my + nz - Vt,$$

so that l, m, n are the direction-cosines of the wave-normal, and V the velocity of propagation. Also let θ denote the actual displacement in the plane of the wave, and λ, μ, ν its direction. Thus

$$\xi = \lambda\theta, \quad \eta = \mu\theta, \quad \zeta = \nu\theta;$$

and

$$\left. \begin{aligned}
\lambda\rho_x \frac{d^2\theta}{dt^2} &= \frac{dp}{dx} + b^2\lambda\nabla^2\theta, \\[2mm]
\mu\rho_y \frac{d^2\theta}{dt^2} &= \frac{dp}{dy} + b^2\mu\nabla^2\theta, \\[2mm]
\nu\rho_z \frac{d^2\theta}{dt^2} &= \frac{dp}{dz} + b^2\nu\nabla^2\theta.
\end{aligned} \right\} \quad \ldots\ldots\ldots\ldots\ldots(2)$$

Now let

$$\theta = \theta_0 \epsilon^{i(lx+my+nz-Vt)}, \quad p = p_0 \epsilon^{i(lx+my+nz-Vt)}, \ldots\ldots\ldots(3)$$

where θ_0 and p_0 are complex constants. On substitution in (2),

$$\left. \begin{aligned}
\lambda\theta_0 (V^2\rho_x - b^2) &= -ip_0 l, \\
\mu\theta_0 (V^2\rho_y - b^2) &= -ip_0 m, \\
\nu\theta_0 (V^2\rho_z - b^2) &= -ip_0 n;
\end{aligned} \right\} \quad \ldots\ldots\ldots\ldots\ldots(4)$$

and since $l\lambda + m\mu + n\nu = 0$,

$$\frac{l^2}{V^2\rho_x - b^2} + \frac{m^2}{V^2\rho_y - b^2} + \frac{n^2}{V^2\rho_z - b^2} = 0;$$

or if, as in the ordinary notation, a, b, c are the principal velocities of propagation*

$$\frac{l^2}{V^2/a^2 - 1} + \frac{m^2}{V^2/b^2 - 1} + \frac{n^2}{V^2/c^2 - 1} = 0. \ldots\ldots\ldots\ldots(5)$$

The equations determining the directions of vibration are

$$\left. \begin{aligned}
&\frac{l}{\lambda}\left(\frac{1}{b^2} - \frac{1}{c^2}\right) + \frac{m}{\mu}\left(\frac{1}{c^2} - \frac{1}{a^2}\right) + \frac{n}{\nu}\left(\frac{1}{a^2} - \frac{1}{b^2}\right) = 0, \\[2mm]
&l\lambda + m\mu + n\nu = 0.
\end{aligned} \right\} \quad \ldots\ldots\ldots(6)$$

* The meaning of b is here changed.

Equations (5) and (6) constitute the analytical solution of the problem. I had originally expected to reproduce in their integrity the beautiful laws of Fresnel; but a slight examination will show that, in order to reconcile (5) and (6) with Fresnel's equations, we must write, for V^2, a^2, b^2, c,

$$V^{-2}, \quad a^{-2}, \quad b^{-2}, \quad c^{-2},$$

respectively. The directions of vibration are parallel to the axes of the section of the ellipsoid

$$x^2/a^2 + y^2/b^2 + z^2/c^2 = 1$$

by the plane of the wave; and the velocities of propagation are *directly* proportional to the lengths of the axes. Accordingly the wave-surface is the envelope of planes drawn parallel to the central sections of the ellipsoid $x^2/a^2 + y^2/b^2 + z^2/c^2 = 1$ at distances directly proportional to the lengths of the axes. Fresnel's surface is the locus of points situated on the normals to the sections of the same ellipsoid and at the same distances. We see, therefore, that the new surface is related to Fresnel's in the following way:—Through any point of the latter draw a plane perpendicular to the line joining it to the centre; the envelope of these planes is the former surface. In the principal planes of a biaxal crystal the new surface agrees with Fresnel's as regards the section which is a circle; but the other is not a true ellipse. Within a uniaxal crystal one ray always follows the ordinary law.

In ordinary media the transversal vibrations can be propagated without any tendency to produce dilatation (positive or negative). But it is not so here. Suppose in our illustration that the centre of the ellipsoid is constrained to move in a certain plane. We should find two directions of possible vibration and two corresponding periods, just as for light in a crystal. The question presents itself, What in the latter case takes the place of the external constraint? *The resistance of the æther to compression*— is the answer. Any part of the æther during the passage of a transverse wave over it tends (except in particular cases) to move normally; but the tendency is shared by all the other parts in the same sheet parallel to the wave-front. The motion, therefore, cannot be actually performed, because it would involve a compression of the medium, which by hypothesis requires an infinite force. The pressure p, however, is not without effect; for it modifies the reflection and refraction when light enters a crystal, and it is probably closely connected with the oblique propagation of a ray in the interior. The actual direction of a ray is to be found from the wave-surface, just as in Fresnel's theory.

I had got about as far as this in my original work when, on reference to Professor Stokes's report, I was greatly surprised to find allusions to a theory of double refraction mathematically, if not physically, identical with

that here advanced.   After insisting on the importance of precise measurements, he says :—"To make my meaning clearer, I will refer to Fresnel's
construction, in which the laws of polarization and wave-velocity are
determined by the sections, by a diametral plane parallel to the wave-front,
of the ellipsoid

$$a^2x^2 + b^2y^2 + c^2z^2 = 1, \dots\dots\dots\dots\dots(11)$$

where a, b, c denote the principal wave-velocities.   The principal semiaxes
of the section determine by their direction the normals to the two planes of
polarization, and by their magnitude the reciprocals of the corresponding
wave-velocities.   Now a certain other physical theory which might be
proposed leads to a construction differing from Fresnel's only in this, that
the planes of polarization and wave-velocities are determined by the section,
by a diametral plane parallel to the wave-front, of the ellipsoid

$$x^2/a^2 + y^2/b^2 + z^2/c^2 = 1, \dots\dots\dots\dots\dots(12)$$

the principal semiaxes of the section determining by their direction the
normals to the two planes of polarization, and by their magnitudes the
corresponding wave-velocities.   The law that the planes of polarization of
the two waves propagated in a given direction bisect respectively the two
supplemental dihedral angles made by planes passing through the wave-
normal and the two optic axes, remains the same as before ; but the positions
of the optic axes themselves, as determined by the principal indices of
refraction, are somewhat different ; the difference, however, is but small
if the differences between a^2, b^2, c^2 are a good deal smaller than the quantities
themselves. Each principal section of the wave-surface, instead of being
a circle and an ellipse, is a circle and an oval, to which an ellipse is a near
approximation.   The difference between the inclinations of the optic axes
and between the amounts of extraordinary refraction in the principal planes,
on the two theories, though small, are quite sensible in observation, but
only on condition that the observations are made with great precision.   We
see from this example of what great advantage for the advancement of theory
observations of this character may be."

And again :—

"The curious and unexpected phenomenon of conical refraction has
justly been regarded as one of the most striking proofs of the general
correctness of the conclusions resulting from the theory of Fresnel.   But
I wish to point out that the phenomenon is not competent to decide between
several theories leading to Fresnel's construction as a near approximation....
We see, therefore, that the limitation of the number of tangent planes to
the wave-surface which can be drawn in a given direction on one side of
the centre to two, or at the most three, is intimately bound up with the
number of dimensions of space ; so that the existence of the phenomenon of

internal conical refraction is no proof of the truth of the particular form of wave-surface assigned by Fresnel rather than that to which some other theory would conduct. Were the law of wave-velocity expressed, for example, by the construction already mentioned having reference to the ellipsoid (12), the wave-surface (in this case a surface of the 16th degree) would still have plane curves of contact with the tangent-plane, which in this case also, as in the wave-surface of Fresnel, are, as I find, circles, though that they should be circles could not have been foreseen.

" The existence of external conical refraction depends upon the existence of a conical point in the wave-surface, by which the interior sheet passes to the exterior. The existence of a conical point is not, like that of a plane curve of contact, a necessary property of a wave-surface. Still it will readily be conceived that if Fresnel's wave-surface be, as it undoubtedly is, at least a near approximation to the true wave-surface, and if the latter have, moreover, plane curves of contact with the tangent plane, the mode by which the exterior sheet passes within one of these plane curves into the interior will be very approximately by a conical point; so that in the impossibility of operating experimentally on mere rays the phenomena will not be sensibly different from what they would have been had the transition been made rigorously by a conical point."

Between the theory here advanced and that of Fresnel observation ought to decide; but it does not appear that any experiments hitherto made are competent to do so. As Professor Stokes points out, all the measurements which are to be combined in one calculation should refer to the same specimen of the crystal; otherwise an element of uncertainty is introduced sufficient to render the application of the test ambiguous. Should the verdict go against the view of the present paper, it is hard to see how any consistent theory is possible, which shall embrace at once the laws of scattering, regular reflection, and double refraction.

[1899. Shortly after the appearance of this paper Sir G. Stokes published the results of his measurements which were sufficient " *absolutely to disprove* the law resulting from the theory which makes double refraction depend on a difference of inertia in different directions." (*Proc. Roy. Soc.*, vol. 20, p. 443, 1872).]

11.

ON THE REFLECTION OF LIGHT FROM TRANSPARENT MATTER.

[*Phil. Mag.* XLII. pp. 81—97, 1871.]

IN connexion with other investigations on light I had occasion to consider the problem of reflection, in order to see how far the facts might be accounted for by the different hypotheses which have been made as to the condition of the æther in transparent matter. Although, as I now find, some of the results then arrived at have been already given by Lorenz, of Copenhagen, the publication of the present paper may not be without use, as I cannot agree with him on many important points, and great misapprehension seems to prevail on the subject generally.

Starting with the assumption that the rigidity is the same in the two media, and that the vibrations of light are normal to the plane of polarization, Fresnel was led to the conclusion that if the incident vibration be represented by unity, the reflected vibration is given by the expressions

$$\frac{\sin(\theta, - \theta)}{\sin(\theta, + \theta)}, \qquad \frac{\tan(\theta, - \theta)}{\tan(\theta, + \theta)},$$

according to whether the plane of primitive polarization coincides with or is perpendicular to the plane of incidence. The process by which the first (sine) formula is obtained is rigorous, or at least may be made so by additional explanations. With regard to the second, the reasoning cannot be considered demonstrative, although, as a matter of fact, the arbitrary principle assumed (that the vibrations in the two media* when resolved parallel to the surface of separation are equal) and the conclusion are approximately true. Fresnel did not contemplate the possibility of a change of phase which, as we now know from Jamin's experiments†, accompanies reflection in one, if not in both of the principal cases.

---

* No account being taken of surface-waves.
† *Ann. de Chimie*, t. XXIX. p. 31.

Green's important work "On the Laws of Reflection and Refraction of Light at the Common Surface of Two Non-crystallized Media," was read before the Cambridge Philosophical Society on December 11, 1837, and published in the Transactions for 1838*. In this paper, which has never received on the Continent the attention which it deserves, Green investigates the equations of motion of an elastic medium, setting out, as we should now say, from the principle of energy. By Lagrange's method he deduces both the general equations applicable throughout the interior, and the conditions which must be satisfied at the surface of separation of two media. The statical properties of an isotropic medium are defined by two constants A and B, the second expressing the *rigidity*, and the first depending, though not in the simplest manner, on the *compressibility*. For the luminiferous æther it is shown that A must be indefinitely great, or that the medium resists change of volume with an infinite force. During motion the inertia of the medium comes into play, and a constant expressing the density must be added to the two statical constants already mentioned. In all this there seems to be nothing to which exception can be taken, unless it be to the assumption (expressly stated by Green) that the sphere of sensible action of the molecular forces, or, as I should prefer to say, the range of the mutual influence of the parts of the medium, is insensible in comparison with the length of the wave, and that the transition from the one state of things to the other at the bounding surface is so rapid that it may be treated as abrupt.

But in the application to the question of reflection further assumptions are made whose significance has been strangely misunderstood. When light passes from air into a denser medium, it propagates itself slower than before in the ratio of $\mu : 1$, but this consideration alone is not sufficient to lead to a definite solution. From the equation

$$\frac{B_{\prime}}{\rho_{\prime}} = \frac{1}{\mu^2} \frac{B}{\rho}$$

we can infer nothing as to the relation between B and $B_{\prime\prime}$, which must be known before further progress (other than tentative) can be made. From the fact that in all the known gases A is independent of the nature of the gas, Green argues that we may assume the same for B, "at least when we consider those phenomena only which depend merely on different states of the same medium, as is the case with light"—an inference which certainly appears very precarious. In a note he says, "Though for all known gases A is independent of the nature of the gas, perhaps it is extending the analogy rather too far to assume that in the luminiferous æther the constants A and B must always be independent of the state of the æther as found

---

* Reprinted in Green's *Mathematical Papers*, edited by Ferrers, Macmillan and Co. 1871.

in different refracting substances. However, since the hypothesis greatly simplifies the equations due to the surface of junction of the two media, and is itself the most simple that could be selected, it seemed natural first to deduce the consequences which follow from it before trying a more complicated one, and, as far as I have yet found, these consequences are in accordance with observed facts."

In a very wild criticism of this theory, at the end of an otherwise sound paper*, Kurz, having mistaken the meaning of A, B, attributes to Green the absurd assumption that the wave-velocities are the same in the two media, and metaphorically holds up his hands in amazement. I need hardly point out that Green's conditions $A = A_{,}$, $B = B_{,}$ are something quite different, and imply simply an identity of statical properties in the case of the two media. It may be shown, however, that the first ($A = A_{,}$) is unnecessary, a fact which Green does not seem to have perceived. The cause of the refraction is a variation of the dynamical property (density). The rest of Green's reasoning is rigorous, admitting of no cavil. When the vibrations are normal to the plane of incidence, the amplitude of the reflected vibration is expressed accurately by Fresnel's sine-formula; but the tangent-formula is only applicable to vibrations in the plane of incidence as a first approximation. It is evident that, in order that theory may at all agree with observation, the vibrations of light must be supposed to be performed normally to the plane of polarization; indeed the two assumptions of constant rigidity and normal vibrations are closely bound up together in all parts of optics. The effect of the hypothetical relations $A = A_{,}$, $B = B_{,}$ is greatly to simplify the bounding conditions which then express the equality of the component displacements *and their derivatives* on the two sides of the separating surface. In this form they become identical with the so-called Principle of Continuity of Movement stated by Cauchy, who does not appear to have seen that a continuity of strain implies necessarily a continuity of statical properties across the surface of separation, as is evident in a moment from D'Alembert's principle. So far there is absolute agreement between Green and Cauchy, the only difference being that Green went deeper into the matter and gave the interpretation, if not the justification, of the principle assumed straight off by Cauchy. The divergence which exists between the results of the two theories takes its rise in their treatment of the longitudinal wave produced when the vibrations are in the plane of incidence, whose consideration cannot be dispensed with, although its direct effect is confined to within a few wave-lengths of the surface. Green merely supposes that the velocity of propagation of disturbances depending on change of volume is infinite in both media, and accordingly arrives at a result which contains only one constant—the refractive index; while Cauchy,

* *Pogg. Ann.* vol. cviii. p. 396.

on the other hand, imagines a sort of opacity to longitudinal vibrations, in virtue of which the waves are damped, and introduces a new constant called the coefficient of extinction. Cauchy, I believe, never published the proof of his formulæ; but the want has been supplied by German physicists*. Whatever may be thought of the processes by which they are obtained, there can be no doubt that Cauchy's formulæ agree very well with the observations of Jamin; while the same cannot be said of Green's as they stand in his original memoir. A modified form of the latter, however, has been given by Haughton†, to which I am inclined to adhere. He thought that, by supposing the incompressibility, though great, to be still finite, the second constant might be introduced, without which an agreement with observation is impossible. Apart from the difficulty of explaining what becomes of the longitudinal wave when the incidence is nearly normal, in which case it must be propagated in the ordinary way, his reasoning is entirely vitiated by an oversight already remarked on by Eisenlohr. The difference between Cauchy's formulæ and Green's, as modified by Haughton, is barely sensible in the experiments of Jamin, which are for the most part confined to the neighbourhood of the polarizing angle; but according to Kurz‡, whose observations extended over a wider range, the latter has a decided advantage as an empirical representation of the facts.

Quite different from the foregoing is the theory of MacCullagh and Neumann, which is given in an accessible form in Lloyd's 'Wave-Theory of Light.' The following principles are laid down as the basis of investigation :—

I.   The vibrations of polarized light are *parallel* to the plane of polarization.

II.   The density of the æther is the same in all bodies as *in vacuo*.

III.   The *vis viva* is preserved; from which it follows that the masses of the æther put in motion, multiplied by the squares of the amplitudes of vibration, are the same before and after reflection.

IV.   The resultant of the vibrations is the same in the two media; and therefore in singly refracting media the refracted vibration is the resultant of the incident and reflected vibrations.

When the vibrations are normal to the plane of incidence, and therefore parallel in all three waves, the application of these principles gives rigorously Fresnel's tangent expression. If the vibrations are in the plane of incidence, the fourth principle alone leads to Fresnel's sine-formula. This only shows that the fourth principle is inconsistent with the others; for, as we shall

* Beer, *Pogg. Ann.* vols. XCI. and XCII. Eisenlohr, *Pogg. Ann.* vol. CIV. p. 346.
† *Phil. Mag.* S. 4, vol. VI. p. 81.
‡ *Pogg. Ann.* vol. CVIII.

see, unexceptionable reasoning founded on I. and II. leads to an altogether different result. The very particular case of IV. required when the vibrations are normal to the plane of incidence happens to be correct. In order to prevent misapprehension, I should say there is a sense in which IV. is perfectly true. If the vibrations belonging to the longitudinal surface-waves be included, it expresses merely the continuity of displacement, a condition which must necessarily be fulfilled according to any view of the subject. But understood in this true sense, it does not carry the con-sequences deduced from it. It remains then to be seen what the magnitude of the reflected wave would be according to principles I. and II., when the light is polarized in the plane of incidence. Let us take up the question after the method of Green, and inquire what are the consequences of the various suppositions which may be made: and first for light vibrating normally to the plane of incidence.

The plane of separation of the media being $x = 0$, let the axis of z be parallel to the fronts of the waves, so that $z = 0$ is the plane of incidence. The displacements in the two media are in general denoted by ξ, η, ζ; $\xi_{,}$, $\eta_{,}$, $\zeta_{,}$; but in this case ξ, η, $\xi_{,}$, $\eta_{,}$, all vanish. For the general equation of motion we have

$$\frac{d^2\zeta}{dt^2} = \frac{n}{D}\left(\frac{d^2\zeta}{dx^2} + \frac{d^2\zeta}{dy^2}\right), \qquad \frac{d^2\zeta_{,}}{dt^2} = \frac{n'}{D'}\left(\frac{d^2\zeta_{,}}{dx^2} + \frac{d^2\zeta_{,}}{dy^2}\right); \quad \dots\dots(1)$$

and for the bounding conditions, when $x = 0$,

$$\zeta = \zeta_{,}, \qquad n\frac{d\zeta}{dx} = n'\frac{d\zeta_{,}}{dx}; \quad \dots\dots\dots\dots\dots(2)$$

n, n' are the rigidities; D, D' the densities.

Assume

$$\zeta = f(ax + by + ct) + F(-ax + by + ct),$$
$$\zeta_{,} = f_{,}(a_{,}x + by + ct),$$

the coefficients b and c being necessarily the same for all three waves, since their traces on the surface must move together. Hence from (2)

$$f' + F' = f_{,}', \qquad n(af' - aF') = n'a_{,}f_{,}',$$

and

$$\frac{F'}{f'} = \frac{a/a_{,} - n'/n}{a/a_{,} + n'/n};$$

or, since $b/a = \tan\theta$, $b/a_{,} = \tan\theta_{,}$,

$$\frac{F'}{f'} = \frac{\dfrac{\tan\theta_{,}}{\tan\theta} - \dfrac{n'}{n}}{\dfrac{\tan\theta_{,}}{\tan\theta} + \dfrac{n'}{n}}, \quad \dots\dots\dots\dots\dots\dots (3)$$

an equation giving the ratio of the reflected and incident vibrations.

Case I.   (Green's) $n = n'$:

$$\frac{F'}{f'} = \frac{\cot \theta - \cot \theta_{,}}{\cot \theta + \cot \theta_{,}} = \frac{\sin (\theta_{,} - \theta)}{\sin (\theta_{,} + \theta)}.$$

Case II.   (MacCullagh's) $D = D'$.

Since generally $n/D : n'D' = \mu^2 : 1$, we have

$$\frac{n'}{n} = \frac{1}{\mu^2} = \frac{\sin^2 \theta_{,}}{\sin \theta},$$

and then (3) gives

$$\frac{F'}{f'} = \frac{\tan (\theta_{,} - \theta)}{\tan (\theta_{,} + \theta)}.$$

If we assume the complete accuracy of Fresnel's expressions, either case agrees with observation; only, if $n = n'$, light vibrates normally to the plane of polarization; while if $D = D'$, the vibrations are parallel to that plane. But we know that Fresnel's tangent-formula is not accurate, and that there is in general no angle of complete polarization, so that already the presumption is in favour of Case I.; but I would not lay much stress upon this, as the phenomena investigated by Jamin are of a secondary character, and might be due to the action of disturbing causes.

Case III.   We may suppose that n and D both vary.   Here we should obtain something between Fresnel's two expressions, which could hardly be reconciled with observation, unless one variation were very subordinate to the other.   Other considerations seem to exclude this case; for if n and D both vary, there is nothing to prevent their varying proportionally, so as to leave the wave-velocity unchanged, or $\mu = 1$.   The transmitted wave would then not be turned, although there would be a finite reflection.   Nothing of this kind is known in nature, whichever way the light may be polarized. But the most satisfactory argument against the joint variation is derived from the theory of the diffraction of light from very small particles, whose diameter does not exceed a small fraction of the wave-length.   Hitherto there has been no theoretical difficulty.   Case I. is only a translation into analysis of the reasoning of Fresnel, and Case II. of the reasoning of MacCullagh.   But when we pass on to the consideration of the problem when the vibrations are in the plane of incidence, our footing is no longer so sure.   However close the analogy may be between the phenomena of light and the transverse vibrations of an elastic solid, one cannot but feel that it may not extend to those motions which are independent of rigidity, and of which in the case of the æther we have no direct knowledge.   Still, in the absence of all others, we cannot do better than follow the guide which has already served us so well.

Since the displacement is entirely in the plane of incidence, $\zeta = 0$, and ξ, η are independent of z. The equations to be satisfied in the interior of the first medium are*,

$$\left. \begin{aligned} \frac{d^2\xi}{dt^2} &= g^2 \frac{d}{dx}\left(\frac{d\xi}{dx}+\frac{d\eta}{dy}\right) + \gamma^2 \frac{d}{dy}\left(\frac{d\xi}{dy}-\frac{d\eta}{dx}\right), \\ \frac{d^2\eta}{dt^2} &= g^2 \frac{d}{dy}\left(\frac{d\xi}{dx}+\frac{d\eta}{dy}\right) + \gamma^2 \frac{d}{dx}\left(\frac{d\xi}{dy}-\frac{d\eta}{dx}\right), \end{aligned} \right\} \quad \cdots\cdots\cdots (4)$$

where

$$g^2 = (m+n)/D, \qquad \gamma^2 = n/D.$$

Putting, with Green,

$$\xi = \frac{d\phi}{dx}+\frac{d\psi}{dy}, \qquad \eta = \frac{d\phi}{dy}-\frac{d\psi}{dx}, \quad \cdots\cdots\cdots\cdots(5)$$

we find

$$\frac{d^2\phi}{dt^2} = g^2\left(\frac{d^2\phi}{dx^2}+\frac{d^2\phi}{dy^2}\right), \qquad \frac{d^2\psi}{dt^2} = \gamma^2\left(\frac{d^2\psi}{dx^2}+\frac{d^2\psi}{dy^2}\right). \quad \cdots\cdots(6)$$

Two similar equations apply to the lower medium.

The boundary conditions are

$$\xi = \xi_{,}, \quad \eta = \eta_{,},$$

$$(m+n)\frac{d\xi}{dx} + (m-n)\frac{d\eta}{dy} = (m'+n')\frac{d\xi_{,}}{dx} + (m'-n')\frac{d\eta_{,}}{dy},$$

$$n\left(\frac{d\xi}{dy}+\frac{d\eta}{dx}\right) = n'\left(\frac{d\xi_{,}}{dy}+\frac{d\eta_{,}}{dx}\right);$$

of which the first pair express the continuity of displacement, and the second the continuity of *stress*.   Assume

$$\left. \begin{aligned} \psi &= \psi' \epsilon^{i(ax+by+ct)} + \psi'' \epsilon^{i(-ax+by+ct)}, \\ \psi &= \phi \epsilon^{i(a'x+by+ct)}, \end{aligned} \right\} \quad \text{upper medium;}$$

$$\left. \begin{aligned} \psi_{,} &= \psi_{,} \epsilon^{i(a_{,}x+by+ct)}, \\ \phi_{,} &= \phi_{,} \epsilon^{i(a_{,}'x+by+ct)}, \end{aligned} \right\} \quad \text{lower medium.}$$

The coefficient of t must be the same for all the waves on account of the periodicity, and b must be the same because the traces of all the waves on the plane of separation $x = 0$ must move together.   The constants $\psi', \psi'', \dots$ are complex.   From (6) we get the following relations,

$$c^2 = \gamma^2(a^2+b^2) = \gamma_{,}^2(a_{,}^2+b^2) = g^2(a'^2+b^2) = g_{,}^2(a_{,}'^2+b^2).$$

Since g and $g_{,}$ are indefinitely great,

$$a'^2 + b^2 = 0, \quad a_{,}'^2 + b^2 = 0; \quad \cdots\cdots\cdots\cdots\cdots(7)$$

whence we obtain

$$a' = ib, \quad a_{,}' = -ib,$$

---

* See Green, or Thomson and Tait, § 698.

if the upper medium correspond to the positive x. Equations (7) express the incompressibility of the æther in the two media, for

$$\delta = \frac{d\xi}{dx} + \frac{d\eta}{dy} = \frac{d^2\phi}{dx^2} + \frac{d^2\phi}{dy^2} = -(a'^2 + b^2)\,\phi.$$

It is therefore hardly correct to call the surface-waves expressed by ϕ *longitudinal*. They are more allied to those motions with which we have so much to do in hydrodynamics, which involve neither rotation nor yet change of volume.

Since $b/a = \tan\theta$, $b/a_{,} = \tan\theta_{,}$,

$$\frac{a^2 + b^2}{a_{,}^2 + b^2} = \frac{\sin^2\theta_{,}}{\sin^2\theta} = \frac{\gamma_{,}^2}{\gamma^2} = \frac{1}{\mu^2},$$

which expresses the ordinary law giving the direction of the refracted wave.

We have now to satisfy the boundary conditions. From the continuity of displacement,

$$a'\phi + b\,(\psi' + \psi'') = a_{,}'\phi_{,} + b\psi_{,}.$$
$$b\phi - a\,(\psi' - \psi'') = b\,\phi_{,} - a_{,}\psi_{,};$$

or, on introducing the values of a', $a_{,}'$, and putting $\psi' + \psi'' = X$, $\psi' - \psi'' = Y$,

$$i\,(\phi + \phi_{,}) = \psi_{,} - X, \qquad b\,(\phi - \phi_{,}) = aY - a_{,}\psi_{,}. \quad\ldots\ldots(8)$$

Were we to ignore the surface-wave altogether and put $\phi = \phi_{,} = 0$, equations (8) would give us

$$X = \psi_{,,} \quad Y = \frac{a_{,}}{a}\psi_{,,}$$

whence

$$\frac{\psi'}{\psi'} = \frac{X - Y}{X + Y} = \frac{1 - a_{,}/a}{1 + a_{,}/a} = \frac{\sin(\theta_{,} - \theta)}{\sin(\theta_{,} + \theta)},$$

Fresnel's first expression. This is exactly what has been done by Zech[*], and is in fact merely a translation into analysis of MacCullagh's fourth principle. The worthlessness of the argument is sufficiently shown by the consideration that no assumption has yet been made as to the relations between n, n', D, D', other than that implied in taking the ratio of the wave-velocities equal to μ. It is as necessary to satisfy the second pair of boundary conditions, expressing the continuity of stress, as the first; and this cannot be done without the introduction of finite surface-waves. Expressed in terms of ϕ, ψ, they take the form

$$(m + n)\frac{d^2\psi}{dx^2} + (m - n)\frac{d^2\phi}{dy^2} + 2n\frac{d^2\psi}{dx\,dy} = \text{similar expression},$$

$$n\left\{2\frac{d^2\phi}{dx\,dy} + \frac{d^2\psi}{dy^2} - \frac{d^2\psi}{dx^2}\right\} = \text{similar expression};$$

* Pogg. *Ann.* vol. cix. p. 60.

or on substitution of the values of ϕ, ψ, with regard to (8),

$$\phi\,\{m\,(a'^2+b^2)-2nb^2\}+2nab\,Y=\phi,\ \{m'\,(a_{,}'^2+b^2)-2n'b^2\}+2n'a_{,}b\psi_{,}, \quad (9)$$

$$n\,\{b^2\psi_{,}-a^2X+ib\,(aY-a_{,}\psi_{,})\}=n'\,\{b^2X-a_{,}^2\psi_{,}+ib\,(aY-a_{,}\psi_{,})\}\ldots\ldots(10)$$

Although a'^2+b^2 is vanishingly small, we are not at liberty to leave it out, because $m\,(a'^2+b^2)$ is finite.   In fact

$$m\,(a'^2+b^2)=Dc^2, \qquad m'\,(a_{,}'^2+b^2)=D'c^2, \quad\ldots\ldots\ldots(11)$$

for we may neglect n in comparison with m.   Using these we obtain

$$D\phi-D'\phi_{,}=\frac{b^2\,(n-n')}{c^2}\left\{\frac{\psi_{,}-X}{i}-\frac{aY+a_{,}\psi_{,}}{b}\right\}. \quad\ldots\ldots(9)'$$

Equations (8), (9'), and (10) contain the solution of the problem.

Case 1.   Let $n=n'$; (9') and (10) give

$$D\phi=D'\phi_{,}, \qquad X\,(a^2+b^2)=\psi_{,}\,(a_{,}^2+b^2).$$

Now $D':D=\mu^2$; so that, from (8),

$$\mu^2\left\{\frac{\psi_{,}-X}{i}-\frac{aY-a_{,}\psi_{,}}{b}\right\}=\frac{\psi_{,}-X}{i}+\frac{aY-a_{,}\psi_{,}}{b};$$

or since

$$X=\frac{a_{,}^2+b^2}{a^2+b^2}\,\psi_{,}=\mu^2\psi_{,}, \quad\ldots\ldots\ldots\ldots\ldots\ldots\ldots\ldots\ldots\ldots (12)$$

$$Y=\left\{\frac{a_{,}}{a}+i\,\frac{b}{a}\frac{(\mu^2-1)^2}{\mu^2+1}\right\}\psi_{,}=\left\{\frac{\cot\theta_{,}}{\cot\theta}+i\tan\theta\,M\,(\mu^2-1)\right\}\psi_{,}, \quad\ldots (13)$$

if we put

$$\frac{\mu^2-1}{\mu^2+1}=M. \quad\ldots\ldots\ldots\ldots\ldots\ldots\ldots (14)$$

From (13) and (14),

$$2\psi'=X+Y=\left\{\mu^2+\frac{\cot\theta_{,}}{\cot\theta}+i\tan\theta\,M\,(\mu^2-1)\right\}\psi_{,},$$

$$2\psi''=X-Y=\left\{\mu^2-\frac{\cot\theta_{,}}{\cot\theta}-i\tan\theta\,M\,(\mu^2-1)\right\}\psi_{,}.$$

The quantities within the brackets are complex, and may be exhibited in the forms Re^{ie}, $R'\epsilon^{ie'}$; e and e' then denote the difference of phase between the incident and refracted, the reflected and refracted waves respectively, and are given by

$$\cot e=\frac{1}{M\,(\mu^2-1)}\,\{\mu^2\cot\theta+\cot\theta_{,}\}=\frac{1}{M}\cot\,(\theta-\theta_{,}); \quad\ldots\ldots(15)$$

by trigonometrical transformation, with use of relation $\sin\theta=\mu\sin\theta_{,}$;

$$\cot e'=\frac{1}{M\,(\mu^2-1)}\,\{-\mu^2\cot\theta+\cot\theta_{,}\}=-\frac{1}{M}\cot\,(\theta+\theta_{,}).\ \ldots (16)$$

We have seen that when the vibrations are normal to the plane of incidence there is no difference of phase between the incident and reflected waves, unless the change of sign, when the second medium is the denser, be considered such. Now what is observed in experiments is the acceleration or retardation of the one polarized component with regard to the other, and is therefore given simply by $e - e'$. The ambiguity must be removed by the consideration that when the incidence is normal there is no relative change of phase, though throughout Jamin's papers it is assumed that there is in that case a phase-difference of half a period. I am at a loss to understand how Jamin could have entertained such a view, which is inconsistent with continuity, inasmuch as when $\theta = 0$ the distinction between polarization in the plane of reflection and polarization in the perpendicular plane disappears.

The ratio of the amplitudes of the reflected and incident vibrations is given by

$$\frac{R'^2}{R^2} = \frac{(-\mu^2 \cot \theta + \cot \theta_{,})^2 + M^2 (\mu^2 - 1)^2}{(\mu^2 \cot \theta + \cot \theta_{,})^2 + M^2 (\mu^2 - 1)^2} = \frac{\cot^2 (\theta + \theta_{,}) + M^2}{\cot^2 (\theta - \theta_{,}) + M^2} \cdot \quad \dots (17)$$

The corresponding quantity when the light is polarized in the plane of incidence is

$$\frac{R''^2}{R^2} = \frac{\sin^2 (\theta - \theta_{,})}{\sin^2 (\theta + \theta_{,})},$$

and therefore

$$\frac{R'^2}{R''^2} = \frac{\cos^2 (\theta + \theta_{,}) + M^2 \sin^2 (\theta + \theta_{,})}{\cos^2 (\theta - \theta_{,}) + M^2 \sin^2 (\theta - \theta_{,})} \cdot \quad \dots (18)$$

Equations (14), (15), (16), (17), (18) constitute the solution of the problem on the hypothesis that $n = n'$, and are equivalent to results given by Green.

Case 2. Let $D = D'$; $n' : n = 1 : \mu^2$. (9′) and (10) assume the form

$$\mu^2 (a^2 + b^2)(aY - a_{,}\psi_{,}) = b^2 (\mu^2 - 1) \left\{ \frac{b}{i} (\psi_{,} - X) - aY - a_{,}\psi_{,} \right\},$$

$$\mu^2 (b^2\psi_{,} - a^2 X) - b^2 X + a_{,}^2\psi_{,} = -(\mu^2 - 1) ib (aY - a_{,}\psi_{,}),$$

the value of $\phi - \phi_{,}$ being substituted from (8); or, on expressing a, b, &c. in terms of the angles of incidence and refraction,

$$\cot \theta Y - \cot \theta_{,}\psi_{,} = \frac{\mu^2 - 1}{\mu^2} \sin^2 \theta \left\{ \frac{\psi_{,} - X}{i} - \cot \theta Y - \cot \theta_{,}\psi_{,} \right\},$$

$$\mu^2 (\psi_{,} - \cot^2 \theta X) - X + \cot^2 \theta_{,}\psi_{,} = -i (\mu^2 - 1)(\cot \theta Y - \cot \theta_{,}\psi_{,}).$$

From these two equations the values of X and Y as functions of the angle of incidence might be tabulated with any given value of μ. One particular case is very remarkable. At the polarizing angle ($\tan^{-1} \mu$) the *amplitude* of the reflected wave is the same as it would be given by Fresnel's sine-formula—a coincidence for which I have not been able to see any reason.

My object in bringing forward the present hypothesis is to disprove it, and is sufficiently attained by the disproof of a particular case. Let us therefore suppose that the difference of refrangibility between the two media is so small that the square and higher powers of $(\mu^2 - 1)$ may be neglected. In the small terms we are to put

$$X = Y = \psi_{,,} \quad \cot \theta = \cot \theta_{,.}$$

The second equation gives

$$X = \frac{\mu^2 + \cot^2 \theta_,}{1 + \mu^2 \cot^2 \theta} \psi_{,,},$$

while from the first

$$Y = \frac{\cot \theta_,}{\cot \theta} \psi_, - 2 \frac{\mu^2 - 1}{\mu^2} \sin^2 \theta \, \psi_,.$$

Now

$$\cot^2 \theta_, = \cot^2 \theta + (\mu^2 - 1)/\sin^2 \theta,$$

$$\cot \theta_, = \cot \theta \left(1 + \frac{\mu^2 - 1}{2 \cos \theta} \right) \text{ approx.}$$

Hence

$$X = \{ 1 + 2 (\mu^2 - 1) \sin^2 \theta \} \, \psi_{,,},$$

$$Y = \left\{ 1 + (\mu^2 - 1) \left[\frac{1}{2 \cos^2 \theta} - 2 \sin^2 \theta \right] \right\} \psi_{,,},$$

$$\frac{\psi_{,,}}{\psi_,} = \frac{X - Y}{X + Y} = (\mu^2 - 1) \frac{2 \sin^2 2\theta - 1}{4 \cos^2 \theta} = -\frac{\mu^2 - 1}{2} \frac{\cos 4\theta}{1 + \cos 2\theta} . \quad \text{...(19)}$$

From (19) we see that the reflected wave vanishes when $\cos 4\theta = 0$; that is, when

$$\theta = \pi/8, \quad \text{or} \quad \theta = 3\pi/8.$$

It appears, then, that on the hypothesis $D = D'$, there would be two polarizing angles ($\pi/8$, $3\pi/8$ respectively) whenever the difference of refrangibility between the two media is small. Since nothing of the sort is observed, we conclude that D cannot be equal to D', and are driven to adopt Green's original view that the rigidity of the æther is the same in all media.

Results substantially equivalent to (19) have been already given in a different form by Lorenz*, who, however, has not discussed them, but simply states that they cannot be reconciled with Fresnel's formulæ. Curiously enough he has taken the same particular case for disproof which I, without a knowledge of his work, had hit upon. Those who have done me the honour of reading my papers on the action of small particles on light will understand

* Pogg. Ann. vol. cxiv.

how I anticipated the two polarizing angles by the very different process there employed. Lorenz draws the conclusion that the elastic force of the æther is the same in all transparent uncrystalline substances as *in vacuo*, and that the vibrations of light are performed normally to the plane of polarization. He might, I think, have omitted the word *uncrystalline**.

There is also another paper† by Lorenz on this subject, in which he endeavours to account for the correction to Fresnel's tangent-formula required by experiment, by supposing that the transition from the one medium to the other, instead of being sudden as we have hitherto considered it, occupies a distance not immeasurably less than the wave-length,—certainly a very reasonable supposition. But there are two objections to his view which are, to my mind, fatal. In the first place, Fresnel's tangent-formula does not express the result of a sudden transition; and what is more, Green's formula (17), which does express it, deviates from the truth on the other side. The difficulty is not to explain why Fresnel's formula is not accurately correct, but why the divergences from it are not greater than we actually find them. According to (17), the light reflected at the polarizing angle from the diamond or any other substances of high refractive index would be a very considerable fraction of the whole, very much greater than what is observed. Another objection to the view that the light reflected at the polarizing angle is due to the want of abruptness in the transition, seems to be contained in the consideration that, if this were really its origin, it ought to show a colour corresponding to the blue of the first order in Newton's scale, being to all intents and purposes reflected from a thin plate. Observation, so far as I am aware, gives no support to such an idea.

Cauchy's formulæ, which differ from (15), (16), (17) merely by the substitution of $- \epsilon \sin \theta$ for M, agree very well with experiment; but I cannot regard them as having a sound dynamical foundation. The introduction of evanescent waves of the kind used by Cauchy involves, as Lorenz remarks, a theory of imperfectly elastic media. But the case is even worse than this; for it may, I believe, be shown that no reasonable theory could lead to the peculiar form of evanescence assumed by Cauchy. Let us examine this point.

If, in the investigation of Cauchy's formulæ as given by Beer, we introduce the functions ϕ and ψ used by Green, we find that ϕ is still expressed by an exponential function of the same form as before, viz. $\epsilon^{i(a'x+by+ct)}$. The only difference is that, whereas in Green's theory $a'^2 + b^2 = c^2 \div g^2$, the relation between a', b, c, according to Cauchy, is

$$a'^2 + b^2 = - k^2, \dots\dots\dots\dots\dots\dots(20)$$

* *Phil. Mag.* S. 4, vol. XLI. p. 519. [Art. X.]
† *Pogg. Ann.* vol. CXI.

where k is the so-called coefficient of extinction. The working out is nearly the same as before. Instead of (8) we have

$$a'\phi - a_{,}'\phi_{,} = b(\psi_{,} - X), \qquad b(\phi - \phi_{,}) = aY - a_{,}\psi_{,}. \quad\text{........}(21)$$

Again, since, according to Cauchy's principle, $m' = m$, $n' = n$, (9') becomes, in virtue of (20),

$$k^2\phi = k_{,}^2\phi_{,}, \quad\text{..............................}(22)$$

(10) is replaced by

$$2a'b\phi + (b^2 - a^2)X = 2a_{,}'b\phi_{,} + (b^2 - a_{,}^2)\psi_{,},$$

or, by (21),

$$X = \frac{a_{,}^2 + b^2}{a^2 + b^2}\psi_{,} = \mu^2\psi_{,} \text{............................}(23)$$

From (21), (22),

$$\frac{Y}{\psi_{,}} = \frac{a_{,}}{a} - \frac{b^2(\mu^2 - 1)}{a} \frac{k^2 - k_{,}^2}{a_{,}'k^2 - a'k_{,}^2}. \quad\text{..................}(24)$$

From (24) we may fall back on Green's corresponding equation (13) by putting $k = 0$, $k_{,} = 0$, $k_{,} : k = \mu : 1$; but Cauchy supposes, on the contrary, that k^2, $k_{,}^2$ are very large in comparison with b^2, and writes

$$a' = ik, \quad a_{,}' = -ik,$$

which convert (24) into

$$\frac{Y}{\psi_{,}} = \frac{a_{,}}{a} + i\frac{(\mu^2 - 1)b^2}{a}\left(\frac{1}{k} - \frac{1}{k_{,}}\right).$$

Cauchy further takes

$$(k^{-1} - k_{,}^{-1})\,2\pi/\lambda = -\epsilon;$$

so that, since $2\pi/\lambda . \sin\theta = b$, the solution of the problem is

$$\left.\begin{aligned}X &= \mu^2\psi_{,}, \\ Y &= \left\{\frac{a_{,}}{a} - i(\mu^2 - 1)\frac{b}{a}\epsilon\sin\theta\right\}\psi_{,}.\end{aligned}\right\}\text{..............}(25)$$

It may, however, be remarked that Cauchy has no right to suppose that ϵ is a constant for the rays of different wave-lengths. In fact if k and $k_{,}$ are constants, ϵ varies inversely as λ; so that the same objection arises here as in the theory of Lorenz. The only difference between (25) and (12), (13) lies in the substitution of $-\epsilon\sin\theta$ for M. It is therefore unnecessary to write down the results corresponding to (15), (16), (17), (18).

But what I wish particularly to point out is the extraordinary differential equation satisfied by ϕ. By differentiating the expression for ϕ and substitution in (20), we find

$$\frac{d^2\phi}{dx^2} + \frac{d^2\phi}{dy^2} = -\frac{k^2}{c^2}\frac{d^2\phi}{dt^2}.$$

I am at a loss to understand how any mechanical theory of imperfect elasticity could lead to such an equation*. If we were to speculate as to the most probable form of the equation of motion, we should perhaps give the preference to

$$\frac{d^2\phi}{dt^2} + h\frac{d\phi}{dt} = g^2\left(\frac{d^2\phi}{dx^2} + \frac{d^2\phi}{dy^2}\right);$$

but the form of ϕ so determined is different from Cauchy's, and leads to a more complicated solution. On the whole I cannot see that Cauchy's theory of reflection has any claim to be considered dynamical, although his formulæ are, beyond doubt, very good empirical representations of the facts.

I now come to the modification of Green's theory proposed by Haughton. If M were an arbitrary constant instead of a definite function of μ, there would be but little difference between the two sets of formulæ; for the factor $\sin\theta$ would not vary greatly in the neighbourhood of the polarizing angle, where alone the correction to Fresnel's original expression is sensible. So far as the question has been treated experimentally, the balance of evidence seems to be rather against than for the factor $\sin\theta$. I have already remarked that Haughton's reasons for considering M as an independent constant cannot be sustained, but at the same time I think that others of considerable force may be given.

In a supplement to his memoir "On the Reflection of Light†," Green says:—" Should the radius of the sphere of sensible action of the molecular forces bear any finite ratio to λ, the length of a wave of light, as some philosophers have supposed in order to explain the phenomena of dispersion, instead of an abrupt termination of our two media we should have a continuous though rapid change of state of the ætherial medium in the immediate vicinity of their surface of separation. And I have here endeavoured to show by probable reasoning that the effect of such a change would be to diminish greatly the quantity of light reflected at the polarizing angle, even for highly refractive substances, supposing the light polarized perpendicular to the plane of incidence." The contrast between this view and that of Lorenz is remarkable.

Referring to equation (9), we see that when $n' = n$, it reduces to

$$m\left(a'^2 + b^2\right)\phi = m'\left(a_{\prime}'^2 + b^2\right)\phi_{\prime}.$$

Reasoning from the analogy of elastic solids, we found

$$m\left(a'^2 + b^2\right) : m'\left(a_{\prime}'^2 + b^2\right) = D : D'. \quad\dots\dots\dots(11)$$

Now although the transition between the two media is so sudden that the principal waves of transverse vibrations are affected nearly in the same way

---

* [1898. See however Art. XVI. footnotes, pp. 142, 146.]
† *Cambridge Trans.* 1839, or Green's works.

as if it were instantaneous, yet we may readily imagine that the case is
different for the surface-waves, whose existence is almost confined to the
layer of variable density. It is probable that the ratio of

$$m \left(a'^2 + b^2\right) \; : \; m' \left(a_{,}'^2 + b^2\right),$$

instead of being equal to $1 : \mu^2$, approaches much more nearly to a ratio of
equality. We may therefore take

$$\phi \; : \; \phi_{,} = \mu_0{}^2 \; : \; 1,$$

where $\mu_0{}^2$ is less than μ^2. The solution is the same as before, except that
now $M = (\mu_0{}^2 - 1)/(\mu_0{}^2 + 1)$.

This explanation of the deviation of M from Green's value seems to me
the most probable; but the ground might be taken that the densities
concerned in the propagation of the so-called longitudinal waves are unknown,
and may possibly not be the same as those on which transverse vibrations
depend. For sulphuret of arsenic, Jamin's experiments give

$$\mu = 2\cdot454, \quad \mu_0 = 1\cdot083,$$

showing that μ_0 is very considerably less than μ.

One of the most remarkable of Jamin's results shows that in many cases
M is negative, or μ_0 less than unity. There are a few substances of an
intermediate character for which $M = 0$; and then Fresnel's original formulæ
express the laws of the phenomena. The value of μ is usually about 1·45.
No adequate explanation has hitherto been given of the singular law; and
in the remarks which follow I wish to be understood as merely throwing out
a suggestion which may or may not contain the germ of an explanation.

It is known that many solid bodies have the power of condensing gases
on their surfaces, a property on which the action of Grove's gas-battery
seems to depend. Now, if we were to suppose that at the surfaces of solid
and liquid bodies there exists a sheet of condensed air, which need not
extend to a distance greater than the wave-length, but is of an optical
density corresponding to about $\mu = 1\cdot5$, the occurrence of negative values of
M would, I think, be explained. There is nothing à priori very improbable
in the existence of such a sheet, so far as I am able to see; but it is
for experiment to decide whether the phenomena observed near the polarizing
angle depend in any manner on the nature of the gas with which the
reflecting body is in contact, and whether the sign of M may change from
negative to positive when vacuum is substituted for atmospheric air. The
fact that the value of M for the surface of separation of (say) glass and water
cannot be calculated from the values of M corresponding respectively to
glass and air, water and air, seems to indicate that the phenomenon is, so to
speak, of an accidental character.

[1899. On the general experimental question see *Phil. Mag.* vol. XXXIII.
p. 1, 1892.]

12.

ON A CORRECTION SOMETIMES REQUIRED IN CURVES PROFESSING TO REPRESENT THE CONNEXION BETWEEN TWO PHYSICAL MAGNITUDES.

[*Phil. Mag.* XLII. pp. 441—444, 1871.]

THE nature of the correction which is the subject of the present paper, and of not infrequent application in experimental inquiry, will be best understood from an example, as it is a little difficult to state with full generality. Suppose that our object is to determine the distribution of heat in the spectrum of the sun or any other source of light. A line thermopile would be placed in the path of the light, and the deflection of the galvanometer noted for a series of positions. But the observations obtained in this way are not *sharp*—that is, they do not correspond to *definite* values of the wave-length or refractive index. In the first place, the spectrum cannot be absolutely pure; at each point there is a certain admixture of neighbouring rays. Further, even if the spectrum were pure, it would still be impossible to operate with a mathematical line of it; so that the result, instead of belonging to a simple definite value of the independent variable, is really a kind of average corresponding to values grouped together in a small cluster.

For the sake of simplicity, let us suppose that the spectrum is originally pure, and that the true curve giving the relations between the two quantities is PQR. Also let MN be the range over which the independent variable changes in each observation—in our case the width of the thermopile. Then the observed curve is to be found from the true by taking m, the middle point of MN, and erecting an ordinate pm, such that

$$pm \,.\, MN = \text{area of curve } PQNM.$$

The locus of p will give the curve expressing the result of the observations. It remains to find a convenient method of passing from the one curve to the other.

In the figure PRQ represents the true curve, MN the *range* as before; $Mm = mN = h$; p is the point on the observed curve found in the manner described; $Om = x_0$, $Rm = y_0$, $pm = y'$. Now

$$\text{area } MPRQNM = \int_{-h}^{+h} y\, dh$$

$$= \int_{-h}^{+h} \left(y_0 + \frac{dy}{dx_0} h + \tfrac{1}{2} \frac{d^2y}{dx_0^2} h^2 \right) dh = 2h \left(y_0 + \frac{h^2}{6} \frac{d^2y}{dx_0^2} \right).$$

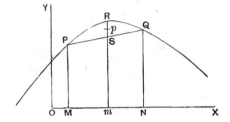

Thus

$$y' = pm = y_0 + \frac{h^2}{6} \frac{d^2y}{dx_0^2}, \quad\dotfill(A)$$

which shows how to deduce y' from y.

To pass backwards, we observe that

$$\frac{d^2y'}{dx^2} = \frac{d^2y}{dx^2} + \frac{h^2}{6} \frac{d^4y}{dx^4};$$

so that

$$\frac{h^2}{6} \frac{d^2y}{dx^2} = \frac{h^2}{6} \left(\frac{d^2y'}{dx^2} - \frac{h^2}{6} \frac{d^4y}{dx^4} \right) = \frac{h^2}{6} \frac{d^2y'}{dx^2},$$

if h^4 be neglected. Thus

$$y = y' - \frac{h^2}{6} \frac{d^2y'}{dx^2} \quad\dotfill(B)$$

(A) and (B) give the analytical solution of the problem; but for practical purposes the following interpretation is important:

$$Sm = \tfrac{1}{2} \left\{ y_0 - \frac{dy}{dx_0} h + \frac{d^2y}{dx_0^2} \frac{h^2}{2} \right\} + \tfrac{1}{2} \left\{ y_0 + \frac{dy}{dx_0} h + \frac{d^2y}{dx_0^2} \frac{h^2}{2} \right\}$$

$$= y_0 + \frac{d^2y}{dx_0^2} \frac{h^2}{2};$$

so that

$$y' = y_0 - \tfrac{1}{3} RS.$$

In passing from the observed to the true curve, the curvature is everywhere to be increased instead of diminished, and

$$pm = Rm + \tfrac{1}{3} RS.$$

It may be remarked that while it is always possible to pass accurately from the true curve to that derived from it with any prescribed range, the inverse problem is not determinate unless it be understood that the range is small, so that its fourth power may be neglected. The practical utility of the solution obtained is scarcely affected by this consideration; indeed it is only when the curvature of the curve is considerable that the correction itself is of much importance.

It often happens that the connexion between the two curves is not so simple, at least at first sight. Suppose, for example, that in the case taken as an illustration the spectrum is impure from the sensible width (2k) of the image of the slit. The observed curve is then connected with true by a *double* integration,

$$y' = \frac{1}{4hk} \int_{x-k}^{x+k} dk \left\{ \int_{x-h}^{x+h} y\, dh \right\}.$$

Now

$$\frac{1}{2h} \int_{x-h}^{x+h} y\, dh = y + \frac{h^2}{6} \frac{d^2y}{dx^2}, \qquad \frac{1}{2k} \int_{x-k}^{x+k} y\, dk = y + \frac{k^2}{6} \frac{d^2y}{dx^2},$$

$$\frac{1}{2k} \int_{x-k}^{x+k} \frac{h^2}{6} \frac{d^2y}{dx^2} dk = \frac{h^2}{6} \left(\frac{d^2y}{dx^2} + \frac{k^2}{6} \frac{d^4y}{dx^4} \right) = \frac{h^2}{6} \frac{d^2y}{dx^2},$$

if the term in $h^2 k^2$ may be neglected. Thus

$$y' = y + \frac{h^2 + k^2}{6} \frac{d^2y}{dx^2}.$$

The rules remain just as before, except that instead of h we now have $\sqrt{(h^2 + k^2)}$. Similarly, when the want of sharpness is due to more than two causes, we must replace h by $\{\Sigma h^2\}^{\frac{1}{2}}$. When, as often happens, the product of the quantities $hk\ldots$ is to be considered as given, the experiments are best arranged so as to make the independent quantities equal; for then the agreement between the two curves is the closest.

The practical rule to which we are led by the considerations explained in this paper is therefore as follows:—

Construct the curve representing the immediate results of the observations in the ordinary way. Let Rm *be any ordinate. Draw parallels* PM, QN

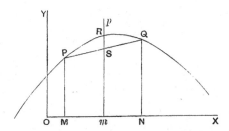

at distances equal to h, *or* $\{\Sigma h^2\}^{\frac{1}{2}}$, *and join* PQ *cutting* Rm *in* S. *The point* p *on the true curve corresponding to the abscissa* Om *is to be found by taking* pR *equal to one-third of* RS, *and so that* p *and* S *lie on opposite sides of* R.

13.

ON 'THE VIBRATIONS OF A GAS CONTAINED WITHIN A RIGID SPHERICAL ENVELOPE.

[*London Math. Soc. Proc.* IV. pp. 93—103, 1872.]

SEE *Theory of Sound,* §§ 330, 331.

[The principal results are contained in the following table, giving the wave-lengths of free plane waves whose periods are the same as those of the proper vibrations of a spherical mass of radius unity. The gravest vibration, in which the gas sways from side to side much as in a cylindrical pipe closed at both ends, corresponds to the harmonic term of order unity.]

Order of Harmonic.

Number of Internal Spherical Nodes.	0	1	2	3	4	5	6
0	1·3983	3·0186	1·8800	1·392	1·113	·9300	·8002
1	·81334	1·0577	·86195	·7320	·6385		
2	·57622	·68251	·59208	·5248			
3	·44670	·50653	·45380				
4	·36485	·40330					
5	·30833	·33523					

14.

INVESTIGATION OF THE DISTURBANCE PRODUCED BY A SPHERICAL OBSTACLE ON THE WAVES OF SOUND.

[*London Math. Soc. Proc.* IV. pp. 253—283, 1872.]

SEE *Theory of Sound*, §§ 296, 334, 335.

[One principal result of the investigation may be thus stated. If the velocity-potential (ϕ) of primary waves incident upon a gaseous spherical obstacle at the origin be denoted by

$$\phi = \cos k\,(at + x), \quad\dots\dots\dots\dots\dots\dots (1)$$

in which a represents the velocity of sound, then the velocity-potential (ψ) of the disturbance due to the obstacle, reckoned at the point whose polar coordinates are r, $\mu\,(=\cos\theta)$, is given by

$$\psi = -\frac{\pi T}{\lambda^2 r}\left\{\frac{m'-m}{m'} + 3\,\frac{\sigma'-\sigma}{\sigma + 2\sigma'}\,\mu\right\}\cos k\,(at - r), \quad\dots\dots (2)$$

where T represents the volume of the obstacle; m', m, σ', σ the compressibilities and the densities of the obstacle and of the rest of the medium respectively. Further $k = 2\pi/\lambda$.

By supposing in (2) that σ', m' are both infinite, we may obtain the corresponding solution for a fixed rigid obstacle, viz.

$$\psi = -\frac{\pi T}{\lambda^2 r}\,(1 + \tfrac{3}{2}\mu)\cos k\,(at - r). \quad\dots\dots\dots\dots (3)]$$

15.

NOTES ON BESSEL'S FUNCTIONS.

[*Phil. Mag.* XLIV. pp. 328—344, 1872.]

[MOST of the results are embodied in *Theory of Sound*. See especially §§ 331, 332, 339 ; Appendix A. The following problem may be mentioned. A rigid cylinder contains incompressible fluid, which has been once at rest and is set in motion in such a manner that at a certain section (perpendicular to the axis) the velocity parallel to the axis is expressed by $1 + \mu r^2$, where r is the distance of any point from the centre of the section. It is required to determine the motion and the energy thereof.

Taking the axis of the cylinder as that of z, and $z = 0$ for the plane section, while $r = 1$ is the equation of the cylinder, we have the following conditions to which the potential ϕ is subject :—

(1)   that when $r = 1$, $d\phi/dr = 0$, for all positive values of z ;

(2)   that when $z = 0$, $d\phi/dz = 1 + \mu r^2$ from $r = 0$ to $r = 1$.

The complete value of ϕ is found to be

$$\phi = (1 + \tfrac{1}{2}\mu)\, z - 4\mu \, \Sigma \, \frac{e^{-pz} J_0(pr)}{p^3 J_0(p)} \, ;$$

where J_0 denotes the Bessel's function of zero order, and the summation extends to all the finite roots of $J_0'(p) = 0$.

If l be the (great) length of the tube,

$$2 \text{ Kinetic Energy} = \pi l (1 + \tfrac{1}{2}\mu)^2 + 16\pi\mu^2 \Sigma p^{-5},$$

in which $\Sigma p^{-5} = \cdot 0012822.$]

16.

ON THE REFLECTION AND REFRACTION OF LIGHT BY INTENSELY OPAQUE MATTER.

[*Phil. Mag.* XLIII. pp. 321—338, 1872.]

It is, I believe, the common opinion, that a satisfactory mechanical theory of the reflection of light from metallic surfaces has been given by Cauchy, and that his formulæ agree very well with observation. The result, however, of a recent examination of the subject has been to convince me that, at least in the case of vibrations performed in the plane of incidence, his theory is erroneous, and that the correspondence with fact claimed for it is illusory, and rests on the assumption of inadmissible values for the arbitrary constants. Cauchy, after his manner, never published any investigation of his formulæ, but contented himself with a statement of the results and of the principles from which he started. The intermediate steps, however, have been given very concisely and with a command of analysis by Eisenlohr (*Pogg. Ann.* vol. CIV. p. 368), who has also endeavoured to determine the constants by a comparison with measurements made by Jamin. I propose in the present communication to examine the theory of reflection from thick metallic plates, and then to make some remarks on the action on light of a *thin* metallic layer, a subject which has been treated experimentally by Quincke.

The peculiarity in the behaviour of metals towards light is supposed by Cauchy to lie in their *opacity*, which has the effect of stopping a train of waves before they can proceed for more than a few wave-lengths within the medium. There can be little doubt that in this Cauchy was perfectly right; for it has been found that bodies which, like many of the dyes, exercise a very intense selective absorption on light, reflect from their surfaces in excessive proportion just those rays to which they are most opaque. Permanganate of potash is a beautiful example of this, given by Professor Stokes. He found (*Phil. Mag.* vol. VI. p. 293) that when the light reflected from a crystal at the polarizing angle is examined through a Nicol held so as to extinguish the rays polarized in the plane of incidence, the residual light is green, and that, when analyzed by the prism, it shows bright bands just where the absorption-spectrum shows dark ones. This very instructive experiment can be repeated with ease by using sunlight, and instead of a crystal a piece of ground glass sprinkled with a little of the powdered salt, which is then well rubbed in and burnished with a glass stopper or otherwise. It can without difficulty be so arranged that the two spectra are seen from the same slit one over the other, and compared with accuracy.

With regard to the chromatic variations, it would have seemed most natural to suppose that the opacity may vary in an arbitrary manner with the wave-length, while the optical density (on which alone in ordinary cases the refraction depends) remains constant, or is subject only to the same sort of variations as occur in transparent media. But the aspect of the question has been materially changed by the observations of Christiansen and Kundt (*Pogg. Ann.* vols. CXLI., CXLIII., CXLIV.) on anomalous dispersion in Fuchsin and other colouring-matters, which show that on either side of an absorption-band there is an abnormal change in the refrangibility (as determined by prismatic deviation) of such a kind that the refraction is *increased* below (that is, on the red side of) the band and *diminished* above it. An analogy may be traced here with the repulsion between two periods which frequently occurs in vibrating systems. The effect of a pendulum suspended from a body subject to horizontal vibration is to increase or diminish the virtual inertia of the mass according as the natural period of the pendulum is shorter or longer than that of its point of suspension*. This may be expressed

---

* [1898. If ξ be the displacement of the bob of the pendulum, whose length is l, x the displacement of the point of support, the equation for ξ is

$$\frac{d^2\xi}{dt^2} = -\frac{g}{l}(\xi - x).$$

Thus if x and ξ vary as $\cos pt$ and $g/l = n^2$,

$$\xi = -\frac{n^2 x}{p^2 - n^2}.$$

For the reaction of the pendulum upon its point of support we have

$$\frac{mg}{l}(\xi - x) = -m\frac{d^2\xi}{dt^2} = \frac{n^2 m}{p^2 - n^2}\frac{d^2 x}{dt^2}.$$

The effect of the pendulum upon its point of support is therefore the same as if the mass of the latter were changed from M to

$$M - \frac{n^2 m}{p^2 - n^2},$$

i.e. in the ratio

$$\frac{p^2 - n^2(M+m)/M}{p^2 - n^2}.$$

In the optical application the vibrations are to be regarded as *forced*, so that p is given. If the medium, whose displacements are represented by x, be such that the square of the velocity of propagation is inversely as the density, the last written ratio gives the square of the refractive index of the medium (as altered by the pendulums) referred to its original state.]

by saying that if the point of support tends to vibrate more rapidly than the pendulum, it is made to go faster still, and *vice versâ*. Below the absorption-band the material vibration is naturally the higher, and hence the effect of the associated matter is to increase (abnormally) the virtual inertia of the æther, and therefore the refrangibility. On the other side the effect is the reverse*. It would be difficult to exaggerate the importance of these facts from the point of view of theoretical optics; but it lies beside the object of the present paper to go further into the question here.

That a sufficient opacity is as competent as a high optical density to produce an abundant reflection is evident without any analysis. So long as the medium into which the light seeks to penetrate remains nearly at rest, the greater part of the motion must be thrown back without any regard to the *cause* of the approximate quiescence. Whether the sluggishness be due to a great inertia or a correspondingly great friction is in this respect of no importance. In order, however, to account for the reflection from silver (90 or 95 per cent.) without opacity, a very high optical density would be required, much higher than we have any reason to think at all likely. On the other hand, we know that the opacity of metals to light is very great.

In this connexion it is interesting to note that some, and probably many, non-metallic substances possess a quasi-metallic reflecting-power for dark radiation. De la Provostaye and Desains long ago remarked on the large percentage of dark heat reflected from glass, which was much in excess of that calculated from Fresnel's formulæ with the known refractive index. The observation seems to have remained uninterpreted; but we cannot well be wrong in attributing the extra reflection to an opacity to the rays of dark heat, which, always great, rises somewhere in the spectrum to such a magnitude as to damp the entering rays within a few wave-lengths of the surface. Nothing but direct experiment can inform us what substances are sufficiently opaque to exercise an abnormal reflection; for the stoppage of radiant heat by a plate of ordinary thickness may well be complete to sense, and yet not sufficiently sudden to give any material assistance in reflection. I am glad therefore to be able to refer to the experiments of the late Professor Magnus (*Pogg. Ann.* vol. CXXXIX.), in which he investigates the proportion of heat reflected by plates of various substances, the incident radiation being derived from moderately heated plates of the same or of a different material.

First let us see what fraction of the incident radiation (unpolarized) would be reflected from the surface of a substance having a refractive index of 1·5—about that of glass. If θ be the angle of incidence, and I the

---

* See Sellmeier, *Pogg. Ann.* vol. CXLIII. p. 272, 1871.

corresponding fraction, I find by calculation from Fresnel's formulæ the following :—

$\theta = 0$	$\theta = 33°$	$\theta = 45°$	$\theta = 62°$
$I = \cdot040$	$I = \cdot042$	$I = \cdot050$	$I = \cdot100$

This is for one surface. If the plate be quite transparent, the reflection may be nearly the double of the above. Now for glass at an angle of 45°, Magnus found no smaller value of I than ·084; and as this must be attributed almost, if not entirely, to the first surface, it is clear that something not taken account of in Fresnel's theory must have come into operation. But by far the most remarkable result was with fluor-spar for the reflecting, and rock-salt for the radiating plate. The reflection at 33° was no less than ·23, at 45° ·242, and at 62° ·335; and to this the second surface cannot contribute sensibly. Unquestionably therefore the reflecting-power of fluor-spar for a certain kind of dark radiation is greatly in excess of what can be accounted for without an extreme opacity—a result which is the more remarkable because for dark radiation in general fluor-spar is one of the most transparent things known *. The reflection from a plate of rock-salt was found to be much the same as from glass; but here, I presume, we may consider both surfaces to be operative, in which case the result is normal†. It is curious that opacity first diminishes the reflection from a plate, and then when extreme increases it again, and that without limit.

The effect of opacity is represented mathematically by the introduction into the differential equation of a term proportional to the velocity. If we suppose that $x = 0$ is the surface of separation, and that the vibrations are parallel to z and perpendicular to the plane of incidence xy, the differential equation in the opaque medium is

$$D, \frac{d^2\zeta_{,}}{dt^2} + h \frac{d\zeta_{,}}{dt} = n \left(\frac{d^2\zeta_{,}}{dx^2} + \frac{d^2\zeta_{,}}{dy^2} \right),$$

where h is a *positive* constant depending on the opacity, and D denotes the optical density. In the upper medium $h = 0$, and the equation is

$$\frac{d^2\zeta}{dt^2} = \frac{n}{D} \left(\frac{d^2\zeta}{dx^2} + \frac{d^2\zeta}{dy^2} \right).$$

Both h and $D_{,}$ are subject to extensive chromatic variations; and the equations are therefore not to be regarded as general equations of motion applicable to all possible cases. It is probable, however, that they represent

---

* [1899. The remarkable results in this direction recently found by Rubens are well known.]

† According to the experiments of Masson and Jamin, the transmission of a perfectly transparent plate is always about 92 per cent., whether the material be glass, rock-salt, or alum. This is in agreement with the calculation in the text, as about 8 per cent. would be reflected.

with sufficient, if not absolute, accuracy the laws of motion for a system of plane waves of given period, provided that suitable values of h and D, are assumed. The boundary conditions according to Green and Cauchy (*Phil. Mag.* August 1871 [Art. XI.]), expressing that there is no discontinuity in the displacement or strain, are

$$\zeta = \zeta_{,}, \quad d\zeta/dx = d\zeta_{,}/dx,$$

when $x = 0$. The system of waves is given by

$$\zeta = \zeta' \epsilon^{i(ax+by+ct)} + \zeta'' \epsilon^{i(-ax+by+ct)},$$

$$\zeta_{,} = \zeta_{,} \epsilon^{i(a_{,}x+by+ct)},$$

where

$$a = \frac{2\pi}{\lambda} \cos \theta, \quad b = \frac{2\pi}{\lambda} \sin \theta, \quad c = \frac{2\pi V}{\lambda} = \frac{2\pi}{\tau}.$$

The coefficients b, c are necessarily the same all through; the multipliers ζ', ζ'', $\zeta_{,}$ are complex. The boundary conditions being the same as for transparent media, we have (*Phil. Mag.* August 1871) $\dfrac{\zeta''}{\zeta'} = \dfrac{a - a_{,}}{a + a_{,}}$, only that now $a_{,}$ is not real. In fact if $n/D = \gamma$, $n/D_{,} = \gamma_{,}$, we obtain from the differential equations

$$c^2 = \gamma^2 (a^2 + b^2), \qquad c^2 - i\, hc/D_{,} = \gamma_{,}^2 (a_{,}^2 + b^2),$$

whence

$$\frac{a_{,}^2 + b^2}{a^2 + b^2} = \frac{\gamma^2}{\gamma_{,}^2} \left(1 - i\, \frac{h}{D_{,}c}\right) = \mu^2 \text{ (say).}$$

We see that $a_{,}$ is determined as a function of a and b by an equation of precisely the same form as for transparent media, the only difference being that μ^2 is now no longer real. If we suppose $\theta_{,}$ to be *defined* by the equation $\sin \theta_{,} = \mu^{-1} \sin \theta$, we may use the forms previously investigated.

From the physical interpretation of μ^2, we see *that its real part is positive, and imaginary part negative.* If we write $\mu^2 = R^2 \epsilon^{2i\alpha}$, 2α must lie between 0 and $-\tfrac{1}{2}\pi$. This remark will be found to be of great importance. For instance, the assumption by Cauchy and others of a real negative value of μ^2 in their treatment of the so-called longitudinal waves produced when light vibrating in the plane of incidence is reflected from the surface of transparent matter, really corresponds to an unstable medium in which the forces resulting from a displacement tend still further to increase it.

The value of μ is $R\epsilon^{i\alpha}$, and at perpendicular incidence ($b = 0$),

$$\frac{a_{,}}{a} = \pm \mu, \text{ say } +\mu,$$

as the sign of μ is in our power. Now, since the refracted wave is $\psi_{,} \epsilon^{i(a_{,}x+by+ct)}$, wherein x is negative, we see that the real part of $a_{,}$ must

10

be positive, while the imaginary part is negative. The same is true of $R\epsilon^{i\alpha}$; or, as R is taken positive, α lies between 0 and $-\frac{1}{2}\pi$. Since, as we have seen, 2α is situated in the same quadrant, α must lie between 0 and $-\frac{1}{4}\pi$. The value of α is determined by $\tan 2\alpha = -h/D_{,}c$. It vanishes with h, and, on the other hand, when $h : cD_{,}$ is very great, approximates to $-\frac{1}{4}\pi$. In this extreme case the real and imaginary parts of μ are numerically equal; the imaginary part is never the greater*.

I have been thus particular to examine the limits between which α may lie, because it appears to me that there is on this point a serious omission, not to say error, in Eisenlohr's paper. In that investigation the necessity of a limitation on the magnitudes of the real and imaginary parts of μ does not appear, mainly because the author has assumed at starting expressions for the incident, reflected, and refracted waves without reference to the differential equations tacitly implied. To suppose, as he does for silver, that $\alpha = 83°$, and therefore $2\alpha = 166°$, is tantamount to the assumption of a medium essentially unstable†.

We may now proceed to transform the expression for the reflected wave

$$\frac{\zeta''}{\zeta'} = \frac{a - a_{,}}{a + a_{,}}.$$

In terms of $\theta_{,}$, $a_{,}/a = \tan\theta/\tan\theta_{,}$, where $\sin\theta_{,} = \mu^{-1}\sin\theta$. To simplify the expressions, it is convenient, following Cauchy and Eisenlohr, to introduce two auxiliary variables, defined by the equation

$$c\epsilon^{iu} = \cos\theta_{,} = \sqrt{(1 - \sin^2\theta/\mu^2)},$$

whence

$$c^2\cos 2u = 1 - \frac{\cos 2\alpha \sin^2\theta}{R^2}, \qquad c^2\sin 2u = -\frac{\sin 2\alpha \sin^2\theta}{R^2}.$$

Thus

$$\frac{a_{,}}{a} = \frac{\sin\theta}{\cos\theta} \cdot \frac{\mu c\epsilon^{iu}}{\sin\theta} = \frac{Rc\,\epsilon^{i(u+\alpha)}}{\cos\theta},$$

$$\frac{\zeta''}{\zeta'} = \frac{\cos\theta - Rc\,\epsilon^{i(u+\alpha)}}{\cos\theta + Rc\,\epsilon^{i(u+\alpha)}}.$$

---

* I apprehend that this conclusion is not limited to the particular form of the differential equation which has been assumed. Whatever that may be, μ^2 will always consist of a real and an imaginary part, of which the former cannot be supposed negative without compromising the stability of the medium. [1898. This conclusion applies in general only when the frequency of vibration is very low. We have already seen (p. 142) that there are cases of finite frequencies for which μ^2 is a real negative quantity. Compare Larmor, *Phil. Trans.* vol. cxc. p. 242, 1897.]

† In Eisenlohr's paper the incident wave travels in the direction of the positive x, while I here suppose the opposite. The change amounts to a reversal of the sign of c; and thus, on Eisenlohr's supposition, the real part of μ^2 ought to be positive and the imaginary part also positive; his result requires that the real part should be negative.

If this quantity be called P, and that obtained from it by changing the sign of i be called Q,

$$\frac{\zeta''}{\zeta'} = \sqrt{PQ} \cdot \epsilon^{id},$$

where

$$\tan d = \frac{P - Q}{i(P + Q)}.$$

The intensity of the reflected light is therefore

$$PQ = \frac{\cos^2 \theta - 2Rc \cos \theta \cos (u + \alpha) + R^2 c^2}{\cos^2 \theta + 2Rc \cos \theta \cos (u + \alpha) + R^2 c^2} = \tan \left(f - \frac{\pi}{4} \right),$$

if

$$\cot f = \cos (u + \alpha) \frac{2Rc/\cos \theta}{1 + R^2 c^2/\cos^2 \theta} = \cos (u + \alpha) . \sin 2 \tan^{-1} \left(\frac{\cos \theta}{Rc} \right).$$

d represents the change of phase; its tangent is given by

$$\tan d = \sin (\alpha + u) . \tan 2 \tan^{-1} \left(\frac{\cos \theta}{Rc} \right).$$

These are Cauchy's formulæ for light polarized in the plane of incidence. At perpendicular incidence,

$$\theta = 0, \quad \theta_{,} = 0, \quad c = 1, \quad u = 0,$$

whence

$$\tan f = \frac{1 + R^2}{2R \cos \alpha} = \frac{1}{2 \cos \alpha} \left(R + \frac{1}{R} \right),$$

which may be expressed in terms of $\gamma_{,}$ and h by means of:—

$$R^2 \cos 2\alpha = \frac{\gamma^2}{\gamma_{,}^2}; \quad R^2 \sin 2\alpha = -\frac{\gamma^2}{\gamma_{,}^2} \frac{h}{cD_{,}}.$$

In the case of metal the reflected light is a large percentage, and therefore $\tan f$ is considerable. This can only be when R is great; and then $1/R$ is relatively small. Approximately therefore

$$\tan f = \frac{R}{2 \cos \alpha} = \frac{\gamma}{\gamma_{,} \sqrt{2}} \frac{\sqrt{(1 + h^2/c^2 D_{,}^2)}}{\sqrt{\{1 + \sqrt{(1 + h^2/c^2 D_{,}^2)}\}}};$$

or, if $h/cD_{,}$ is considerable,

$$\tan f = \frac{\gamma}{\sqrt{2} . \gamma_{,}} \sqrt[4]{1 + \frac{h^2}{c^2 D_{,}^2}} = \frac{\gamma}{\gamma_{,}} \sqrt{\frac{h}{2cD_{,}}} \quad \text{nearly.}$$

Also

$$\gamma/\gamma_{,} = \sqrt{D_{,}}/\sqrt{D};$$

and thus, when the opacity is so great that the reflected light is a large fraction of the whole, its intensity will be

$$\frac{\sqrt{h} - \sqrt{2Dc}}{\sqrt{h} + \sqrt{2Dc}}.$$

If we suppose that h is constant for the different waves of light, we find that the reflection is better and better the longer the wave, since c varies inversely as the period of vibration. Most metals, it would appear, reflect the red rays in greater quantity than the more refrangible, and dark heat better than any.

The wave entering the metal is represented by $\zeta_{,}\epsilon^{i(a_{,}x+ct)}$, or, on substitution of its value for $a_{,}$,

$$\zeta_{,}\epsilon^{-R\sin a\,ax} \cdot \epsilon^{i\,(R\cos a\,ax+ct)}.$$

The velocity of wave-propagation is $c/R\cos\alpha\,a$ against c/a in air. Thus in a certain sense $R\cos\alpha$ (that is, the *real part* of μ) may be regarded as the refractive index of the metal for the kind of light under consideration; but I wish to remark that great confusion has arisen in the use of the expression "refractive index" as applied to metallic or quasi-metallic bodies, the same name being given to quantities which, though coincident for transparent matter, may here differ widely.

Expressed in terms of $\gamma_{,}$ and h,

$$R\cos\alpha = \frac{\gamma}{\gamma_{,}\sqrt{2}}\left\{1 + \sqrt{(1 + h^2/c^2 D_{,}^2)}\right\}^{\frac{1}{2}};$$

or if $h/cD_{,}$ be very large,

$$R\cos\alpha = \frac{\gamma}{\gamma_{,}}\sqrt{\left(\frac{h}{2cD_{,}}\right)} = \sqrt{\left(\frac{h}{2Dc}\right)},$$

which, we have seen, presumably increases with the period of vibration. In this approximation we have supposed that the influence of opacity is paramount, so that $\sin 2\alpha = -1$, and

$$R\cos\alpha = -R\sin\alpha = R/\sqrt{2}.$$

The wave-length within the medium may be taken to be

$$\lambda' = \frac{\lambda}{R\cos\alpha} = \sqrt{\frac{2Dc}{h}} \cdot \lambda = 2\sqrt{\frac{\pi VD}{h}} \cdot \lambda^{\frac{1}{2}} \text{ approx.,}$$

on substitution of the value of c. Hence, if h be constant, the wave-length in the metal varies as the square root of the wave-length in air. The quantity here called the internal wave-length is that which physically best deserves the name; and it is connected with what we have called the refractive index by the usual relation,

Internal wave-length = external wave-length ÷ refractive index;

but it must be remembered that, from an analytical point of view, the internal wave-length and refractive index are imaginary, being denoted by $\lambda \div \mu$ and μ respectively.

The factor expressing the absorption is

$$\epsilon^{-R \sin \alpha . ax} = \epsilon^{-2\pi x/\lambda . R \sin \alpha},$$

or, in terms of λ',

$$\epsilon^{-2\pi x/\lambda' . \tan \alpha},$$

where, it will be remembered, both x and $\tan \alpha$ are negative, showing that, if α be constant, the penetration expressed in terms of λ' is always the same. In cases where the influence of opacity is overwhelming, $\tan \alpha = -1$.

In order to form an idea of the sort of magnitudes with which we are dealing, let us take silver—an extreme case. Exact measurements of the percentage of light reflected at perpendicular incidence are wanting (so far as I know); but De la Provostaye and Desains found in some cases a reflection of dark heat amounting to 95 per cent. Using this in our formulæ, we find

$$R + R^{-1} = 2 \cos \alpha . 39.$$

Now, since $\cos \alpha$ can never be less than $1/\sqrt{2}$, it follows that $1/R$ can be neglected in comparison with R; and thus $R = 80 \cos \alpha$;

$$R \cos \alpha = 80 \cos^2 \alpha; \quad R \sin \alpha = 40 \sin 2\alpha.$$

If we further suppose that the great value of R is due to opacity, we may put $\cos \alpha = 1/\sqrt{2}$, and

$$R = 40 \sqrt{2}, \quad R \cos \alpha = 40, \quad R \sin \alpha = -40.$$

Thus $\lambda' = \lambda/40$; otherwise the ratio of $\lambda' : \lambda$ is still smaller.

For the metals it is probable that of the total reflection the greater part is due to opacity; in other cases it often happens that the effect of opacity is only a slight increase of the reflection that would otherwise take place. Let us inquire what the strength of absorption must be.

If μ_0 denote the refractive index which the medium would possess in virtue of its density alone (γ/γ_i), we have

$$R^2 \cos 2\alpha = \mu_0^2, \quad R^2 \sin 2\alpha = - \mu_0^2 \frac{h}{D_i c};$$

while the reflection is given by

$$\text{Reflection} = \frac{\tan f - 1}{\tan f + 1}$$

and

$$\tan f = \frac{1}{2 \cos \alpha} \left(R + \frac{1}{R} \right) = \frac{1}{2 \cos \alpha} \left(\frac{\mu_0}{\sqrt{\cos 2\alpha}} + \frac{\sqrt{\cos 2\alpha}}{\mu_0} \right);$$

from which it appears that, when α is small, its effect depends on α^2.

On the other hand the factor representing the absorption is

$$\epsilon^{-2\pi x/\lambda . R \sin \alpha}, \text{ or approximately } \epsilon^{-2\pi x/\lambda . \mu_0 \alpha},$$

in which the coefficient of x varies as α. For instance, let $\alpha^2 = \frac{1}{100}$. The effect on reflection would be insensible to ordinary observation, though the opacity is so great as to halve the light within a distance equal to the wavelength in air. Thus it is evident that, in order to aid in reflection, opacity must be very extreme.

We have hitherto supposed that the reflection takes place at the bounding surface of the opaque medium and air; but it is easy to adapt our formulæ so as to express the result when the first medium, still supposed transparent, or at least not very opaque, has a refractive index μ' different from unity. The only change required is to write $R \div \mu'$ for R. Thus at perpendicular incidence,

$$\tan f = \frac{2}{\cos \alpha}\left(\frac{R}{\mu'} + \frac{\mu'}{R}\right).$$

If the reflection be still so good as to allow of the neglect of the second term, we have

$$\tan f = \frac{2R}{\mu' \cos \alpha}.$$

The reflection when light strikes from glass on silver would be considerably less perfect than when the first medium is air; in fact the percentage not reflected

$$= \frac{2}{1 + \tan f} = \frac{\mu' \cos \alpha}{R} \text{ approx.}$$

So much for vibrations perpendicular to the plane of incidence. When we pass to the consideration of vibrations in that plane, we are embarrassed by difficulties of the kind met with in the theory of ordinary reflection, here presenting themselves in an aggravated form. If, following Green, we assumed the equations of motion applicable to elastic solids with the addition of terms proportional to the velocity to represent the frictional loss, and further supposed that the rigidity is the same in the two media, while the compressibility is indefinitely small, we should arrive at results differing only from his by the substitution of an imaginary for a real refractive index. But we know from experiment that Green's results are not verified for transparent media without a modification of doubtful significance, and of magnitude *increasing rapidly with* μ. It is therefore useless to attempt to apply Green's results. The only other course appears to be to start from Fresnel's tangent-formula, and transform that, as we have done the one involving sines, by the introduction of a complex refractive index (and angle of refraction). This is what has been done by Cauchy and Eisenlohr. Following a process similar to that used for vibrations normal to the plane of incidence and with the same notation, we find that the intensity of the reflected light is represented by $\tan(g - \frac{1}{4}\pi)$, where

$$\cot g = \cos(\alpha - u) \cdot \sin 2 \tan^{-1}\left(\frac{c}{R \cos \theta}\right),$$

while the change of phase d' is given by

$$\tan d' = \sin(\alpha - u) \cdot \tan 2 \tan^{-1}\left(\frac{c}{R \cos \theta}\right).$$

However, what we should most require for comparison with experiment relates to the *relative* intensities and changes of phase of the two polarized components; and these are directly obtained by Eisenlohr by transforming Fresnel's corresponding expression* after the introduction of the complex refractive index. If the ratio of the *amplitudes* be called $\tan \beta$, we have

$$\cos 2\beta = \cos(\alpha + u) \cdot \sin 2 \tan^{-1}\left(\frac{\sin^2 \theta}{cR \cos \theta}\right),$$

$$\tan(d' - d) = \sin(\alpha + u) \cdot \tan 2 \tan^{-1}\left(\frac{\sin^2 \theta}{cR \cos \theta}\right),$$

c and u being determined as before. Eisenlohr has compared these formulæ with measurements made by Jamin relating to the so-called principal angle of incidence (making $d' - d$ equal to $\frac{1}{4}\pi$) and the corresponding ratio of amplitudes, and has deduced, as I have already remarked, values of the constants which make the real part of μ^2 negative, and are therefore in-admissible. Another argument leading to the same conclusion is as follows.

Consider a case in which μ^2 is so considerable that c is sensibly equal to unity and α to zero, or, in other words, so refractive that the entering ray is always sensibly parallel to the normal of the surface; and let the incident ray strike the surface at that particular angle which gives a relative phase-difference of a quarter of a period. The angle in question is that determined by $\sin^2 \theta / cR \cos \theta = 1$; so that $\cos 2\beta = \cos \alpha$. Since α must lie between 0 and $-\frac{1}{4}\pi$, $\cos 2\beta$ must be comprised between the limits 1 and $1/\sqrt{2}$. Now in order that the reflection may be perfect at all angles of incidence, it is only necessary that the density or opacity or both should be sufficiently large; and then $\cos 2\beta$ must be sensibly equal to zero. And yet if the formulæ under consideration were true, there would always be a certain angle of incidence making the ratio of the two polarized components very different from unity—a state of things incompatible with a nearly complete reflection. I do not think that the failure of the formulæ for light vibrating in the plane of incidence need cause surprise, when it is considered that Fresnel's tangent-formula, which forms the starting-point of the investigation, is not verified even with transparent media, and differs more and more from the truth as the refracting-power increases. The failure of theory is the more unfortunate, because the relative change of phase and ratio of ampli-tudes for the two polarized components are precisely the quantities best adapted to experimental measurement. As it is, we must conclude, I fear,

$$* \quad \frac{\cos(\theta_, + \theta)}{\cos(\theta_, - \theta)}.$$

that the careful investigations of Jamin on the subject are at present un-available for the purpose of forming an estimate of the values of the density and opacity of the various metals. Experiments on the absolute reflecting-power of the metals for the different parts of the spectrum at perpendicular incidence would be valuable and probably easy; but they do not appear to have been attempted.

It has hitherto been supposed that there is no interruption in the con-tinuity of the metallic medium within such a distance from the surface that the intromitted wave is still sensible. This is a very different thing from assuming, as it has been asserted that the theory does*, that the reflection takes place entirely from the surface, if indeed such an assumption could have any meaning. When, as in the experiments of Quincke, the metallic layer is so thin as to transmit a sensible quantity of light, it is clear that the theory requires modification. If the media on the two sides of the metal are optically similar, a sufficient reduction of the thickness of the layer must at last result in a destruction of the reflection, just as with thin transparent plates.

One of the most remarkable of Quincke's results relates to the influence of a thin metallic layer on the *phase* of the transmitted light. In many cases the phase was accelerated so as to be in advance of what would correspond to a layer of air in place of the silver—an effect which, according to ordinary ideas, would imply a refractive index less than unity. However, it is not difficult to see that, in regard to the effect on the phase of the transmitted wave, the influence of opacity is altogether different from that of density. According to a method used by me in the investigation of the light scattered from very small particles (as in the sky)†, we may suppose that the incident wave passes on undisturbed, if suitable forces are imagined to act on the æther in the metal in order to compensate for the alteration of optical pro-perties. In the case when the thickness of the metallic plate is small compared to the internal wave-length, this point of view possesses con-siderable advantages, and gives a clear insight into the peculiarities of the question. The forces which we must suppose to act in the region of the metallic layer divide themselves into two groups—one dependent on variation of density, and the other on opacity. The first set correspond in phase with the acceleration of the æther, the second with the velocity. There is thus a difference of a quarter of a period. In my former paper I showed that the effect of a force acting at any point is to produce at another, distant r from it, a disturbance whose retardation relative to the force is r simply. In our case each particle of the plate must be considered to be a centre from which a disturbance emanates; and it will readily appear that the phase due

* Wüllner, *Lehrbuch der experimentalen Physik*, vol. II. p. 471.
† *Phil. Mag.* February 1871. [Art. VIII.]

to the whole system of forces is just a quarter of a period behind that corresponding to that element of disturbance which suffers least retardation. In fact, if we divide the whole plate into Huyghens's zones, we know that the effect of the whole is the half of that of the first zone. Now the phase of the disturbance due to the first zone is the mean between that corresponding to its centre and circumference, of which the latter is half a period behind the former. Thus the wave produced by the set of forces due to the alteration of density is a quarter of a period behind the direct wave which has been supposed to pass through undisturbed. The effect of the disturbance is accordingly a maximum on the phase (calculated without allowance for the metallic plate), and a minimum (to the order of approximation here considered, vanishing) on the intensity. It is just the opposite with the second group of forces due to the opacity, which are originally a quarter of a period behind the first. The disturbance due to them will be half a period behind the direct wave with which it has to be compounded, and therefore produces a maximum effect on the intensity *and a vanishing one on the phase*. A very thin film can produce no effect on the phase of the transmitted light in virtue of its opacity, however great, but acts just as if it were deprived of its absorbent power and reduced to the condition of an ordinary transparent plate. Of course this does not explain the *acceleration* of phase found by Quincke; but it shows at least that we must be prepared to distinguish between the effects of density and opacity, though these are in the same direction so far as regards the magnitude of the reflection &c.

Let us then consider analytically the behaviour of a thin metallic plate when light is incident normally upon it. Above let the disturbance be

$$\zeta = \epsilon^{i(ax+ct)} + B_, \epsilon^{i(-ax+ct)} ;$$

below the plate,

$$\zeta = A_2 \, \epsilon^{i(ax+ct)}.$$

In the interior we must introduce both kinds of exponentials, in order to represent the reflection from the second surface. Thus

$$\zeta' = A' \epsilon^{i(a,x+ct)} + B' \epsilon^{i(-a,x+ct)},$$

where $a_, = \mu a$ as before.

The conditions to be satisfied are the continuity of ζ and $d\zeta/dx$ at the two surfaces of separation, viz. when $x = 0$ and when $x = -\delta$, which give four simple equations for the determination of $B_,, A', B', A_2$. On elimination of A', B', we obtain

$$B_, = - \frac{(\mu^2 - 1) \, i \sin \mu a\delta}{2\mu \cos \mu a\delta + (\mu^2 + 1) \, i \sin \mu a\delta},$$

$$A_2 = \frac{\epsilon^{ia\delta}}{\cos \mu a\delta + (\mu^2 + 1)/2\mu \cdot i \sin \mu a\delta}.$$

These expressions contain the ordinary results for transparent plates. Considering μ real, the reflected wave is

$$\frac{(\mu^2 - 1)\sin \mu a\delta}{\sqrt{4\mu^2 \cos^2 \mu a\delta + (\mu^2 + 1)^2 \sin^2 \mu a\delta}} \epsilon^{i(-ax+ct+e)},$$

where

$$\tan e = -\frac{\mu^2 + 1}{2\mu} \tan \mu a\delta.$$

Similarly the transmitted wave is

$$\frac{1}{\sqrt{\cos^2 \mu a\delta + \frac{1}{4}(\mu^2 + 1)^2/\mu^2 . \sin^2 \mu a\delta}} \epsilon^{i(ax+ct+a\delta+e')},$$

where

$$\tan e' = -\frac{\mu^2 + 1}{2\mu} \tan \mu a\delta.$$

If $\mu a\delta$ be very small, the expression for the wave becomes approximately

$$\epsilon^{i\{ax+ct+a\delta-\frac{1}{2}(\mu^2+1)a\delta\}} = \epsilon^{i\{ax+ct-\frac{1}{2}(\mu^2-1)a\delta\}}.$$

In this case there is no loss of intensity in the transmitted light, and the *retardation* is $\frac{1}{2}(\mu^2 - 1)\delta$.

But if μ be complex (equal to Re^{ia}),

$$B_I = \frac{-(R^2\epsilon^{2ia} - 1)i\sin(Re^{ia}a\delta)}{2Re^{ia}\cos(Re^{ia}a\delta) + i(R^2\epsilon^{2ia} + 1)\sin(Re^{ia}a\delta)}.$$

The intensity of the reflected light is to be found by multiplying B_I by the quantity derived from it by changing the sign of i. The numerator of the resulting fraction is

$$(R^4 - 2R^2 \cos 2\alpha + 1)\sin(Re^{ia}a\delta)\sin(Re^{-ia}a\delta).$$

The product of sines is the half of

$$\cos\{2R\sin\alpha . ia\delta\} - \cos\{2R\cos\alpha . a\delta\}$$

$$= \tfrac{1}{2}\{\epsilon^{2R\sin a . a\delta} + \epsilon^{-R\sin a . a\delta}\} - \cos\{2R\cos\alpha . a\delta\}.$$

We may infer that the intensity of the reflected light is nearly proportional to

$$1 - \frac{2\cos\{2R\cos\alpha . a\delta\}}{\epsilon^{2R\sin a . a\delta} + \epsilon^{-2R\sin a . a\delta}}.$$

For transparent media the sum of exponentials reduces to the constant 2, but for opaque media it increases rapidly with δ. After the first, corresponding to $\delta = 0$, the minima are no longer zero, and soon all fluctuation becomes insensible.

Another effect of the exponential terms is to displace the position of the maxima and minima with respect to δ. They tend to occur earlier than they otherwise would do. In our ignorance of the values of the constants it seems hardly worth while to follow the result more minutely.

The transformation of A_2 when μ is complex leads to a long expression; and I will therefore confine myself to the particular case of a very thin layer, whose thickness does not amount to more than a small fraction of the wavelength.

Let the reduced expression for the transmitted wave be

$$A_2 \epsilon^{i(ax+ct)} = (\text{amplitude})\ \epsilon^{i(ax+a\delta+ct+e')}.$$

Then e' is given by the equation

$$\tan e' = \frac{D' - D}{i(D' + D)},$$

if we denote the denominator of the expression for A_2 by D, that derived from it by changing the sign of i by D'. Now

$$D = 1 - \frac{R^2 \epsilon^{2ia}(a\delta)^2}{2} + \frac{1 + R^2 \epsilon^{2ia}}{2}\ ia\delta + \text{cubes in } \delta,$$

$$D' = 1 - \frac{R^2 \epsilon^{-2ia}(a\delta)^2}{2} - \frac{1 + R^2 \epsilon^{-2ia}}{2}\ ia\delta + \text{cubes in } \delta;$$

$$\therefore\ D' - D = R^2 a^2 \delta^2 i \sin 2\alpha - ia\delta(1 + R^2 \cos 2\alpha),$$

$$D' + D = 2 - R^2 a^2 \delta^2 \cos 2\alpha - R^2 \sin 2\alpha . a\delta;$$

$$\therefore\ \tan e' = \frac{-a\delta(1 + R^2 \cos 2\alpha) + R^2 \sin 2\alpha . a^2\delta^2}{2 - R^2 \cos 2\alpha . a^2\delta^2 - R^2 \sin 2\alpha . a\delta}$$

$$= -\frac{a\delta(1 + R^2 \cos 2\alpha)}{2}\left\{1 + \frac{R^2 \sin 2\alpha . a\delta}{2}\left(\frac{1}{2} - \frac{1}{1 + R^2 \cos 2\alpha}\right)\right\}.$$

As a first approximation,

$$e' = -\tfrac{1}{2}a\delta(1 + R^2 \cos 2\alpha);$$

and the transmitted wave is

$$(\text{amplitude})\ \epsilon^{ia\{x - \frac{1}{2}(R^2 \cos 2\alpha - 1)\delta\} + ict};$$

so that the retardation is $\tfrac{1}{2}(R^2 \cos 2\alpha - 1)\delta$, independent of the opacity, as we have already seen it ought to be.

The second term in the approximate value of e' has a contrary effect to the first, because $\sin 2\alpha$ is negative. Moreover $\sin 2\alpha$ is numerically large. This may account for the acceleration of phase observed by Quincke—though if this explanation be correct, there must always be a retardation when the film is thin enough. It may happen that, in virtue of the great opacity of silver, its elimination by a reduction of the thickness may be impracticable without at the same time bringing the retardation due to density below the point at which it could be detected.

Postscript.—Since the above was written there has appeared a paper by O. E. Meyer, entitled "An Attempt to account for Anomalous Dispersion of Light"*, in which the author arrives at an expression for the refractive index equivalent to that found above as the value of $R \cos \alpha$, namely

$$R \cos \alpha = \frac{\gamma}{\gamma_{,} \sqrt{2}} \left\{ 1 + \sqrt{(1 + h^2/c^2 D_r^2)} \right\}^{\frac{1}{2}}.$$

Considering h constant, he sees in this an explanation of anomalous dispersion, inasmuch as $R \cos \alpha$ increases with λ, $(c \propto \lambda^{-1})$. In this view I cannot at all agree. Meyer seems to have overlooked the fact that h (in his notation κ), the constant of opacity, is subject to enormous chromatic variations, in comparison with which those of λ may be treated as quite insignificant. But this is not all. It has been laid down by Kundt as the result of his observations—and a very remarkable and suggestive result it is—that the anomalous effect is not confined to those rays which are intensely absorbed. Probably indeed the effect vanishes at that part of the spectrum which corresponds to the centre of the absorption-band, where, according to Meyer's formula (though not his interpretation of it), it should be a maximum. I have already indicated what appears to me to be the true mechanical character of the phenomenon by an illustration derived from ordinary dynamics. The mathematical analysis of the problem referred to, which turns up in almost all branches of physics dealing with vibrations, is well known and therefore need not be given here †.

---

* *Pogg. Ann.* vol. CXLV. p. 80, translated in *Phil. Mag.* vol. XLIII. p. 295.

† [1899. See footnote on p. 142. I have lately discovered that Maxwell had (earlier than Sellmeier) considered the problem of anomalous dispersion. His results were given in the Mathematical Tripos Examination, Jan. 21, 1869 (see *Cambridge Calendar* for that year), and they may have been in my mind when the text of this paper was written.]

17.

PRELIMINARY NOTE ON THE REPRODUCTION OF DIF-FRACTION-GRATINGS BY MEANS OF PHOTOGRAPHY.

[*Proceedings of the Royal Society*, Vol. xx. pp. 414—417, 1872.]

DURING the last autumn and winter I was much engaged with experiments on the reproduction of gratings by means of photography, and met with a considerable degree of success. A severe illness has prevented my pursuing the subject for some months, and my results are in consequence still far from complete; but as I may not be able immediately to resume my experiments, I think it desirable to lay this preliminary note before the Royal Society, reserving the details and some theoretical work connected with the subject for another opportunity.

It is some years since the idea first occurred to me of taking advantage of the minute delineating power of photography to reproduce with facility the work of so much time and trouble. I thought of constructing a grating on a comparatively large scale, and afterwards reducing by the lens and camera to the required fineness. I am now rather inclined to think that nothing would be gained by this course, that the construction of a grating of a given number of lines and with a given accuracy would not be greatly facilitated by enlarging the scale, and that it is doubtful whether photographic or other lenses are capable of the work that would be required of them.

However this, may be, the method that I adopted is better in every respect, except perhaps one. Having provided myself with a grating by Nobert with 3000 lines ruled over a square inch, I printed from it on sensitive dry plates in the same way as transparencies for the lantern are usually printed from negatives.

In order to give myself the best chance of success, I took as a source of light the image of the sun formed by a lens placed in the shutter of a dark room. I hoped in this way that, even if there should be a, small interval between the lines of the grating and the sensitive surface, still a *shadow* of the lines would be thrown across it. Results of great promise were at once obtained, and after a little practice I found it possible to produce copies comparing not unfavourably with the original. A source of uncertainty lay in the imperfect flatness of the glass on which the sensitive film was prepared, though care was taken to choose the flattest pieces of patent plate. The remedy is, of course, to use worked glass, which is required in any case if the magnifying-power of a telescope is to be made available.

Almost any of the dry processes known to photographers may be used. I have tried plain albumen, albumen on plain collodion, and Taupenot plates. The requirements of the case differ materially from those of ordinary photography, sensitiveness being no object, and hardness rather than softness desirable in the results. After partial development, I have found a treatment with iodine, in order to clear the transparent parts, very useful. In proceeding with the intensifying, the deposit falls wholly on the parts that are to be opaque. It is more essential that the transparent parts should be quite clear than that the dark parts should be very opaque.

The performance of these gratings is very satisfactory. In examining the solar spectrum, I have not been able to detect any decided inferiority in the defining-power of the copies. With them, as with the original, the nickel line between the *D*'s is easily seen in the third spectrum. I work in a dark room, setting up the grating at a distance from the slit fastened in the shutter, and using no collimator. The telescope is made up of a single lens of about thirty inches focus for object-glass, and an ordinary eyepiece held independently. I believe this arrangement to be more efficient than a common spectroscope, with collimator and telescope all on one stand; at any rate, the magnifying-power is considerably greater, and it seems to be well borne.

I have also experimented on the reproduction of gratings by a very different kind of photography. It will be remembered that a mixture of gelatine with bichromate of potash is sensitive to the action of light, becoming insoluble, even in hot water, after exposure. In ordinary carbon printing the colouring-matter is mixed with the gelatine and the print is developed with warm water, having been first transferred so as to expose to the action of the water what was during the operation of the light the hind surface. In my experiments the colouring-matter was omitted, and the bichromated gelatine poured on the glass like collodion and then allowed to dry in the dark. A few minutes' exposure to the direct rays of the sun then sufficed to produce such a modification under the lines of the gratings that, on treatment with warm water, a copy of the original was produced capable of giving brilliant spectra. In these gelatine-gratings all parts are alike transparent, so that the cause of the peculiar effect must lie in an alternate elevation and depression of the surface. That this is the case may be proved by pressing soft sealing-wax on the grating, when an impression appears on the wax, giving it an effect like that of mother-of-pearl. It is known that the effect of water on a gelatine print is to make the protected parts project in consequence of their greater absorption, but it might have been expected that on drying the whole would have come flat again. It is difficult to say exactly what does happen; and I am not even sure whether the part protected by the scratch on the original is raised or sunk. Gelatine can scarcely be actually dissolved away, because the uppermost layer must have become insoluble under the influence of light. I do not at present see my way to working by transfer, as in ordinary carbon printing.

I have not yet been able to reduce the production of these gelatine-gratings to a certainty, but can hardly doubt the possibility of doing so. One or two of considerable perfection have been made, capable of showing the nickel line between the D's, and giving spectra of greater brightness than the common photographs. Not only so, but the gelatine copy surpasses even the original in respect of brightness. The reason is that, on account of the broadening of the shadow of the scratch, a more favourable ratio is established between the breadths of the alternate parts.

Theory shows that with gratings composed of alternate transparent and opaque parts the utmost fraction of the original light that can be concentrated in one spectrum is only about $\frac{1}{10}$, and that this happens in the first spectrum when the dark and bright parts are equal. But if instead of an opaque bar stopping the light, a transparent bar capable of retarding the light by half an undulation can be substituted, there would be a fourfold increase in the light of the first spectrum. I accordingly anticipate that the gelatine-gratings are likely to prove ultimately the best, if the conditions of their production can be sufficiently mastered.

With regard to the application of the photographs, I need not say much at present; it is evident that the use of gratings would become more general if the cost were reduced in the proportion, say, of 20 to 1, more particularly if there were no accompanying inferiority of performance.

The specimens sent with this paper are both capable of showing the nickel line and give fairly bright spectra, but they must not be supposed to be the limit of what is possible. From their appearance under the microscope I see no reason to doubt that lines 6000 to the inch can be copied by the same method, a point which I hope shortly to put to the test of experiment.

18.

ON THE APPLICATION OF PHOTOGRAPHY TO COPY DIFFRACTION-GRATINGS.

[British Association Report, 1872, p. 39.]

GREAT interest has always attached itself to the beautiful phenomena discovered by Fraunhofer, which present themselves when a beam of light falls on a surface ruled with a great number of parallel and equidistant lines. Their unexpected character, the brilliant show of colour, and the ready explanation of the main points on the principles of the Wave Theory recommend them to all, while the working physicist recognizes in them the key to the exact measurement of wave-lengths, which has been so splendidly used by Ångström and others.

The production, however, of gratings of sufficient fineness and regularity is a matter of no ordinary difficulty. Indeed the exactness required and obtained is almost incredible. The wave-lengths of the two sodium lines differ by about the thousandth part. If in two gratings, or two parts of the same grating, the average interval between the divisions differed by this fraction, the less refrangible sodium line of one would be superposed on the more refrangible corresponding to the other. In point of fact the gratings ruled by Nobert of Barth, to whom the scientific world has been greatly indebted, are capable of distinguishing a difference of wave-length probably of a tenth part of that above mentioned. But in order that the D-lines may be resolved at all, there must be no average error (running over a large part of the grating) of $\frac{1}{1000}$ part of the interval between consecutive lines. When it is remembered what the interval is (from $\frac{1}{1000}$ to $\frac{1}{6000}$ of an inch, or even less), the degree of success which has been reached seems very remarkable.

A work requiring so much accuracy is necessarily costly—the reason, probably, why gratings fit to be used with the telescope for the purpose of showing the fixed lines are comparatively rare. The hope of being able to perfect a process for the reproduction of gratings at a comparatively cheap rate has induced the author to return at the first opportunity to the experiments described in a Preliminary Note read before the Royal Society in June last. Although the subject is as yet by no means exhausted, the author thought it worth while to bring before the Association an account of the progress that has been made, with specimens of the results.

The method of procedure is very simple. A dry plate prepared by any photographic process on a flat surface of glass, or other transparent material not affected by the fluid media employed, is brought into contact with the ruled surface of the grating in a printing-frame, and exposed to light. In the author's first experiments he used exclusively as a source of light the image of the sun in a lens of short focus placed in the shutter of a darkened room; but so small a source is not necessary. The light from the clouds or sky reflected by a mirror through a hole several inches in aperture will be sufficiently concentrated if the frame be a few feet distant. The author has not as yet specially investigated the point, but he believes that if the light were too much diffused, the experiment would fail. Much would, no doubt, depend on the perfection of the contact—an element very likely to vary. The variable intensity of diffused daylight, which it is almost impossible to estimate with precision, has induced him to use exclusively in his later experiments with ordinary photographic plates the light of a moderator lamp. This, with globe removed, is placed at a distance of 1 or 2 feet from the printing-frame, the distance being carefully measured. Working in this way there is little difficulty in giving consecutive plates any relative exposure that may be required. A collateral advantage is the possibility of operating at any time of the day or night.

With regard to the preparation of the plates, the author has latterly been using the tannin process introduced by Major Russell. A preliminary coating with dilute albumen is generally advisable, as any loosening of the film from the glass must be avoided on account of the distortion that it might introduce. In some states of the collodion an edging of black varnish put on after the exposure is sufficient to hold the film down. The glasses, after being coated with collodion (Mawson's was used), are immersed as usual in the silver bath, and then allowed to soak in distilled water, best contained in a dipping-bath. They are then washed under a tap for about half a minute, and put into the tannin solution (about 15 grains to the ounce) held, in the author's practice, in a small dish. The author usually prepares his plates in the evening, standing them up to dry on blotting paper. In the morning they are in a fit state for use. Artificial heat might no doubt be used if a more rapid drying were desired.

At a distance of about 1 foot from the lamp the exposure required is four or five minutes. The development is the most critical part of the process. The pyrogallic solution should contain plenty of acid (acetic or citric), and its action must not be pushed too far—the mistake which a photographer accustomed to negative work is most likely to make. At this stage the spectra given by a candle-flame are not very brilliant, on account of the iodide of silver still covering the parts which are to be transparent. Any trace of fog is especially .to be avoided. The author has experienced advantage in many cases from a solution of iodine in iodide of potassium applied to the film previously to fixing; but its action must be carefully watched, or too much silver will be converted. The iodide of silver is then cleared away with hyposulphite of soda or cyanide, followed by a careful washing under the tap.

With regard to the gelatine copies, the author has not much to add to the account read before the Royal Society. The process is very simple and some of the results very perfect, but he has not hitherto succeeded in sufficiently mastering the details. Plates apparently treated in precisely the same manner turned out very differently. That difficulties should arise is not very extraordinary, considering the novelty of the method; but it is curious that some of the very first batch prepared are among the best yet produced. The value of the results is so great that the author has no intention of abandoning his attempts, and perseverance must at last secure success.

The author then said a few words about the performance and prospects of the new copies. Their defining power on the fixed lines in the solar spectrum is all that could be desired, being, so far as he can see, in no way inferior to the originals. In the third spectrum the 3000 to the inch gratings show the line between the D's, if the other optical arrangements are suitable. The fourth line of the group b is distinguished with the utmost ease. The author is not sufficiently familiar with spectroscopic work to make an exact comparison, but presumes that two prisms of 60° at least would be required to effect as much. The author is here speaking of photographs on worked glass. With ordinary patent plate, although very good results may be obtained if tested by the naked eye only, it is a great chance whether the magnifying power of a telescope will not reveal the imperfect character of the surface.

With direct sunlight the light is abundantly sufficient; but it is here in all probability that the weak point of gratings lies. It should be distinctly understood that where light is deficient gratings will not compete with prisms. There are cases, however, where the scale might be turned by the opacity of all highly dispersive substances to the rays under examination. Even if glass be retained as the substratum, it may be used in a very thin layer, while prisms are essentially thick. The immense advantage of a diffraction-spectrum for the investigation of dark heat need not here be insisted on. Taking all things into consideration, it is probable that photographed gratings will supersede prisms for some purposes, though certainly not for all.

The specimens exhibited by Mr Ladd are copies of two gratings by Nobert, each of a square inch in surface, the one containing 3000 and the other 6000 lines. The latter cost about £20.

19.

ON THE DIFFRACTION OF OBJECT-GLASSES.

[*Astron. Soc. Month. Not.* XXXIII. pp. 59—63, 1872.]

IN observing the Sun with a telescope astronomers have to adopt some device in order to obviate the injurious effects which the intense light and heat would otherwise have on the eye. The most obvious way of doing this would be to contract the aperture of the object-glass, until the amount of light was reduced to within the necessary limit. But, as is well known, such a course cannot be followed without an enormous sacrifice of definition. The image, in the focus of the object-glass, of a mathematical point is a patch of light surrounded by rings, the dimensions of the system for a given wave-length varying inversely with the diameter of the lens. If this be reduced by a diaphragm, the patches dilate, those whose centres are within a small distance overlap, and the resolving power of the telescope suffers.

It has occurred to me that the result would be quite otherwise if, instead of the marginal, the central parts of the glass were stopped off, so that the light, coming from the lens to the focus, formed a hollow cone of rays. In this case the peculiar advantage of a large aperture would not be lost, while any imperfections arising from outstanding spherical aberration would be much diminished.

The general dependence of the diffraction phenomena which occur at the focus of a telescope on the aperture and wave-length may be explained without mathematical analysis. Consider the centre of the image given by a well-corrected object-glass, as illuminated by secondary waves coming from every part. Since the phases of all these elements agree at the point in question, the illumination is a maximum, and varies as the *square* of the area of the object-glass. Now take a neighbouring point in the focal plane. If the difference of its distances from the nearest and furthest point of the lens be but a small fraction of the wave-length [λ] of the light, the illumination must be sensibly the same as before.   The distance which it is necessary to go from the centre, in order that the illumination may be diminished in a given proportion, will evidently vary directly with λ, and inversely with the diameter of the glass.   If, then, we cover up the edge of the glass, the image of a point dilates, but no such effect ensues if we obstruct the central parts.

Another point seems to deserve a passing remark.   Red glass is, I believe, often used to diminish the solar glare.   From an optical point of view, this is the worst that could be chosen, for the ring system, being proportional to λ, is the largest for red light.   The wave-lengths for the fixed lines C and F are in the ratio $655 : 486$.   The substitution of a green-blue light for red would therefore be equivalent to an enlargement of about one-third in the diameter of the object-glass.

My immediate object in the present communication is, if possible, to induce some astronomer in possession of a good telescope to make a few observations on the defining power of an object-glass provided with a central stop.   The subject might be either the Sun himself, or as in Foucault's experiments (Verdet, *Leçons d'Optique Physique*, tom. I. p. 308), a scale illuminated by sunlight and placed at a sufficient distance.   Any results that might be obtained could not fail to interest physicists generally.

I append a mathematical statement of the two most contrasted cases, namely, (1) when the object-glass is completely uncovered, (2) when a narrow marginal rim is alone left to act.

Let dI denote the amplitude of the vibration corresponding to a ring $(2\pi R dR)$ of the object-glass at a point whose distance from the centre of the image subtends at the lens an angle θ.   Then, if $x = 2\pi \sin\theta\, R/\lambda$,

$$dI = 2\pi R dR \cdot J_0(x),$$

where
$$J_0(x) = \frac{1}{\pi} \int_0^\pi \cos(x\cos\phi)\, d\phi,$$

is the Bessel's function of zero order.

For the complete aperture from 0 to R, we have

$$I = 2\pi \int_0^R R dR J_0(2\pi \sin\theta\, R/\lambda).$$

The intensity is represented in the two cases by the squares of dI and I.

The expression for I may be written

$$I = \frac{2\pi R^2}{x^2} \int_0^x x\, dx J_0(x).$$

Now, differentiation being denoted by accents, $J_0(x)$ satisfies the equation

$$J_0'' + \frac{1}{x} J_0' + J_0 = 0,$$

whence

$$\int_0^x x\, dx\, J_0(x) = - \int_0^x dx\, \{x J_0'' + J_0'\} = - \int_0^x dx\, \frac{d}{dx} \{x J_0'\} = -x J_0',$$

so that

$$I = -\pi R^2 \cdot \frac{2 J_0'(x)}{x},$$

or since $J_0'(x) = -J_1(x)$,

$$I = \pi R^2\, \frac{2 J_1(x)}{x}.$$

The intensities in the two cases are accordingly in the ratio

$$\{2x^{-1} J_1(x)\}^2 \;:\; \{J_0(x)\}^2.$$

Let us consider first the positions of the dark rings. Their radii are given by the roots of the equations $J_1(x) = 0$ and $J_0(x) = 0$.

In case (2) $\{J_0(x) = 0\}$ the values of x corresponding to the dark rings are

2·41,   5·52,   8·66,   11·8,   14·9,   18·1,   &c.,

being ultimately of the form $(m - \frac{1}{4})\pi$ for the mth dark ring. These are for the pattern given by the marginal rim alone. For the whole lens we require the roots of $J_1(x) = 0$, which are

3·83,   7·02,   10·17,   13·3,   16·5,   19·6,   &c.,

the mth being ultimately $(m + \frac{1}{4})\pi$. So far the advantage lies entirely with case (2). If the size of the image be regarded as the space included within the first dark ring, the rim above gives a smaller image than the whole lens in the ratio of 2·41 : 3·83, or about 1 : 1·6. From Foucault's experiments (see Verdet) it would appear that the effective image does not extend so far as even to the first dark circle. Something, however, may depend upon the distribution of brightness. The functions $2x^{-1} J_1(x)$, $J_0(x)$ have their greatest values when $x = 0$, and are then both equal to unity. But the maxima after the first are less for $2x^{-1} J_1(x)$ than for $J_0(x)$. The positions of the maxima of $\{J_0(x)\}^2$ are given by the equation $J_1(x) = 0$, and are the same as those just found for the vanishing of $\{2x^{-1} J_1(x)\}^2$.

Corresponding to the values of x,

3·83,   7·02,   10·17,   &c.,

we have for $J_0(x)$ without regard to sign

403,   ·300,   ·250,   &c.

The squares of these give the maxima illuminations. To find when $2x^{-1}J_1(x)$ is a maximum, we have by differentiation,

$$J_1'(x) - x^{-1}J_1(x) = 0.$$

But by a property of these functions,

$$J_2(x) = x^{-1}J_1(x) - J_1'(x);$$

and therefore our conditions are simply $J_2(x) = 0$.

This gives for x approximately

$$5\cdot1, \quad 8\cdot4, \quad 11\cdot6, \quad \&c.,$$

the series being ultimately the same as for $J_0(x)$, so that the $(m-1)$th root is $(m-\frac{1}{4})\pi$.

The corresponding values of $2x^{-1}J_1(x)$, without regard to sign, are

$$\cdot13, \quad \cdot064, \quad \cdot040, \quad \&c.$$

For convenience of comparison I have drawn up a table of the maximum intensities, and of the places where the maxima and minima occur in the two cases. The calculations were made by means of the tables of Bessel's functions computed by Hansen (Lommel, *Studien ueber die Bessel'schen Functionen*).

| | Whole Lens | | Marginal Rim | |
	x	Intensity	x	Intensity
1st max.	0·00	1·0000	0·00	1·000
1st min.	3·83	0·0000	2·41	0·000
2nd max.	5·1	0·0170	3·83	0·160
2nd min.	7·02	0·0000	5·52	0·000
3rd max.	8·4	0·0041	7·02	0·090
3rd min.	10·17	0·0000	8·66	0·000
4th max.	11·6	0·0016	10·17	0·057
4th min.	13·3	0·0000	11·8	0·000

P.S. Since the above paper was written, Mr Dunkin has called my attention to a paper by the Astronomer Royal on "The Diffraction of an Annular Aperture" (*Phil. Mag.*, Jan. 1841), with which I was previously unacquainted. The purport of the two papers is, however, quite different. My object was not to solve a problem in physical optics, but to direct the attention of the possessors of telescopes to the theoretical consequences of using a central stop, and to suggest that its application may possibly be found advantageous in observations where light would otherwise be in excess. There is no pretence of originality in the mathematics; indeed, the solution of the problem for an annulus is a [usual] step in the treatment of the question of a lens with full aperture, the details of which are well known.

20.

AN EXPERIMENT TO ILLUSTRATE THE INDUCTION ON ITSELF OF AN ELECTRIC CURRENT.

[*Nature*, VI. p. 64, 1872.]

IT is well known that the sudden development of a current in a conductor is opposed by an influence analogous to the *inertia* of ordinary matter. A powerful movement of electricity cannot be suddenly produced ; neither can it be suddenly stopped. One consequence is that a periodic interruption of a circuit in which a constant electromotive force acts is sufficient, when the self-induction is great, to stop all sensible current, even although the interruptions themselves may be of very short duration. Before any copious flow can be produced the circuit is broken, and the work has to be begun over again. Whether in any particular case the influence of self-induction is paramount, or not, will depend also on the *resistance* of the circuit, and on the rapidity of the intermittence. The magnitudes which really come into direct comparison are the interval between the breaks, and the time which would elapse while a current, generated in the circuit and then left to itself, falls to a specific fraction (such as one half) of its original magnitude. In ordinary cases the duration of transient currents is but a small part of a second of time, so that, in order to bring out the effects of self-induction, the breaks must recur with considerable rapidity.

There is, however, one remarkable exception to the general rule, which occurs when, alongside of the principal coil to which the sluggishness is due, there exists an independent course along which the electricity can circulate. For instance, suppose that a coil with two wires, such as is often used for electro-magnets, is so arranged that one wire is included in the principal circuit, while the ends of the other are joined. The effect of the second circuit is then to neutralise the self-induction of the first, and so to increase largely the current that passes through it. Let us trace the progress of the phenomenon ; supposing that the first circuit has been closed for a sufficient time to allow of the development of the full current which can be excited by the actual electromotive force.

The moment the rupture is complete, the current in the first wire must stop, but another of the same magnitude and direction is at once developed in the neighbouring circuit. In fact, in virtue of its *inertia*, the electrical motion tends to continue with as little change as possible, a result which is attained in great degree by the formation of the second current to fill the place of the first. In a short time the induced current would diminish and become insensible under the operation of resistance (analogous to ordinary friction); but we are supposing that before this takes place to any considerable extent the contact is renewed, and the electromotive force again begins, in the first circuit, to push the electricity on. It is now that the peculiarity of the arrangement manifests itself. The current instantly transfers itself back again to the first circuit, which thus, without any delay, has the advantage of the full current which the electromotive force can sustain. If it had not been for the second circuit and its current, the development in the first would only have been gradual, and by supposition so slow that it would be checked by another interruption before any considerable progress could be made. In short, the self-induction of the principal circuit is virtually destroyed*.

In my experiment the principal circuit consisted of a Smee cell and one wire of a coil belonging to a large electro-magnet, and which I may call A_1. The interrupter was a tuning-fork, arranged after Helmholtz, and set into regular vibrations of about 128 per second by an independent current and battery. The fork itself was forged by the village blacksmith, and the whole affair was home-made. Across one prong was placed a sort of rider of copper wire, dipping on either side into a mercury cup, and so arranged that during the vibration its ends should enter and leave the mercury, thereby establishing and interrupting the continuity of the circuit. The current was measured by means of a short-wire galvanometer whose electrodes were connected with two neighbouring points of the circuit in such a manner that a small but constant proportion of the entire current passed through the instrument. The second wire of the coil A_2, which is similar to the first and *put on with it*, formed the second circuit, when its ends were joined by a short wire. In order to increase at pleasure the effects of induction, iron wires or rods of about a quarter of an inch in diameter were provided, whose insertion in the coil materially increased the decisiveness of the result.

In the first place, the deflection produced on the galvanometer when the circuit was permanently completed was 58°, which to fell 39° when the

---

* Mathematicians familiar with the theory of electricity will follow this by putting the three induction coefficients (in Maxwell's notation, L, M, N) equal, and the resistance of the second circuit, S, equal to zero.

interrupter was at work, the circuit of A_2 being open, and without iron. On closing A_2 the deflection rose to 46°. A_2 was again opened, and one iron wire introduced, which gave 30°. Two wires gave 25°, while the introduction of thirty reduced the deflection to 12°. Again closing A_2, the reading was 43°, raised to 44° only by the removal of the iron. It was clear that the second circuit almost secured the first from the influence of induction, which otherwise greatly reduced the electrical circulation. I may add that the arrangement was very efficient, the galvanometer needle remaining perfectly steady, so that the readings could be taken with ease and accuracy.

Another experiment made at the same time (about two years ago) may be noticed, if only for its contrast with the preceding. The coil A_1, being removed from the main circuit, was included in the branch with the galvanometer, as shown in the figure. Here neither the insertion of iron nor the

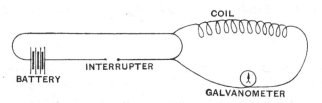

closing of A_2 made any difference; the circuit containing the coil remaining always closed, whatever might be the condition of the other. In such circumstances the average current indicated by the galvanometer is independent of the self-induction of the coil, varying only with the resistance in the branch, and with the *average* difference of potential at the points of derivation.

21.

SOME GENERAL THEOREMS RELATING TO VIBRATIONS.

[*Proceedings of the London Mathematical Society*, Vol. IV. pp. 357—368, 1873.]

THIS paper contains a short account of some general theorems, with which I have lately become acquainted during the preparation of a work on Acoustics. As they seem to possess considerable interest, I take the present opportunity of bringing them before the Society.

SECTION I.

The natural periods of a conservative system, vibrating freely about a configuration of stable equilibrium, fulfil the stationary condition.

Let the system be referred, in the usual manner, to independent co-ordinates ψ_1, ψ_2, ψ_3,... whose origin is taken to correspond with the configuration of equilibrium. Then, the square of the motion being neglected, the kinetic and potential energies are expressible in the form

$$T = \tfrac{1}{2}[11]\,\dot{\psi}_1^2 + \tfrac{1}{2}[22]\,\dot{\psi}_2^2 + \ldots\ldots + [12]\,\dot{\psi}_1\dot{\psi}_2 + \ldots\ldots \quad \ldots(1),$$

$$V = \tfrac{1}{2}\{11\}\,\psi_1^2 + \tfrac{1}{2}\{22\}\,\psi_2^2 + \ldots\ldots + \{12\}\,\psi_1\psi_2 + \ldots\ldots \quad \ldots(2),$$

where $[11]\ldots$, $\{11\}\ldots$ are constants, subject to the condition of making T and V always positive. For the present purpose, it is convenient, though not necessary, to transform the coordinates in the manner explained in Thomson and Tait's *Natural Philosophy*, § 337, so as to reduce T and V to a sum of squares;

$$T = \tfrac{1}{2}[1]\,\dot{\phi}_1^2 + \tfrac{1}{2}[2]\,\dot{\phi}_2^2 + \ldots\ldots \quad \ldots\ldots\ldots(3),$$

$$V = \tfrac{1}{2}\{1\}\,\phi_1^2 + \tfrac{1}{2}\{2\}\,\phi_2^2 + \ldots\ldots \quad \ldots\ldots\ldots(4),$$

where the coefficients are necessarily positive. The natural vibrations of the system are those represented by the separate variation of the coordinates

$\phi_1, \phi_2, \ldots$; and the corresponding differential equations obtained by Lagrange's method are of the form

$$[s]\, \ddot{\phi}_s + \{s\}\, \phi_s = 0 \quad \ldots\ldots\ldots\ldots\ldots\ldots\ldots\ldots(5),$$

showing that the period of the natural vibration ϕ_s is given by

$$\tau_s = 2\pi\, [s]^{\frac{1}{2}} \div \{s\}^{\frac{1}{2}} \quad \ldots\ldots\ldots\ldots\ldots\ldots\ldots(6).$$

Let us now suppose that the system is no longer allowed to choose its type of vibration, but that an arbitrary type is imposed upon it by a suitable constraint, leaving only one degree of freedom. Thus, let

$$\phi_1 = A_1 \theta, \quad \phi_2 = A_2 \theta, \quad \&c. \ldots\ldots\ldots\ldots\ldots\ldots(7),$$

where $A_1, A_2, \ldots$ are given real coefficients. The expressions for T and V become

$$T = \{\tfrac{1}{2}\,[1]\, A_1{}^2 + \tfrac{1}{2}\,[2]\, A_2{}^2 + \ldots\ldots\}\, \dot{\theta}^2 \quad \ldots\ldots\ldots\ldots(8),$$

$$V = \{\tfrac{1}{2}\,\{1\}\, A_1{}^2 + \tfrac{1}{2}\,\{2\}\, A_2{}^2 + \ldots\ldots\}\, \theta^2 \quad \ldots\ldots\ldots\ldots(9),$$

whence, if θ varies as $\cos pt$,

$$p^2 = \frac{\{1\}\, A_1{}^2 + \{2\}\, A_2{}^2 + \ldots\ldots}{[1]\, A_1{}^2 + [2]\, A_2{}^2 + \ldots\ldots} \quad \ldots\ldots\ldots\ldots(10).$$

This gives the period of the vibration of constrained type; and it is evident that the period is stationary, when all but one of the coefficients $A_1, A_2, \ldots$ vanish, that is to say, when the type coincides with one of those natural to the system.

By means of this theorem we may prove that an increase in the mass of any part of a vibrating system is attended by a prolongation of all the natural periods, or at any rate that no period can be diminished. Suppose the increment of mass to be infinitesimal. After the alteration, the types of vibration will in general be changed; but, by a suitable constraint, the system may be made to retain any one of the former types. If this be done, it is certain that any vibration which involves a motion of the part whose mass is increased, has its period prolonged. Only as a particular case (as, for example, when a load is placed at the node of a vibrating string) can the period remain unchanged. The theorem now allows us to assert that the removal of the constraint, and the consequent change of type, can only affect the period by a quantity of the second order; and that therefore, in the limit, the free period cannot be less than before the change. By integration we infer that a finite increase of mass must prolong the period of every vibration which involves a motion of the part affected, and that in no case can any period be diminished; but in order to see the correspondence of the two sets of periods, it may be necessary to suppose the alteration made by steps. The converse of this and corresponding theorems relating to an alteration in the potential energy of a system will now be obvious.

A very useful application of the principle may be made to the approximate calculation of the natural periods of a system whose constitution, though complicated, is but slightly different from one of a much simpler nature. The main difficulty of the general problem consists in the determination of the free types, which may involve the solution of a difficult differential equation. We now see that an approximate knowledge of the type may be sufficient for practical purposes, and that, in the class of cases referred to, the adoption of the type natural to the approximate simpler system in the calculation of T and V will entail an error of the second order only in the final result.

To illustrate this question, we may take a case not without interest of its own—namely, the transverse motion of a stretched string of nearly, but not quite, uniform longitudinal density. If the uniformity were exact, the type of the sth component vibration would be

$$y = \phi_s \sin \frac{s\pi x}{l} \quad \dots\dots\dots\dots\dots(11),$$

where l is the length, x the distance of any particle from one end, and y the transverse displacement. In accordance with the plan proposed, we are to calculate the period for the variable string on the supposition that (11) is also applicable to it. We find

$$T = \tfrac{1}{2}\phi_s^2 \int_0^l \rho \sin^2 \frac{s\pi x}{l}\, dx$$

$$= \frac{\rho_0 l}{4}\, \phi_s^2 \left\{ 1 + 2\int_0^l \frac{\Delta\rho}{l\rho_0} \sin^2 \frac{s\pi x}{l}\, dx \right\} \quad \dots\dots\dots(12)^*,$$

if $\rho = \rho_0 + \Delta\rho$, $\Delta\rho$ being small.

For the potential energy we have (T_1 being the tension) the usual expression

$$V = \tfrac{1}{2}T_1 \int_0^l \left(\frac{dy}{dx}\right)^2 dx = \frac{T_1 l}{4}\cdot\frac{s^2\pi^2}{l^2}\, \phi_s^2 \quad \dots\dots\dots(13).$$

Hence, if the solution be

$$\phi_s = A \cos\left(\frac{2\pi t}{\tau_s} - \epsilon\right) \quad \dots\dots\dots\dots(14),$$

the period τ_s is given by

$$\tau_s^2 = \frac{4\rho_0 l^2}{s^2 T_1}\left\{ 1 + 2\int_0^l \frac{\Delta\rho}{l\rho_0}\sin^2 \frac{s\pi x}{l}\, dx \right\}\dots\dots\dots(15).$$

As might be expected, the effect of an alteration of density vanishes at the nodes, and is a maximum midway between them.

* [An erratum is here corrected, 1898.]

A similar method applies to a great variety of problems, and gives the means of calculating the correction due to the necessary deviation of any actual system, on which experiments can be made, from the ideal simplicity assumed in theory.

Another point of importance with reference to this application has yet to be noticed. It appears from (10) that the period of the vibration corresponding to any hypothetical type is included between the greatest and least of the periods natural to the system. In the case of systems like strings and plates, which are treated as capable of continuous deformation, there is no least natural period; but we may still assert that the period calculated from any hypothetical type cannot exceed that belonging to the gravest normal type. When, therefore, the object is to estimate the longest natural period of a system by calculations founded on an assumed type, we have, à priori, the assurance that the result will come out too small. For example, the value for τ_1, given in (15), is certainly less than the truth, while the error is of the second order in $\Delta\rho$.

In the choice of a hypothetical type, judgment must be used, the object being to approach the truth as nearly as can be done without too great a sacrifice of simplicity. The type for a string heavily weighted at any point might suitably be taken from the extreme case of an infinite load, that is to say, the two parts of the string might be supposed to be straight. Even with a uniform unloaded string, the result of the above hypothesis is not so very far from the truth. Taking $y = mx \cos pt$ from $x = 0$ to $x = \frac{1}{2}l$, we find, for the whole string,

$$T = \frac{\rho p^2 m^2 l^3}{24} \sin^2 pt, \qquad V = m^2 \frac{T_1 l}{2} \cos^2 pt;$$

whence

$$p^2 = \frac{12 T_1}{\rho l^2} \quad\dots\dots\dots\dots\dots\dots(16).$$

The correct result for a uniform string is

$$p^2 = \frac{\pi^2 T_1}{\rho l^2} \quad\dots\dots\dots\dots\dots\dots(17);$$

so that the period calculated from the assumed type is too small in the ratio of $\pi : \sqrt{12}$ or $\cdot907 : 1$.

A much closer approximation would be obtained by the assumption of a parabolic form

$$y = m\left(1 - 4x^2/l^2\right) \quad\dots\dots\dots\dots(18).$$

Proceeding in the same way as before, we should find a period too short in the ratio $\pi : \sqrt{10}$, or $\cdot9936 : 1$. In order that the natural type should be parabolic, the density of the string would have to vary as $(l^2 - 4x^2)^{-1}$, being thus a minimum in the middle, and becoming infinite at either end.

The gravest tone of a square plate is obtained when the type of vibration is such that the nodal lines form a cross passing through the centre of the plate and parallel to the edges. The next type in order of importance gives the diagonals for the nodal lines. Chladni found experimentally that the interval between the two tones was about a fifth. It so happens that the second kind of vibration can be completely treated theoretically, being referable to the simpler case of the vibration of bars*; but the first has not hitherto been successfully attacked. I find that if we assume for the type of vibration

$$z = xy \cos pt \quad \dots\dots\dots\dots\dots\dots(19),$$

the nodal lines being taken for axes of x and y, the boundary conditions are satisfied, and the calculated period comes out greater than that corresponding to the diagonal position of the nodal lines in the ratio of $1\cdot37 : 1$. Since this ratio is certainly too small, Chladni's result is about what might have been expected from theoretical considerations.

Before leaving the subject of natural vibrations, I wish to direct the attention of mathematicians to a point which does not appear to have been sufficiently considered: I refer to the expansion of arbitrary functions in series of others of specified types. The best known example of such expansions is that generally called after Fourier, in which an arbitrary periodic function is analysed into a series of harmonics, whose periods are sub-multiples of that of the given function. It is well known that the difficulty of the question is confined to the proof of the *possibility* of the expansion; if this be assumed, the determination of the coefficients is easy enough. What I wish now to draw attention to is, that in this, and an immense variety of similar cases, the possibility of the expansion may be inferred from physical considerations.

To fix our ideas, let us consider the small vibrations of a uniform string stretched between two fixed points. We know, from the general theory, that the whole motion, whatever it may be, can be analysed into a series of harmonic functions of the time, representing component vibrations, each of which can exist by itself. If we can discover these normal types, we shall be in a position to represent the most general possible vibration by combining them, each with arbitrary amplitude and phase. The nature of the normal types is given by the solution of the differential equation

$$\frac{d^2y}{dx^2} + k^2y = 0 \quad \dots\dots\dots \dots\dots\dots\dots(20),$$

whence it appears that they are expressed by

$$y = \sin\frac{\pi x}{l}, \qquad y = \sin\frac{2\pi x}{l}, \qquad y = \sin\frac{3\pi x}{l}, \text{ \&c.}$$

---

* [1898. Only in the case where Poisson's ratio vanishes; see *Theory of Sound*, § 226.]

We infer that the most general position which the string can assume is capable of representation by a series of the form

$$A_1 \sin \frac{\pi x}{l} + A_2 \sin \frac{2\pi x}{l} + \ldots\ldots,$$

which is a particular case of Fourier's theorem. There would be no difficulty in proving it in its most general form.

So far the string has been supposed uniform. But we have only to introduce a variable density, or even a single load at one point of the string, in order completely to alter the expansion whose possibility may be inferred from dynamical theory. It is evident that corresponding to any system, whether string, bar, membrane, plate, or what not, there is an appropriate expansion for an arbitrary function of one or more variables. Thus the expansion in La Place's series may be proved by considering the motion of a thin layer of gas between two concentric spherical surfaces, the expansion in Bessel's functions from the vibrations of a circular membrane, or of the air contained within a rigid cylinder, &c. When the difficulty of a direct analytical proof of even these simple cases is considered, the advantage of the physical point of view will be admitted.

The method of definite integration (or summation, if the system have only finite freedom), by which the constants are determined to suit arbitrary initial circumstances, is well known, and has been applied to a great variety of problems, dealing not only with vibrations, but with other physical questions such as the conduction of heat; but I have never seen the reason of its success distinctly stated. It may be said to depend on the characteristic property of the normal coordinates, namely, their power of expressing the energy of a system as a sum of squares only. In the case of a string, for example, we have

$$y = \phi_1 \sin \frac{\pi x}{l} + \phi_2 \sin \frac{2\pi x}{l} + \ldots\ldots,$$

where ϕ_1, ϕ_2,...... are the normal coordinates. The expression for the energy is

$$\frac{\rho}{2} \int_0^l \dot{y}^2 \, dx, \quad \text{or} \quad \frac{\rho}{2} \int_0^l dx \left\{ \dot{\phi}_1 \sin \frac{\pi x}{l} + \dot{\phi}_2 \sin \frac{2\pi x}{l} + \ldots\ldots \right\}^2.$$

If by the solution of the differential equation, or otherwise, we have assured ourselves as to the nature of the normal types, we may assume, without further proof, that the products of the coordinates will disappear from the expression for the energy,—in the above instance, that

$$\int_0^l \sin \frac{r\pi x}{l} \sin \frac{s\pi x}{l} \, dx$$

will vanish, if r and s be different.

SECTION II.

The Dissipation Function.

The original equation of motion of a system in rectangular coordinates, as obtained at once by an application of the Principle of Virtual Velocities, is

$$\Sigma m \left(\ddot{x}\, \delta x + \ddot{y}\, \delta y + \ddot{z}\, \delta z \right) = \Sigma \left(X \delta x + Y \delta y + Z \delta z \right) \ldots \ldots \ldots (21).$$

When transformed to independent coordinates, and restricted so as to give the equations of vibratory motion in their simplest form, this becomes

$$\frac{d}{dt} \left(\frac{dT}{d\dot{\psi}_1} \right) = \Psi_1, \ \&c. \ \ldots \ldots \ldots \ldots \ldots (22),$$

where $\Psi_1 \delta\psi_1 + \Psi_2 \delta\psi_2 + \ldots$ is the transformation of $\Sigma \left(X \delta x + Y \delta y + Z \delta z \right)$, denoting the work done on the system by the applied forces during the hypothetical displacement.

If we separate from Ψ the forces which depend only on the position of the system, we obtain

$$\frac{d}{dt} \left(\frac{dT}{d\dot{\psi}_1} \right) + \frac{dV}{d\psi_1} = \Psi_1, \ \&c. \ \ldots \ldots \ldots \ldots (23).$$

The principal object of the present section is to show that another group of forces may be advantageously treated in a similar manner.

The forces referred to are those which vary in direct proportion to the component velocities of the parts of the system. It is well known that friction, and other sources of dissipation, may be usefully represented as following this law approximately; and even when the true law is different, the principal features of the case will be brought out. The effect of such forces will be to introduce into the original equation terms of the form

$$\Sigma \left(\kappa_x \dot{x}\, \delta x + \kappa_y \dot{y}\, \delta y + \kappa_z \dot{z}\, \delta z \right) \ \ldots \ldots \ldots \ldots (24),$$

where κ_x, κ_y, κ_z are the coefficients of friction, parallel to the axes, for the particle x, y, z. The transformation to the independent coordinates is effected in a similar manner to that of $\Sigma m \left(\dot{x}\, \delta x + \dot{y}\, \delta y + \dot{z}\, \delta z \right)$, and gives

$$\frac{dF}{d\dot{\psi}_1}\, \delta\psi_1 + \frac{dF}{d\dot{\psi}_2}\, \delta\psi_2 + \ldots \ldots,$$

where

$$F = \tfrac{1}{2} \Sigma \left(\kappa_x \dot{x}^2 + \kappa_y \dot{y}^2 + \kappa_z \dot{z}^2 \right)$$

$$= \tfrac{1}{2}(11)\, \dot{\psi}_1^2 + \tfrac{1}{2}(22)\, \dot{\psi}_2^2 + \ldots \ldots + (12)\, \dot{\psi}_1 \dot{\psi}_2 + \ldots \ldots \ \ldots (25).$$

F, it will be observed, is, like T and V, a necessarily positive quadratic function of the coordinates, and represents the rate at which energy is dissipated.

The above investigation refers to forces proportional to the absolute velocities; but it is equally important to include such as depend on the *relative* velocities of the parts of a system, and this fortunately can be done without any increase of complication. For example, if a force acts on the particle x_1 proportional to $\dot{x}_1 - \dot{x}_2$, there must be at the same moment, by the law of action and reaction, an equal and opposite force acting on x_2. The additional terms in the fundamental equation will be of the form

$$\kappa (\dot{x}_1 - \dot{x}_2) \, \delta x_1 + \kappa (\dot{x}_2 - \dot{x}_1) \, \delta x_2,$$

which may be written

$$\kappa (\dot{x}_1 - \dot{x}_2) \, \delta (x_1 - x_2) = \delta \psi_1 \frac{d}{d\psi_1} \{ \tfrac{1}{2} (\dot{x}_1 - \dot{x}_2)^2 \} + \dots\dots ;$$

and so on for any number of pairs of mutually influencing particles. The only effect is the addition of new terms to F, which still appears in the form (25)*.

The existence of the function F does not seem to have been recognised hitherto, and indeed is expressly denied in the excellent *Acoustics* of the late Prof. Donkin (p. 101). We shall see that its existence implies certain relations between the coefficients in the generalised equations of motion, which carry with them important consequences.

Lagrange's equation, after the separation from Ψ of the forces proportional to the displacements *and velocities*, whether absolute or relative, becomes

$$\frac{d}{dt} \left(\frac{dT}{d\dot{\psi}} \right) + \frac{dF}{d\dot{\psi}} + \frac{dV}{d\psi} = \Psi \quad \dots\dots\dots\dots\dots(26),$$

where

$$\left. \begin{aligned}
T &= \tfrac{1}{2} [11] \, \dot{\psi}_1{}^2 + \dots\dots + [12] \, \dot{\psi}_1 \dot{\psi}_2 + \dots\dots \\
F &= \tfrac{1}{2} (11) \, \dot{\psi}_1{}^2 + \dots\dots + (12) \, \dot{\psi}_1 \dot{\psi}_2 + \dots\dots \\
V &= \tfrac{1}{2} \{11\} \, \psi_1{}^2 + \dots\dots + \{12\} \, \psi_1 \psi_2 + \dots\dots
\end{aligned} \right\} \quad \dots\dots(27).$$

On substitution, we obtain a system of equations, which may be written:

$$\left. \begin{aligned}
\overline{11}\,\psi_1 + \overline{12}\,\psi_2 + \overline{13}\,\psi_3 + \dots\dots &= \Psi_1 \\
\overline{21}\,\psi_1 + \overline{22}\,\psi_2 + \overline{23}\,\psi_3 + \dots\dots &= \Psi_2 \\
\overline{31}\,\psi_1 + \overline{32}\,\psi_2 + \overline{33}\,\psi_3 + \dots\dots &= \Psi_3 \\
\dots\dots\dots\dots\dots\dots\dots\dots\dots\dots\dots\dots\dots
\end{aligned} \right\} \quad \dots\dots\dots\dots(28),$$

where a coefficient such as $\overline{rs}$ is an abbreviation for the quadratic operator

$$[rs] \frac{d^2}{dt^2} + (rs) \frac{d}{dt} + \{rs\}.$$

It is to be carefully noticed that since $[rs] = [sr]$, $(rs) = (sr)$, $\{rs\} = \{sr\}$, it follows that $\overline{rs} = \overline{sr}$.

* The differences referred to in the text may of course pass into differential coefficients in the case of a continuous body.

The small vibrations of a system free from dissipative influences can always be analysed into a series of normal components, each of which is similar in character to that of a system possessing but one degree of freedom. It is, in general, otherwise with the vibrations of a dissipative system. These may, indeed, be analysed into components of the quasi-harmonic type (Thomson and Tait, § 343); but these last are different in character from the vibration of a simple dissipative system. For instance, the system, supposed to be animated by one component, does not pass simultaneously through the configuration of equilibrium. The reason of the difference will appear at once. When there is no friction, a suitable transformation of coordinates will always reduce T and V to a sum of squares, and the equations of motion become

$$[1]\,\ddot{\phi}_1 + \{1\}\,\phi_1 = \Phi_1, \ \&\text{c.,}$$

the same as for a simple system. The presence of friction will not interfere with the reduction of T and V; but the transformation proper for them will not in general suit also the requirements of F. The equation can then only be reduced to the form

$$[1]\,\ddot{\phi}_1 + (11)\,\dot{\phi}_1 + (12)\,\dot{\phi}_2 + \ldots\ldots + \{1\}\,\phi_1 = \Phi_1 \ \ldots\ldots(29),$$

and not to the simple form expressing the vibration of a system of one degree of freedom,

$$[1]\,\ddot{\phi}_1 + (1)\,\dot{\phi}_1 + \{1\}\,\phi_1 = \Phi_1 \ \ldots\ldots\ldots\ldots(30).$$

We may, however, choose which two of the three functions we shall reduce, and the selection would vary according to circumstances.

Cases, however, arise in which, owing to the special character of the system, the same transformation of coordinates will reduce all three functions to sums of squares, and then the motion possesses an exceptional simplicity. Under this head the most important are probably when F is of the same form as T or V. In the problem of the string, if we assume a direct retarding force proportional to the velocity, we have F proportional to T; if the dissipation is due to viscosity, we might have F proportional to V. The same exceptional reduction is possible when F is a linear function of T and V, or when T is itself of the same form as V. In any of these cases the equations of motion for each component are of the same form as for a dissipative system with one degree of freedom, and the elementary types are characterised by the fact that the whole system passes simultaneously through the configuration of equilibrium. It appears that the law of friction usually assumed for a string is of an exceptional character, and leads to results of, in some respects, delusive simplicity.

Section III.

The present section is devoted to the proof and illustration of a very important law of a reciprocal character, connecting the forces and motions of any two types. Particular cases of it have been noticed by previous writers; but the general theorem is, I believe, new, and indeed could not be proved without the results of the preceding section.

The following partial statement will convey an idea of its nature :—

Let a periodic force Ψ_s, equal to $A_s \cos pt$, act on a system either conservative, or subject to dissipation represented by the function F, giving the forced vibration $\psi_r = \kappa A_s \cos(pt - \epsilon)$, where κ is the coefficient of amplitude, and ϵ the retardation of phase. The theorem asserts that if the system be acted on by the force $\Psi_r = A_r \cos pt$, the corresponding forced motion of type s will be

$$\psi_s = \kappa A_r \cos(pt - \epsilon).$$

The solution of the general equations (28) may be expressed in the form

$$\left. \begin{aligned} \nabla \psi_1 &= \frac{d\nabla}{d.\overline{11}} \Psi_1 + \frac{d\nabla}{d.\overline{12}} \Psi_2 + \ldots\ldots \\[2mm] \nabla \psi_2 &= \frac{d\nabla}{d.\overline{21}} \Psi_1 + \frac{d\nabla}{d.\overline{22}} \Psi_2 + \ldots\ldots \\[2mm] \ldots \quad \ldots \quad \ldots \quad \ldots \quad \ldots \end{aligned} \right\} \ldots\ldots\ldots(31),$$

where ∇ denotes the determinant

$$\begin{vmatrix} \overline{11}, & \overline{12}, & \overline{13}, & \ldots\ldots \\ \overline{21}, & \overline{22}, & \overline{23}, & \ldots\ldots \\ \overline{31}, & \overline{32}, & \overline{33}, & \ldots\ldots \\ \ldots & \ldots & \ldots & \\ \ldots & \ldots & \ldots & \end{vmatrix} \ldots\ldots\ldots\ldots(32),$$

and the partial differentiations are made without recognition of the relations $\overline{sr} = \overline{rs}$, &c. By the nature of determinants it follows that, since $\overline{sr} = \overline{rs}$,

$$\frac{d\nabla}{d.\overline{rs}} = \frac{d\nabla}{d.\overline{sr}} \quad \ldots\ldots\ldots\ldots\ldots(33).$$

Thus the component displacement ψ_r due to a force Ψ_s is given by

$$\nabla \psi_r = \frac{d\nabla}{d.\overline{rs}} \Psi_s \ldots\ldots\ldots\ldots\ldots(34).$$

12—2

If, now, we inquire what the effect of a force Ψ_r will be in producing the displacement of type s, we find

$$\nabla\psi_s = \frac{d\nabla}{d.sr}\,\Psi_r \dots\dots\dots\dots\dots\dots(35),$$

so that in virtue of (33) the relation of ψ_s to Ψ_r, in the second case, is the same as the relation of ψ_r to Ψ_s in the first.

Distinguishing the second case by a dash affixed to the corresponding quantities, let us take

$$\Psi_s = A_s\epsilon^{ipt}, \qquad \Psi_r' = A_r'\epsilon^{ipt},$$

where the coefficients A_s, A_r' may, without loss of generality, be supposed to be real. The solution may be expressed in the form

$$\left.\begin{aligned} \psi_r &= A_s\frac{d\log\nabla\,(ip)}{d.\overline{rs}}\,\epsilon^{ipt} \\[2mm] \psi_s' &= A_r'\frac{d\log\nabla\,(ip)}{d.\overline{sr}}\,\epsilon^{ipt} \end{aligned}\right\} \dots\dots\dots\dots(36),$$

where d/dt is replaced by ip in ∇ and its differentials. Hence by (33) we see that

$$A_r'\psi_r = A_s\psi_s' \dots\dots\dots\dots\dots\dots(37),$$

which is the symbolical expression of the reciprocal theorem with respect both to amplitude and phase. If $\Psi_r' = \Psi_s$, then will $\psi_s' = \psi_r$; but it must be remembered that the forces and displacements of different types are not necessarily comparable. The following statement will, however, hold good in all cases :—The force Ψ_r' does as much work on the motion due to Ψ_s, as Ψ_s does on the motion due to Ψ_r'.

There is an important class of cases to which our principle, general as is the proof just given, would not at first sight appear to apply. Among these may be noticed systems in which the cause of the dissipation, or of part of it, is the conduction and radiation of heat. The dissipation cannot always be represented by a function F, which shall be the same in form under all circumstances. I am not at present in a position to discuss this question completely; but there is one consideration which may here be referred to as sufficient to bring a large additional field within the sweep of the demonstration. Since the investigation is concerned only with harmonic motions of period (p), it will be sufficient for the establishment of the theorem if the dissipation function exist *for all vibrations of the given period*.

A few examples may promote the comprehension of a theorem which, on account of its extreme generality, may appear vague.

Let A and B be two points of a uniform or variable stretched string. If a periodic transverse force act at A, the same vibration will be produced at B as would have ensued at A had the force acted at B.

In a space occupied by air, let A and B be two sources of disturbance. The vibration excited at A will have at B the same relative amplitude and phase as if the places were exchanged. Helmholtz (*Crelle*, Band LVII.) has proved this result in the case of a uniform fluid without friction, in which may be immersed any number of rigid fixed solids; but we are now in a position to assert that the reciprocity will not be interfered with, whatever number of strings, membranes, forks, &c. may be present, even though they are subject to damping.

The theorem includes the optical law, that if one point can be seen from a second, the second can also be seen from the first, whatever reflections or refractions the light may have to undergo on its passage.

A last example may be taken from electricity. Let there be two linear conducting circuits A and B, in whose neighbourhood there may be any number of others (either closed or terminating in condensers), or solid conducting masses. The theorem asserts that an electromotive force acting in A gives the same variable current in B as would be produced in A if the electromotive force were transferred to B.

[1899. An application of the principles of this paper to the "Time-moduli of Dissipative Systems" will be found in *British Assoc. Report*, Aberdeen, p. 911, 1885.]

22.

ON THE NODAL LINES OF A SQUARE PLATE.

[*Phil. Mag.* XLVI. pp. 166—171, 246—247, 1873.]

SEE *Theory of Sound*, § 226.

[1899. In general the bending of a square plate with free edges takes place in both planes; but when the material is such that Poisson's ratio vanishes, i.e. when longitudinal extension entails no lateral contraction, the bending may be in one plane only, so that the surface remains cylindrical. In this case the bending follows the law of a simple bar, and the nodal system consists of two straight lines parallel to one pair of edges.

By superposing, after Wheatstone, two such modes of vibration, the phases being the same and the amplitudes equal or opposite, we obtain other nodal systems easily constructed from the known functions expressing the free vibrations of a bar. In the latter case symmetry suffices to shew that the nodal lines are the diagonals of the plate.]

23.

NOTE ON A NATURAL LIMIT TO THE SHARPNESS OF SPECTRAL LINES.

[*Nature*, Vol. VIII. pp. 474, 475, 1873.]

In the explanation usually given of the broadening of the fixed lines with increased pressure, it appears to be assumed that their finite width depends upon the disturbance produced by the mutual influence of the colliding molecules. I desire to point out that even if each individual molecule were allowed to execute its vibrations with perfect regularity, the resulting spectral line would still have a finite width, in consequence of the motion of the molecules in the line of sight. If there be any truth at all in the kinetic theory of gases, the molecules of sodium, or whatever the substance may be, are moving in all directions indifferently and with velocities whose magnitudes cluster about a certain mean. The law of distribution of velocities is probably the same as that with which we are familiar in the theory of errors, according to which, the number of molecules affected with a given velocity increases, the nearer that velocity is to the mean.

By the principles of this theory of gases the mean square of the velocity of the molecules can be deduced from the known pressure and mass. If v denote the velocity whose square is equal to the mean, it is found that for air at the freezing point, $v = 485$ metres per second.

At the temperature of flame the velocity may be about three times greater. For the purposes of a rough estimate it will be accurate enough to take the mean velocity of the molecules at 1500 metres per second, and that of light at 300,000,000 metres per second. The wave-length of the light emitted by a molecule moving with the mean velocity from the eye will therefore be greater by about five millionths than if the molecules were at rest. The double of this will be a moderate estimate of the width of the spectral line, as determined by the cause under consideration. We may conclude that however rare the gas, and however perfect our instrument may be, a fixed line cannot be reduced to within narrower limits than about a hundredth part of the interval between the sodium lines. I must leave it to spectroscopists more practised and skilful than myself, to say whether this result is in agreement with the appearance of the spectrum.

[1899. The application of Doppler's principle to moving molecules had been given two or three years earlier by Lippich (*Pogg. Ann.* t. 139, p. 465, 1870).

The subject is further discussed in a memoir on the Limit to Interference when Light is radiated from Moving Molecules (*Phil. Mag.* vol. 27, p. 298, 1889), and a calculation is given of the actual structure of a spectrum line upon the basis of Maxwell's law for the distribution of velocities in a gas.]

24.

ON THE VIBRATIONS OF APPROXIMATELY SIMPLE SYSTEMS.

[*Phil. Mag.* XLVI. pp. 357—361, 1873; XLVIII. pp. 258—262, 1874.]

SEE *Theory of Sound*, §§ 90, 91, 102.

[1899. The suggestion of Art. XXI, p. 172, is here developed. If the system is varied, we get in place of (3), (4),

$$T + \delta T = \tfrac{1}{2} \left([1] + \delta\,[1] \right) \dot{\phi}_1^{\,2} + \ldots + \delta\,[12]\,\dot{\phi}_1 \dot{\phi}_2 + \ldots,$$
$$V + \delta V = \tfrac{1}{2} \left(\{1\} + \delta\,\{1\} \right) \phi_1^{\,2} + \ldots + \delta\,\{12\}\,\phi_1 \phi_2 + \ldots.$$

In the original system a normal vibration is represented by the sole variation of ϕ_r proportionally to $\cos p_r t$, where $p_r^{\,2} = \{r\} \div [r]$. In the altered system the new type is determined by

$$\phi_s : \phi_r = \frac{\delta\,[rs]\,p_r^{\,2} - \delta\,\{rs\}}{[s]\,(p_s^{\,2} - p_r^{\,2})},$$

and

$$p_r^{\,2} = \frac{\{r\} + \delta\,\{r\}}{[r] + \delta\,[r]} - \Sigma\,\frac{(p_r^{\,2}\delta\,[rs] - \delta\,\{rs\})^2}{[r]\,[s]\,(p_s^{\,2} - p_r^{\,2})},$$

in which the summation extends to all values of s other than r. It is to be observed that the terms under the sign of summation are of the *second* order in $\delta\,[rs]$, $\delta\,\{rs\}$, regarded as small quantities.]

25.

ON THE FUNDAMENTAL MODES OF A VIBRATING SYSTEM.

[*Phil. Mag.* XLVI. pp. 434—439, 1873.]

SEE *Theory of Sound*, § 164.

[1899. The general theory is illustrated by the case of a bar vibrating laterally. The normal functions are usually found from an ordinary differential equation with application of certain terminal conditions, and the conjugate property, represented by $\int uv\,dx = 0$, is usually established in the same way. It is shown that the natural foundation for the conjugate property is the original variational equation from which in the method of Green the ordinary differential equation is deduced.]

26.

VIBRATIONS OF MEMBRANES.

[*London Mathem. Soc. Proceedings*, v. pp. 9—10, 1873.]

Theory of Sound, §§ 210, 211.

[1899. This is an application of a theorem of Art. XXI, to the effect that an increase in the potential energy of a vibrating system is attended by a rise in pitch, extending to all the fundamental modes. If the system consist of a uniformly stretched membrane with a fixed edge, it follows that any contraction of the boundary must cause an elevation of pitch, because the new state of things may be conceived to differ from the old merely by the introduction of an additional constraint. Springs (without inertia) are supposed to urge the line of the proposed boundary towards its equilibrium position, and gradually to become stiffer. At each step the vibrations become more rapid, until they approach a limit, corresponding to infinite stiffness of the springs and absolute fixity of their points of application.

Use is also made of the theorem that if the area of a membrane be given, the pitch is gravest when the form is circular.]

27.

HARMONIC ECHOES.

[*Nature*, VIII. pp. 319, 320, 1873.]

ACCORDING to Dr Brewer* "The harmonic echo repeats in a different tone or key the direct sound. The harmonic is generally either the third, fifth or tenth of the tonic....On the river Nahe, near Bergen, and not far from Coblentz, is an echo thus described by Barthius:—It makes seventeen repetitions at unequal intervals. Sometimes the echo seems to approach the listener, sometimes to be retreating from him; sometimes it is very distinct, at others extremely feeble; at one time it is heard at the right, and the next at the left; now in unison with the direct sound, and presently a third, fifth, or tenth of the fundamental. Occasionally it seems to combine two or more voices in harmony, but more frequently it resembles the voice of a single mimic.

"At Paisley, in Scotland, there is a somewhat similar echo in the burying-place of Lord Paisley, Marquis of Abercorn. Musical notes rise softly, swell till the several echoes have reverberated the sound either in unison or harmony, and then die away in general cadence.

"At the Lake of Killarney, in Ireland, is a very celebrated harmonic echo, which renders an excellent second to any simple air played on a bugle.

"There was formerly, according to the authority of Dr Birch, an harmonic echo no less remarkable, seventeen miles above Glasgow, near a mansion called Rosneath. If a trumpeter played eight or ten notes, the echo would repeat them correctly a third lower. After a short silence another repetition was heard, still lower than the former; and after a similar pause the same notes were repeated a third time, in a lower key and feebler tone, but nevertheless, with the same undeviating fidelity. This echo no longer exists."

* Brewer, *On Sound and its Phenomena* (1864), p. 305.

It is difficult to believe that these descriptions are accurate, but that they have a basis of truth there can be little doubt. My attention was first drawn to the subject by an echo at Bedgebury Park, the country residence of Mr Beresford Hope. The sound of a woman's voice was returned from a plantation of firs, situated across a valley, with the pitch *raised an octave*. The phenomenon was unmistakeable, although the original sound required to be loud and rather high. With a man's voice we did not succeed in obtaining the effect.

At the time I had no idea that such an alteration of pitch had ever been observed, or was possible; but it soon occurred to me that the explanation was similar to that which I had given of the blue of the sky a year or two previously (*Phil. Mag.* Feb. 1871). Strange to say, at the very time of the observation, I had in my portfolio a mathematical investigation* of the problem of the disturbance of the waves of sound by obstacles which are small in all their dimensions relatively to the length of the sound-waves. In such a case (precisely as in the parallel problem for light) it appears that the reflecting, or rather diverting, power of the obstacle varies inversely as the fourth power of the wave-length. When a composite note, such as that proceeding from the human throat, impinges on the obstacle, its components are diverted in very different proportions. A group of small obstacles will return the first harmonic, or octave, sixteen times more powerfully than the fundamental. After this, it is not hard to understand how a wood, which may be considered to be made up of a great number of obstacles, many of which, in two or three of their dimensions, are small in comparison with the wave-length, returns a sound which appears to be raised an octave.

The increased reflection is, of course, at the expense of the direct sound. If we conceive a group of small obstacles to act on a train of plane waves of sound, the effect will be a diffused echo, which may be heard on all sides, appearing to proceed from the group, and the direct waves which maintain their direction. If the original sound be composite, the diffused echo contains the higher elements in excessive proportion, and for the same reason the direct wave, being shorn of these higher elements, will appear duller than the original sound. It is well known that pure tones are liable to be estimated an octave too low, and thus it may be possible that a note in losing its harmonies may appear to fall an octave.

What is here called the direct sound may itself be converted into an echo by regular reflection. For example, if a plane wall were covered with small projections, there would be a diffused echo, due to the projections in which the higher elements preponderated, and an ordinary echo, obeying the law of reflection, in which the lower elements would preponderate.

---

* Since communicated in an amplified form to the Mathematical Society [Art. xiv].

28.

NOTE ON THE NUMERICAL CALCULATION OF THE ROOTS
OF FLUCTUATING FUNCTIONS.

[*Proceedings of the London Mathematical Society*, v. pp. 119—124, 1874.]

THERE is an important class of functions, often occurring in physical investigations, whose numerical calculation is easy when the argument is either small or great. In the first case the function is readily calculated from an ascending series, which is always convergent and might be employed whatever the value of the variable may be, were it not for the length to which the calculations would run. When the argument is great, a series proceeding by descending powers is employed, whose character is quite different. In this case the series is of the kind called semi-convergent, though strictly speaking it is not convergent at all; for, when carried sufficiently far, the sum of the series may be made to exceed any assignable quantity. But, though ultimately divergent, it begins by converging, and when a certain point is reached the terms become very small. It can be proved that, if we stop here, the sum of the terms already obtained represents the required value of the functions, subject to an error which in general cannot exceed the last term included. Calculations founded on this series are therefore only approximate; and the degree of the approximation cannot be carried beyond a certain point. If more terms are included, the result is made worse instead of better. In the class of functions referred to, the descending series is abundantly adequate when the argument is large, but there will usually be a region—often the most interesting part of the whole—where neither series is very convenient. The object of the present note is to point out how a part of the difficulty thence arising may sometimes be met.

Though the method in question is not limited to, it arose out of and is well exemplified by, the case of Bessel's functions, which are required among other purposes to represent the vibrations of a circular membrane, or of air contained within a cylindrical case. The roots of the equation obtained by equating the function to zero, give the possible periods of vibration, and the mechanics of the question show (as in all such cases) that no imaginary or complex root can occur.

If the function of the nth order be denoted as usual by $J_n(z)$, the ascending series is

$$J_n(z) = \frac{z^n}{2^n\,\Gamma(n+1)} \left\{ 1 - \frac{z^2}{2\,.\,2n+2} + \frac{z^4}{2\,.\,4\,.\,2n+2\,.\,2n+4} - \dots \right\} \quad \dots\dots\dots(1),$$

which is always ultimately convergent, whatever may be the values of z and n. When, however, z is considerable, the series becomes perfectly useless for arithmetical purposes. The descending series is

$$J_n(z) = \left(\frac{2}{\pi z}\right)^{\frac{1}{2}} \left\{ 1 - \frac{(1^2 - 4n^2)(3^2 - 4n^2)}{1\,.\,2\,(8z)^2} + \dots \right\} \cos\left(z - \tfrac{1}{4}\pi - \tfrac{1}{2}n\pi\right)$$

$$+ \left(\frac{2}{\pi z}\right)^{\frac{1}{2}} \left\{ \frac{1^2 - 4n^2}{1\,.\,8z} - \frac{(1^2 - 4n^2)(3^2 - 4n^2)(5^2 - 4n^2)}{1\,.\,2\,.\,3\,(8z)^3} + \dots \right\} \sin\left(z - \tfrac{1}{4}\pi - \tfrac{1}{2}n\pi\right).$$

$$\dots\dots\dots\dots(2).$$

This series terminates when n is of the form (integer $+ \frac{1}{2}$), and then of course constitutes the expression of the function in finite terms, but otherwise it runs on to infinity, and becomes ultimately, in all cases, divergent. In numerical calculations we are to include only the convergent part.

When z is very great, it is evident that the roots of the equation $J_n(z) = 0$, tend to the form

$$z = \frac{\pi}{4} + n\frac{\pi}{2} + \frac{2m-1}{2}\pi,$$

where m is an integer.

For purposes of explanation it will be sufficient to take the particular case of $n = 0$, worked out by Prof. Stokes in his paper "On the numerical calculation of a class of definite integrals and infinite series*."

To calculate the roots, we find

$$\frac{z}{\pi} = m - \tfrac{1}{4} + \frac{\cdot 050661}{4m - 1} - \frac{\cdot 053041}{(4m - 1)^3} + \frac{\cdot 262051}{(4m - 1)^5} + \quad \dots\dots\dots(3),$$

which is adequate for all the roots, except the first corresponding to $m = 1$. If we denote the roots in order by p_1, p_2, &c., we get

$$p_2 \div \pi = 1\cdot757098, \qquad p_5 \div \pi = 4\cdot7527,$$
$$p_3 \div \pi = 2\cdot754568, \qquad p_6 \div \pi = 5\cdot7522,$$
$$p_4 \div \pi = 3\cdot7534, \qquad p_7 \div \pi = 6\cdot7519.$$

Cambridge Transactions, vol. IX. 1850.

The higher the order of the root, the more accurately it can be calculated from this series, but an accurate value of the first root can only be derived from the ascending series. To attain this end in the ordinary way requires a considerable expenditure of computation. My object is to show how the difficulty may be evaded.

Returning to the general function $J_n(z)$, and excluding the n zero roots, we see that

$$1 - \frac{z^2}{2 \cdot 2n + 2} + \frac{z^4}{2 \cdot 4 \cdot 2n + 2 \cdot 2n + 4} - \cdots = \left(1 - \frac{z^2}{p_1^2}\right)\left(1 - \frac{z^2}{p_2^2}\right)\left(1 - \frac{z^2}{p_3^2}\right) \cdots$$

must be an identical equation. Taking the logarithms of both sides, expanding, and equating like powers of z, we find that

$$\Sigma p^{-2} = \frac{1}{2^2 \cdot n + 1}, \quad \Sigma p^{-4} = \frac{1}{2^4 (n+1)^2 \cdot n + 2},$$

$$\Sigma p^{-6} = \frac{1}{2^5 (n+1)^3 \cdot n + 2 \cdot n + 3}, \quad \Sigma p^{-8} = \frac{5n + 11}{2^8 (n+1)^4 (n+2)^2 \cdot n + 3 \cdot n + 4},$$

$$\Sigma p^{-10} = \frac{2 \cdot 7n + 19}{2^{10} (n+1)^5 (n+2)^2 \cdot n + 3 \cdot n + 4 \cdot n + 5},$$

the last result requiring a good deal of reduction.

If $n = 0$,

$$\Sigma p^{-2} = \tfrac{1}{4}, \quad \Sigma p^{-4} = \tfrac{1}{32}, \quad \Sigma p^{-6} = \tfrac{1}{192}, \quad \Sigma p^{-8} = \tfrac{11}{12288}, \quad \Sigma p^{-10} = \tfrac{19}{122880}.$$

If $n = 1$,

$$\Sigma p^{-2} = \tfrac{1}{8}, \quad \Sigma p^{-4} = \tfrac{1}{192}, \quad \Sigma p^{-6} = \tfrac{1}{3072}, \quad \Sigma p^{-8} = \tfrac{1}{46080}, \quad \Sigma p^{-10} = \tfrac{13}{8847360}.$$

Now Σp^{-10} must depend mainly on the first root, and being known numerically may be used to derive an accurate value of that root with the assistance of approximate values of the other roots. We have, when $n = 0$,

$$
\begin{aligned}
\pi^{10} \div p_2^{10} &= \cdot 00356484 \\
\pi^{10} \div p_3^{10} &= \cdot \quad\quad 3976 \\
\pi^{10} \div p_4^{10} &= \cdot \quad\quad\; 180 \\
\pi^{10} \div p_5^{10} &= \cdot \quad\quad\quad 17 \\
\pi^{10} \div p_6^{10} &= \cdot \quad\quad\quad\; 3 \\
\pi^{10} \div p_7^{10} &= \cdot \quad\quad\quad\; 1 \\
\hline
& \quad\quad\; \cdot 00360661
\end{aligned}
$$

and

$$\frac{\pi^{10}}{p_1^{10}} = \frac{19\pi^{10}}{122880} - \Sigma_2^\infty \frac{\pi^{10}}{p^{10}},$$

so that
$$p_1^{-10} = \cdot 0001546224 - \cdot 000000038513 = \cdot 0001545839.$$

Seven significant figures in p_1^{-10} require only three in $\Sigma_2^{\infty} \pi^{10}/p^{10}$, so that all the roots after the third might be neglected, and the values of the second and third themselves are only required to be known roughly. The resulting value of p_1 is

$$p_1 = 2\cdot404826.$$

The advantage of the present method would appear to depend upon the combined use of both the ascending and descending series.

The value of p_1 might of course be calculated from Σp^{-2}, or the sum of any other inverse powers. The advantage of using a high power is that the result becomes approximately independent of the other roots, and the tenth power was chosen as facilitating the numerical calculations. If a higher power than the tenth were required the algebraical reduction would become long, but this might easily be compensated by an advantage in the arithmetic.

As far as the 20th power inclusive, the expansion may be made without much difficulty by the aid of the table given on p. 775 of De Morgan's *Differential and Integral Calculus*. In this way the value of p_1 (for $n = 0$) might be obtained to 6 or 7 significant figures without *any* allowance for p_2, p_3, &c.

The method is equally successful in its application to the case of $n = 1$. We find

$$p_1^{-10} + \cdot000000003553 = \tfrac{13}{8847360} = \cdot000001469365,$$

whence $p_1 = 3\cdot831706 = 1\cdot219670\,\pi.$

In cases where there is a difficulty about the *two* first roots, it might be possible to obtain the desired results by using the values of two of the sums of powers, e.g. Σp^{-10} and Σp^{-2}. In calculating the value of $\Sigma_3^{\infty} p^{-2}$, a large number of roots would have to be included, but the calculation could generally be facilitated by the use of approximate formulæ. For example, in the case of $J_0(z)$, the roots after a certain point, say p_r, might be adequately represented by $z = m\pi$. Then

$$\Sigma_1^{\infty} p^{-2} = \Sigma_1^{r} p^{-2} - \Sigma_1^{r} \frac{1}{m^2\pi^2} + \Sigma_1^{\infty} \frac{1}{m^2\pi^2},$$

where the last term is $\tfrac{1}{6}$, by a known formula, which is in fact derivable from our general expression

$$\Sigma p^{-2} = \frac{1}{2^2 \cdot n + 1},$$

inasmuch as $J_{\frac{1}{2}}(z) = (2/\pi z)^{\frac{1}{2}} \sin z$.

A more important question arises as to whether this method can be applied to calculate the argument corresponding to a prescribed value (ϕ) of the function, other than zero. It might appear that this could always be done, inasmuch as when $J_0(z) = \phi$, another function, viz. $J_0(z) - \phi$, vanishes. But then the roots of this last function are not all real, and therefore cannot be calculated by the same method as before. Nevertheless, if ϕ were small, so that the first three or four roots were real, it would seem that the complex roots would be large enough to be dispensed with, as not sensibly influencing the value of Σp^{-10}, and then the first root could be deduced with sufficient precision from the values of the other real roots as calculated by aid of the descending series. I am not sufficiently acquainted with numerical calculations to say whether this application could ever be practically useful. If too laborious for the systematic tabulation of a function, it might perhaps be occasionally available as a control.

Nov. 22nd.

Prof. Cayley, to whom the preceding note was referred, has pointed out that a similar result may be attained by a method given in a paper by Encke, "Allgemeine Auflösung der numerischen Gleichungen," *Crelle*, t. XXII. (1841), pp. 193—248.

"Taking the equation

$$0 = 1 - ax + bx^2 - cx^3 + dx^4 - ex^5 + fx^6 - gx^7 + hx^8 - \dots;$$

if the equation whose roots are the squares of these is

$$0 = 1 - a_1 x + b_1 x^2 - c_1 x^3 + \dots,$$

then $a_1 = a^2 - 2b,$ $b_1 = b^2 - 2ac + 2d,$ $c_1^2 = c^2 - 2bd + 2ae - 2f,$

$$d_1^2 = d^2 - 2ce + 2bf - 2ag + 2h, \ \&c.;$$

and we may in the same way derive $a_2, b_2, c_2,$ &c., from $a_1, b_1, c_1,$ &c., and so on.

As regards the function

$$J_n(z) = \frac{z^n}{2^n \cdot \Gamma(n+1)} \left\{ 1 - \frac{z^2}{2 \cdot 2n+2} + \frac{z^4}{2 \cdot 4 \cdot 2n+2 \cdot 2n+4} - \dots \right\},$$

we have as follows:

$a^{-1} = 2^2 \cdot n + 1,$

$b^{-1} = 2^5 \cdot n + 1 \cdot n + 2,$

$c^{-1} = 2^7 \cdot 3 \cdot n + 1 \dots n + 3,$

$d^{-1} = 2^{11} \cdot 3 \cdot n + 1 \dots n + 4,$

$e^{-1} = 2^{13} \cdot 3 \cdot 5 \cdot n + 1 \dots n + 5,$

$f^{-1} = 2^{16} \cdot 3^2 \cdot 5 \cdot n + 1 \dots n + 6,$

$g^{-1} = 2^{18} \cdot 3^2 \cdot 5 \cdot 7 \cdot n + 1 \dots n + 7,$

$h^{-1} = 2^{23} \cdot 3^2 \cdot 5 \cdot 7 \cdot n + 1 \dots n + 8,$

$a_1^{-1} = 2^4 \cdot (n+1)^2 \cdot n + 2,$

$b_1^{-1} = 2^9 \cdot (n+1 \cdot n+2)^2 \cdot n+3 \cdot n+4,$

$c_1^{-1} = 2^{13} \cdot 3 \cdot (n+1 \dots n+3)^2 \cdot n+4 \dots n+6,$

$d_1^{-1} = 2^{19} \cdot 3 \cdot (n+1 \dots n+4)^2 \cdot n+5 \dots n+8,$

$$a_2 = \frac{5n + 11}{2^8 \cdot (n+1)^4 (n+2)^2 \, n+3 \cdot n+4},$$

$$b_2 = \frac{25n^2 + 231n + 542}{2^{17} \cdot (n+1 \cdot n+2)^4 (n+3 \cdot n+4)^2 \, n+5 \ldots n+8},$$

$$a_3 = \frac{429n^5 + 7640n^4 + 53752n^3 + 185430n^2 + 311387n + 202738}{2^{16} (n+1)^8 (n+2)^4 (n+3 \cdot n+4)^2 \, n+5 \cdot n+6 \cdot n+7 \cdot n+8} *.$$

If $n = 0$, $\quad \Sigma p^{-16} = a_3 = \dfrac{101369}{2^{27} \cdot 3^3 \cdot 5 \cdot 7} = p_1^{-16}$, suppose ;

whence $\qquad\qquad\qquad p_1 = 2{\cdot}404825$."

It will be seen that the expression a_3 for Σp^{-16} gives a very accurate value of p_1, without any allowance for the other roots.

If $n = 1$, we have $\quad \Sigma p^{-16} = \alpha^{-16} = a_3 = \dfrac{761376}{2^{30} \cdot 3^7 \cdot 10^2 \cdot 7}$;

whence $\qquad\qquad\qquad \alpha = 3{\cdot}831690,$

and this is the approximate value of p_1. To get a corrected value, we may take

$$p_1 = \alpha \left(1 + \tfrac{1}{16} \alpha^{16} \Sigma_2^\infty \, p^{-16}\right) ;$$

or, if all but p_2 may be omitted,

$$p_1 = \alpha \left\{1 + \tfrac{1}{16} \left(\frac{\alpha}{p_2}\right)^{16}\right\}$$

In the present case $p_2 = 7{\cdot}015$, so that

$$p_1 = \alpha \times 1{\cdot}0000039 = 3{\cdot}831705.$$

It may be observed that the value obtained without correction for the higher roots is always an under-estimate.

* I have altered the coefficient of n^3 in the numerator from 53572 to 53752.

29.

A HISTORY OF THE MATHEMATICAL THEORIES OF ATTRAC-
TION AND THE FIGURE OF THE EARTH FROM THE
TIME OF NEWTON TO THAT OF LAPLACE. By I. Todhunter, M.A., F.R.S. Two Volumes. (London, Macmillan & Co.
1873.)

[*The Academy*, v. pp. 176—177, 1874.]

Scientific men must often experience a feeling not far removed from
alarm, when we contemplate the flood of new knowledge which each year
brings with it. New societies spring into existence, with their Proceedings
and Transactions, laden with the latest discoveries, and new journals continually appear in response to the growing demand for popular science.
Every year the additions to the common stock of knowledge become more
bulky, if not more valuable ; and one is impelled to ask, Where is this to
end ? Most students of science who desire something more than a general
knowledge, feel that their powers of acquisition and retention are already
severely taxed. It would seem that any considerable addition to the burden
of existing information would make it almost intolerable.

It may be answered that the tendency of real science is ever towards
simplicity ; and that those departments which suffer seriously from masses of
undigested material are also those which least deserve the name of science.
Happily, there is much truth in this. A new method, or a new mode of
conception, easily grasped when once presented to the mind, may supersede
at a stroke the results of years of labour, making clear what was before
obscure, and binding what was fragmentary into a coherent whole. True
progress consists quite as much in the more complete assimilation of the old,
as in the accumulation of new facts and inferences which in many cases
ought to be regarded rather as the raw materials of science than as science
itself. Nevertheless, it would be a mistake to suppose that the present
generation can afford to ignore the labours of its predecessors, or to assume

that so much of them as is really valuable will be found embodied in recent memoirs and treatises. Of the dangers of such a course, History gives ample warning. The case of Young will at once suggest itself as that of a man who from various causes did not succeed in gaining due attention from his con- temporaries. Positions which he had already occupied were in more than one instance reconquered by his successors at a great expense of intellectual energy.

It is one of the objects of books like Mr Todhunter's to check this deplorable waste of labour, by bringing together all the writings of the older authors which bear on certain selected subjects. No one who has not tried it, can imagine how much time is lost in hunting backwards and forwards through endless Transactions and periodicals in various tongues, many of them difficult of access, for memoirs of which after all the value may prove very trifling. When the problem in hand is of no great difficulty, the student may even find an independent attack the shortest in the end. There cannot be two opinions as to the great importance of the work that Mr Todhunter has undertaken. It is one demanding much clear-sightedness and patience, and we are not surprised to learn that it occupied seven years. Some may think that the same talents and industry would be better devoted to original work; but it must be allowed that to elucidate and render accessible the labours of others may be a service as valuable as the addition of new material to the common store. To deny this would be an error parallel to that of some economists, who glorify the labourer and manufac- turer at the expense of the merchant.

The theory of Attraction and of the Figure of the Earth is a subject to which most of the greatest mathematicians have contributed. In itself of great interest, it was the occasion of the invention of the mathematical weapons which have since been so successfully used in almost all branches of Physics. The first steps or rather strides were made by Newton. His theorems with respect to the attraction of spheres—that a spherical shell exerts no force on an internal particle, and attracts an external one exactly as if its mass was concentrated at the centre—are the foundation of the whole subject, and it is difficult to imagine anything more simple and beautiful than his exposition of them. To him we owe the first investigation of the earth's figure. A mass of uniform attracting fluid, if at rest, would evidently shape itself into a sphere. The question is, How will this form be altered when the whole revolves? What will be the effect of the centrifugal force? Newton's solution of this important problem was not complete; but on the assumption that the form might be that of an oblate spheroid, or as Mr Todhunter calls it, an *Oblatum*, he investigated the degree of eccentricity and the law of variation of gravity at the surface. Though progress had been made by Stirling and Clairaut, the gap in Newton's work was not fully filled up until the time of Maclaurin, who proved conclusively that the conditions of relative equilibrium were satisfied in the case of an oblatum.

The period embraced in Mr Todhunter's history extends to the first quarter of the present century. Perhaps this was the best point at which to stop, though a slight sketch of more recent discoveries would have been acceptable. The most important part of the work, considered as a book of reference, is probably the analysis of the memoirs of Legendre and Laplace; but for the genuine student of scientific history the earlier efforts are of equal, if not superior, interest. The whole work bears evidence of its author's well-known care; and the claims of the various mathematicians, whose labours are reviewed, appear to be discussed with perfect impartiality. D'Alembert and Ivory are perhaps those whose reputations suffer most in Mr Todhunter's hands, while Laplace takes a position even higher than had been assigned to him by previous writers. Without a complete survey of earlier memoirs it was difficult to know how much of the *Mécanique Céleste* was original, and how much borrowed; for Laplace, like too many modern French writers, was not in the habit of acknowledging his obligations.

In such a work as that before us accuracy and completeness are almost everything, and minor defects may well be passed over. Of course many points are discussed which admit of some difference of opinion. In estimating the value of various contributions to his subject, Mr Todhunter shows perhaps a tendency to prefer rigour of treatment to originality of conception. But the strictest proof is not always the most instructive or even the most convincing. To deserve the name of demonstration an argument should make its subject-matter plain, and not merely force an almost unwilling assent.

30.

ON THE MANUFACTURE AND THEORY OF DIFFRACTION-GRATINGS.

[*Phil. Mag.* XLVII. pp. 81—93, 193—205, 1874.]

IN a "Preliminary Note on the Reproduction of Diffraction-gratings by means of Photography," published in the *Proceedings of the Royal Society* for 1872, and in the *Philosophical Magazine* for November of the same year [Art. XVII], I gave a short account of experiments with which I had been for some time occupied. A few further details were communicated to the British Association at Brighton (*Brit. Assoc. Report*, p. 39) [Art. XVIII]. I now propose to give the results of more recent experience in the practical manufacture of gratings, as well as some theoretical conclusions which have been in manuscript since the subject first engaged my attention.

There are two distinct methods of copying practised by the photographer —(1) by means of the camera, (2) by contact-printing. The first, if it were practicable for our purpose, would have the advantage of leaving the scale arbitrary, so that copies of varying degrees of fineness might be taken from the same original. By this method I have obtained a photograph of a piece of striped stuff on such a scale that there was room for about 200 lines in front of the pupil of the eye, capable of showing lateral images of a candle; but I soon found that the inherent imperfections of our optical appliances, if not the laws of light themselves, interposed an almost insuperable obstacle to obtaining adequate results.

However perfect a lens may be, there is a limit to its powers of condensing light into a point. Even if the source from which the light proceeds be infinitely small, the image still consists of a spot of finite size surrounded by dark and bright rings. That this must be so may be shown by general considerations without any calculations. If a lens is absolutely free from aberration, the secondary waves issuing from the different parts of its hinder surface agree perfectly in phase at the focal point. Let us consider the illumination at a neighbouring point in the focal plane. If the distance between the two points is so small that the difference of the distances between the point under consideration and the nearest and furthest parts of the object-glass is but a small fraction of the wave-length (λ), the group of secondary waves are still sensibly in agreement, and therefore give a resultant illumination the same as before. At a certain distance from the focal point the secondary waves divide themselves into two mutually destructive groups, corresponding to the nearer and further parts of the object-glass. There is therefore here a dark ring. Further out there is again light, then another dark ring, and so on, the intensity of the bright rings, however, rapidly diminishing.

The radius r of the first dark ring subtends at the centre of the lens an angle θ given by

$$\sin \theta = {\cdot}61 \, \frac{\lambda}{R} \, *,$$

where R is the radius of the lens. If f be the focal length, we have

$$r = {\cdot}61 \frac{f\lambda}{R}.$$

Let us now suppose that the problem is to cover a square inch with 3000 lines. On account of the curvature of the field it would be impossible to obtain extreme definition over the surface of a square inch with a less focal distance than (say) four inches. If we take $f = 4$ and $\lambda = \frac{1}{40,000}$, we find

$$R = \frac{{\cdot}61}{10,000 \, r},$$

which gives $R = {\cdot}2$ for $r = \frac{1}{3000}$. That is to say, if the focal length were 4 inches and aperture ·4 inch, the first dark ring corresponding to one of the lines would fall on the focal point of the neighbouring one—a state of things apparently inconsistent with good definition. It is true that the aperture might well be greater than half an inch, so that it may seem possible to satisfy the requirements of the case. But the result of the above calculation, being founded on the supposition of entire freedom from aberration, both spherical and chromatic, is subject in practice to a large modification. In astronomical telescopes, where everything is sacrificed to the requirement of extreme definition at the centre of the field, the theoretical limit is sometimes closely approached; but the case is very different with a photographic lens. In fact the very first thing it occurs to a photographer to do, when he wishes to improve the definition, is to contract the aperture of his lens by means of a stop—a course which would be attended with the opposite result in the case of a perfect object-glass, or even a good astronomical telescope. While,

* Verdet, *Leçons d'Optique Physique*, vol. I. p. 305.

therefore, it might be too much to say that the reproduction of 3000 lines in an inch by lens and camera is impossible, the attempt to do so without very special appliances appears in a high degree unpromising. It would certainly require a lens more than usually free from spherical aberration, and unlike either a telescopic or a photographic object-glass*, achromatic (if the expression may be allowed) for the chemical rays, unless indeed the latter requirement could be evaded by using approximately homogeneous light. It must be understood that nothing is here said against the practicability of covering a small space with lines *at the rate of* 3000 to the inch, a feat probably well within the powers of a good microscopic object-glass.

The method of contact-printing, on the other hand, is free from optical difficulties. The photographic film prepared on a flat piece of glass (or other support) may be brought by moderate pressure in a printing-frame within a very short distance of the lines of the original grating; and if the source of light be moderately small and the rays fall perpendicularly, the copy rarely fails in definition, unless through some photographic defect. When direct processes not depending on development are employed, the unclouded light of the sun is necessary. To avoid too much diffused light, I usually place the printing-frame on the floor of a room into which the sun shines, and adjust its position until the light reflected from the plate-glass front is sent back approximately in the direction of the sun. Too much time should not be lost in this operation, which requires no particular precision. Usually I cut off part of the extraneous light by partially closing the shutters; but I cannot say whether this makes any difference in the result. Those who are accustomed to this kind of experimenting will know that it is often less trouble to take a precaution than to find out whether it is really necessary. In an early stage of an investigation, when the causes of failure are numerous and unknown, it is best to exclude everything that can possibly be supposed to be prejudicial. When the principal difficulties have been overcome, it will be time enough to determine what precautions are necessary, if the question has not been already settled by accidental experience.

In the case of developed plates there is more choice of lights in consequence of the higher sensitiveness. I have used successfully cloud or skylight reflected horizontally from the zenith by a mirror through a hole of two or three inches diameter in the shutter of a darkened room, the frame being set up in a vertical plane at a few feet distance. The principal objection to this plan is the difficulty of estimating the exposure with proper precision— a difficulty which is more felt than in ordinary photography, as it is convenient to develop a good many copies at once. On a really fine day the image of the sun formed by a condensing lens of short focus placed in the

---

* Photographic lenses are corrected on the principle of making the "visual and chemical foci" coincident, which leads to a different construction from what would be adopted were the chemical rays alone attended to.

shutter (as used in diffraction experiments) constitutes a very convenient source of light.   As the exposure is only a few seconds, there is no difficulty in dodging isolated clouds, whose progress may be watched from within by examining their *image* with a coloured glass.   When there is any haze this method is not more satisfactory than the other.

With the more sensitive processes artificial light may be employed.   I have done a good many copies by the aid of a moderator-lamp (without the globe) at two feet distance from the frame.   An Argand gas-flame would probably be still better.

The printing-frame I employ has a thick plate-glass front, against which the original grating and the prepared plate are pressed by screws.   These are more under control than the springs generally used in the common printing-frames.   When everything is ready, the original is placed on the glass front of the frame with the engraved face upwards, care being taken to exclude all grit by means of a camel's-hair brush.   The prepared plate is then placed face downwards on the grating, then a pad to equalize the pressure (I have used one of india-rubber), and on the pad the rigid back of the frame, on which the screws are made to press with a moderate force. When the film is delicate, care should be taken to place it in the proper position at once without sliding.

The two surfaces of the plate-glass front of the printing-frame and the back of the original grating may be cleaned in the ordinary way with a soft cloth or wash-leather; but the engraved face of the grating requires more delicate treatment. If touched at all with a solid (wash-leather), the greatest care should be used.   I prefer to wash it, when soiled, with a stream of water from a tap, afterwards flooding it with pure alcohol and setting it up to drain and dry spontaneously.   Sometimes I have found nitric acid useful; but I always try to avoid the rubbing contact of a solid.   These precautions have been so successful that, after several hundred copies have been taken, the originals have scarcely, if at all, deteriorated.

For the support of the photographic film it is no doubt most satisfactory to use optically worked "parallel" glass.   Ordinary glass would fail, for two reasons.   In the first place it would generally be impossible to secure a sufficiently close contact in the printing.   But even if this difficulty could be surmounted, the spectrum given by the copy would not bear the magnifying-power which it is generally desirable to apply.   It is indeed evident that the glass support of the grating requires the same precision of workmanship as the object-glass of the telescope used in conjunction with it.

Although ordinary glass taken at random is inadmissible, I have done a great number of excellent gratings on selected pieces of patent plate.   In order to choose the best, I lay the plates on a table in such a position that the bars of a window or skylight are seen reflected in them. Each bar appears in general double, one image corresponding to each surface. By sliding the plate about, while the head is kept still, irregularities are easily detected by the shifting or curvature of the images. From a package of two dozen 5 × 4 plates as issued by photographic dealers, three or four, often lying together, may usually be selected as flat enough for the purpose, or at any rate decidedly superior to the remainder. It is worth notice that the object aimed at is *flatness* of the two faces, exact *parallelism* being of much less consequence; for it is evident that the interposition of a truly worked prism of very acute angle would produce no evil result. A glass is therefore not to be rejected merely because the two images of the bar seen reflected in it are decidedly separated. The question is rather whether this separation remains constant as the plate is moved about without rotation. I have never seen a piece of patent plate that could not be at once distinguished from worked glass in the way described; so that the test is abundantly sufficient for the purpose. The more delicate methods by which worked glass is examined would be less practically useful.

Whatever kind of glass be used, if the photographic process be at all complicated, there is considerable economy of labour in preparing comparatively large pieces, to be afterwards cut with the diamond to the required size. A 5 × 4, or even a $4\frac{1}{4} \times 3\frac{1}{4}$, plate will do very well for four gratings. In the case of worked glass economy is an object; but when patent plate is used I should recommend 5 × 4 glasses, as a margin is convenient. Even when, as in the collodio-chloride process to be presently described, the plate for each grating is prepared separately, it is convenient to perform the preliminary operations of cleaning and albumenizing on larger pieces. The cutting of prepared plates requires a little care. I place them face downwards on a sheet of clean paper, make the diamond cuts on the back, and then, before breaking, remove as much as possible of the glass powder. As it is important to prevent any grit from getting between the film and the engraved face of the original, I usually brush the surface with a large camel's-hair brush kept scrupulously clean.

In the preparation of the plates I have used a considerable variety of methods. The process with gelatine and bichromate of potash described in my previous papers has decided advantages; but all my efforts to obtain a mastery of it have been unavailing. Plates prepared to all appearance in precisely the same manner, and even at the same time, turned out differently, while modifications purposely introduced seemed to be for the most part without effect. It required a strong scientific prejudice to hold the uniformity of nature in the face of so much adverse evidence. The uncertainty of this method is provoking, as some of the results are exceedingly good; but I gave up my attempts sooner than I might otherwise have done in consequence of the discovery of another method by which most of the advantages of the gelatine process, namely simplicity of manipulation and brilliancy of results, might be attained with much less risk of failure.

It is very possible that a photographer skilled in the employment of gelatine might succeed where I failed. In case any such should wish to make the attempt, I will mention a few points that seemed important. The solution of gelatine should be carefully filtered. For thick liquids containing gelatine, albumen, &c., the best filtering material that I know of is tow. The tow should be cleaned from grease by boiling with soda and subsequent washing, and a small plugget of it pushed with moderate force into the neck of the funnel. Some arrangement must be adopted for keeping the gelatine hot, or the operation will hardly succeed. It is important that the coat of gelatine should be even, for which object the glass must be free from grease, and the plate on which the prepared glasses are put away to set perfectly level. Even then a good deal depends on the manipulation; but this is soon learned. The uniformity of the coat may be tested by the colour when the plate is placed on a sheet of white paper and examined in a weak white light. By candle-light the colour of weak bichromate of potash is scarcely visible. The exposure may be from two to six minutes to the direct rays of the sun. I have not been able to detect any deterioration when the plates were kept a few days in the dark before being used.

A photographer accustomed to either the plain albumen or the Taupenot process will find it very suitable for gratings. The hardness of the surface, which allows varnish to be dispensed with, is a great advantage. In my experiments with plain albumen, the principal difficulty was the purely photographic one of avoiding stains. It must be observed, however, that in actual use the gratings are not seen in focus, and that excellent spectra may be obtained from copies which a photographer would be inclined to throw away at once as hopelessly faint and dirty. The objection to the Taupenot process is the trouble of preparing the plates; but this is much mitigated when the plan is adopted of preparing large pieces to be afterwards cut up.

Among those requiring development, the tannin process is the one with which I have been most successful. In order to counteract the well-known tendency of the film to loosen, a preliminary coating of dilute albumen or gelatine is generally necessary. In the production of gratings the photographer must not be satisfied with merely keeping the film on the glass; the slightest tendency to looseness must be considered highly objectionable. The plates are coated with Mawson's collodion, excited in an ordinary silver-bath, washed first in distilled water and afterwards under the tap, and then immersed for a minute in a well-filtered 15-grain solution of tannin. On removal from the tannin, they are set up cornerwise on blotting-paper to drain and dry.

For the development of these plates I prefer a solution of gallic acid employed in a dish, though I have obtained very good results by the ordinary method with pyrogallic acid. Prepare the two following :—

(1)   Gallic acid   .   .   .   100 grains.
      Alcohol   .   .   .   . $2\frac{1}{2}$ oz.
      Distilled water .   .   . $2\frac{1}{2}$ oz.

(2)   Nitrate of silver   .   .   100 grains.
      Glacial acetic acid   .   .   2 oz.
      Distilled water .   .   .   16 oz.

The dish used for developing should be of glass, and is best cleaned with a little strong nitric acid, which may be used over and over again. The developing solution is prepared by mixing (1) and (2) in equal parts and diluting with water to half the strength. The alcohol helps to keep the film tight; and the development is well under control. In warm weather the operation may take an hour; but much depends upon the exposure, and still more upon the temperature. The proper point to which to carry the development can only be learned by experience; but the beginner is most likely to err on the side of excess, particularly if he uses pyrogallic acid. If, as is desirable, the film be creamy and thick, the spectra of a candle do not appear to advantage at this stage, in consequence of the unaltered iodide of silver. For fixing, "hypo" is the safest, though cyanide may be used if the film will bear it.

Tannin plates when finished are hardly secure without varnish; but there is considerable risk of spoiling gratings in the operation if an ordinary negative varnish be used. The crystal (benzole) varnish, which is applied cold, is much easier to use and gives adequate protection.

But the process which I am now most inclined to recommend is that introduced by Mr Wharton Simpson, and known as the collodio-chloride process. The collodion, which may be procured ready for use from Messrs Mawson and Swan, of Newcastle, consists of an emulsion of finely divided chloride of silver held in suspension by the dissolved gun-cotton, together with a carefully adjusted excess of free nitrate. After a time the chloride of silver is precipitated and the preparation becomes useless; but if properly mixed in the first instance, it will remain fit for use for weeks or even months. In the production of gratings the consumption is very small; so that, if required for this purpose alone, it is well to order it in small quantities.

In order to secure a proper adhesion, I have found a preliminary coating of albumen absolutely necessary. The white of an egg beaten up with a pint of distilled water gives a solution of sufficient strength. The plates, previously cleaned, are coated in any way that may be found convenient, and then set up on blotting-paper to drain and dry. The principal precautions necessary are to filter the albumen very carefully and to work in a room free from dust. It will generally be convenient to prepare a considerable number of plates at a time. Though of almost infinitesimal thickness, the film of albumen produces a very marked effect. Without the albumen the skin of collodion will usually come right away from the glass when washed under the tap ; with it the adhesion is remarkably good, and the film so tough as even to bear rubbing with the finger while wet.

The plate is coated with collodion in the ordinary way, and, after resting a few seconds, is dried by heat over a spirit-lamp or otherwise. After the plate has been made quite warm, it is put aside in the dark to cool and to absorb a certain amount of moisture from the atmosphere. This may take five or ten minutes. If the plate is used too soon the result is unsatisfactory; but, on the other hand, it will not do to leave it long enough to become sensibly moist. Something will probably depend on the particular sample of collodio-chloride.

The exposure required is about five or seven minutes to the autumn sun. On a hazy day something more may be required ; but if there are many clouds about, the experimenter will do well to postpone operations.

On removal from the frame, the plates may be placed in a dish of water until it is convenient to finish them. They are fixed, without any toning, in a dilute solution of hyposulphite of soda, such as is used for paper prints, and then carefully washed. The most effective washing is a combination of rinsing and soaking. My practice is to rinse the plates under the tap for half a minute in order to remove the greater part of the hyposulphite of soda, and then to allow them to soak for an hour or two in water changed two or three times. After a final rinsing the plates may be set up to dry.

Gratings finished in this way give excellent definition, but the spectra are rather deficient in brilliancy. This defect is of less importance than might be supposed ; for in order to see the finer fixed lines, sunlight is in any case indispensable, and with sunlight there is usually illumination to spare. Nevertheless, as gratings are likely to be largely used for the purpose of popular illustration under circumstances where artificial light must be employed, I am glad to be in a position to recommend a simple mode of treatment by means of which the brilliancy of the spectra may be materially enhanced. For this purpose it is only necessary to treat the fixed and washed impression with a solution of corrosive sublimate. When the whitening effect is complete, the plate must be again washed and then set up to dry. Considered as a photographic transparency, the grating is reduced rather than heightened in intensity by this process. The cause of the improvement of the spectra will be touched upon presently. These mercury-treated gratings cannot be varnished without sacrificing most of the advantage of the method. I have occasionally applied the same treatment successfully to tannin plates.

When not in use, the finished gratings should be kept in a dry place and protected from dust and other atmospheric deposit. For this purpose they may be put away wrapped in paper. For a short time there is no objection to leaving them standing, face inwards, against a wall; but a better plan is to place them, face downwards, on a flat and thoroughly clean piece of plate-glass.

The originals from which I have hitherto taken copies are three in number. Two are by Nobert, and contain respectively 3000 and 6000 lines, in each case covering a square inch (Paris). On a casual inspection the second, apart from the greater number of its lines, would be preferred as presenting a more even appearance. The 3000-line grating is divided into three parts, giving spectra of differing degrees of brightness, corresponding no doubt to a variation in the cut of the diamond or other stone employed, a peculiarity which is faithfully preserved in the copies. But on actual trial it is found that the spectra of the 3000-line grating are much the best in respect of definition; and the same difference is observed in the copies. The superior brilliancy of the closer-ruled grating is thus of little or no advantage for the investigation of the solar spectrum. In order to make good use of it, a higher degree of magnifying-power would be necessary than the definition of the spectra will bear.

The other original grating was engraved by Mr Rutherfurd, of New York, and was kindly lent me by Mr Browning; it contains 6000 lines to the inch. Owing to a change of residence, I have not hitherto had an opportunity of testing either the original or the copies on the solar spectrum; but I may observe that in respect of brightness they fall far short of Nobert's. This, as I have already remarked, is not always an objection; and the accuracy of division, on which definition depends, is said to be very superior*.

In testing gratings I prefer to work in a dark room. The slit is fastened in the window-shutter, outside which is placed the heliostat or *porte-lumière*. As slits are frequently required in optical experiments, and as usually made are rather expensive, I may be allowed to mention a very simple method by which serviceable slits may often be obtained. A piece of glass is covered with tinfoil, which must be made to adhere well; I have found a weak shellac-varnish a suitable cement. The alcohol is allowed to evaporate, and the thin layer of shellac softened by heat. In order to make a slit, it is only necessary to lay a straight-edge on the tinfoil and to draw a line with a sharp knife, afterwards wiping the line of the cut with a rag moistened with alcohol. The width and regularity of the slit may be judged of by holding it close to

---

* Draper, "On Diffraction-Spectrum Photography," *Phil. Mag.* Dec. 1873, p. 419.

the eye, and observing the appearance presented by a distant candle. The narrower the aperture the more dilated (in the direction of the width of the slit) the image will appear. Broader slits may be made by removing the foil between two parallel cuts.

At a distance of 12 feet or more from the shutter are placed the grating and the object-glass of the observing-telescope. In making the preliminary adjustments, it is convenient to use a slit so wide that the spectra and the light reflected from the grating can be seen on a screen. By the second the aspect of the plane of the grating can be judged of; and when the line of spectra is horizontal, it will be known that the lines of the grating are vertical and parallel to the slit. As object-glass, I am in the habit of using a single lens of about 24 inches solar focus. The eyepiece is a high-power achromatic, supplied by Mr Browning, and forms, with the object-glass, a telescope of much higher magnifying-power than is ordinarily used in spectroscopes. Without a high power it is impossible to bring out the full value of the grating. In order to obtain the best definition, it is necessary to adjust carefully the aspect of the object-glass; and I find that the best aspect is not always the same. It is possible that the performance of other optical instruments might occasionally be improved if means were provided for a slight alteration in the direction of the optic axis of one of the lenses employed. The grating itself I usually place approximately in the position of minimum deviation.

The copies on worked glass by the ordinary photographic processes and by the modification of the collodio-chloride last described rarely fail in definition. With the original (3000) grating, or with the copies, I can make out nearly, but not quite, all that is shown in Ångström's map. With this grating the third spectrum is generally the most serviceable. When the picked patent plate is employed, there will generally be a proportion whose performance is less satisfactory, though few which would not give very fair results when tested by a low power only. Some cannot be considered inferior to the worked glass, at least when the object-glass is specially adjusted for them. In many cases the definition may be considerably improved by the use of a diaphragm in the form of a *horizontal* slit, so placed that only the central parts of the lines of the grating are operative. In respect of brilliancy, gratings may be more quickly judged of; it is sufficient merely to examine the spectra of a candle placed in a dark room.

The lines themselves are of course too close to be seen without a microscope; but their presence may be detected, and even the interval between them measured, without optical aid, by a method not depending on the production of spectra or requiring a knowledge of the wave-length of light. If two photographic copies containing the same number of lines to the inch be placed in contact, film to film, in such a manner that the lines are nearly parallel in the two gratings, a system of parallel bars develops itself, whose direction bisects the external angle between the directions of the original lines, and whose distance increases as the angle of inclination diminishes. The cause of the phenomenon will be readily understood by drawing on paper two sets of equally distant and not too thin bars inclined at a small angle. Where the opaque and transparent parts severally overlap, the obstruction of light is, on the average, less than the double of that due to each set separately*, and consequently these places appear by comparison bright. The interval between the bars is evidently half the long diagonal of the rhombus formed by two pairs of consecutive lines, and is expressed by $a \cos \frac{1}{2} \theta \div \sin \theta$, or approximately $a \div \theta$, where a is the interval between the primary lines, and θ the mutual inclination of the two sets.

When parallelism is very closely approached, the bars become irregular, in consequence of the imperfection of the ruling. This phenomenon might perhaps be made useful as a test.

If the planes of the films be not quite parallel, bars parallel to the original lines may appear when the line of intersection of the planes is in the same direction. This arises from a fore-shortening of one of the sets, making it equivalent to a grating of a somewhat higher degree of fineness.

When examined under the microscope, the opaque bar on the copy, which corresponds to the shadow of the groove of the original, is seen to be composite, being not unfrequently traversed along its length by several fine lines of transparency. In one case, where the copy was on common glass, this effect went so far that at certain parts of the grating the periodicity was altered by each line splitting into two, the first spectrum altogether disappearing. In order to make this observation, the eye should be placed at the point where the pure spectra are formed and be focussed on the grating. The places in question will then appear as irregular dark bands.

The disappearance of the first spectrum is very unusual; but it is common for bands to appear when the eye is adjusted to the place of the fourth and higher spectra. When the order is high, the bands will not be black, but coloured with light belonging to one of the other spectra. There is no difficulty in understanding how this occurs. In the process of copying, the groove of the original is widened into a bar, whose width depends on the closeness of contact, an element which necessarily varies at different parts of the plate. The dark bands are the locus of points at which the relation of the alternate parts is such as to destroy the spectrum in question.

I have not had an opportunity of trying the method of copying on lines closer than 6000 to the inch; but I have no doubt that the limit of fineness was not attained. I should expect to find no difficulty with lines 10,000 or

---

* The mathematical reader will easily prove this from the law of absorption.

12,000 to the inch; but beyond that point it is possible that the method would fail, or require special precautions, such as the use of extra-flat glass and greater pressure to ensure close contact in the printing. For preliminary experiments I should be inclined to try mica as a support, whose flexibility would facilitate a close contact. I may mention that I have done copies of the 3000-line grating on sheets of mica, such as may be obtained very thin and smooth from the photographic dealers. For more convenient manipulation in the preliminary stages of preparation, the mica should be mounted on a sheet of glass of the same size as itself. A small drop of water interposed will ensure a sufficiently close adhesion.

I have tried to take copies of copies, but with indifferent success, even when the performance of the first was not perceptibly inferior to that of the original.

Gratings may be copied without the aid of photography by simply taking a *cast*. Following Brewster, I have obtained a fair result by allowing filtered gelatine to dry after being poured on the 3000 Nobert. This method, however, is attended with much more risk to the original, and is besides open to other objections, sufficient, I think, to prevent its competing with photography.

The remainder of this paper is principally occupied with theoretical considerations relating to the performance of gratings considered as light-analyzing apparatus. The more popular works on the theory of light give only the main outlines of the subject, and pass over almost in silence the important questions of illumination and definition. On the other hand, the mathematical treatises, such as Airy's "Tracts" and Verdet's *Leçons*, though they give analytical results involving most of the required information, are occupied rather with explaining the production of spectra as a diffraction-phenomenon than with investigating on what conditions their perfection depends. On examining the question for myself, I came to the conclusion that the theory of gratings, as usually presented, is encumbered with a good deal that may properly be regarded as extraneous.

One of the first things to be noticed is the extraordinary precision required in the ruling. The difference of wave-length of the two sodium-lines is about a thousandth part. If, therefore, we suppose that one grating has 1000 lines in the space where another has 1001, it is evident that the first grating would produce the same deviation for the less-refrangible D line that the second would produce for the more-refrangible D line. We have only to suppose the two combined into one in order to see that, in a grating required to resolve the D line, there must be no systematic irregularity to the extent of a thousandth part of the small interval. Single lines may, of course, be out of position to a much larger amount. It is easy to see, too, that the same accuracy is required, whatever be the order of the spectrum examined.

The precision of ruling actually attained in gratings is very great. In the 3000 Nobert it is certain that the average interval between the lines does not vary by a six-thousandth part in passing from one half of the grating to the other; for the D lines, when well defined, do not appear so broad as a sixth part of the space separating them.

In considering the influence of the number of lines (n) and the order of the spectrum (m), we will suppose that the ruling is accurate, and that plane waves are incident perpendicularly upon the face of the grating whose width is represented in the figure by AB. But inasmuch as a large part of the phenomenon covered by the usual mathematical investigation depends upon the *limitation* of the grating at A and B, we 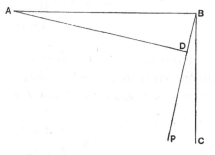 shall find it convenient to take first the simple case of an aperture represented by AB, and afterwards to consider the influence of the ruling.

In the perpendicular direction BC all the secondary waves emanating from AB are in complete agreement of phase, and their resultant accordingly attains its highest possible value. In a direction BP, making with BC a very small angle, the agreement of phase will be disturbed. If BP be so drawn that the projection of AB upon it is equal to λ, the phases of the secondary waves will be distributed uniformly over a complete period, and the resultant will therefore be *nil*. The same result must ensue whenever BD is an exact multiple of λ.

For the intermediate directions we require a little more calculation.

The phase of the resultant will always correspond with that of the secondary wave which issues from the middle of the aperture. If x denotes the retardation of any element with respect to this one, the amplitude of the resultant is given by

$$\int_{-\frac{1}{2}R}^{+\frac{1}{2}R} \cos x \, dx \div R,$$

where R is the relative retardation of the extreme parts A and B, or, on integration,

$$\frac{\sin \frac{1}{2}R}{\frac{1}{2}R}.$$

This expression gives the magnitude of the resultant amplitude compared with that in the principal direction BC, where all the components agree in phase.

The composition of elementary vibrations whose phases vary uniformly within certain limits may be illustrated by a mechanical analogy. Each

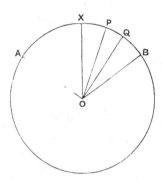

elementary vibration is represented by a force proportional to the element of circular arc PQ and acting at O along a direction OP, making with a fixed line of reference OX an angle corresponding to the phase of the vibration. The force may be supposed to be due to the attraction of the arc on a particle placed at O. The group of vibrations is thus represented by the group of forces whose directions are distributed uniformly through the angle AOB; and the resultant of the forces, found by resolving in the ordinary way, represents on the same system the resultant of the vibrations. In the present case AOB corresponds to R, and the integrated expression $\sin \frac{1}{2}R \div \frac{1}{2}R$ denotes the ratio of the resultant force to the aggregate of its components calculated without allowance for the difference of direction—that is, as if the whole attracting mass were concentrated at X.

According to what has been already explained, $(\sin \frac{1}{2}R \div \frac{1}{2}R)^2$ vanishes when $\frac{1}{2}R$ is an even multiple of $\frac{1}{2}\pi$. The positions of the maxima (determined by $\tan \frac{1}{2}R = \frac{1}{2}R$) do not exactly bisect the angles between the vanishing directions; but it will be sufficient for our present purpose to note that the principal maximum ($R = 0$) is unity, and that the others do not differ greatly from $(2/3\pi)^2$, $(2/5\pi)^2$, $(2/7\pi)^2$, &c. It is evident that on either side of the principal direction the illumination falls off with great rapidity. If AB is 1 (inch) and $\lambda = \frac{1}{40,000}$, the angle CBP corresponding to the first minimum is only about $5''$.

The image of an infinitely narrow line of light (whose length is perpendicular to the plane of the diagram) as formed by an object-glass with aperture AB, is thus a series of parallel stripes, composed of a central narrow band whose illumination varies from a maximum in the middle to zero at the edges, enclosed by parallel bands of rapidly decreasing illumination. We have now to examine the effect of the ruling.

For the sake of simplicity, we will take first the case of a grating composed of transparent bars of width a, alternating with opaque bars of width d, and consider the central image or spectrum of zero order. In the principal direction, BC, the secondary waves are, as before, in complete agreement, but the amplitude is diminished by the ruling in the ratio $a : a + d$. In another direction, making a small angle with BC such that the relative retardation of A and B amounts to a few wave-lengths, it is easy to see that the mode of interference is the same as if there were no ruling. For example, when the direction is such that the projection of AB upon it amounts to one wave-length, the elementary components neutralize one another, because their phases are on the whole distributed symmetrically, though discontinuously, round the entire circumference. The only effect of the ruling is to diminish the amplitude in the ratio $a : a + d$; and except for the difference in illumination, the appearance of a line of light is exactly the same as if the aperture were perfectly free.

The lateral images occur in such directions that the projection of the element $a + d$ of the grating upon them is an exact multiple of λ. The effect of each element of the grating is then the same; and unless this vanishes on account of a particular adjustment of the ratio $a : d$, the resultant amplitude becomes comparatively very great. These directions, in which the retardation between A and B is exactly $mn\lambda$, may be called the principal directions. On either side of any one of them the illumination is distributed according to the same law as for the central image ($m = 0$), vanishing, for example, when the retardation amounts to $(mn \pm 1)\lambda$. In considering the relative brightness of the different spectra, it is therefore sufficient to attend merely to the principal directions, provided that the whole deviation be not so great that its cosine differs considerably from unity*.

Under the restriction just stated, the intensity of the secondary waves may be supposed not to be diminished by the obliquity; and thus we obtain for the ratio of brightnesses :—

$$B_m : B_0 = \left[\int_{-\frac{am\pi}{a+d}}^{+\frac{am\pi}{a+d}} \cos x \, dx \div \frac{2am\pi}{a+d} \right]^2 = \left(\frac{a+d}{am\pi} \right)^2 \sin^2 \frac{am\pi}{a+d} ;$$

where B_m denotes the brightness of the mth spectrum, and B_0 of the central image.

If B denotes the brightness of the central image when the whole of the space occupied by the grating is transparent, we have

$$B_0 : B = a^2 : (a+d)^2,$$

and thus

$$B_m : B = \frac{1}{m^2\pi^2} \sin^2 \frac{am\pi}{a+d}.$$

The sine of an angle can never be greater than unity; and consequently under the most favourable circumstances only $1/m^2\pi^2$ of the original light can be obtained in the mth spectrum. We conclude that, with a grating composed of transparent and opaque parts, the utmost light obtainable in any one spectrum is in the first (the central image not being included), and there amounts to $1/\pi^2$, or about $\frac{1}{10}$, and that for this purpose a and d must be equal.

---

* This point is perhaps made clearer by supposing the original light to be always incident at such an angle that the diffracted spectrum under consideration occurs in the normal direction.

When $d = a$, the general formula becomes

$$B_m : B = \frac{1}{m^2\pi^2} \sin^2(\tfrac{1}{2}m\pi),$$

showing that, when m is even, B_m vanishes, and that, when m is odd,

$$B_m : B = \frac{1}{m^2\pi^2}.$$

The third spectrum has thus only one-ninth of the brilliancy of the first.

It is here supposed that the light is homogeneous; but if it be composed of elements of continuously varying wave-length, another factor, $1/m$, must be introduced, due to the varying elongation of the spectra. The deficiency of light accompanying an increased dispersion, however, is nothing peculiar to gratings, and may be met by a proportional widening of the slit, without lowering the original standard of purity.

Another particular case of interest is obtained by considering a small relatively to $a + d$. Unless the spectrum be of high order, we have simply

$$B_m : B = \left(\frac{a}{a+d}\right)^2,$$

so that the brightness of all the (lateral) spectra is the same. According to this, the spectra of a sodium-flame observed through a grating should be of equal brilliancy when a is small relatively to $a + d$. In the grating with 3000 lines this condition is fulfilled; but the theoretical result is contradicted by observation, the second and third lateral spectra being much brighter than the first. I am not in a position to explain the discrepancy, which I only noticed while drawing up this paper for the press. Unless due to some mathematical blunder, the cause would appear to be deep-seated. The effect of an insufficient narrowness of grooves would be in the opposite direction.

Our expressions have been obtained without taking into account the reflection from the face of the grating; and therefore the light stopped by the opaque parts, together with that distributed in the central image and lateral spectra, ought to make up the original brightness. Thus, if $a = d$, we ought to have

$$\frac{1}{2} = \frac{1}{4} + \frac{2}{\pi^2}\left(1 + \frac{1}{9} + \frac{1}{25} + \dots\right),$$

which is true by a known theorem. In the general case,

$$\frac{a}{a+d} = \left(\frac{a}{a+d}\right)^2 + \frac{2}{\pi^2}\sum_{m=1}^{m=\infty}\frac{1}{m^2}\sin^2\left(\frac{m\pi a}{a+d}\right),$$

a result which may be verified by Fourier's theorem.

By Babinet's principle it is a matter of indifference, so far as the brightness of the lateral spectra are concerned, whether a represent the width of the transparent or of the opaque part, inasmuch as the secondary waves from the transparent and opaque parts together would give a zero resultant. The same conclusion may be derived from the expression for the ratio $B_m : B$.

From the value of $B_m : B_0$ we see that no lateral spectrum can surpass the central image in brightness; but this result depends upon the hypothesis that the lines of the grating act by opacity, which is generally very far from being true. In an engraved glass grating or in a gelatine copy there is no opaque material present by which light could be absorbed, and the effect depends on a difference of retardation due to the alternate parts. It is remarkable that this point is never alluded to in the ordinary treatises on optics, and, so far as I know, was first noticed by Quincke (*Pogg. Ann.* vol. CXXXII. p. 321, 1867), who made a theoretical and experimental examination of the phenomena presented when light is diffracted at the edge of a transparent obstacle. My attention was first drawn to it, before I was acquainted with Quincke's work, by observing that, contrary to my anticipations, it was possible for the lateral spectra of a soda-flame to exceed the central image in brilliancy. When once the question is raised, the explanation is easy enough; for if the grating were composed of equal alternate parts, both alike transparent but giving a relative retardation of half a wavelength, it is evident that the central image would be entirely extinguished, while the first spectrum would be four times as bright as if the alternate parts were opaque. If it were possible to introduce at every part of the aperture of the grating an arbitrary retardation, *all* the light might be concentrated in any desired spectrum. By supposing the retardation to vary uniformly and continuously, we fall on the case of the ordinary prism; but there would then be no analysis of light, except such as depends on the variation of retardation with wave-length. To obtain a diffraction-spectrum in the ordinary sense containing all the light, it would be necessary that the retardation should gradually alter by a wave-length in passing over each element of the grating and then fall back to its previous value, thus springing suddenly over a wave-length. It is not likely that such a result will ever be obtained in practice; but the case was worth stating, in order to show that there is no theoretical limit to the concentration of light of assigned wave-length in one spectrum, and as illustrating the frequently observed unsymmetrical character of the spectra on either side of the central image.

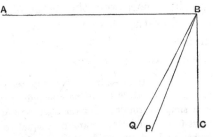

We have now to consider the dependence of the *resolving-power* of a grating on the number of its lines (n) and the order of the spectrum observed (m).

Let BP be the direction of the principal maximum for the wavelength λ in the mth spectrum; then the projection of AB on BP is $mn\lambda$

[representing the relative retardation of the extreme rays which pass through A and B]. If BQ be the direction corresponding to the first minimum, the projection of AB on BQ is $(mn + 1)\lambda$. Suppose now that $\lambda + \delta\lambda$ is the wavelength for which BQ gives the principal maximum, then

$$(mn + 1)\lambda = mn(\lambda + \delta\lambda);$$

whence

$$\frac{\delta\lambda}{\lambda} = \frac{1}{mn},$$

which shows that the resolving-power varies directly as m and n^*.

It is not possible to say precisely under what circumstances a double line would appear to be resolved—something no doubt would depend on the intensity of the light—but it seems probable that there would be no distinct resolution when the two images are separated by only half the width of the central band of either. If this be so, we may take $\lambda/m\delta\lambda$ as the least number of grooves capable of resolving in the mth spectrum a double line whose wave-lengths are λ and $\lambda + \delta\lambda$. In the case of the soda-lines $\delta\lambda/\lambda$ is about $\frac{1}{1000}$; and therefore to resolve them in the first spectrum would require 1000 grooves, in the second spectrum 500, and so on. It is evident that if the ruling be perfectly accurate and the illumination sufficient, the work may be accomplished with comparatively few lines by using a spectrum of elevated order.

The result of an attempt to determine experimentally the number of lines necessary to resolve D in the solar spectrum may here be recorded. The 3000-to-the-inch Nobert was used, its horizontal aperture being limited by the jaws of an adjustable slit. From the width of the slit found to be necessary the number of lines in operation was calculated.

In the first spectrum 1200 lines were required, in the second 630, and in the third 375. These numbers should be in the ratio $6 : 3 : 2$; but the last, which presents the greatest deviation, was difficult of exact measurement.

The number of lines necessary in the first spectrum is .very much what might have been expected. The effect of a limitation of the aperture of the grating by a slit whose length is *horizontal*, is, of course, quite different. As the slit is narrowed, the image of a *point* would be dilated in a vertical direction; but this is of no moment when the subject of observation is itself a uniform vertical line. As was mentioned in the first part, the definition is often materially improved.

* [1899. If we define as the *dispersion* in a particular part of the spectrum the ratio of the angular interval $d\theta$ to the corresponding increment of wave-length $d\lambda$, we may express it by a very simple formula. For the alteration of wave-length entails, at the two limits of a diffracted wave-front, a relative retardation equal to $mn\,d\lambda$. Hence, if a be the width of the diffracted beam, and $d\theta$ the angle through which the wave-front is turned, $a\,d\theta = mn\,d\lambda$, or

$$\text{dispersion} = d\theta/d\lambda = mn/a.]$$

Since a grating resolves in proportion to the total number of its grooves, it might be supposed that the defining-power depends on different principles in the case of gratings and prisms; but the distinction is not fundamental. The limit to definition arises in both cases from the impossibility of representing a line of light otherwise than by a band of finite though narrow width, the width in both cases depending only on the horizontal aperture (for a given λ). If a grating and a prism have the same horizontal aperture and dispersion, they will have equal resolving-powers on the spectrum; the greater dispersion is the only cause of the superiority of the diffraction-spectra of high order.

In estimating the value of light-analyzing apparatus, there are three things to be considered—the brightness, the purity, and the apparent magnitude of the resulting spectrum. In the case of a prism, where the loss of light by reflection and absorption may in a rough approximation be neglected, the first two characteristics are inseparably connected, whether a telescope be used or not, so long as the pupil of the eye is filled with light. In whatever degree the purity be enhanced, whether by increasing the dispersion or narrowing the slit, in the same degree must the brightness suffer. The angular magnitude of the spectrum is merely a question of magnifying-power. No matter how small the dispersion may be, the spectrum may yet be made to appear as long as we please by sufficiently increasing the focal length of the object-glass and the power of the eyepiece. But if the brilliancy is not to suffer, the size of the prism and the aperture of the telescope must be proportionally increased; for otherwise the condition will be violated of keeping the pupil filled with light. There are thus two ways of obtaining a powerful spectroscope. The first is to procure a great dispersion by multiplication of prisms; the second is to be satisfied with a small dispersion, and attain the necessary length of spectrum by a high-power telescope, which may involve a larger aperture. It may be questioned whether the second method has received as much attention as it deserves. When there is light to spare, a higher power than usual may often be employed advantageously without an augmentation of aperture.

In the case of gratings the question is complicated by the choice of spectra; but some remarks may perhaps be useful. Much misapprehension appears to exist as to the nature of the advantage derived from close ruling. It is generally supposed that the closer the ruling the greater the resolving-power of the instrument; but this is for the most part a mistake. When there is no limitation on the order of the spectrum to be observed, resolving-power depends not on the *closeness* but on the *accuracy* of the ruling. Let us take the case of a grating with 3000 lines in an inch, and consider the effect of interpolating an additional 3000 lines. The effect of the addition will be to destroy by interference the first, third, and the odd spectra generally; while the advantage gained in the spectra of even order is not in dispersion, nor in resolving-power, but simply in brilliancy, which is increased to four times. If we now suppose half the grating to be cut away, so as to leave 3000 lines in half an inch, the dispersion will evidently not be altered, while the brightness and resolving-power are halved. If, therefore, resolving-power be our object, we should aim at covering a considerable breadth with very accurately placed lines, rather than at extreme closeness of the lines themselves. On the other hand, for experiment on dark heat, or whenever a narrow slit is not available, resolving-power is of less importance, and the best grating will be one that covers the largest space with the finest lines.

I have already mentioned that my 6000-to-the-inch Nobert defines not only not better, but decidedly worse, than the one with 3000 lines in the inch. This inferior definition is probably an accident; for there seems to be no theoretical reason for it. In brightness the closer-ruled grating has greatly the advantage.

The preceding investigations are founded on the principles ordinarily adopted in explaining diffraction-phenomena, and not on a strict dynamical theory. In the present state of our knowledge with respect to the nature of light and its relations to ponderable matter, vagueness in the fundamental hypotheses is rather an advantage than otherwise; a precise theory is almost sure to be wrong. Nevertheless it is often instructive to examine optical questions from a more special point of view; and therefore I hope that an investigation of an ideal grating on dynamical principles will not be out of place, though not very closely connected with the preceding portion of the paper.

In actual gratings the lines or grooves occur at the boundary of two media of different refrangibilities; but, for the sake of simplicity, we shall here suppose the medium on both sides to be the same. The grating will thus consist simply of bars (infinitely long) whose optical properties differ from those of the rest of the medium; and we further suppose

(1) that the variation of optical properties depends upon a difference of inertia, and is small in amount;

(2) that the diameter of the bars is very small in relation to the wave-length of light.

The supposition that refraction depends upon a difference of inertia is that of Fresnel and Green, and has been shown by the latter to lead to Fresnel's laws. In several papers in this magazine*, I have shown that, if the analogy with an elastic solid holds good at all, no other supposition is reconcilable with the facts of the reflection of light from surfaces and its

* *Phil. Mag.* 1871. [Arts. VIII, IX, XI.]

diffraction by small particles. Whether true or not, it is at any rate mechanically possible.

Since the bars are very small, the effect of each is quite independent of the rest; and so the dynamical theory need only concern itself with one. In my paper "On the Light from the Sky*," I proved that the effect of a body in disturbing the waves of light incident upon it may be calculated by ordinary integration from those of its parts, provided that the square of the alteration of mechanical properties may be neglected. This proposition, though true as stated, requires some caution in use, and is practically inapplicable when the body is elongated in the direction of original propagation, because the dimension of the body in this direction divided by λ may occur as a factor in the terms omitted. In the present case, however, where the light is incident normally to the plane of the grating, this difficulty does not arise.

Let the bar under consideration be taken for axis of z, and let the axis of x be parallel to the direction of propagation of the original light. The original vibration is thus, according to the polarization, parallel to either z or y. We will take first the former case, where the disturbance due to the bar must be symmetrical in all directions round OZ, and parallel to it.

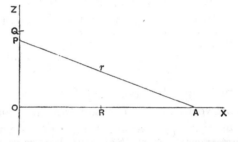

The element of the disturbance at A due to PQ (dz) will be proportional to dz in amplitude, and will be retarded in phase by an amount corresponding to the distance r. In calculating the effect of the whole bar, we have to consider the integral

$$\int_0^\infty \frac{dz}{r} \cos \frac{2\pi}{\lambda} (bt - r) = \int_R^x \frac{dr \cos \dfrac{2\pi}{\lambda} (bt - r)}{\sqrt{r^2 - R^2}}.$$

Now the denominator $= \sqrt{r - R}\,\sqrt{r + R}$, a product of which the second factor may be treated as constant in the integration in view of the fact that the parts for which r differs much from R destroy each other's effects. After this simplification the integral may be evaluated by means of the formula

$$\int_0^\infty \frac{\sin x}{\sqrt{x}} \, dx = \int_0^x \frac{\cos x}{\sqrt{x}} \, dx = \sqrt{\frac{\pi}{2}}.$$

The result is

$$\int_0^\infty \frac{dz}{r} \cos \frac{2\pi}{\lambda} (bt - r) = \frac{\sqrt{\lambda}}{2\sqrt{R}} \cos \frac{2\pi}{\lambda} \left(bt - R - \frac{\lambda}{8} \right),$$

showing that the total effect is retarded $\frac{1}{8}\lambda$ behind that due to the element at O. This result is analogous to, though different from, that of the ordinary integration by Huyghens's zones. In that case the effect of each zone is very nearly the same, and therefore the whole is the half of that of the first zone. If the first zone be divided into rings by circles drawn so that r increases in arithmetical progression, the rings will be of equal area, and therefore the phase of the resultant vibration will be halfway between that corresponding to the centre and circumference—that is, will be retarded relatively to the centre by one *fourth* of λ. In the present case, if the bar be divided on the same principle so that each piece gives a result retarded $\frac{1}{2}\lambda$ behind its predecessor, the lengths will rapidly diminish from the centre outwards, and therefore the same argument does not apply. The retardation of the resultant relatively to the central element is less than before, on account of the preponderance of the central parts.

By the result investigated in my paper previously referred to, if T be the volume of the element PQ, D and D' the original and altered densities, the disturbance at A due to the element is

$$\frac{D' - D}{D} \frac{\pi \cdot T}{r\lambda^2} \sin\alpha \, \cos\frac{2\pi}{\lambda}(bt - r)^*,$$

the original vibration at PQ being denoted by $\cos(2\pi bt/\lambda)$. α is the angle between the ray PA and the direction of original vibration OZ; but in the present application we may put $\sin\alpha = 1$, since only the central parts are really operative. If we replace T by $A\,dz$, A being the sectional area of the bar, and use the integral just investigated, we find for the effect of the whole bar

$$\frac{D' - D}{D} \frac{A\pi}{\lambda^2} \sqrt{\frac{\lambda}{R}} \cos\frac{2\pi}{\lambda}\left(bt - R - \frac{\lambda}{8}\right),$$

corresponding to $\cos(2\pi bt/\lambda)$ for the incident light.

When the original vibration is parallel to y, the disturbance due to the bar will no longer be symmetrical round OZ. If α be the angle between Ox and the direction of the scattered ray, it is only necessary to introduce the factor $\cos\alpha$ in order to make the preceding expression applicable.

If the direction of the original vibration be inclined at an angle θ to OZ, and that of the scattered ray at an angle ϕ, we have on resolution

$$\tan\phi = \cos\alpha \tan\theta,$$

which expresses the important law enunciated by Stokes†.

---

* The factor π was inadvertently omitted in the original memoir.
† "Dynamical Theory of Diffraction," *Camb. Trans.* vol. IX.

The preceding investigation depends upon the smallness of $D' - D$ as well as of A; but where the original vibrations are parallel to the bar, the result is correct to all powers of $D' - D$. I find that, if the bar be circular and composed of material for which the density is D' and the rigidity n' (the corresponding quantities for the rest of the medium being D and n), the expression for the scattered vibration is

$$\gamma = \frac{2\pi A}{\lambda^{\frac{3}{2}} R^{\frac{1}{2}}} \left\{ \frac{D' - D}{2D} - \frac{n' - n}{n' + n} \cos \alpha \right\} \times \cos \frac{2\pi}{\lambda} \left(bt - R - \frac{\lambda}{8} \right),$$

corresponding to

$$\gamma = \cos \frac{2\pi}{\lambda} bt$$

for the incident light at the centre of the bar*. If we suppose that $n' = n$, this agrees with the result already found; and it is correct if the bar be small enough, whatever may be the magnitude of $(D' - D)/D$.

[For further work upon this subject see *Phil. Mag.* vol. XI. p. 196, 1881.]

* [An analogous investigation is given in *Theory of Sound*, § 343.]

31.

INSECTS AND THE COLOURS OF FLOWERS.

[*Nature*, XI. p. 6, 1874.]

THERE is one point connected with Mr Darwin's explanation of the bright colours of flowers which I have never seen referred to. The assumed attractiveness of bright colours to insects would appear to involve the supposition that the colour-vision of insects is approximately the same as our own. Surely this is a good deal to take for granted, when it is known that even among ourselves colour-vision varies greatly, and that no inconsiderable number of persons exist to whom, for example, the red of the scarlet geranium is no bright colour at all, but almost a match with the leaves.

32.

A STATICAL THEOREM.

[*Phil. Mag.* XLVIII. pp. 452—456, 1874; XLIX. pp. 183—185, 1875.]

IN a paper "On some General Theorems relating to Vibrations," published in the *Mathematical Society's Proceedings* for 1873 [Art. XXI], I proved a very general reciprocal property of systems capable of vibrating, with or without dissipation, about a position of stable equilibrium. The principle may be shortly, though rather imperfectly, stated thus:—If a periodic force of harmonic type and of given amplitude and period act upon the system at the point P, the resulting displacement at a second point Q will be the same both in amplitude and phase as it would be at the point P were the force to act at Q.

If we suppose the period of the force to be very great, the effects both of dissipation and inertia will ultimately disappear, and the system will be in a condition of what is called moveable equilibrium; that is to say, it will be found at any moment in that configuration in which it would be maintained at rest by the then acting forces, supposed to remain unaltered. The statical theorem to which the general principle then reduces is so extremely simple that it can hardly be supposed to be altogether new; nevertheless it is not to be found in any of the works on mechanics to which I have access, and was not known to the physicists to whom I have mentioned it. In any case, I think, two or three pages may not improperly be devoted to the consideration of it.

Let the system be referred to the independent coordinates ψ_1, ψ_2, &c., reckoned in each case from the configuration of equilibrium. Since only small displacements are contemplated, ψ_1, &c. are small quantities whose squares are to be neglected. Then, if Ψ_1, Ψ_2, &c. are the impressed forces of the corresponding types, the equations of equilibrium are of the form

$$
\left.
\begin{aligned}
a_{11}\psi_1 + a_{12}\psi_2 + a_{13}\psi_3 + \ldots &= \Psi_1, \\
a_{21}\psi_1 + a_{22}\psi_2 + a_{23}\psi_3 + \ldots &= \Psi_2, \\
a_{31}\psi_1 + a_{32}\psi_2 + a_{33}\psi_3 + \ldots &= \Psi_3,
\end{aligned}
\right\} \ldots\ldots\ldots\ldots\ldots(A)
$$

being the same in number as the degrees of freedom. It may be observed that forces of a constant character need not be included in Ψ_1, &c.; for the effect of such is only to alter the configuration of equilibrium, and they may be supposed to be already accounted for in the estimation of that configuration.

If the system be conservative, as is here supposed, the coefficients in the equations (A) are not all independent; for in order that an energy-function may exist, any coefficient such as a_{rs} must be equal to the corresponding a_{sr}.

The solution of equations (A) is

$$
\left.
\begin{aligned}
\nabla\psi_1 &= A_{11}\Psi_1 + A_{12}\Psi_2 + \cdots, \\
\nabla\psi_2 &= A_{21}\Psi_1 + A_{22}\Psi_2 + \cdots, \\
\nabla\psi_3 &= A_{31}\Psi_1 + A_{32}\Psi_2 + \cdots,
\end{aligned}
\right\} \quad\cdots\cdots\cdots\cdots\cdots\cdots(B)
$$

where ∇ denotes the determinant

$$
\begin{vmatrix}
a_{11} & a_{12} & a_{13} \cdots \\
a_{21} & a_{22} & a_{23} \cdots \\
a_{31} & a_{32} & a_{33} \cdots \\
 & \cdots & \cdots
\end{vmatrix}
$$

and

$$
A_{11} = \frac{d\nabla}{da_{11}}, \qquad A_{12} = \frac{d\nabla}{da_{12}}, \quad \&c.
$$

in which, therefore, by a property of determinants,

$$
A_{rs} = A_{sr}.
$$

In the application that we are about to make it will be supposed that all the forces but two vanish, for example that Ψ_3, Ψ_4, &c. vanish. Under these circumstances we obtain from (B)

$$
\nabla\psi_1 = A_{11}\Psi_1 + A_{12}\Psi_2, \qquad \nabla\psi_2 = A_{21}\Psi_1 + A_{22}\Psi_2, \cdots\cdots\cdots\cdots(C)
$$

equations which determine the displacements ψ_1, ψ_2 when the forces are given. The consequences which follow from the fact that $A_{12} = A_{21}$ may be exhibited in three ways.

First Proposition.—Suppose $\Psi_2 = 0$. From the second equation, we see that $\nabla\psi_2 = A_{21}\Psi_1$. Similarly if we had supposed $\Psi_1 = 0$, we should get $\nabla\psi_1 = A_{12}\Psi_2$, *showing that the relation of ψ_1 to Ψ_2 in the second case is the same as the relation of ψ_2 to Ψ_1 in the first.*

In order to fix our ideas we will take the case of a rod, not necessarily uniform, supported in any manner in a horizontal position—for example, with one end clamped and the other free. Then, if P and Q be any two points of its length, we assert that a pound weight hung on at P will give the same linear deflection at Q as is observed at P when the weight is hung at Q; and the only thing on which our conclusion depends is the proportionality of strains and stresses. If we take angular instead of linear displacements, the theorem will run:—A given couple at P will produce the same rotation at Q as the couple at Q would give at P. Or if one displacement be linear and the other angular, the result may be stated thus:—A couple at P would do as much work in acting over the rotation at P due to a simple force at Q, as the force at Q would do in acting over the linear displacement at Q due to the couple at P. In the last case the statement is more complicated, since the forces, being of different kinds, cannot be made equal.

Second Proposition.—Suppose that $\psi_1 = 0$. Then, from (C),

$$A_{12}\nabla\psi_2 = (A_{12}{}^2 - A_{11}A_{22})\,\Psi_1.$$

From this we conclude that *if ψ_2 is given, it requires the same force Ψ_1 to keep $\psi_1 = 0$, as would be required in Ψ_2 to keep $\psi_2 = 0$, if ψ_1 had the given value.*

Thus, if the rod be supported at P so that that point cannot fall, while Q is depressed one inch by a force there acting, the reaction on the support at P is the same as it would have been on a support at Q if P had been depressed one inch*.

Third Proposition.—Suppose, first, that $\Psi_1 = 0$. Then, from (C),

$$\psi_1 : \psi_2 = A_{12} : A_{22}.$$

Secondly, suppose $\psi_2 = 0$. Then

$$\Psi_2 : \Psi_1 = - A_{12} : A_{22}.$$

Thus, when Ψ_2 alone acts, the ratio of displacements $\psi_1 : \psi_2$ is the negative of the ratio of the forces $\Psi_2 : \Psi_1$ necessary to keep $\psi_2 = 0$.

If the rod is supported at P and bent by a force acting downwards at Q, the reaction bears the same ratio to the force as the displacement at Q would bear to the displacement at P when the unsupported rod is bent by a force applied at P.

In this proposition the interchange of P and Q gives a different, though of course an equally true statement. The first two propositions are themselves reciprocal in form.

The second and third propositions, as well as the first, admit of the extension to the vibrations of systems subject to inertia and dissipation; but I do not here pursue this part of the subject.

---

* The verification of these results with rods variously supported, or more complicated structures, gives a very good experimental exercise.

Our fundamental equations (C) may be arrived at with less analysis and perhaps equal rigour by a somewhat modified process. The conditions that the forces Ψ_3, Ψ_4, &c. vanish, impose linear relations on the coordinates, and virtually reduce the degrees of freedom enjoyed by the system to two. But for only two independent coordinates we have at once

$$\Psi_1 = b_{11}\psi_1 + b_{12}\psi_2, \qquad \Psi_2 = b_{21}\psi_1 + b_{22}\psi_2, \dots\dots\dots\dots\dots(D)$$

where the coefficients b_{12}, b_{21} are equal. The equality of the coefficients b_{12}, b_{21} is a consequence of the existence of an energy-function, or may be proved *de novo* by taking the system round the cycle of configurations represented by the square whose angular points are

$$\left.\begin{matrix}\psi_1 = 0 \\ \psi_2 = 0\end{matrix}\right\} \qquad \left.\begin{matrix}\psi_1 = 0 \\ \psi_2 = 1\end{matrix}\right\} \qquad \left.\begin{matrix}\psi_1 = 1 \\ \psi_2 = 1\end{matrix}\right\} \qquad \left.\begin{matrix}\psi_1 = 1 \\ \psi_2 = 0\end{matrix}\right\}.$$

From (D) we may deduce the three propositions directly, or mediately with the aid of (C), which is merely the algebraic solution of (D).

Finally, I would remark that essentially the same method, though with a somewhat different interpretation, is applicable to systems other than those contemplated in the preceding demonstrations. In thermodynamics the condition of a body is regarded as depending on two independent coordinates such as the temperature and volume; and by the principles of that subject it is known that a function of that condition exists, representing the work that can be got out of the system in reducing it to a standard condition of volume and temperature, any communication or abstraction of heat being made at the standard temperature. The simplest course that can be taken is along an adiabatic up to the standard temperature, and then along the isothermal until the standard volume is attained. If the actual condition of the body be defined by $v + dv$, $t + dt$, while the standard condition corresponds to v, t, we have for the available energy, or entropy ($d\epsilon$),

$$2d\epsilon = dp\, dv + d\phi\, dt,$$

where dp is the variation of pressure, and $d\phi$ the variation of the thermodynamic function.

In this equation dv, dt correspond to ψ_1, ψ_2; dp, $d\phi$ to Ψ_1, Ψ_2; and $d\epsilon$ corresponds to the potential energy of the purely mechanical system. Our first proposition shows that, if $d\phi = 0$, dt/dp has the same value as that of $dv/d\phi$ when there is no variation of pressure, the interpretation of which is that the heat (measured as work) necessary to increase the volume by unity at constant pressure, is numerically equal to the product of the absolute temperature into the increase of pressure required to raise the temperature one degree when no heat is allowed to escape. (See Maxwell's *Heat*, p. 167.) In like manner the other propositions may be interpreted.

Since the publication of the paper in the December Number of the *Philosophical Magazine*, entitled "A Statical Theorem," I have made some tolerably careful experimental measurements in illustration of one of the results there given, which are perhaps worth recording. The "system" consisted of a strip of plate-glass 2 feet long, 1 inch broad, and about $\frac{1}{10}$ inch in thickness, supported horizontally at its ends on two very narrow ledges. In the first experiment two points, A and B, were marked upon it, A near the centre, and B about 5 inches therefrom, for which the truth of the theorem was to be tested. When a weight W is suspended at A, the deflection in a vertical direction at B should be the same as is observed at A when W is attached at B.

The weight was suspended from a hook whose pointed extremity rested on the upper surface of the bar at the marked points. In this way there was no uncertainty as to the exact point at which the weight was applied. The measurements of deflection were made with a micrometer-screw reading to the ten-thousandth of an inch, the contact of the rounded extremity of the screw with its image in the upper surface of the glass being observed with a magnifier. The reading in each position was repeated four times, with the following results.

Case 1. W hung at A. Deflection observed at B:—

	W on.		*W* off.
	79		1473
	82		1473
	79		1476
	76		1474
Mean . .	79	Mean . . .	1474

The deflection at B due to W at A was therefore 1395.

Case 2. W hung at B. Deflection observed at A:—

	W on.		*W* off.
	50		1447
	47		1449
	45		1445
	45		1446
Mean . .	47	Mean . . .	1447

Accordingly the deflection at A due to W at B was 1400.

The difference of the two deflections, amounting to only about $\frac{1}{3}$ per cent., is quite as small as could be expected, and is almost within the limits of experimental error.

In the second experiment the test was more severe, B being replaced by another point B' $7\frac{1}{2}$ inches distant from A, instead of only 5 inches. The deflections in the two cases here came out identical and equal to 993 divisions, W being the same as before. With W at A, the deflection at A was about 1700; and with W at B', the deflection at B' was 760.

The theorem here verified might sometimes be useful in determining the curve of deflection of a bar when loaded at any point A. Instead of observing the deflection at a number of points P, it might be simpler to measure the deflections at the fixed point A, while the load is shifted to the various points P.

For the benefit of those whose minds rebel against the vagueness of generalized coordinates, a more special proof of the theoretical result may here be given. The equation of equilibrium of a bar (whose section is not necessarily uniform) is

$$\frac{d^2}{dx^2}\left(B\frac{d^2y}{dx^2}\right) = Y^*, \quad \dots\dots\dots\dots(A)$$

in which for the present application Y denotes the impressed force, *not* including the weight of the bar itself, and y is the vertical displacement due to Y.

Let y, y' denote two sets of displacements corresponding to the forces Y, Y'. Then

$$\int\left\{y'\frac{d^2}{dx^2}\left(B\frac{d^2y}{dx^2}\right) - y\frac{d^2}{dx^2}\left(B\frac{d^2y'}{dx^2}\right)\right\}dx = \int\{y'Y - yY'\}\,dx, \dots\dots(B)$$

where the integration extends over the whole length of the bar. Now, integrating by parts,

$$\int y'\frac{d^2}{dx^2}\left(B\frac{d^2y}{dx^2}\right)dx = y'\frac{d}{dx}\left(B\frac{d^2y}{dx^2}\right) - \frac{dy'}{dx}B\frac{d^2y}{dx^2} + \int B\frac{d^2y}{dx^2}\frac{d^2y'}{dx^2}\,dx,$$

in which the integrated terms always vanish in virtue of the terminal conditions. In the present case, for example, y, y', d^2y/dx^2, d^2y'/dx^2 vanish at each extremity. Thus the left-hand member of (B) vanishes, and we derive

$$\int\{y'Y - yY'\}\,dx = 0. \quad \dots\dots\dots\dots\dots(C)$$

Let us now suppose that Y vanishes at all points of the bar except in the neighbourhood of A, and also that Y' vanishes except in the neighbourhood of B. Then from (C),

$$y_A'\int Y dx = y_B\int Y' dx;$$

or if $\int Y dx = \int Y' dx$,

$$y_A' = y_B, \quad \dots\dots\dots\dots\dots\dots(D)$$

as was to be proved.

* Thomson and Tait's *Natural Philosophy*, § 617.

A similar method is applicable to all such cases.

I may here add that, corresponding to each of the statical propositions in my former paper, there are others relating to initial motions in which *impulses* and *velocities* take the place of *forces* and *displacements*. Thus, to take an example from electricity, if A and B represent two circuits, the sudden generation of a given current in one of them gives rise to an electro-motive impulse in the other, which is the same, whether it be A or B in which the current is generated. Or, to express what is really the same thing in another way, the ratio of the currents in A and B due to an electro-motive impulse in B is the negative of the ratio of the impulses in B and A necessary in order to prevent the development of a current in B. These statements are not affected by the presence of other circuits, C, D, &c., in which induced currents are at the same time excited.

[1899. The following presentation of the matter is due to Prof. Everett.

A weight W, corresponding to $\int Y dx$ above, applied at the point A produces a drop a at A and a drop b at B; while a weight W' applied at B produces a drop a' at A and b' at B. If both weights be in operation, the potential energy of deformation is definite, and its value may be arrived at in two ways by applying the weights in different orders. If we first apply W, the potential energy so far is $\frac{1}{2}Wa$; and if we then apply W' the additional potential energy is $\frac{1}{2}W'b' + Wa'$, making altogether

$$\tfrac{1}{2}Wa + \tfrac{1}{2}W'b' + Wa'.$$

In like manner if the order of application be reversed, we find

$$\tfrac{1}{2}Wa + \tfrac{1}{2}W'b' + W'b.$$

Accordingly

$$Wa' = W'b,$$

as was to be proved.]

33.

MR HAMILTON'S STRING ORGAN.

[*Nature*, XI. pp. 308, 309, 1875.]

IN the *Philosophical Magazine* for February there is a paper by Mr R. Bosanquet on the mathematical theory of this instrument, in which, however, as it appears to me, the principal points of interest are not touched upon. As the remarks that I have to offer will not require any analysis for their elucidation, I venture to send them to *Nature* as more likely than in the *Philosophical Magazine* to meet the eyes of those interested.

The origin of the instrument has led, as I cannot but think, to considerable misconception as to its real acoustical character. The object of Mr Hamilton and his predecessors was to combine the musical qualities of a string with the sustained sound of the organ and harmonium. This they sought to effect by the attachment of a reed, which could be kept in continuous vibration by a stream of air. Musically, owing to Mr Hamilton's immense enthusiasm and perseverance, the result appears to be a success, but is, I think, acoustically considered, something very different from what was originally intended. I believe that the instrument ought to be regarded rather as a modified reed instrument than as a modified string instrument.

Let us consider the matter more closely. The string and reed together form a system capable of vibrating in a number, theoretically infinite, of independent fundamental* modes, whose periods are calculated by Mr Bosanquet. The corresponding series of tones could only by accident belong to a harmonic scale, and certainly cannot coexist in the normal working of Mr Hamilton's instrument, one of whose characteristics is great sweetness and smoothness of sound. I conceive that the vibration of the system is rigorously or approximately simple harmonic, and that accordingly the sound

---

* In the mechanical, not the musical sense.

emitted directly from the reed, or string, or from the resonance-board in connection with the string, is simple harmonic. On the other hand, it is certain that the note actually heard is compound, and capable of being resolved into several components with the aid of resonators.

The explanation of this apparent contradiction is very simple. Exactly as in the case of the ordinary free reed, whose motion, as has been found by several observers, is rigorously simple harmonic, the intermittent stream of air, which does not take its motion from the reed, gives rise to a highly compound musical note, whose gravest element is the same as that of the pure tone given by the string and resonance-board. One effect of the string, therefore, and that probably an important one, is to intensify the gravest tone of the compound note given by the intermittent stream of air.

The fact that the *pitch* of the system is mainly dependent upon the string, seems to have distracted attention from the important part played by the stream of air, and yet it is obvious that wind cannot be forced through such a passage as the reed affords without the production of sound. A few very simple experiments would soon decide whether the view I am advocating is correct, but I have not hitherto had an opportunity of making them properly. I may mention, however, that I have noticed on one or two occasions an immediate falling off in the sound when the wind was cut off, although the string and reed remained in vibration for a second or two longer. A resonator tuned to one of the principal overtones was without effect when held to the string, but produced a very marked alteration in the character of the sound when held to the reed.

It will be seen that according to my explanation the principal acoustical characteristic of the string—that its tones form a harmonic scale—does not come into play, the office of the string being mainly to convey the vibration of the reed itself (as distinguished from the wind) to the resonance-board and thence through the air to the ear of the observer. A second advantage due to the string appears to be a limitation of the excursion of the reed, whereby the peculiar roughness of an ordinary reed is in great measure avoided.

34.

GENERAL THEOREMS RELATING TO EQUILIBRIUM AND INITIAL AND STEADY MOTIONS.

[*Phil. Mag.* XLIX. pp. 218—224, 1875.]

IF a material system start from rest under the action of given impulses, the energy of the actual motion exceeds that of any other which the system might have been guided to take by the operation of mere constraints; and the difference is equal to the energy of the motion which must be compounded with either to produce the other (Bertrand). A proof of this interesting theorem is given in Thomson and Tait's *Natural Philosophy*, § 311—by a slight modification of which a more general result may be arrived at, giving rise to important corollaries.

Let P, Q, R denote the components of impulse on the particle m, and $\dot{x}$, $\dot{y}$, $\dot{z}$ the component velocities assumed. Then, if $\dot{x}'$, $\dot{y}'$, $\dot{z}'$ denote any other velocities consistent with the connexions of the system, the Principle of Virtual Velocities gives

$$\Sigma \left\{ (P - m\dot{x}) \dot{x}' + (Q - m\dot{y}) \dot{y}' + (R - m\dot{z}) \dot{z}' \right\} = 0,\ldots\ldots\ldots\ldots(1)$$

by means of which the initial velocities $\dot{x}$, &c. are completely determined.

In equation (1) the hypothetical velocities $\dot{x}'$, &c. are any whatever consistent with the constitution of the system; but if they are limited to be such as the system could acquire under the operation of the given impulses with the assistance of mere constraints, we have

$$2T' = \Sigma m (\dot{x}'^2 + \dot{y}'^2 + \dot{z}'^2) = \Sigma (P\dot{x}' + Q\dot{y}' + R\dot{z}'). \ldots\ldots\ldots\ldots(2)$$

This includes the case of the actual motion.

Returning now to the general case, suppose that E' denotes the function

$$E' = \Sigma (P\dot{x}' + Q\dot{y}' + R\dot{z}') - \tfrac{1}{2} \Sigma m (\dot{x}'^2 + \dot{y}'^2 + \dot{z}'^2), \ldots\ldots\ldots\ldots(3)$$

becoming for the actual motion

$$E = T = \tfrac{1}{2} \Sigma m \, (\dot{x}^2 + \dot{y}^2 + \dot{z}^2), \quad \dots\dots\dots\dots\dots (4)$$

or, for any motion of the kind considered in (2),

$$E' = T' = \tfrac{1}{2} \Sigma m \, (\dot{x}'^2 + \dot{y}'^2 + \dot{z}'^2). \quad \dots\dots\dots\dots\dots (5)$$

For the difference between E' and E in (3) and (4), we get

$$E - E' = \tfrac{1}{2} \Sigma m \, (\dot{x}^2 + \dot{y}^2 + \dot{z}^2) + \tfrac{1}{2} \Sigma m \, (\dot{x}'^2 + \dot{y}'^2 + \dot{z}'^2) - \Sigma \, (P\dot{x}' + Q\dot{y}' + R\dot{z}'),$$

in which by (1),

$$\Sigma \, (P\dot{x}' + Q\dot{y}' + R\dot{z}') = \Sigma m \, (\dot{x}\dot{x}' + \dot{y}\dot{y}' + \dot{z}\dot{z}') \, ;$$

so that

$$E - E' = \tfrac{1}{2} \Sigma m \, \{ (\dot{x} - \dot{x}')^2 + (\dot{y} - \dot{y}')^2 + (\dot{z} - \dot{z}')^2 \}, \quad \dots\dots\dots (6)$$

which shows that E is a maximum for the actual motion (in which case it is equal to T), and exceeds any other value E' by the energy of the difference of the real and hypothetical motions. From this we obtain Bertrand's theorem, if we introduce the further limitation that the hypothetical motion is such as the system can be guided to take by mere constraints; for then by (5)

$$E - E' = T - T'.$$

By means of the general theorem (6) we may prove that the energy due to given impulses is increased by any diminution (however local) in the inertia of the system. For whatever the motion acquired by the altered system may be, the value of E corresponding thereto (viz. T) is greater than if the velocities had remained unchanged; and this, again, is evidently greater than the actual E (viz. T) of the original motion. The total increase of energy is equal to the decrease which the alteration of mass entails in the energy of the former motion, together with the energy (under the new conditions of the system) of the difference of the old and new motions. If the change be small, the latter part is of the second order.

On the other hand, of course, any addition to the mass lessens the effect of the given impulses.

A similar deduction may be made from Thomson's theorem, which stands in remarkable contrast with that above demonstrated. The theorem is, that if a system be set in motion with prescribed velocities by means of applied forces of corresponding types, the whole energy of the motion is *less* than that of any other motion fulfilling the prescribed velocity-conditions. "And the excess of the energy of any other such motion, above that of the actual motion, is equal to the energy of the motion which would be generated by the action alone of the impulse which, if compounded with the impulse producing the actual motion, would produce this other supposed motion."

From this it follows readily that, with given velocity-conditions, the energy of the initial motion of a system rises and falls according as the inertia of the system is increased or diminished*.

We now pass to the investigation of some statical theorems which stand in near relation to the results we have just been considering. The analogy is so close that the one set of theorems may be derived from the other almost mechanically by the substitution of "force" for "impulse," and "potential energy of deformation" for "kinetic energy of motion." A similar mode of demonstration might be used, but it will be rather more convenient to employ generalized coordinates.

Consider then a system slightly displaced by given forces from a position of stable equilibrium, from which configuration the coordinates are reckoned. The potential energy of the displacement V is a quadratic function of the coordinates ψ_1, ψ_2, &c.

$$V = \tfrac{1}{2} a_{11}\psi_1^2 + \tfrac{1}{2} a_{22}\psi_2^2 + \dots + a_{12}\psi_1\psi_2 + a_{23}\psi_2\psi_3 + \dots \dots \dots (7)$$

If, then,

$$E' = \Psi_1\psi_1' + \Psi_2\psi_2' + \dots - V', \dots \dots \dots \dots (8)$$

where Ψ_1, Ψ_2, &c. are the forces, E' will be an absolute maximum for the position actually assumed by the system. In equation (8), V is to be understood merely as an abbreviation for the right-hand member of (7), and the displacements ψ_1, &c. are any whatever.

In the position of equilibrium, since then $\Psi_1 = dV/d\psi_1$, &c.,

$$E = V = \tfrac{1}{2}(\Psi_1\psi_1 + \Psi_2\psi_2 + \dots); \dots \dots \dots \dots (9)$$

and thus

$$E - E' = \tfrac{1}{2}(\Psi_1\psi_1 + \dots) + V' - (\Psi_1\psi_1' + \dots)$$
$$= \tfrac{1}{2}\Psi_1(\psi_1 - \psi_1') + \dots + V' - \tfrac{1}{2}(\Psi_1\psi_1' + \dots).$$

Now, by a reciprocal property readily proved†,

$$\Psi_1\psi_1' + \Psi_2\psi_2' + \dots = \Psi_1'\psi_1 + \Psi_2'\psi_2 + \dots, \dots \dots \dots (10)$$

and also

$$V' = \tfrac{1}{2}\Psi_1'\psi_1' + \tfrac{1}{2}\Psi_2'\psi_2' + \dots, \dots \dots \dots \dots (11)$$

where Ψ_1', &c. is the set of forces necessary to maintain the configuration ψ_1', &c. Thus by (10) and (11),

$$E - E' = \tfrac{1}{2}(\Psi_1 - \Psi_1')(\psi_1 - \psi_1') + \dots, \dots \dots \dots \dots (12)$$

a positive quantity representing the potential energy of the deformation $(\psi_1 - \psi_1')$, &c. Thus E' attains its greatest value E in the case of the actual

---

* See a paper by the author on Resonance, *Phil. Trans.* 1871, p. 94 [Art. v. p. 51].
† By substituting $\Psi_1 = dV/d\psi_1$, &c., $\Psi_1' = dV'/d\psi_1'$, &c.

configuration, and the excess of this value E over any other is the potential energy of the displacement which must be compounded with either to produce the other. So far the displacement represented by ψ_1', &c. is any whatever; but if we confine ourselves to displacements due to the given forces and differing from the actual displacements only by reason of the introduction of constraints limiting the freedom of the system, then $E' = V'$; and the theorem as to the maximum value of E' may be stated with the substitution of V' for E'. Thus the introduction of a constraint has the effect of diminishing the potential energy of deformation of a system acted on by given forces; and the amount of the diminution is the potential energy of the difference of the deformations*.

For an example take the case of a horizontal rod clamped at one end and free at the other, from which a weight may be suspended at the point Q. If a constraint is applied holding a point P of the rod in its place (e.g. by a support situated under it), the potential energy of the bending due to the weight at Q is less than it would be without the constraint by the potential energy of the difference of the deformations. And since the potential energy in either case is proportional to the descent of the point Q, we see that the effect of the constraint is to diminish this descent.

The theorem under consideration may be placed in a clearer light by the following interpretation of the function E.

In forming the conditions of equilibrium, we are only concerned with the forces which act upon the system when in that position; but we may, if we choose, attribute any consistent values to the forces for other positions. Suppose, then, that the forces are constant, as if produced by weights. Then, in any position, E denotes the work, positive or negative, which must be done upon the system in order to bring it into the configuration defined by $V = 0$. Thus, to return to the rod with the weight suspended from Q, E represents the work which must be done in order to bring the rod from the configuration to which E refers into the horizontal position. And this work is the difference between the work necessary to raise the weight and that gained during the unbending of the rod. Further, if the configuration in question is one of equilibrium with or without the assistance of a constraint (such as the support at P), the work gained during the unbending is exactly the half of that required to raise the weight; so that E is the same as the potential energy of the bending, or *half* the work required to raise the weight.

When the rod, unsupported at P, is bent by the weight at Q, the point P drops. The energy of the bending is the same as the total work required to restore the rod to a horizontal position. Now this restoration may be effected in two steps. We may first, by a force applied at P, raise that point into its

* Compare Maxwell's *Theory of Heat*, p. 131.

proper position, a process requiring the expenditure of work. The system will now be in the same condition as that in which it would have been found if the point P had been originally supported; and therefore it requires less work to restore the configuration $V = 0$ when the system is under constraint than when it is free. Accordingly the potential energy of deformation is also less in the former case.

We may now prove that any relaxation in the stiffness of a system equilibrated by given forces is attended by an *increase* in the potential energy of deformation. For if the original configuration be maintained, E will be greater than before, in consequence of the diminution in the energy of a given deformation. *À fortiori*, therefore, will E be greater when the system adjusts itself to equilibrium, when the value of E is as great as possible. Conversely, any increase in V as a function of the coordinates entails a diminution in the actual value of V corresponding to equilibrium. Since a loss of freedom may be regarded as an increase of stiffness, we see again how it is that the introduction of a constraint diminishes V.

The statical analogue of Thomson's theorem for initial motions refers to systems in which given deformations are produced by the necessary forces of corresponding types—for example, the rod of our former illustration, of which the point P is displaced through a given distance, as might be done by raising the support situated under it. The theorem is to the effect that the potential energy V of a system so displaced and in equilibrium is as small as it can be under the circumstances, and that the energy of any other configuration exceeds this by the energy of that configuration which is the difference of the two.

To prove this, suppose that the conditions are that $\psi_1, \psi_2, \psi_3, \ldots \psi_r$ are given, while the forces of the remaining types Ψ_{r+1}, Ψ_{r+2}, &c. vanish. The symbols ψ_1, &c., Ψ_1, &c. refer to the actual equilibrium-configuration, and $\psi_1 + \Delta\psi_1$, $\psi_2 + \Delta\psi_2$, &c., $\Psi_1 + \Delta\Psi_1$, $\Psi_2 + \Delta\Psi_2$, &c. to any other configuration subject to the same *displacement*-conditions. For each suffix, therefore, *either* $\Delta\psi$ *or* Ψ vanishes. Now for the potential energy of the hypothetical deformation we have

$$
\begin{aligned}
2(V + \Delta V) &= (\Psi_1 + \Delta\Psi_1)(\psi_1 + \Delta\psi_1) + \ldots \\
&= 2V + \Psi_1\Delta\psi_1 + \Psi_2\Delta\psi_2 + \ldots \\
&\quad + \Delta\Psi_1 \cdot \psi_1 + \Delta\Psi_2 \cdot \psi_2 + \ldots \\
&\quad + \Delta\Psi_1 \cdot \Delta\psi_1 + \Delta\Psi_2 \cdot \Delta\psi_2 + \ldots . \quad \ldots\ldots\ldots\ldots(13)
\end{aligned}
$$

But by the reciprocal relation,

$$
\Psi_1 \cdot \Delta\psi_1 + \Psi_2 \cdot \Delta\psi_2 + \ldots = \Delta\Psi_1 \cdot \psi_1 + \Delta\Psi_2 \cdot \psi_2 + \ldots,
$$

of which the former by hypothesis is zero. Thus

$$
2\Delta V = \Delta\Psi_1 \cdot \Delta\psi_1 + \Delta\Psi_2 \cdot \Delta\psi_2 + \ldots, \quad \ldots\ldots\ldots\ldots(14)
$$

as was to be proved.

The effect of a relaxation in stiffness must clearly be to diminish V; for such a diminution would ensue if the configuration remained unaltered, and therefore still more when the system returns to equilibrium under the altered conditions. It will be understood that in particular cases the diminution spoken of may vanish.

The connexion between the two statical theorems, dealing respectively with systems subject to given displacements and systems displaced by given forces, will be perhaps brought out more clearly by another demonstration of the latter. We have to show that the removal of a constraint is attended by an increase in the potential energy of deformation. By a suitable choice of coordinates the conditions of constraint may be expressed by the vanishing of the first r coordinates $\psi_1 \dots \psi_r$. The relation of the two cases to be compared is expressed by supposing the forces of the remaining types $\Psi_{r+1}, \dots$ to be the same, so that $\Delta\Psi_{r+1}$, &c. vanish. Thus for every suffix either ψ vanishes or else $\Delta\Psi$. Accordingly $\Sigma\psi\Delta\Psi$ is zero, and therefore also, by the law of reciprocity, $\Sigma\Psi\Delta\psi$. Hence, as above,

$$2\Delta V = \Sigma\Delta\Psi\Delta\psi, \quad \dots\dots\dots\dots\dots\dots\dots(15)$$

showing that the removal of the constraint increases the potential energy by the potential energy of the difference of the deformations.

Corresponding to the above theorems for T and V, there are two more relating to the function F introduced by me in a paper printed in the *Proceedings of the Mathematical Society* for 1873 [Art. XXI.], expressing the effects of viscosity. We have here to consider systems destitute both of kinetic and potential energy, of which probably the best example is a combination of electrical conductors, conveying currents, whose inductive effects, dependent on inertia, may be neglected. The equations giving the magnitudes of the steady currents are of the form

$$dF/d\dot{\psi} = \Psi, \quad \dots\dots\dots\dots\dots\dots\dots\dots\dots(16)$$

where F is a quadratic function (in this case with constant coefficients) of the velocities $\dot{\psi}_1$, &c., representing half the dissipation of energy in the unit of time, and Ψ_1, &c. are the electromotive forces. It is scarcely necessary to go through the proofs, as they are precisely similar to those already given with the substitution of F for T, and steady forces for impulses. The analogue of Bertrand's theorem tells us that, if *given electromotive forces* act, the development of heat in unit time is diminished by the introduction of any constraint, as, for example, breaking one of the contacts. And by comparison with Thomson's theorem for initial motions we learn that, if *given currents* be maintained in the system by forces of corresponding types, the whole development of heat is the least possible under the circumstances (Maxwell's *Electricity and Magnetism*, § 284). And precisely as before, we might deduce corollaries relating to the effect of altering the resistance of any part of the combination.

35.

ON THE DISSIPATION OF ENERGY.

[*Roy. Instit. Proc.* VII. pp. 386—389, 1875; *Nature*, XI. pp. 454—455, 1875.]

THE second law of thermodynamics, and the theory of dissipation founded upon it, has been for some years a favourite subject with mathematical physicists, but has not hitherto received full recognition from engineers and chemists, nor from the scientific public. And yet the question under what circumstances it is possible to obtain work from heat is of the first importance. Merely to know that when work is done by means of heat, a so-called equivalent of heat disappears is a very small part of what it concerns us to recognize.

A heat-engine is an apparatus capable of doing work by means of heat supplied to it at a high temperature and abstracted at a lower, and thermodynamics shows that the fraction of the heat supplied capable of conversion into work depends on the limits of temperature between which the machine operates. A non-condensing steam-engine is not, properly speaking, a heat-engine at all, inasmuch as it requires to be supplied with water as well as heat, but it may be treated correctly as a heat-engine giving up heat at 212° Fahr. This is the lower point of temperature. The higher is that at which the water boils in the boiler, perhaps 360° Fahr. The range of temperature available in a non-condensing steam-engine is therefore small at best, and the importance of working at a high pressure is very apparent. In a condensing engine the heat may be delivered up at 80° Fahr.

It is a radical defect in the steam-engine that the range of temperature between the furnace and the boiler is not utilized, and it is impossible to raise the temperature in the boiler to any great extent, in consequence of the tremendous pressure that would then be developed. There seems no escape from this difficulty but in the use of some other fluid, such as a hydrocarbon oil, of much higher boiling point. The engine would then consist of two parts—an oil-engine taking in heat at a high temperature, and doing work by means of the fall of heat down to the point at which a steam-engine becomes available, and secondly a steam-engine receiving the heat given out by the oil-engine and working down to the ordinary atmospheric temperature.

Heat-engines may be worked backwards, so as by means of work to raise heat from a colder to a hotter body. This is the principle of the air or ether freezing machines now coming into extensive use. In this application a small quantity of work goes a long way, as the range of temperature through which the heat has to be raised is but small.

If the work required for the freezing machine is obtained from a steam-engine, the final result of the operation is that a fall of heat in the prime mover is made to produce a rise of heat in the freezing machine, and the question arises whether this operation may be effected without the intervention of mechanical work. The problem here proposed is solved in Carré's freezing apparatus, described in most of the text-books on heat. There are two communicating vessels, A and B, which are used alternately as boiler and condenser. In the first part of the operation aqueous ammonia is heated in A, until the gas is driven off and condensed under considerable pressure in B, which is kept cool with water. Here we have a fall of heat, the absorption taking place at the high temperature and the emission at the lower. In the second part of the operation A is kept cool, and the water in it soon recovers its power of absorbing the ammonia gas, which rapidly distils over. The object to be cooled is placed in contact with B, and heat passes from the colder to the hotter body. Finally the apparatus is restored to its original condition, and therefore satisfies the definition of a heat-engine. M. Carré has invented a continuously working machine on this principle, which is said to be very efficient.

Other freezing arrangements depending on solution or chemical action may be brought under the same principle, if the cycle of operations be made complete.

When heat passes from a hotter to a colder body without producing work, or some equivalent effect such as raising other heat from a colder to a hotter body, energy is said to be dissipated, and an opportunity of doing work has been lost never to return. If on the other hand the fall of heat is fully utilized, there is no dissipation, as the original condition of things might be restored at pleasure; but in practice the full amount of work can never be obtained, in consequence of friction and the other imperfections of our machines.

The prevention of unnecessary dissipation is the guide to economy of fuel in industrial operations. Of this a good example is afforded by the regenerating furnaces of Siemens, in which the burnt gases are passed through a passage stacked with fire-bricks, and are not allowed to escape until their temperature is reduced to a very moderate point. After a time the products of combustion are passed into another passage, and the unburnt gaseous fuel and air are introduced through that which has previously been heated. The efficiency of the arrangement depends in great degree on the fact that the cold fuel is brought first into contact with the colder parts of the flue, and does not take heat from the hotter parts until it has itself become hot. In this way the fall of heat is never great, and there is comparatively little dissipation.

The principal difficulty in economy of fuel arises from the fact that the whole fall of heat from the temperature of the furnace is seldom available for one purpose. Thus in the iron smelting furnaces heat below the temperature of melting iron is absolutely useless. But when the spent gases are used for raising steam, the same heat is used over again at another part of its fall*. There is no reason why this process should not be carried further. All the heat discharged from non-condensing steam-engines, which is more than nine-tenths of the whole, might be used for warming, or drying, or other operations in which only low temperature heat is necessary.

The chemical bearings of the theory of dissipation are very important, but have not hitherto received much attention. A chemical transformation is impossible, if its occurrence would involve the opposite of dissipation (for which there is no convenient word); but it is not true, on the other hand, that a transformation which would involve dissipation must necessarily take place. Otherwise the existence of explosives like gunpowder would be impossible. It is often stated that the development of heat is the criterion of the possibility of a proposed transformation, though exceptions to this rule are extremely well-known. It is sufficient to mention the solution of a salt in water. This operation involves dissipation, or it would not occur, and it is not difficult to see how work might have been obtained in the process. The water may be placed under a piston in a cylinder maintained at a rigorously constant temperature, and the piston slowly raised until all the water is evaporated, and its tension reduced to the point at which the salt would begin to absorb it at the temperature in question. After the salt and vapour are in contact the piston is made to descend until the solution is effected. In this process work is gained, since the pressure under the piston during the expansion is greater than at the corresponding stage of the contraction. If the salt is dissolved in the ordinary way energy is dissipated, an opportunity of doing work at the expense of low temperature heat has been missed and will not return.

* [1899. This illustration is inappropriate, inasmuch as the blast furnace gases are still combustible, and raise steam in virtue of chemical action.]

The difficulty in applying thermodynamical principles to chemistry arises from the fact that chemical transformations cannot generally be supposed to take place in a reversible manner, even although unlimited time be allowed. Some progress has, however, recently been made, and the experiments of Debray on the influence of pressure on the evolution of carbonic anhydride from chalk throw considerable light on the matter. By properly accommodating the pressure and temperature, the constituents of chalk may be separated or recompounded without dissipation, or rather dissipation may theoretically be reduced without limit by making the operation slowly enough.

The possibility of chemical action must often depend on the density of the reacting substances. A mixture of oxygen and hydrogen in the proper proportions may be exploded by an electric spark at the atmospheric pressure, and energy will be dissipated. In this operation the spark itself need not be considered, as a given spark is capable of exploding any quantity of gas. Suppose, now, that previously to explosion the gas is expanded at constant temperature, and then after explosion brought back to the former volume. Since in the combination there is a condensation to two-thirds, the pressure required to compress the aqueous vapour is less than that exercised at the same volume by the uncombined gases, and accordingly work is gained on the whole. Hence the explosion in the expanded state involves less dissipation than in the condensed state, and the amount of the difference could be increased without limit by carrying the expansion far enough. It follows that beyond a certain point of rarity the explosion cannot be made, as [according to what has been said its occurrence] could not then involve any dissipation. But although the tendency to combine diminishes as the gas becomes rarer, the heat developed during the combination remains approximately constant.

It must be remembered that the heat of combination is generally developed at a high temperature, and that therefore work may be done during the cooling of the products of combustion. If, therefore, it is a necessity of the case that the act of combustion should take place at a high temperature, the possibility of explosion will cease at an earlier point of rarefaction than would otherwise have been the case.

It may probably be found that many mixtures which show no tendency to explode under ordinary conditions will become explosive when sufficiently condensed.

36.

ON THE WORK THAT MAY BE GAINED DURING THE MIXING OF GASES.

[*Phil. Mag.* XLIX. pp. 311—319, 1875.]

THE well-known fact that hydrogen tends to escape through fine apertures more rapidly than air enters to supply its place, even although the advantage of the greater pressure may be on the side of the air, proves that the operation of mixing the two gases has a certain mechanical value. In a common form of the experiment a tube containing hydrogen and closed at the upper end with a porous plug of plaster of Paris stands over water. In a short time the escape of hydrogen creates a partial vacuum in the tube, and the water rises accordingly. Whenever then two gases are allowed to mix without the performance of work, there is dissipation of energy, and an opportunity of doing work at the expense of low temperature heat has been for ever lost. The present paper is an attempt to calculate this amount of work.

The result at which I have arrived is extremely simple. It appears that the work that may be done during the mixing of the volumes v_1 and v_2 of two different gases is the same as that which would be gained during the expansion of the first gas from volume v_1 to volume $v_1 + v_2$, together with the work gained during the expansion of the second gas from v_2 to $v_1 + v_2$, the expansions being supposed to be made into vacuum. Now these expansions may be considered actually to take place; and thus the rule is brought under Dalton's principle that each gas behaves to the other as a vacuum. It is understood that the gases follow the common law of independent pressures, so that the total pressure is always the sum of those which would be exerted by each gas in the absence of the other.

We will take first the case when one gas is condensable, and estimate how much work must be done in order to separate the components of a mixture. Suppose, then, that a long cylinder, closed at the bottom, contains a uniform mixture of (for example) hydrogen and steam confined under a piston, and that the walls of the cylinder are maintained at a constant temperature. When the piston descends, heat will be generated; but the operation is supposed to proceed so slowly, that not only is the temperature rigorously constant throughout, but also the mixture is at any time in that condition which it would finally attain were the descent of the piston arrested. The pressure on the piston resisting the descent is by hypothesis the sum of those which it would experience from the hydrogen and steam separately*. When the space under the piston is reduced to that which the given quantity of steam is capable of saturating at the given temperature, condensation commences and continues as the steam-space is gradually diminished.

By carrying this process sufficiently far, the condensation of the steam may be effected with any desired degree of completeness, and thus the water and hydrogen separated. A second moveable piston may now be inserted immediately over the condensed water, and a very gradual expansion allowed until the original total volume is recovered. If the second piston be allowed free motion, the constituents of the original mixture are now separated, under equal pressures, and occupying the same total volume as before; and the question is, how much work has been expended in arriving at this state of things?

In view of the fact that during the first part of the operation the hydrogen and steam press independently, it is clear that the total work done is the same as that which would be required to compress the hydrogen from the original volume $v_1 + v_2$ to the volume v_1 if no steam were present, together with the work necessary to compress the steam from $v_1 + v_2$ to v_2 if no hydrogen were present. And since every step of this process is reversible, the same amount of work might be gained in making the mixture, and is dissipated if the mixture is allowed to take place by free diffusion.

The same argument will apply when the condensation of one of the gases is effected by chemical means. Suppose, for example, that we have a mixture of carbonic anhydride and hydrogen at a red heat, and that it is proposed to absorb the carbonic gas with quicklime. It has been proved by Debray that at every temperature above a certain point carbonic gas in contact with quicklime and carbonate has a definite tension; any excess will be absorbed by the lime, and any deficiency supplied by a decomposition of the carbonate.

If the tension of the carbonic gas in the given mixture be higher than that proper to the temperature, absorption will take place in an irreversible manner. In order to prevent dissipation, the mixture of gases must be first

---

* For the sake of simplicity we may suppose a vacuum on the other side of the piston, though, of course, any constant pressure would give finally no result.

expanded until the tension of the carbonic gas is no higher than that corresponding to the temperature at which it is proposed to work. When the contact is made, the mixture may be very slowly condensed, so that after the point is passed at which chemical action commences, the tension of the carbonic anhydride remains constant. This process may be continued until nearly all the carbonic anhydride is absorbed. The hydrogen may then be separated. The space over the carbonate of lime must next be slowly increased until the original quantity of carbonic gas has been again evolved, when the connexion with the quicklime must be cut off. It now only remains to reduce the separated gases to the same pressure and to a total volume equal to that of the original mixture.

From the preceding considerations we may, I think, infer that the law above stated is general whenever the gases really press independently; for it is difficult to see how its truth could depend on what would seem to be the accident of the existence or non-existence of a chemical capable of absorbing one or other of the gases.

It is worthy of notice that exactly the same rule applies for the mechanical value of the separation of two gases, even when the pressures are different; for we get the same result whether we first before mixing allow the pressures to become equal and add the work gained in this process to that due to the subsequent mixing, or whether we calculate at once the work due to the separate expansion of the two gases from their original volumes to the total volume of the mixture.

In like manner the work that can be gained during the mixing of any number of pure and different gases, which press independently, is the sum of those due to the expansions of the several gases from their original to their final volumes, where the volume of a gas is understood to mean the space in which the gas is confined.

The next problem which presents itself is that of finding the work that may be done during the mixture of two quantities of mixed gases—for example, oxygen and hydrogen. Suppose the two mixtures to be contained in a cylinder, and separated from one another by a piston which moves freely. The rule is that the work required to be estimated is that which would be gained during the equalization of the oxygen-pressures if the hydrogen were annihilated, together with that which would be gained during the equalization of the hydrogen-pressures if the oxygen were annihilated.

If the proportions of the gases are the same in the two mixtures, and also the total pressures, there is, of course, no possibility of doing work. If, on the other hand, the gas on the one side of the piston be pure oxygen, and on the other side pure hydrogen, the more general rule reduces to that already given for pure gases.

I now pass to another proof of the fundamental rule, depending on the possibility of separating two gases of different densities by means of gravity. In a vertical column maintained at a uniform temperature, two gases which press independently will arrange themselves each as if the other were absent. Consequently, if there be any difference in density, the percentage composition will vary at different heights, and a partial separation of the gases is thus effected.

Imagine now a large reservoir containing gas at sensibly constant pressure, on which is mounted a tall narrow vertical tube; and first, in order to understand the operation more easily, let there be only one kind of gas present. If p be the pressure and ρ the density, $p = k\rho$, since the temperature is constant; and if z be the height measured from the reservoir in which the pressure is P,

$$dp = -g\rho\,dz = -\mu p\,dz, \qquad \text{if } \mu = g/k;$$

whence, by integration,

$$p = P\epsilon^{-\mu z} \quad \dots\dots\dots\dots\dots\dots\dots\dots\dots(1)$$

expresses the law of variation of pressure with height. Suppose now that a small quantity of gas of volume v is (1) removed from the top of the tube, (2) compressed to volume v_0 until it is of the same pressure as the gas in the reservoir, (3) allowed to fall through the height z to the level of the reservoir, and (4) forced into the reservoir. The effect of this series of operations is *nil*, and there can be neither gain nor loss of work. The work gained in the first operation is consumed in the fourth, since $pv = Pv_0$, so that attention may be confined to the second and third operations. Now the work consumed in the compression from v to v_0 is

$$\int_{v_0}^{v} p\,dv = Pv_0 \int_{v_0}^{v} \frac{dv}{v} = Pv_0 \log \frac{v}{v_0} = Pv_0 \log \frac{P}{p}; \quad \dots\dots\dots(2)$$

and the work gained in the descent

$$= gz\rho_0 v_0 = \mu z v_0 P.$$

And these are equal, since $\log(P/p) = \mu z$; so that on the whole no work is lost or gained.

The case is different when there are two kinds of gases present. Although, as before, the work gained in the first operation is consumed in the fourth, there is no longer compensation in the second and third operations. If P_1 and P_2 are the partial pressures in the reservoir, the work required for the compression from v to v_0 is

$$(P_1 + P_2)\,v_0 \log \frac{v}{v_0}.$$

On the other hand, the work gained in the descent is

$$gzv_0 \left(\frac{Q_1}{k_1} + \frac{Q_2}{k_2} \right) = zv_0 \left(\mu_1 Q_1 + \mu_2 Q_2 \right),$$

if Q_1, Q_2 are the partial pressures of the abstracted gas after condensation to volume v_0. Thus, on the whole, if W be the work done on the gases, since $Q_1 + Q_2 = P_1 + P_2$,

$$\frac{W}{v_0} = Q_1 \left(\log \frac{v}{v_0} - \mu_1 z \right) + Q_2 \left(\log \frac{v}{v_0} - \mu_2 z \right).$$

Now $p_1 = P_1 \epsilon^{-\mu_1 z}$, and $vp_1 = v_0 Q_1$; so that $vP_1 \epsilon^{-\mu_1 z} = v_0 Q_1$. Accordingly

$$\log \frac{v}{v_0} - \mu_1 z = \log \left(\frac{Q_1}{P_1} \epsilon^{+\mu_1 z} \right) - \mu_1 z = \log \frac{Q_1}{P_1}.$$

Similarly

$$\log \frac{v}{v_0} - \mu_2 z = \log \frac{Q_2}{P_2};$$

and thus

$$W = v_0 \left\{ Q_1 \log \frac{Q_1}{P_1} + Q_2 \log \frac{Q_2}{P_2} \right\}. \quad \dots\dots\dots\dots\dots (3)$$

Since the process is reversible, (3) gives the work which may be done during the mixing of a volume v_0 of two gases under the partial pressures Q_1 and Q_2, with a large quantity of the same gases under an equal total pressure, but with partial pressures P_1 and P_2.

The quantity denoted by W can never be negative. To verify this from (3), write it in the form

$$- \frac{W}{v_0 (P_1 + P_2)} = x' \log \frac{x}{x'} + y' \log \frac{y}{y'} = \log \left\{ \left(\frac{x}{x'} \right)^{x'} \left(\frac{y}{y'} \right)^{y'} \right\}, \quad \dots\dots (4)$$

where

$$x = \frac{P_1}{P_1 + P_2}, \quad y = \frac{P_2}{P_1 + P_2}, \quad x' = \frac{Q_1}{Q_1 + Q_2}, \quad y' = \frac{Q_2}{Q_1 + Q_2};$$

so that $x + y = x' + y' = 1$.

Now (Todhunter's *Algebra*, p. 392) if $a, b, c, \dots$ be any positive quantities,

$$\frac{a + b + c + d + \dots}{n} > (abcd \dots)^{\frac{1}{n}}.$$

Suppose that $a, b, c, \dots$ consist of p equal quantities α and q equal quantities β; then

$$\frac{p}{n} \alpha + \frac{q}{n} \beta > \alpha^{p/n} \beta^{q/n}.$$

If now we take

$$p/n = x', \quad q/n = y', \quad \alpha = x/x', \quad \beta = y/y',$$

we see that

$$x + y > \left(\frac{x}{x'}\right)^{x'} \left(\frac{y}{y'}\right)^{y'},$$

and therefore, since $x + y = 1$, that W is always positive, unless $\alpha = \beta$, in which case the composition of the two mixtures is the same, and W vanishes.

We have now to show how the formula for the mixture of two pure gases may be derived from (3). Let v_1 be the volume of the first gas and v_2 of the second, at the constant pressure $P_1 + P_2$. The value of the interdiffusion of v_1 and v_2 must be the same as that of their diffusion into a large quantity of a mixture whose composition is identical with that of the mixture of v_1 and v_2. For, on this supposition, the separation of the two mixtures spoken of would have no mechanical value. Now by (3) the value of W for the diffusion of a quantity v_0 of pure gas into a large quantity of a mixture whose partial pressures are P_1 and P_2 is (since Q_2 and $Q_2 \log Q_2$ vanish)

$$W = (P_1 + P_2) v_0 \log \frac{P_1 + P_2}{P_1}; \quad \dots\dots\dots\dots\dots(5)$$

and hence the value of W for the interdiffusion of the quantities v_1 and v_2 is

$$W = v_1 (P_1 + P_2) \log \frac{P_1 + P_2}{P_1} + v_2 (P_1 + P_2) \log \frac{P_1 + P_2}{P_2},$$

or, since by hypothesis $P_2 : P_1 = v_2 : v_1$,

$$W = (P_1 + P_2) \left\{ v_1 \log \frac{v_1 + v_2}{v_1} + v_2 \log \frac{v_1 + v_2}{v_2} \right\}$$

$$= (P_1 + P_2) \log \frac{(v_1 + v_2)^{v_1 + v_2}}{v_1^{v_1} v_2^{v_2}}. \quad \dots\dots\dots\dots\dots(6)$$

This equation agrees with the rule enunciated at the beginning of this paper, inasmuch as $(P_1 + P_2) v_1 \log \dfrac{v_1 + v_2}{v_1}$ represents the work gained in the expansion of the first gas from volume v_1 to volume $v_1 + v_2$, and

$$(P_1 + P_2) v_2 \log \frac{v_1 + v_2}{v_2}$$

represents the corresponding quantity for the second gas.

The significance of equation (5) may perhaps be more fully brought out by the following investigation of it. Whatever the relative proportions of the two gases in the reservoir may be, it will always be possible by going high enough to obtain a small quantity of the lighter gas in any required degree of purity. The removal of this at the top of the tube, its condensation to the pressure in the reservoir, the fall to the level of the reservoir, and the introduction into the reservoir would, on the whole, require no work to be done if this kind of gas had alone been present. The only effect of the heavier gas is to render necessary a greater condensation in the third operation; and thus W is the work that is required to condense the gas from the partial pressure P_1 to the total pressure in the reservoir $P_1 + P_2$, whence equation (5) follows at once. If it is desired to isolate a small quantity of the heavier gas, the tube must be taken downwards.

It is to be observed that the work required to force a given quantity of gas into a large reservoir containing gas at the same pressure is independent of this pressure, since, according to Boyle's law, v is diminished in the same proportion that p is increased.

The principle of dissipation may be employed to prove that the pressure in a vertical column of mixed gases is greater when there is free diffusion than when the gases are uniformly mixed; for if the gases be allowed to rise from the reservoir tolerably quickly (or if a series of moveable pistons be interpolated), the composition in the tube will be the same as in the reservoir. If free diffusion be now allowed, there must be dissipation. The original state of things will be restored if the mixture be slowly forced back into the reservoir; and accordingly the work consumed in condensation must be greater than that gained in the expansion. In fact it may be proved algebraically by a process somewhat similar to that applied to equation (3), that the pressure of the gases under free diffusion p, where

$$p = P_1 \epsilon^{-\mu_1 z} + P_2 \epsilon^{-\mu_2 z}, \quad \dots\dots\dots\dots\dots\dots(7)$$

is greater than the pressure of a uniform mixture p', where

$$p' = (P_1 + P_2)\, \epsilon^{-z \frac{\mu_1 P_1 + \mu_2 P_2}{P_1 + P_2}}. \quad \dots\dots\dots\dots\dots(8)$$

It is, however, possible to imagine other distributions which shall give a pressure greater than (7). The mechanical equilibrium gives one equation involving the two quantities p_1 and p_2, viz.

$$\frac{dp_1}{dz} + \frac{dp_2}{dz} + \mu_1 p_1 + \mu_2 p_2 = 0; \quad \dots\dots\dots\dots\dots\dots(9)$$

and the subsidiary conditions are that $p_1 = P_1$, $p_2 = P_2$, when $z = 0$. Hence we may take as the most general solution,

$$p_1 = P_1 \epsilon^{-\mu_1 z} + \epsilon^{-\mu_1 z} \int_0^z \epsilon^{\mu_1 x} X\, dx,$$

$$p_2 = P_2 \epsilon^{-\mu_2 z} - \epsilon^{-\mu_2 z} \int_0^z \epsilon^{\mu_2 x} X\, dx,$$

where X is an arbitrary function of x. Thus the total pressure

$$p_1 + p_2 = P_1 \epsilon^{-\mu_1 z} + P_2 \epsilon^{-\mu_2 z} + \int_0^z \left\{ \epsilon^{-\mu_1 (z-x)} - \epsilon^{-\mu_2 (z-x)} \right\} X dx. \quad ......(10)$$

For free diffusion $X = 0$; but it could always be taken so as to make the integral either positive or negative, as might be desired.

The work required to decompose a mixture of gases is in general small, and could scarcely be of much importance from an industrial point of view. When, however, the proportion of one ingredient is very insignificant, more work is required. Thus the separation of the carbonic anhydride from the atmosphere would require, relatively to the quantity obtained, a much larger expenditure of-work than the separation of the oxygen. Thus consideration shows that extreme purity in any gas will always be attained and maintained with difficulty. Even when the necessary work is small, as in the separation of oxygen from the atmosphere, it is well to bear in mind that some work is absolutely essential. The reversible absorption of the oxygen of air may be effected by a substance like baryta; but we must not expect to recover the pure oxygen at the same temperature and under a pressure equal to the total pressure at which it was absorbed. Either the temperature must be raised, or the gas must be exhausted at a pressure less than that under which it existed in the mixture during the absorption*. It is just possible that this point might be found to be of practical importance in the solution of the problem of extracting oxygen from the air.

* [1899. This is the method employed in the Brin process, subsequently introduced.]

37.

VIBRATIONS OF A LIQUID IN A CYLINDRICAL VESSEL.

[*Nature*, XII. p. 251, 1875.]

See *Phil. Mag.* I. pp. 257—279, 1876; Art. XXXVIII. below.

The fact that the period of a given mode of vibration of liquid in a cylindrical vessel of infinite depth and of section always similar to itself (*e.g.* always circular) is proportional to the square root of the linear dimension of the section, follows from the theory of dimensions without any calculation. For the only quantities on which the period (τ) could depend are (i) ρ the density of the liquid, (ii) g the acceleration of gravity, and (iii) the linear dimension d. Now, as in the case of the common pendulum, it is evident that τ cannot depend upon ρ. If the density of the liquid be doubled, the force which acts upon it is also doubled, and therefore the motion is the same as before the change. Thus τ, a time, is a function of d, a length, and of g. Since g is of -2 dimensions in time, $\tau \propto g^{-\frac{1}{2}}$, and therefore in order to be independent of the unit of length, it must vary as $d^{\frac{1}{2}}$, inasmuch as g is of one dimension in length. Hence $\tau \propto d^{\frac{1}{2}} g^{-\frac{1}{2}}$. This reasoning, it will be observed, only applies when the depth may be treated as infinite, [or when it bears some constant finite ratio to the linear dimension of the section].

38.

ON WAVES.

[*Phil. Mag.* I. pp. 257—279, 1876.]

THE theory of waves in a uniform canal of rectangular section, in the case when the length of the wave is great in comparison with the depth of the canal and when the maximum height of the wave is small in comparison with the same quantity, was given long ago by Lagrange, and is now well known. A wave of any form, subject to the above conditions, is propagated unchanged, and with the velocity which would be acquired by a heavy body in falling through half the depth of the canal. The velocity of propagation here referred to is of course relative to the undisturbed water. If we attribute to the water in the canal a velocity equal and opposite to that of the wave, the wave-form, having the same relative velocity as before, is now fixed in space, and the problem becomes one of *steady* motion. It is under this aspect that I propose at present to consider the question; and we will therefore suppose that water is flowing along a tube, whose section undergoes a temporary and gradual alteration in consequence of a change in the vertical dimension of the tube. The principal question will be how far the pressure at the upper surface can be made constant by a suitable adjustment of the velocity of flow to the force of gravity.

That the two causes which tend to produce variation of pressure at the upper surface act in opposition to each other is at once evident. If there were no gravity, the pressure would vary on account of the alteration in the velocity of the fluid. Since there must be the same total flow across all sections of the pipe, the fluid which approaches an enlargement must lose velocity, and the change of momentum involves an augmented pressure. On this account, therefore, there is an increased pressure at a place of enlargement and a diminished pressure at a contraction. On the other hand the effect of gravity is in the opposite direction, tending to produce a loss of pressure at the upper surface where that surface is high, and a gain of pressure where the surface is low. This effect of gravity is independent of the velocity; but the changes of pressure due to acceleration and retardation depend on the velocity of flow, and we can therefore readily understand that, with a certain definite velocity of flow, compensation may take place, at least approximately. When this happens, the condition of a free surface is satisfied, the constraint may be removed, and we are left with a stationary wave-form.

In the theory of long waves it is assumed that the length is so great in proportion to the depth of the water, that the velocity in a vertical direction can be neglected, and that the horizontal velocity is uniform across each section of the canal. This, it should be observed, is perfectly distinct from any supposition as to the height of the wave. If l be the undisturbed depth, and h the elevation of the water at any point of the wave, u_0, u the velocities corresponding to l, $l + h$ respectively, we have, by the condition of continuity,

$$u = \frac{l u_0}{l + h}, \qquad \text{so that } u_0^2 - u^2 = u_0^2 \frac{2lh + h^2}{(l + h)^2}.$$

By the principles of hydrodynamics, the increase of pressure due to retardation will be

$$\frac{\rho}{2}(u_0^2 - u^2) = \frac{\rho u_0^2}{2} \frac{2lh + h^2}{(l + h)^2}.$$

On the other hand, the loss of pressure due to height will be $g\rho h$; and therefore the total gain of pressure over the undisturbed parts is

$$\left(\frac{\rho u_0^2}{l} \frac{1 + h/2l}{(1 + h/l)^2} - g\rho \right) h.$$

If now the ratio $h : l$ be very small, the coefficient of h becomes

$$\rho \, (u_0^2/l - g),$$

and we conclude that the condition of a free surface is satisfied provided $u_0^2 = gl$. This determines the rate of flow in order that a stationary wave may be possible, and gives of course at the same time the velocity of a wave in still water.

If we suppose the condition $u_0^2 = gl$ satisfied, the change of pressure is, to a second approximation,

$$\delta p = g\rho h \left\{ \frac{1 + h/2l}{(1 + h/l)^2} - 1 \right\} = -\frac{3}{2} \frac{g\rho h^2}{l},$$

which shows that the pressure is defective at all parts of the wave where h differs from zero. Unless, therefore, h^2 can be neglected, it is impossible to satisfy the condition of a free surface for a stationary long wave—which is the same as saying that it is impossible for a long wave of finite height to be propagated in still water without change of type. If, however, h be every-where positive, a better result can be obtained with an increased value of u_0; and if h be everywhere negative, with a diminished value. We infer that a positive wave moves with a somewhat higher, and a negative wave with a somewhat lower velocity than that due to half the undisturbed depth.

[1899. The approximation may be continued as follows. In a small positive progressive wave the relation between the particle-velocity u at any point (now reckoned relatively to the parts outside the wave) and the elevation h is

$$u = \sqrt{(g/l)} \cdot h. \quad\dots\dots\dots\dots\dots\dots\dots(1)$$

If this relation be violated anywhere, a wave will emerge, travelling in the negative direction. Let us now picture to ourselves the case of a positive progressive wave in which the changes of velocity and elevation are very gradual but become important by accumulation, and let us inquire what conditions must be satisfied in order to prevent the formation of a negative wave. (*Theory of Sound*, § 251.) It is clear that the answer to the question whether or not a negative wave will be generated at any point will depend upon the state of things in the immediate neighbourhood of the point, and not upon the state of things at a distance from it, and will therefore be determined by the criterion applicable to small disturbances. In applying this criterion we are to consider the velocities and elevations, not absolutely, but relatively to those prevailing in the neighbouring parts of the medium, so that the form of (1) proper for the present purpose is

$$du = \sqrt{\{g/(l+h)\}} \cdot dh ; \quad\dots\dots\dots\dots\dots(2)$$

whence

$$u = 2\sqrt{g}\{\sqrt{(l+h)} - \sqrt{l}\} . \quad\dots\dots\dots\dots(3)$$

This is the relation between u and h necessary for a long positive progressive wave.

Consider now the velocity of the crest of a wave of elevation h. It is propagated relatively to the surrounding water with the velocity due to the depth $(l + h)$; but in addition to this there is the particle-velocity defined above. The velocity of the crest relatively to still water is accordingly

$$2\sqrt{g} \{\sqrt{(l+h)} - \sqrt{l}\} + \sqrt{\{g(l+h)\}},$$

or approximately

$$\sqrt{\{g(l+3h)\}} , \quad\dots\dots\dots\dots\dots\dots\dots(4)$$

as given by Airy.]

Although a constant gravity is not adequate to compensate the changes of pressure due to acceleration and retardation in a long wave of finite height, it is evident that complete compensation is attainable if gravity be a function of height; and it is worth while to inquire what the law of force must be in order that long waves of unlimited height may travel with type unchanged. If f be the force at height h, the condition of constant pressure is

$$\tfrac{1}{2}\rho u_0^2 \left\{ 1 - \frac{l^2}{(l+h)^2} \right\} = \rho \int_0^h f\, dh\,;$$

whence

$$f = -\frac{u_0^2}{2}\frac{d}{dh}\frac{l^2}{(l+h)^2} = u_0^2 \frac{l^2}{(l+h)^3}\,,$$

which shews that the force must vary inversely as the cube of the distance from the bottom of the canal. Under this law the waves may be of any height, and they will be propagated unchanged with the velocity $\sqrt{(f_1 l)}$, where f_1 is the force at the *undisturbed level.*

The same line of thought may be applied to the case of a long wave in a canal whose section is uniform but otherwise arbitrary. Let A be the area of the section below the undisturbed level, b the breadth at that level. Then, as before,

$$(A + bh)\, u = A u_0$$

if h be small; whence

$$u_0^2 - u^2 = \frac{2bh}{A}\, u_0^2\,.$$

Now by dynamics

$$u_0^2 - u^2 = 2gh,$$

if the upper surface be free; and thus

$$u_0^2 = gA/b,$$

which gives the velocity of propagation. In the case of a rectangular section we have the same result as before, since $A = bl$.

The energy of a long wave is half potential and half kinetic. If we suppose that initially the surface is displaced, but that the particles have no velocity, we shall evidently obtain (as in the case of sound) two equal waves travelling in opposite directions, whose total energies are equal, and together make up the potential energy of the original displacement. Now the elevation of the derived waves must be half of that of the original displacement, and accordingly the potential energies less in the ratio of $4:1$. Since therefore the potential energy of each derived wave is one quarter, and the total energy one half of that of the original displacement, it follows that in the derived wave the potential and kinetic energies are equal.

We may now investigate the effect on a long wave of a gradual alteration in the breadth of the canal and the area of the section. The potential energy of the wave varies directly as the length, breadth, and square of the height; and, by what has been proved above, the same is true of the total energy. Now the length of the wave in various parts of the canal is obviously proportional to the velocity of propagation, viz. $\sqrt{(A/b)}$; and we may therefore write

$$E \propto \sqrt{(A/b)} \cdot (\text{height})^2 \cdot b.$$

But when the alteration in the canal is very gradual, there is no sensible reflection and the energy of the wave continues constant; so that

$$\text{height} \propto A^{-\frac{1}{4}}b^{-\frac{1}{4}}.$$

In the case of a rectangular section,

$$\text{height} \propto l^{-\frac{1}{4}}b^{-\frac{1}{4}}.$$

These results are due to Green, Kelland, and Airy*. The same method may be even more easily applied to the sound-wave moving in a pipe of gradually varying section.

The theory of long waves may be applied in many cases to ascertain the effect on a stream of a contraction or enlargement of its channel. If the section of the channel up to the natural level of the stream be altered from A_0 to A, the equation of continuity gives

$$(A + bh)\,u = A_0 u_0,$$

where b, the breadth at the surface of the water, is supposed not to vary with height. The condition of a free surface is

$$u_0^2 - u^2 = u_0^2 \left\{ 1 - \frac{A_0^2}{(A + bh)^2} \right\} = 2gh,$$

or

$$\frac{A_0^2}{(A + bh)^2} = 1 - \frac{2gh}{u_0^2},$$

which shows that h can never exceed the height due to the velocity u_0, as is indeed otherwise obvious.

If the variations in A and b are small as well as gradual, and if we put $A = A_0 + \delta A$, we find

$$h = \frac{\delta A}{b} \div \frac{gA_0}{bu_0^2} - 1.$$

When the velocity u_0 is less than that of a free wave, $gA_0 > bu_0^2$, and h has the same sign as δA; viz. a contraction of the channel produces a depression of the surface, and an enlargement produces an elevation. But if the velocity of the stream exceed that of a free wave, these effects are reversed, and an enlargement and contraction of the section entail respectively a depression and an elevation of the surface.

If the velocity of the stream is nearly the same as that of a free wave, a state of things is approached in which a wave can sustain itself in a stationary

* Stokes, *Brit. Assoc. Report on Hydrodynamics*, 1846.

position without requiring a variation in the channel; and then the effects of such a variation are naturally much intensified.

We must not forget that these calculations proceed on the supposition that a steady motion is possible. It would appear that the motion thus obtained is unstable in the case where the velocity of the stream exceeds that of a free wave. If we suppose the upper surface to consist of a moveable envelope, it would indeed be in equilibrium when disposed according to the law above investigated; but if a displacement be made and steady motion be conceived to be re-established, the pressure of the fluid will be less than before if the displacement be downwards, but will be increased if the displacement be upwards; so that the forces brought to bear on the envelope are in both cases in the direction of displacement. The expression for the variation of pressure at the envelope is

$$\frac{\delta p}{\rho} = u_0{}^2 \frac{\delta A}{A_0} + \left(\frac{u_0{}^2 b}{A_0} - g\right) h.$$

The Solitary Wave.

This is the name given by Mr Scott Russell to a peculiar wave described by him in the *British Association Report* for 1844. Since its length is about six or eight times the depth of the canal, this wave is, to a rough approximation, included under the theory of long waves; but there are several circumstances observed by Mr Russell which indicate that it has a character distinct from that of other long waves. Among these may be mentioned the very different behaviour of solitary waves according as they are positive or negative, viz. according as they consist of an elevation or a depression from the undisturbed level. In the former case the wave has a remarkable permanence, being propagated to great distances without much loss; but a negative wave is soon broken up and dissipated.

Airy, in his treatise on Tides and Waves, still probably the best authority on the subject, appears not to recognize anything distinctive in the solitary wave. He says:—" We are not disposed to recognize this wave as deserving the epithets 'great' or 'primary,'......and we conceive that ever since it was known that the theory of shallow waves of great length was contained in the equation $d^2X/dt^2 = g\kappa \, d^2X/dx^2$,......the theory of the solitary wave has been perfectly well known." And again, "Some experiments were made by Mr Russell on what he calls a *negative* wave—that is, a wave which is in reality a progressive hollow or depression. But (we know not why) he appears not to have been satisfied with these experiments, and has omitted them in his abstract. All the theories of our IVth Section, without exception, apply to these as well as to *positive* waves, the sign of the coefficient only being changed."

On the other hand, Professor Stokes says* :—" It is the opinion of Mr Russell that the solitary wave is a phenomenon *sui generis,* in no wise deriving its character from the circumstances of the generation of the wave. His experiments seem to render this conclusion probable. Should it be correct, the analytical character of the solitary wave remains to be discovered."

The theory of the solitary wave has been considered by Earnshaw (*Camb. Trans.* vol. VIII.), who, distrusting what he calls analytical approximations, bases his calculation on a supposed result of experiment, namely that the horizontal velocity is uniform over each section. This, as we have seen, is the fundamental assumption in the theory of long waves; but when the length of the wave is moderate, such a state of things is impossible in a frictionless fluid which has been once at rest; for it involves molecular rotation. In fact if there be a velocity-potential, the horizontal velocity u satisfies Laplace's equation $d^2u/dx^2 + d^2u/dy^2 = 0$, and therefore cannot be a function of x without being also a function of y. The motion investigated by Earnshaw has therefore molecular rotation; and the rotation remains constant for each particle; otherwise the equations of fluid motion would not be satisfied. This is the explanation of the difficulty with which Earnshaw meets,—that while the necessary conditions are satisfied in the wave itself, there is discontinuity in passing from the wave to the undisturbed water. The discontinuity arises from the fact that, as there is no rotation outside the wave, it is necessary to suppose finite rotations imparted to the particles as the wave reaches and leaves them. It is evident that, except in the case of very long waves, u must be treated as a function of y as well as of x.

In considering the theory of long waves (reduced to rest by imparting an opposite motion to the water), we saw that it was impossible to satisfy the condition of a free surface if the height of the wave were finite. It occurred to me to inquire whether there might not be compensation in certain cases between the variation of pressure at the upper surface due to a finiteness of height, and the variation due to a departure from the law of uniform horizontal velocity proper to very long waves. It was conceivable that the surface-condition in the case of a wave of given finite height might be better satisfied by a moderate than by a very great wave-length. In this way I have obtained what seems to be a perfectly satisfactory approximate theory of the solitary wave.

If u and v be the horizontal and vertical velocities in a stream moving in two dimensions without molecular rotation, and ϕ, ψ the potential and stream functions, we have

$$u = \frac{d\phi}{dx} = \frac{d\psi}{dy}, \qquad v = \frac{d\phi}{dy} = -\frac{d\psi}{dx}. \dots\dots\dots\dots(A)$$

* *British Association Report,* 1846.

Hence, if the bottom of the canal be taken for axis of x, we may take for u and v, since they satisfy Laplace's equation,

$$u = \cos\left(y\frac{d}{dx}\right)f(x) = f - \frac{y^2}{1\cdot 2}f'' + \frac{y^4}{1\cdot 2\cdot 3\cdot 4}f^{iv} - \&\text{c.},$$

$$-v = \sin\left(y\frac{d}{dx}\right)f(x) = yf' - \frac{y^3}{1\cdot 2\cdot 3}f''' + \cdots,$$

$$\Bigg\}\quad\ldots\ldots(B)$$

where $f(x)$ is the slowly variable value of u at the bottom when $y = 0$, and accents indicate differentiation with respect to x. The corresponding expression for ψ is

$$\psi = yf - \frac{y^3}{1\cdot 2\cdot 3}f'' + \frac{y^5}{1\cdot 2\cdot 3\cdot 4\cdot 5}f^{iv} - \ldots\ldots(C)$$

This equation applies to the upper boundary, if we understand by ψ the there constant value of the stream-function, and it gives us a relation between the ordinate of the boundary and the function f.

If p be the pressure at the upper surface, we have

$$-2\frac{p-C}{\rho} = 2gy + u^2 + v^2,$$

where C is some constant. We will write for brevity,

$$u^2 + v^2 = \varpi - 2gy;\quad\ldots\ldots\ldots\ldots\ldots\ldots(D)$$

and the object of the investigation is to examine how far it is possible to make ϖ constant by varying the form of y as a function of x. Since

$$u^2 + v^2 = (1 + y'^2)\,u^2,$$

our equation becomes

$$yu = \sqrt{\frac{\varpi y^2 - 2gy^3}{1 + y'^2}},$$

or, on substituting for u its value,

$$fy - \frac{y^3}{1\cdot 2}f'' + \frac{y^5}{1\cdot 2\cdot 3\cdot 4}f^{iv} - \&\text{c.} = \sqrt{\frac{\varpi y^2 - 2gy^3}{1 + y'^2}}.$$

Between this equation and (C), f may be eliminated by successive approximation; and we obtain as the relation between y and ϖ,

$$\psi\left\{1 - \frac{y^3}{3}\left(\frac{1}{y}\right)'' - \frac{y^5}{45}\left(\frac{1}{y}\right)^{iv} - \cdots\right\} = \sqrt{\frac{\varpi y^2 - 2gy^3}{1 + y'^2}}.\quad\ldots\ldots\ldots(E)$$

In this investigation y is regarded as a function of x, which varies slowly, or (as we may put it) a function of ωx, where ω is a small quantity. If we agree to neglect the fourth power of ω, the third and following terms on the left-hand side of (E) may be omitted, and we obtain

$$\psi^2 \left\{ 1 + y'^2 - \tfrac{2}{3} y^3 \, \frac{d^2 y^{-1}}{dx^2} \right\} = \varpi y^2 - 2g y^3,$$

or

$$\psi^2 \left\{ 1 - \tfrac{1}{3} y'^2 + \tfrac{2}{3} y y'' \right\} = \varpi y^2 - 2g y^3, \quad \dots \dots \dots \dots (\text{F})$$

by which the value of ϖ is determined approximately in terms of the form of the upper surface. If we suppose ϖ constant and integrate (F) on that hypothesis, we shall obtain a form of upper surface for which the pressure varies very slightly, provided of course that the solution so obtained satisfies the suppositions on which the differential equation (F) is founded.

To integrate (F) we may write it in the form

$$\psi^2 \left\{ 1 + \tfrac{4}{3} y^{\frac{3}{2}} \, \frac{d^2 y^{\frac{1}{2}}}{dx^2} \right\} = \varpi y^2 - 2g y^3,$$

or

$$\frac{d^2 y^{\frac{1}{2}}}{dx^2} = \frac{3}{4 \psi^2} \left(\varpi y^{\frac{1}{2}} - 2g y^{\frac{3}{2}} - \psi^2 y^{-\frac{3}{2}} \right),$$

which becomes a complete differential when multiplied by $2 \, \dfrac{dy^{\frac{1}{2}}}{dx} \, dx$. Thus we find

$$\tfrac{1}{3} y'^2 = C y + \frac{\varpi y^2 - g y^3}{\psi^2} + 1,$$

C being the constant of integration. Suppose now that in the undisturbed parts of the canal the depth is l and the velocity u_0. Then

$$\varpi = u_0^2 + 2gl, \qquad \text{and} \qquad \psi = \int_0^l u_0 \, dy = u_0 l.$$

Substituting these, we get

$$\tfrac{1}{3} y'^2 = 1 + C y + \frac{u_0^2 + 2gl}{u_0^2 l^2} \, y^2 - \frac{g}{u_0^2 l^2} \, y^3. \quad \dots \dots \dots \dots (\text{G})$$

In this equation g and l are given, while u_0 and C are at our disposal; and thus the cubic expression on the right may be made to vanish for $y = l$ and $y = l'$, where l' is the distance between the summit of the wave and the bottom of the canal. If we substitute these values of y and eliminate C, we find

$$u_0^2 = g l' \quad \dots \dots \dots \dots \dots \dots \dots \dots \dots (\text{H})$$

as the relation between u_0 and l'. The constants C and u_0 being now determined so as to make y' vanish when $y = 0$ and when $y = l'$, it will be found that the third root of the cubic is also l, so that our equation may be put into the form

$$y'^2 + \frac{3}{l^2 l'} (y - l)^2 (y - l') = 0. \dots \dots \dots \dots \dots \dots \dots (\text{I})$$

From this result it appears that there is only one maximum or minimum value of y (besides l); and since $y - l'$ is necessarily negative, it follows that the surface-condition cannot be satisfied to this order of approximation by a solitary wave of depression. Differentiating (I), we get

$$y'' = \frac{3(y - l)}{2l^2 l'} \{2l' + l - 3y\},$$

which shows that the points of zero curvature occur when $y = l$ and when $y = \frac{1}{3}(2l' + l) = l + \frac{2}{3}(l' - l)$. Thus the curvature changes sign at two-thirds of the height of the wave above the undisturbed level, and at these points only. The nature of the wave is sufficiently defined by (I); but we may readily integrate again, so as to obtain the relation between x and y. Thus, if $l' - l = \beta$, $y - l = \eta$,

$$\pm x = \sqrt{\frac{l^2 l'}{3\beta}} \, \log_e \left\{ \frac{2\beta}{\eta} - 1 + 2\sqrt{\frac{\beta^2}{\eta^2} - \frac{\beta}{\eta}} \right\}, \quad \ldots\ldots\ldots\ldots (J)^*$$

the constant being taken so that $x = 0$ when $\eta = \beta$. This equation gives the height η at any point x in terms of one constant, viz. the maximum height of the wave. There is therefore (in a given canal) only one form of solitary wave of given maximum height. On either side the height diminishes without interruption, but does not (according to (J)) absolutely vanish at any finite distance. Accordingly there is no definite wave-length; but if we inquire what value of x corresponds to a given ratio of $\eta : \beta$, we get

$$x \propto \sqrt{\{(l + \beta)/\beta\}}$$

being greatest for the smallest waves.

Suppose, for example, that we regard the wave as ending where the height is one-tenth of the maximum. Then

$$x : 1 = 2\cdot1 \, \sqrt{(1 + l/\beta)}.$$

The shortest wave-length is when $\beta = l$; and then $2x/l = 5\cdot96$. If $\beta = \frac{1}{3}l$, $2x/l = 8\cdot4$. If $\beta = \frac{1}{5}l$, $2x/l = 12\cdot6$. These results are in agreement with Russell's observations.

The form of the wave as determined by (J) is shown in the figure, half the wave only being drawn :—

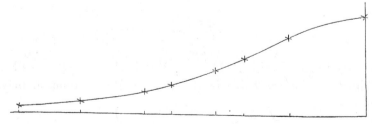

* [1899. The expression for η in terms of x is $\eta = \beta \, \mathrm{sech}^2 (x/2b)$, where $b^2 = l^2(l + \beta)/3\beta$.]

The velocity of propagation is given by (H), which is Scott Russell's formula exactly. In words, the velocity of the wave is that due to half the greatest depth of the water.

Another of Russell's observations is now readily accounted for:—"It was always found that the wave broke when its elevation above the general level became equal, or nearly so, to the greatest depth. The application of mathematics to this circumstance is so difficult, that we confine ourselves to the mention of the observed fact*." When the wave is treated as stationary, it is evident from dynamics that its height can never exceed that due to the velocity of the stream in the undisturbed parts; that is, $l' - l$ is less than $u_0^2/2g$. But $u_0^2 = gl'$, and therefore $l' - l$ is less than $\frac{1}{2}l'$, or $l' - l$ is less than l. When the wave is on the point of breaking, the water at the crest is moving with the velocity of the wave.

Periodic Waves in Deep Water.

The best known theory of these waves is that of Gerstner, Rankine, and Froude, in which the profile is trochoidal. The motion of each particle of the fluid is in a circle, which is described with uniform velocity. If h, k be the coordinates of the centre of one of these circles, measured horizontally and downwards respectively, the position of the particle at time t is given by

$$\xi = h + Re^{-\frac{k}{R}} \sin\left(at + \frac{h}{R}\right), \qquad \eta = k + Re^{-\frac{k}{R}} \cos\left(at + \frac{h}{R}\right).$$

It is not difficult to show that the motion represented by these equations satisfies the condition of continuity, and is consistent with the principles of fluid mechanics; but it involves a molecular rotation, whose amount is

$$ae^{-2k/R} \div (1 - e^{-2k/R}).$$

This molecular rotation, being constant for each particle, is not inconsistent with the properties of frictionless fluid when the motion is once set up; but it is known that a motion of this kind could not be generated from rest in such fluid by any natural force. We proceed to consider the theory of periodic waves in deep water when there is no molecular rotation.

As in the case of long waves, the problem may be reduced to one of steady motion by attributing to the water a velocity equal and opposite to that of the waves. If x be measured horizontally and y downwards from the surface, the conditions of continuity and of freedom from rotation are satisfied by

$$\phi = cx + \alpha e^{-ky} \sin kx, \qquad \psi = cy - \alpha e^{-ky} \cos kx; \quad \ldots\ldots\ldots(A)$$

* Airy, *Tides and Waves*, Art. 401.

where ϕ and ψ are the equipotential and stream functions, c the velocity at a great depth, α a constant depending on the amplitude of the waves, and $k = 2\pi/\lambda$, λ being the wave-length or distance from crest to crest. The motion represented by (A) passes into a uniform horizontal flow at a great depth; and we have only to inquire how far the surface-condition of constant pressure can be satisfied.

If U be the resultant velocity at any point,

$$U^2 = (d\phi/dx)^2 + (d\phi/dy)^2 = c^2 + 2ck\alpha e^{-ky}\cos kx + k^2\alpha^2 e^{-2ky}$$
$$= c^2 + 2ck\,(cy - \psi) + k^2\alpha^2 e^{-2ky},$$

and therefore

$$p/\rho = \text{const.} + (g - kc^2)\,y + ck\psi - \tfrac{1}{2}k^2\alpha^2 e^{-2ky}. \quad\ldots\ldots\ldots\ldots(B)$$

Hence, when ψ is constant and α is so small that α^2 can be neglected, p will also be constant, provided that

$$c = \sqrt{\left(\frac{g}{k}\right)} = \sqrt{\left(\frac{g\lambda}{2\pi}\right)}. \quad\ldots\ldots\ldots\ldots\ldots(C)$$

If c has this value, the surface-condition is satisfied approximately, and (A) may be understood to represent a train of free periodic stationary waves, or, if the motion relatively to deep water be considered, a train of periodic waves advancing without change of type and with a uniform velocity c.

The profile of the wave is determined by the second of equations (A), in which ψ is made constant. By successive approximation we may deduce the value of y in terms of x. If ψ be taken so that the mean value of y is zero, we get

$$y = \frac{\alpha}{c}\left(1 + \frac{5}{8}\frac{k^2\alpha^2}{c^2}\right)\cos kx - \frac{k\alpha^2}{2c^2}\cos 2kx + \frac{3}{8}\frac{k^2\alpha^3}{c^3}\cos 3kx,$$

which is correct as far as α^3. Let

$$\frac{\alpha}{c}\left(1 + \frac{5}{8}\frac{k^2\alpha^2}{c^2}\right) = a\,;$$

then

$$\left.\begin{array}{l}\phi = cx + ca\,(1 - \tfrac{5}{8}k^2a^2)\,e^{-ky}\sin kx, \\[4pt] \psi = cy - ca\,(1 - \tfrac{5}{8}k^2a^2)\,e^{-ky}\cos kx\,;\end{array}\right\} \quad\ldots\ldots\ldots\ldots(D)$$

and for the equation of the surface,

$$y = a\cos kx - \tfrac{1}{2}ka^2\cos 2kx + \tfrac{3}{8}k^2a^3\cos 3kx. \quad\ldots\ldots\ldots\ldots(E)$$

From (B) we may obtain a closer approximation to the value of c. Expanding the exponential, we have (approximately)

$$p/\rho = \text{const.} + (g - kc^2 + k^3\alpha^2)\,y + \ldots\,;$$

so that

$$c^2 = g/k + k^2\alpha^2 = g/k + k^2a^2c^2,$$

or $$c^2 = g/k \,.\, (1 + k^2 a^2), \quad \dots\dots\dots\dots\dots\dots\dots\dots\dots\dots\dots\dots(F)$$

where $k = 2\pi/\lambda$.

Formulæ (E) and (F) are given by Professor Stokes in a memoir published in the *Cambridge Philosophical Transactions*, Vol. VIII.

So long as the depth is everywhere sufficiently great in comparison with the length of the waves, uniformity of depth is immaterial. For waves in water of constant finite depth l, the expression for ψ is

$$\psi = cy - a \cos kx \left\{ e^{-k(y-l)} - e^{k(y-l)} \right\},$$

and the velocity c is determined by

$$c^2 = \frac{g}{k} \,.\, \frac{e^{kl} - e^{-kl}}{e^{kl} + e^{-kl}} = g/k \,.\, \tanh kl, \quad \dots\dots\dots\dots\dots\dots(G)$$

which passes into (C) when l is considerable in comparison with λ. When l is small, we get from (G) $c^2 = gl$, which is the formula proper for long waves. When obtained thus, it is applicable in the first instance only to waves of a particular type; but the fact that it is independent of k or λ would lead us to the conclusion that the same formula would apply to a long wave of any type.

In one respect the theory of irrotational waves may be considered inferior to that of Rankine, which last is *exact*, in the sense that it is independent of any supposition as to the smallness of the waves. So far as I am aware, writers on this subject appear to think that it is only a question of mathematics to determine the form of irrotational waves of finite amplitude to any degree of approximation. But it seems to me by no means certain that any such type exists, capable of propagating itself unchanged with uniform velocity. I see no reason why the possibility of such waves in deep water should be taken for granted, when we know that in shallow water waves of finite height cannot be propagated without undergoing a gradual alteration of type*.

One of the most interesting results of Professor Stokes's theory is the existence of a slow translation of the water near the surface in the direction of the wave. I propose to show that this superficial motion is an immediate consequence of the absence of molecular rotation, and that it is independent of the condition of constant pressure at the bounding surface.

Let AB be the surface from crest to hollow, and CD a neighbouring stream-line. Draw $A'B'$, $C'D'$, two stream-lines at such a depth that the

---

* [1899. This supposed impossibility is connected with the assumption that the waves in shallow water are very long. The more recent researches of Sir G. Stokes (*Collected Papers*, vol. I. p. 314) and of Prof. Korteweg and Dr De Vries (*Phil. Mag.* vol. xxxix. p. 422, 1895) seem to establish the existence of absolutely permanent waves of finite height.]

steady motion of the fluid is uniform, and so as to include a total stream equal to that which flows between AB and CD. Then we have to show that a particle at A will take longer to reach B, than a particle at A' takes to arrive at B'. Now if σ denotes the small breadth of the tube AD at any

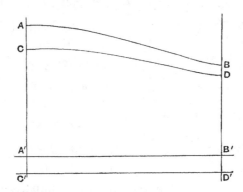

point, and v the velocity, the total stream is σv and is constant. Denoting it by K, we have

$$v = K \div \sigma.$$

The time t occupied by a particle in moving from A to B is therefore

$$t = \int \frac{ds}{v} = \int \frac{\sigma ds}{K} = \text{area } AD \div K.$$

And if t' represents the time between A' and B',

$$t' = \text{area } A'D' \div K,$$

K being the same in both cases, since the total streams are by supposition equal. Thus

$$t : t' = \text{area } AD : \text{area } A'D';$$

and it remains to prove that AD is greater than $A'D'$.

If we draw equipotential lines in such a manner that the small spaces cut off between them and AB, CD are squares, then we know that the same series of equipotential lines will divide the space between $A'B'$, $C'D'$, into small squares also. Now if a line be divided into a given number of parts, the sum of the squares of the parts will be a minimum when the parts are all equal. Hence the space AD is greater than if the squares described on the parts of AB were all equal, and therefore à *fortiori* greater than the space $A'D'$, which consists of the sum of the squares of the same number of equal parts of a shorter line.

It follows that when a particle starting from A' has arrived at B', another particle starting at the same moment from A will fall short of B. Thus in a progressive wave the water near the surface has on the whole a motion of translation in the direction in which the waves advance.

Oscillations in Cylindrical Vessels.

If liquid contained in a cylindrical vessel of any section, whose generating lines are vertical and whose depth is uniform, be disturbed from the position of equilibrium, oscillations will ensue in consequence of the tendency of the fluid to recover its horizontal boundary.

Let us consider in the first place the small vibrations in two dimensions of a compressible fluid such as air when contained within a cylindrical rigid boundary. If x and y be the rectangular coordinates of any point, and ϕ the velocity potential, it is known that ϕ will satisfy over the whole area

$$\frac{d^2\phi}{dt^2} = a^2 \left(\frac{d^2\phi}{dx^2} + \frac{d^2\phi}{dy^2} \right), \quad \dots\dots\dots\dots\dots\dots(A)$$

a being the velocity of sound; while round the contour

$$\frac{d\phi}{dn} = 0, \quad \dots\dots\dots\dots\dots\dots\dots\dots(B)$$

where $d\phi/dn$ denotes the rate of variation of ϕ in a normal direction.

Whatever the motion of the air may be, it can be analyzed into components of the harmonic type. Suppose that for one of these ϕ varies as $\cos kat$; then, from (A),

$$\frac{d^2\phi}{dx^2} + \frac{d^2\phi}{dy^2} + k^2\phi = 0 \dots\dots\dots\dots\dots\dots\dots(C)$$

is an equation which ϕ must satisfy for the component vibration in question. The equations (C) and (B) can only be satisfied with certain definite values of k; and the functions ϕ corresponding to these values are proportional to what may be called the *normal* functions of the air-system. We may denote these functions by u_k. Any function arbitrary over the area can be expanded in a series of the functions u *.

Returning to the liquid-problem, we see that the elevation h of the surface at any point above the undisturbed position may be expressed by the series

$$h = \Sigma \alpha_k u_k(x, y), \quad \dots\dots\dots\dots\dots\dots\dots(D)$$

the quantities α being constants with respect to space, but dependent upon time. The potential energy of the displacement, *calculated on the hypothesis of a constant pressure on the surface*, will clearly be

$$\left. \begin{aligned} V &= g\rho \iint . \int_0^h z\,dz . dx\,dy = \tfrac{1}{2} g\rho \iint h^2 dx\,dy \\ &= \tfrac{1}{2} g\rho \iint (\Sigma \alpha_k u_k)^2 dx\,dy = \tfrac{1}{2} g\rho \, \Sigma \alpha_k^2 \iint u_k^2 dx\,dy \end{aligned} \right\}, \quad \dots\dots\dots(E)$$

by the conjugate property of the functions u. This is the potential energy.

---

* See on this subject several papers by the author, especially "General Theorems relating to Vibrations," *Math. Society Proceedings*, vol. IV. No. 63, and *Phil. Mag.* 1873 [Arts. XXI., XXV.].

The motion of the fluid throughout the interior depends, according to a known theorem, only upon the motion of the surface; and the surface normal velocity

$$- d\phi/dz = \dot{h} = \Sigma \dot{\alpha}_k u_k.$$

If l be the depth, the complete value of ϕ is given by

$$- \phi = \Sigma \frac{e^{k(z-l)} + e^{-k(z-l)}}{k(e^{-kl} - e^{kl})} \dot{\alpha}_k u_k(x, y). \quad \dots\dots\dots\dots(F)$$

For, in the first place, this value of ϕ satisfies Laplace's equation, inasmuch as each term u_k satisfies the equation (C);

Secondly, (F) satisfies the condition imposed by the rigid cylindrical boundary, since $du_k/dn = 0$;

Thirdly, (F) makes $d\phi/dz = 0$, when $z = l$;

And fourthly, when $z = 0$, $- d\phi/dz = \dot{h}$.

The kinetic energy T may now be readily calculated:

$$T = \frac{\rho}{2} \iiint \left\{ \left(\frac{d\phi}{dx}\right)^2 + \left(\frac{d\phi}{dy}\right)^2 + \left(\frac{d\phi}{dz}\right)^2 \right\} dx\,dy\,dz = \frac{\rho}{2} \iint \frac{d\phi}{dn} \, dS,$$

by Green's theorem, dS denoting an element of the surface bounding the mass, and $d\phi/dn$ the rate of variation of ϕ in a normal direction outwards. The surface S consists of three parts—the bottom of the vessel, the cylindrical side of the vessel, and the upper surface of the fluid. Over the first two of these, $d\phi/dn = 0$, and thus

$$T = - \frac{\rho}{2} \iint \left(\phi \frac{d\phi}{dz} \right)_{z=0} dx\,dy.$$

Now when $z = 0$,

$$\phi = \Sigma \dot{\alpha}_k k^{-1} \coth kl \,.\, u_k, \qquad d\phi/dz = - \Sigma \dot{\alpha}_k u_k,$$

so that

$$T = \tfrac{1}{2}\rho \, \Sigma \dot{\alpha}_k^2 k^{-1} \coth kl \,.\, \iint u_k^2 \, dx\,dy, \quad \dots\dots\dots\dots (G)$$

the product of any two functions u_k, $u_{k'}$ vanishing when integrated over the area.

We have now to calculate the work done by impressed forces corresponding to the displacement represented by $\delta \alpha_k$. It must be remembered that these forces are limited to be such as have a potential. Let δp denote the variable part of the pressure at the surface, supposed to remain in its position of rest, whether applied directly or due to impressed body-forces, then

$$\text{work done on system} = - \iint \delta p \, \delta h \, dx \, dy.$$

If δp be expanded in the series, $\delta p = \Sigma \beta_k u_k(x, y)$,

$$\text{work} = - \iint \Sigma \beta_k u_k \,.\, \Sigma \delta \alpha_k u_k \,.\, dx \, dy = - \Sigma \beta_k \delta \alpha_k \iint u_k^2 \, dx \, dy.$$

We can now form the equations of motion in terms of the generalized coordinates α_k. By Lagrange's method,

$$\ddot{\alpha}_k k^{-1} \coth kl + g\alpha_k = -\beta_k/\rho \quad \dots\dots\dots\dots\dots(H)$$

is the equation determining the variation of the coordinate α_k, where

$$\beta_k = \iint \delta p \, u_k \, dx \, dy \div \iint u_k^2 \, dx \, dy.$$

When the oscillations are free, $\beta_k = 0$. If the period be τ_k, and the corresponding period for the air-vibration τ_k',

$$\tau_k = 2\pi \div \sqrt{(gk \tanh kl)}, \qquad \tau_k' = 2\pi \div ka. \quad \dots\dots\dots\dots(I)$$

If λ' be the wave-length of plane aerial vibrations having the period τ_k',

$$k = 2\pi \div \lambda'.$$

If kl be very small, the ratio of periodic times is

$$\tau_k : \tau_k' = a : \sqrt{(gl)}, \quad \dots\dots\dots\dots\dots\dots(J)$$

and is independent of k. Hence the two problems of the vibrations of air and liquid are mathematically analogous whatever the initial circumstances may be ; so that if the condensation in the first follows the same law initially as the elevation in the second, the correspondence will be preserved throughout the subsequent motion, if $a^2 = gl$. The initial circumstances, however, must be such as not to give prominence to the higher components, for which kl would no longer be small.

When kl is not negligible, we learn from formula (I) that the period increases with l until kl is moderately great, when it becomes sensibly

$$\tau_k = 2\pi \div \sqrt{(gk)}. \quad \dots\dots\dots\dots\dots\dots(K)$$

In any case the period is independent of the density of the liquid.

Some careful observations on liquid vibrations have been recently made by Professor Guthrie[*], with which it may be interesting to compare the results of theory. Professor Guthrie used troughs whose horizontal section was rectangular and circular. We will take the rectangular section first.

Confining ourselves to those modes of vibration which depend on only one horizontal coordinate, we may take for the normal functions

$$u = \cos(n\pi x/L),$$

L being the length of the trough, n integral, and x being measured from one end. The corresponding value of k is $n\pi/L$. Hence, from (I), the length of the simple equivalent pendulum is

$$\frac{L}{n\pi} \coth \frac{n\pi l}{L}. \quad \dots\dots\dots\dots\dots\dots(L)$$

---

[*] *Phil. Mag.* October and November 1875.

When $n\pi l/L$ is considerable, (L) becomes

$$L/n\pi , \quad \dots\dots\dots\dots\dots\dots\dots\dots\dots\dots\dots(M)$$

or, for a closer approximation,

$$\frac{L}{n\pi}\left(1 + 2e^{-\frac{2n\pi l}{L}}\right). \quad \dots\dots\dots\dots\dots\dots\dots\dots(N)$$

Formula (M) was found by Professor Guthrie to agree with observation when $n = 1$ or 2. The periods in the two cases are in the ratio $1 : \sqrt{2}$, if the depth be sufficient.

If the depth bear a constant ratio to the length, (L) or (I) shows that the period is directly proportional to the square root of the linear dimension; and the same law will obtain when the depth is great, whatever the absolute value may be.

If $n = 1$, the points of constant elevation occur when $x = \frac{1}{2}L$ (that is, in the middle of the length); and if $n = 2$, when $x = \frac{1}{4}L$ or $\frac{3}{4}L$. The maximum elevations (or depressions) are equal.

These results take into account inertia and gravity only. From some expressions in his paper Professor Guthrie would appear to attribute the effect of shallowness in increasing the period to friction. No doubt friction must act in this direction; but its immediate effect is on the amplitude, and not on the period. In all ordinary cases the action of insufficient depth may be sufficiently accounted for by the increase of the effective inertia due to the contraction of the channels along which the liquid flows, in the same way as the pitch of an organ-pipe is lowered by an obstruction at the mouth. In such vessels as those used by Professor Guthrie it may be doubted whether friction and capillarity have any sensible influence on the periodic time.

The theory for the circular trough depends on the class of functions named after Bessel, which are an extreme case of Laplace's spherical functions. For the symmetrical vibrations we have

$$u = J_0(kr), \quad \dots\dots\dots\dots\dots\dots\dots\dots\dots\dots(O)$$

r being the radius vector; and if R be the radius of the vessel, k is a root of

$$J_0'(kR) = 0. \quad \dots\dots\dots\dots\dots\dots\dots\dots\dots\dots(P)$$

If $x = kR$, the values of x satisfying (P) are 3·832, 7·015, 10·174, &c., of which only the first belongs to the cases experimented on by Professor Guthrie. The approximate formula for the length of the simple equivalent pendulum corresponding to (N) is

$$\frac{R}{3\cdot832}\left(1 + 2e^{-7\cdot7\frac{l}{R}}\right), \quad \dots\dots\dots\dots\dots\dots\dots\dots(Q)$$

or, when l is considerable,

$$R \div 3\text{·}832 \text{ simply.} \dots\dots\dots\dots\dots\dots\text{(R)}$$

Professor Guthrie compares his observations with a pendulum of length $R \div 4$, and finds a fair agreement, which, however, would be improved by the substitution of the theoretical formula (R).

According to (O) the place of zero elevation and depression occurs when

$$r = \frac{2\text{·}405}{3\text{·}832} R = \text{·}6277\ R.$$

According to observation,

$$r = \tfrac{2}{3} R = \text{·}6667\ R.$$

From the Tables of Bessel's functions it appears that the amplitude at the edge of the vessel is ·403 of that at the centre. Professor Guthrie makes this ·5.

For the next set of vibrations in a circular dish u is of the form

$$u = \sin\theta\, J_1(kr),$$

where the admissible values of kR are 1·841, 5·332, 8·536, &c. Hence for the gravest of this group the length of the equivalent pendulum is

$$R \div 1\text{·}841. \dots\dots\dots\dots \dots\dots\dots\text{(S)}$$

In this group of modes the elevation vanishes at all points along a certain diameter $(\theta = 0)$.

In the third group we have

$$u = \sin 2\theta\, J_2(kr),$$

and the admissible values of kR are 3·054, 6·705, 9·965, &c. For the gravest of these the length of the equivalent pendulum is

$$R \div 3\text{·}054, \dots\dots\dots\dots\dots\dots\dots\text{(T)}$$

if the depth be sufficient. The elevation vanishes along two perpendicular diameters $(\theta = 0,\ \theta = \tfrac{1}{2}\pi)$.

In the fourth group there would be three diameters for which $u = 0$; and the length of the pendulum isochronous with the gravest mode will be

$$R \div 4\text{·}201. \dots\dots\dots\dots\dots\dots\text{(U)}$$

The frequencies of vibration in the three gravest modes, being inversely as the square roots of the corresponding pendulum-lengths, are in the ratio

$$1 : 1\text{·}29 : 1\text{·}44.$$

Professor Guthrie's observations give for the value of these ratios

$$1 : 1\cdot31 : 1\cdot48.$$

Possibly too low a frequency is attributed to the gravest vibration owing to the effect of insufficient depth.

When the complete theory of the free vibrations of any system is thoroughly known, it is in general easy to investigate the effect of periodic forces. If u_1, u_2, &c. are the normal functions, and $2\pi/n_1$, $2\pi/n_2$, &c. the periods of the corresponding free vibrations, the effect of forces whose period is $2\pi/p$ can be expressed in terms of the effect produced by similar forces of infinite period, which last can be calculated statically. Thus, if the solution of the problem according to the equilibrium theory is

$$A_1 u_1 \cos pt + A_2 u_2 \cos pt + \dots,$$

the true solution as modified by the inertia of the system will be

$$\frac{A_1 n_1^2}{n_1^2 - p^2} u_1 \cos pt + \frac{A_2 n_2^2}{n_2^2 - p^2} u_2 \cos pt + \dots.$$

Let us calculate in this way the motion in a circular cylindrical basin due to a small horizontal force, acting uniformly throughout the mass of liquid, but variable with the time according to the harmonic law. The equilibrium value of h (the elevation) is evidently

$$h = r \cos \theta \cos pt;$$

and the only difficulty consists in expressing r by a series of Bessel's functions J_1. It may be proved that

$$r = \frac{2 J_1 (kr)}{(k_1^2 - 1) J_1 (k_1)} + \frac{2 J_1 (k_2 r)}{(k_2^2 - 1) J_1 (k_2)} + \dots, \quad \dots\dots\dots\dots(V)$$

where k_1, k_2, &c. are the roots of $J_1'(k) = 0$, and the radius R is taken as unity. Thus the true value of h (after the motion has been going on long enough to be independent of initial circumstances) is

$$h = \frac{2 n_1^2 \cos pt \cos \theta J_1 (k_1 r)}{(n_1^2 - p^2)(k_1^2 - 1) J_1 (k_1)} + \dots, \quad \dots\dots\dots\dots(W)$$

the summation being extended to all the admissible values of k. The value of n^2 is given by

$$n^2 = gk \tanh kl.$$

If the system be at rest at $t = 0$, and displaced according to the law $h = r \cos \theta$ (that is, with an inclined plane surface), the subsequent motion is given in rapidly converging series by

$$h = \frac{2 J_1 (k_1 r) \cos \theta}{(k_1^2 - 1) J_1 (k_1)} \cos n_1 t + \frac{2 J_1 (k_2 r) \cos \theta}{(k_2^2 - 1) J_1 (k_2)} \cos n_2 t + \dots. \quad \dots\dots(X)$$

P.S.—Some recent observations on the periods of the oscillations of water in a large circular tank may be worth recording. The radius of the tank is 60·3 inches [153·2 cm.], and the depth about 43 inches [109 cm.]. The oscillations were excited by dipping one or more buckets synchronously with the beats of a metronome set approximately beforehand. Soon after the withdrawal of the buckets the vibrations were counted (in most cases for five minutes), and the results reduced for a space of one minute.

Gravest symmetrical mode.—Frequency by observation 47·3 [complete oscillations per minute]. The theoretical result for an infinite depth is 47·32, and for actual depth 47·13.

Next highest symmetrical mode.—By observation, frequency $= 64·1$, by theory 64·02. In this case the correction for finite depth is insensible, and the length of the equivalent pendulum $= R \div 7·015$.

Gravest mode with one nodal diameter.—By observation, frequency $= 30·0$. By theory, for infinite depth 32·81, for actual depth 30·48.

One nodal diameter and one nodal circle.—By observation, frequency $= 56·0$; by theory 55·8. The length of equivalent pendulum $= R \div 5·332$.

Two nodal diameters.—By observation, frequency $= 41·5$. By theory, for infinite depth 42·09, for actual depth 41·59.

The agreement between theory and observation is as close as could be expected.

I have lately seen a memoir by M. Boussinesq (1871, *Comptes Rendus,* Vol. LXXII.), in which is contained a theory of the solitary wave very similar to that of this paper. So far as our results are common, the credit of priority belongs of course to M. Boussinesq.

39.

ON THE APPROXIMATE SOLUTION OF CERTAIN PROBLEMS RELATING TO THE POTENTIAL.

[*Proceedings of the London Mathematical Society*, VII. pp. 70—75, 1876.]

THE first problem that I propose for consideration is that of the flow of electricity in two dimensions along a strip of uniform metal sheet, such as tinfoil, the strip being bounded by curves symmetrical with respect to a central line, taken as axis of x, and nowhere more than moderately inclined to that axis. The equations of the boundary may be written $y = \pm y_1$, where y_1 is a slowly varying function of x. On account of the symmetry the potential is an even function of y, and the axis of x is a stream-line, so that the problem is the same as if the axis were itself a boundary. The conditions to be satisfied by the potential ϕ are therefore the usual equation of Laplace,

$$\nabla^2\phi = d^2\phi/dx^2 + d^2\phi/dy^2 = 0,$$

which must be true over the whole area of the strip, together with the boundary conditions, that there must be no normal flow across the curves $y = y_1$ and $y = 0$. If we introduce the stream function ψ, which is related to ϕ by the equations

$$d\phi/dx = d\psi/dy, \qquad d\phi/dy = - d\psi/dx , \ldots\ldots\ldots\ldots(1)$$

and itself satisfies $\nabla^2\psi = 0$, the boundary conditions may be expressed by supposing ψ to be zero when $y = 0$, and when $y = y_1$ to assume a second constant value ψ_1.

If the boundary of the strip were straight and parallel to the axis, the current $d\phi/dx$ would be constant at all points, and we should have

$$\phi = f \cdot x, \qquad \psi = f \cdot y,$$

where f represents the constant current.

The object of the present investigation is to determine the values of ϕ and ψ, when the above simple values may be taken as a basis of approximation, and to calculate corrections for the effect of the deviation of y_1 from constancy.

Denoting by f the slowly varying magnitude of the current when $y = 0$, we may take

$$
\left.\begin{aligned}
\phi &= \cos\left(y\,\frac{d}{dx}\right) f_1 = f_1 - \frac{y^2}{1\,.\,2}f' + \frac{y^4}{1\,.\,2\,.\,3\,.\,4}f''' - \&c. \\
\psi &= \sin\left(y\,\frac{d}{dx}\right) f_1 = yf - \frac{y^3}{1\,.\,2\,.\,3}f'' + \&c.
\end{aligned}\right\} ,\ \ \ldots\ldots(2)
$$

where $f_1 = \int f\, dx$.

These values satisfy the general conditions of the potential and stream functions, and when $y = 0$ make

$$
d\phi/dx = f, \qquad \psi = 0.
$$

The second of equations (2) may be regarded as determining the lines of flow (any one of which may be supposed to be the boundary of the sheet) in terms of f. What remains to be effected is the determination of f from the condition that $\psi = \psi_1$, when $y = y_1$, viz.,

$$
\psi_1 = y_1 f - \frac{y_1^3}{1\,.\,2\,.\,3}f'' + \frac{y_1^5}{1\,.\,2\,.\,3\,.\,4\,.\,5}f^{\mathrm{iv}} - \&c.,
$$

an equation which may be treated by the method of successive approximation. We have

$$
f = \frac{\psi_1}{y_1} + \frac{y_1^2}{1\,.\,2\,.\,3}f'' - \frac{y_1^4}{1\,.\,2\,.\,3\,.\,4\,.\,5}f^{\mathrm{iv}} + \&c.; \ \ \ \ldots\ldots (3)
$$

whence we obtain

$$
\begin{aligned}
f &= \frac{\psi_1}{y_1} + \frac{y_1^2}{6}\left[\frac{d^2}{dx^2}\left(\frac{\psi_1}{y}\right) + \frac{1}{6}\frac{d^2}{dx^2}\left\{y_1^2\frac{d^2}{dx^2}\left(\frac{\psi_1}{y_1}\right)\right\}\right] - \frac{y_1^4}{120}\frac{d^4}{dx^4}\left(\frac{\psi_1}{y_1}\right) \\
&= \frac{\psi_1}{y_1} + \frac{\psi_1}{6}\left(2y_1^{-1}y_1'^2 - y_1''\right) + \frac{y_1^2\psi_1}{36}\frac{d^2}{dx^2}\left(2y_1^{-1}y_1'^2 - y_1''\right) - \frac{y_1^4}{120}\frac{d^4}{dx^4}\left(\frac{\psi_1}{y_1}\right), \ \ldots(4)
\end{aligned}
$$

where accents denote differentiation with respect to x.

In order to calculate the electrical resistance of the strip, we must compare the total current with the difference of potentials at the two extremities. If the lamina be of unit specific conductivity, the total current

$$
= \int_0^{y_1}\frac{d\phi}{dx}\,dy = \int_0^{y_1}\frac{d\psi}{dy}\,dy = \psi_1.
$$

For the potential on the axis we have $\phi = \iint f\,dx$, and therefore the resistance of the strip is represented by

$$\iint f\,dx \div \psi_1,$$

where f has the value given by (4).

If the equipotential lines, between which the resistance is required, be situated in parts of the strip, where the edge is straight and parallel to the axis (for a sufficient distance in comparison with the width), our expression may be simplified by integration by parts, since all the differential coefficients of y_1 with respect to x may be supposed to vanish at the limits under the circumstances contemplated.

We have

$$\int y_1^2 \frac{d^2}{dx^2}\left(2y_1^{-1}y_1'^2 - y_1''\right)dx = \int \left(2y_1^{-1}y_1'^2 - y_1''\right)\frac{d^2y_1^2}{dx^2}\,dx$$

$$= 2\int\left(2y_1^{-1}y_1'^2 - y_1''\right)\left(y_1'^2 - y_1 y_1''\right)dx,$$

$$\int y_1^4 \frac{d^4 y_1^{-1}}{dx^4}\,dx = \int \frac{d^2 y_1^4}{dx^2}\frac{d^2 y_1^{-1}}{dx^2}\,dx = 4\int\left(2y_1^{-1}y_1'^2 - y_1''\right)\left(3y_1'^2 + y_1 y_1''\right)dx\,;$$

and the final result for the resistance may be written

$$\int \frac{dx}{y_1}\left\{1 + \frac{1}{3}y_1'^2 - \frac{1}{45}\left(2y_1'^2 - y_1 y_1''\right)^2\right\}. \quad\ldots\ldots\ldots\ldots(5)$$

This expression admits of various forms. For example, instead of the second term we might take

$$\int \frac{dx}{y_1}\left\{\ldots + \frac{1}{6}\left(2y_1'^2 - y_1 y_1''\right) - \ldots\right\},$$

which differs by the addition of a term which is a complete differential, and therefore disappears on integration.

The success of the approximation depends of course upon the degree of slowness with which y_1 varies, but it is not necessary to suppose that the whole variation of y_1 is small within the limits of x under consideration. If we suppose that y_1 is a function of ωx, where ω is a small quantity, our expression for the resistance will include all terms containing as factors lower powers of ω than ω^6.

The same expression (5) gives us the *conductivity* of the strip between the curves $y = y_1$ and $y = 0$, supposed equipotential. For it is a general proposition in the theory of conduction in two dimensions that if two opposite sides of any curvilinear quadrilateral be at uniform (through different) potentials and the other pair of sides be non-conducting, the resistance is the reciprocal of that corresponding to an altered state of things in which the first pair of opposite sides are non-conducting and the second pair are equipotential.

Its truth is easily seen if we fill up the quadrilateral with the intermediate stream and equipotential curves, which will ultimately divide the area into elementary squares. The resistance of any square between opposite edges is independent of the size of the square, any augmentation of resistance due to an increased distance to be travelled being compensated by the effect of increased width. Thus, if the number of interpolated stream and equipotential curves be respectively $m - 1$ and $n - 1$, the whole resistance is to that of a square in the ratio $n : m$; while for the same reason, if the equipotential and stream-lines be interchanged, the resistance is $m : n$ of that of a square.

The same result is applicable to other problems mathematically analogous; for example, the calculation of the capacity of an electrical accumulator.

A similar method applies to the investigation of conduction in three dimensions in a conductor, which is symmetrical round the axis of x, and whose boundary does not differ violently from a cylindrical surface. If r be the distance of any point from the axis, ϕ and ψ the potential and stream functions, we have

$$\frac{d\phi}{dx} = \frac{1}{r}\frac{d\psi}{dr}, \qquad \frac{d\phi}{dr} = -\frac{1}{r}\frac{d\psi}{dx}, \qquad \dots\dots\dots\dots\dots(6)$$

whence by elimination

$$\frac{d^2\phi}{dr^2} + \frac{1}{r}\frac{d\phi}{dr} + \frac{d^2\phi}{dx^2} = 0, \qquad \dots\dots\dots\dots\dots(7)$$

$$\frac{d^2\psi}{dr^2} - \frac{1}{r}\frac{d\psi}{dr} + \frac{d^2\psi}{dx^2} = 0, \qquad \dots\dots\dots\dots\dots(8)$$

which show that ϕ and ψ are not now, as in the case of two dimensions, interchangeable. The series corresponding to (2) are

$$\left.\begin{aligned} \phi &= F - \frac{r^2 F''}{2^2} + \frac{r^4 F^{iv}}{2^2 \cdot 4^2} - \frac{r^6 F^{vi}}{2^2 \cdot 4^2 \cdot 6^2} + \dots\dots \\ \psi &= \frac{r^2 F'}{2} - \frac{r^4 F'''}{2^2 \cdot 4} + \frac{r^6 F^{v}}{2^2 \cdot 4^2 \cdot 6} - \frac{r^8 F^{vii}}{2^2 \cdot 4^2 \cdot 6^2 \cdot 8} + \dots \end{aligned}\right\}, \qquad \dots\dots\dots(9)$$

where F is a function of x so far arbitrary. If the form of the boundary be defined by $r = y$, we have

$$\psi_1 = \frac{y^2 F'}{2} - \frac{y^4 F'''}{2^2 \cdot 4} + \frac{y^6 F^{v}}{2^2 \cdot 4^2 \cdot 6} - \dots\dots\dots,$$

from which F' is to be determined in terms of y. By successive approximation we obtain

$$F' = \frac{2\psi_1}{y^2} + \frac{y^2}{8}\left\{\frac{d^2}{dx^2}\left(\frac{2\psi_1}{y^2}\right) + \frac{1}{8}\frac{d^2}{dx^2} y^2 \frac{d^2}{dx^2}\left(\frac{2\psi_1}{y^2}\right)\right\} - \frac{y^4}{12 \cdot 4^2}\frac{d^4}{dx^4}\left(\frac{2\psi_1}{y^2}\right)\dots\dots(10)$$

The total stream is given by the integral

$$\int_0^y \frac{d\phi}{dx} 2\pi r\, dr = \int_0^y \frac{1}{r}\frac{d\psi}{dr} \cdot 2\pi r\, dr = 2\pi\psi_1;$$

and therefore the resistance between any two equipotential surfaces is represented by

$$\frac{1}{2\pi\psi_1}\int F'dx.$$

On the supposition that at each limit the boundary is cylindrical for a sufficient distance in comparison with the diameter, we may treat this expression by integration by parts in the same manner as for the corresponding problem in two dimensions, and we then obtain

$$\text{resistance} = \int \frac{dx}{\pi y^2}\left\{1 + \frac{1}{2}y'^2 - \frac{(3y'^2 - yy'')^2}{48}\right\}, \quad\ldots\ldots\ldots(11)$$

where, as before, accents denote differentiation with respect to x.

In a memoir on the Theory of Resonance published in the *Phil. Trans.* for 1871*, I had occasion to consider this problem, and I then proved that the first term of (11) represents an *inferior limit* to the resistance, and that the first and second terms together give a *superior limit*, these limits being applicable without any restriction on the form of the symmetrical boundary. It now appears that the superior limit coincides with the accurate value as far as a *second approximation*, and that the correction is approximately given by the third term in (11), which is essentially negative.

Other problems in Electricity and Heat depending upon the potential may be treated by the above method. For example, the conductivity between the plane $z = 0$, and a neighbouring equipotential surface $z = z_1$, where z_1 is a function of x and y differing from a constant over a finite region only, is approximately

$$\iint \frac{dx\,dy}{z_1}\left\{1 + \frac{1}{3}\left(\frac{dz_1}{dx}\right)^2 + \frac{1}{3}\left(\frac{dz_1}{dy}\right)^2\right\}, \quad\ldots\ldots\ldots\ldots(12)$$

where the area of integration includes the whole of the region through which z_1 varies†.

Finally it may be noticed that the preceding methods are applicable in two dimensions, when we replace x and y by any conjugate functions α, β, of x, y. By this transformation the scope of the analysis may be considerably increased, but to enlarge upon this would take us too far from the principal subject of the paper.

---

* [Art. v. p. 35.] See also Maxwell's *Electricity and Magnetism*, vol. I., Art. 307. In connection with this subject I may be allowed here to record that I have proved that the "correction for the end" of a cylindrical conductor is less than ·8242 of the radius, by a calculation based on an assumed motion containing two arbitrary constants; this result is probably very near the truth. The corresponding superior limit obtained from one arbitrary constant is ·8281 (*Phil. Mag.*, Nov. 1872). If the motion at the end be assumed to be uniform, we get ·849 (Maxwell, Art. 309). An inferior limit is ·785.

† Math. Tripos, 1876, Jan. 21, 1½ to 4, Question x. [Art. XLI. p. 286.]

40.

OUR PERCEPTION OF THE DIRECTION OF A SOURCE OF SOUND.

[*Nature*, XIV. pp. 32—33, 1876.]

THE practical facility with which we recognize the situation of a sounding body has always been rather a theoretical difficulty. In the case of sight a special optical apparatus is provided whose function it is to modify the uniform excitation of the retina, which a luminous point, wherever situated, would otherwise produce. The mode of action of the crystalline lens of the eye is well understood, and the use of a lens is precisely the device that would at once occur to the mind of an optician ignorant of physiology. The bundle of rays, which would otherwise distribute themselves over the entire retina, and so give no indication of their origin, are made to converge upon a single point, whose excitation is to us the sign of an external object in a certain definite direction. If the luminous object is moved, the fact is at once recognized by the change in the point of excitation.

There is nothing in the ear corresponding to the crystalline lens of the eye, and this not accidentally, so to speak, but by the very nature of the case. The efficient action of a lens depends upon its diameter being at least many times greater than the wave-length of light, and for the purposes of sight there is no difficulty in satisfying this requirement. The wave-length of the rays by which we see is not much more than a ten-thousandth part of the diameter of the pupil of the eye. But when we pass to the case of sound and of the ear, the relative magnitudes of the corresponding quantities are altogether different. The waves of sound issuing from a man's mouth are about eight feet long, whereas the diameter of the passage of the ear is quite small, and could not well have been made a large multiple of eight feet. It is evident therefore that it is useless to look for anything corresponding to the crystalline lens of the eye, and that our power of telling the origin of a sound must be explained in some different way.

It has long been conjectured that the explanation turns upon the combined use of both ears; though but little seems to have been done hitherto in the way of bringing this view to the test. The observations and calculations now brought forward are very incomplete, but may perhaps help to clear the ground, and will have served their purpose if they induce others to pursue the subject.

The first experiments were made with the view of finding out with what degree of accuracy the direction of sound could be determined, and for this it was necessary of course that the observer should have no other material for his judgment than that contemplated.

The observer, stationed with his eyes closed in the middle of a lawn on a still evening, was asked to point with the hand in the direction of voices addressed to him by five or six assistants, who continually shifted their position. It was necessary to have several assistants, since it was found that otherwise their steps could be easily followed. The uniform result was that the direction of a human voice used in anything like a natural manner could be told with certainty from a single word, or even vowel, to within a few degrees.

But with other sounds the result was different. If the source was on the right or the left of the observer, its position could be told approximately, but it was uncertain whether, for example, a low whistle was in front or behind. This result led us to try a simple sound, such as that given by a fork mounted on a resonance-box. It was soon found that whatever might be the case with a truly simple sound, the observer never failed to detect the situation of the fork by the noises accompanying its excitation, whether this was done by striking or by a violin bow. It was therefore necessary to arrange the experiment differently. Two assistants at equal distances and in opposite directions were provided with similar forks and resonators. At a signal given by a fourth, both forks were struck, but only one was held over its resonator, and the observer was asked to say, without moving his head, which he heard. When the observer was so turned that one fork was immediately in front and the other immediately behind, it was impossible for him to tell which fork was sounding, and if asked to say one or the other, felt that he was only guessing. But on turning a quarter round, so as to have one fork on his right and the other on his left, he could tell without fail, and with full confidence in being correct.

The possibility of distinguishing a voice in front from a voice behind would thus appear to depend on the compound character of the sound in a way that it is not easy to understand, and for which the second ear would be of no advantage. But even in the case of a lateral sound the matter is not free from difficulty, for the difference of intensity with which a lateral sound is perceived by the two ears is not great. The experiment may easily be tried roughly by stopping one ear with the hand, and turning round backwards and forwards while listening to a sound held steadily. Calculation shows, moreover, that the human head, considered as an obstacle to the waves of sound, is scarcely big enough in relation to the wave-length to give a sensible shadow.   To throw light on this subject I have calculated the intensity of sound due to a distant source at the various points on the surface of a fixed spherical obstacle.   The result depends on the ratio (α) between the circumference of the sphere and the length of the wave.   If we call the point on the spherical surface nearest to the source the anterior pole, and the opposite point (where the shadow might be expected to be most intense) the posterior pole, the results on three suppositions as to the relative magnitudes of the sphere and wave-length are given in the following table :—

	$\alpha = 2$	$\alpha = 1$	$\alpha = \frac{1}{2}$
Anterior Pole	·690	·503	·294
Posterior Pole	·318	·285	·260
Equator	·356	·237	·232

When, for example, the circumference of the sphere is but half the wavelength, the intensity at the posterior pole is only about a tenth part less than at the anterior pole, while the intensity is least of all in a lateral direction. When α is less than $\frac{1}{2}$, the difference of the intensities at the two poles is still less important, amounting to about one per cent., when $\alpha = \frac{1}{4}$.

The value of α depends on the wave-length, which may vary within pretty wide limits, and it might be expected that the facility of distinguishing a lateral sound would diminish when the sound is grave.   Experiments were accordingly tried with forks of a frequency of 128, but no greater difficulty was experienced than with forks of a frequency of 256, except such as might be attributed to the inferior loudness of the former. According to calculation the difference of intensity would here be too small to account for the power of discrimination.

41.

QUESTIONS FROM MATHEMATICAL TRIPOS EXAMINATION FOR 1876.

[*Cambridge University Calendar*, 1876.]

January 6. 9—12.

vi. INVESTIGATE the equations of equilibrium of a flexible string acted upon by any tangential and normal forces.

An uniform steel wire in the form of a circular ring is made to revolve in its own plane about its centre of figure. Show that the greatest possible linear velocity is independent both of the section of the wire and of the radius of the ring, and find roughly this velocity, the breaking strength of the wire being taken as 90,000 lbs. per square inch, and the weight of a cubic foot as 490 lbs.

vii. Calculate from the principle of energy the rate at which water will be discharged from a vessel in whose bottom there is a small hole, explaining clearly why the area of the *vena contracta*, and not that of the hole, is to be used.

A cistern discharges water into the atmosphere through a vertical pipe of uniform section. Show that air would be sucked in through a small hole in the upper part of the pipe, and explain how this result is consistent with an atmospheric pressure in the cistern.

viii. Investigate the disturbance in an unlimited atmosphere due to a source of sound which is concentrated at a single point, and whose effect is to produce an alternate production and destruction of air, given in amount and periodic according to the harmonic law.

Show that, if a given source of sound as defined above be situate at the vertex of an infinite conical tube, the energy emitted in a given time is inversely as the solid angle of the cone.

ix. Define a reversible heat-engine, and point out in what respects a condensing steam-engine as ordinarily worked falls short of the definition.

Can the theory of heat-engines be applied to a non-condensing steam-engine?

x. The ends of a coil of insulated wire can be connected with the poles of a constant battery; investigate the gradual establishment of the current after contact is made.

Show that, when contacts are rapidly made and broken, the average current as indicated by a galvanometer of long period may fall much below that due to the duration of the contacts, and explain the increased effect which attends the closing of the circuit of a second coil in the neighbourhood of the first.

January 6. $1\frac{1}{2}$—4.

v. A straight pipe whose material is thin in comparison with the bore is closed at both ends and subjected to internal fluid pressure. Show that the longitudinal tension of the material is the half of the circumferential tension.

Prove that the greatest quantity of air of given pressure which can be held in a long pipe of given weight is independent of the bore of the pipe.

vi. Investigate the small vibrations of a simple pendulum under the action of gravity.

How could the law connecting the length of the pendulum and the periodic time be arrived at without calculation?

Explain how a boy in a swing is able to increase the amplitude of vibration.

vii. State the law of absorption of homogeneous light in a uniform medium, and show that the colour of light originally white transmitted through a layer of the medium may be entirely altered when the thickness of the layer is increased.

Homogeneous light falls upon a plate of absorbing material, whose surfaces partially reflect the light incident upon them. Calculate the total intensity of the light (1) reflected back on the first side, (2) transmitted, (3) absorbed.

viii. Investigate the distribution of electricity on a conductor in the form of an ellipsoid situate in free space.

Deduce the capacity of an infinitely thin elliptic lamina, and show that if the eccentricity e be small, the capacity may be expressed approximately in terms of the area σ by

$$2\sqrt{\frac{\sigma}{\pi^3}}\left(1+\frac{e^4}{64}\right).$$

ix. Explain the Wheatstone's Bridge method of measuring resistances.

How is it applied to find the position of a fault in an otherwise well insulated cable immersed in a tank of water, both the ends being accessible ?

On account of uniform leakage of electricity the apparent position of a fault is apt to be too near the middle of the cable. Show that, if R the normal insulation resistance be diminished to R' in consequence of the fault, and R' be small in comparison with R, the position of the fault may be found from the formula

$$ml - nl' = \frac{m+n}{2} \frac{R'}{R} (l - l'),$$

where l, l' are the distances of the fault from the ends of the cable, and $m : n$ is the ratio of resistances necessary to obtain a balance.

January 17. 9—12.

1. Enunciate Hooke's law for the extension of an elastic string.

An endless elastic string without weight is placed round a smooth fixed pulley in a vertical plane, whose diameter is half the natural length of the string. The lowest point of the string is made fast to the pulley, and to the highest point is attached a heavy particle. Show that the equilibrium will be neutral, if the weight of the particle be the force necessary, according to Hooke's law, to stretch the string to twice its natural length.

2. Show that any system of forces acting on a rigid body may be replaced by a force acting at a given point, and a couple.

If any system of forces be reduced to two, of which one is of given magnitude and passes through a given point, prove that the line of action of the other will envelope a conic.

4. Show that the solution of the differential equation for vibrations resisted by a frictional force proportional to the velocity, but otherwise free, viz.

$$\ddot{u} + \kappa \dot{u} + n^2 u = 0,$$

may be put into the form

$$u = e^{-\frac{1}{2}\kappa t} \left\{ \dot{u}_0 \frac{\sin n't}{n'} + u_0 \left(\cos n't + \frac{\kappa}{2n'} \sin n't \right) \right\},$$

where $n'^2 = n^2 - \frac{1}{4}\kappa^2$, and $\dot{u}_0$, u_0 are the values of the velocity and displacement when $t = 0$.

Deduce the complete solution of

$$\ddot{u} + \kappa \dot{u} + n^2 u = U$$

in the form

$$u = e^{-\frac{1}{2}\kappa t} \left\{ \dot{u}_0 \frac{\sin n't}{n'} + u_0 \left(\cos n't + \frac{\kappa}{2n'} \sin n't \right) \right\} + \frac{1}{n'} \int_0^t e^{-\frac{1}{2}\kappa(t - t')} \sin n' (t - t') U' dt',$$

where U' is the same function of t' as U is of t.

6. Describe the arrangements necessary for obtaining a finite field of view uniformly lit with approximately homogeneous light. On what conditions does the brightness of the field depend? Is there any limit to the possible brightness when a maximum range of refrangibility is prescribed?

7. Explain the method of obtaining the solar parallax from observations of Mars on the meridian, giving the necessary formulæ.

January 19. 1½—4.

vi. Explain the process by which a function is determined so as to make one integral a maximum or minimum, while another integral involving the same variables is constant, and apply it to determine what functions of x, satisfying the conditions $y = 0$, when $x = 0$, and when $x = l$, make

$$\int_0^l \left(\frac{dy}{dx}\right)^2 dx \text{ stationary in value when } \int_0^l y^2\, dx \text{ is given.}$$

vii. Prove that a uniform frictionless incompressible fluid once at rest can never acquire molecular rotation under the action of natural forces, even though the motion of every particle be resisted by a force proportional to its absolute velocity; and that without molecular rotation there can be no motion within a fixed envelope filled with fluid and enclosing a simply-connected space.

Prove that if in fluid moving with a velocity-potential a portion which at any moment occupies a finite spherical space be instantaneously solidified, the solid so formed will have no motion of rotation.

viii. Plane waves of light are incident directly on an infinite opaque screen, in which there is an aperture of any form. Show how the principles of the wave theory lead to the conclusion that there will be in general constant illumination behind the screen at points well within the projection of the aperture, and zero illumination at points well without the projection of the aperture.

By the aid of Huyghens' zones, or otherwise, explain the central bright spot in Poisson's experiment of the shadow of an opaque circular disc.

ix. Show that the condition of continuity in the interior of a fluid is satisfied by the wave-motion in which the co-ordinates of the individual particles at time t are

$$\xi = h + R\epsilon^{-\frac{k}{R}} \sin\left(at + \frac{h}{R}\right),$$

$$\eta = k + R\epsilon^{-\frac{k}{R}} \cos\left(at + \frac{h}{R}\right);$$

h, k being constants for each particle and a and R constants for all; and investigate the amount of molecular rotation involved.

From the general properties of equipotential and stream-curves, prove that in a regular series of waves moving in deep water without molecular rotation there is necessarily in the neighbourhood of the surface a transference of fluid in the direction of the wave's propagation, whether that surface satisfy the condition of a free surface or not.

xi. State the mathematical conditions which determine the magnetization of a mass of soft iron exposed to the action of given electric currents.

Show that the presence of the iron must increase the induction of each of the circuits upon itself, but may diminish the coefficients of mutual induction.

January 20. $1\frac{1}{2}$—4.

iv. Form the differential equation of the moon's motion, viz.

$$\frac{d^2u}{d\theta^2} + u = \frac{\dfrac{P}{h^2u^2} - \dfrac{T}{h^2u^3}\dfrac{du}{d\theta}}{1 + 2\displaystyle\int \frac{T}{h^2u^3}\,d\theta},$$

where u is the reciprocal of the projection on the ecliptic of the moon's radius vector, and integrate it with the omission of the disturbing force, but without any approximation depending upon the smallness of the eccentricity or of the obliquity of the orbit.

v. Explain carefully what is meant by the instantaneous orbit of a planet, and show how to express the momentary change of the major axis in terms of the energy communicated.

A planet bursts into any number of equal fragments, which then describe undisturbed elliptic orbits. Prove that the harmonic mean of the major axes of the orbits is increased by the explosion.

vi. Prove that if a material system start from rest under the action of given impulses, the energy of the actual motion exceeds that of any other motion which the system (under the action of the same impulses) might have been guided to take by the addition of mere constraints; and that the difference is equal to the energy of the motion which must be compounded with either to produce the other.

Verify this theorem in the case of a lamina acted on by a given impulsive couple in its own plane, supposing the constraint to be produced by holding one point of the lamina fixed.

vii. Define the potential of matter attracting according to the law of nature, and prove that in free space it satisfies the equation

$$\frac{d^2V}{dx^2} + \frac{d^2V}{dy^2} + \frac{d^2V}{dz^2} = 0.$$

The distribution of the attracting matter being symmetrical with respect to the axis of z, and also with respect to the plane of xy, examine the nature of the equilibrium of a particle at the origin; and compare the forces called into play by displacements of given magnitude along and perpendicular to the axis respectively.

viii. Supposing the pressure of a fluid to be a given function of the temperature and volume, investigate an expression for the difference between the specific heat at constant pressure and the specific heat at constant volume.

Apply the result to the case of a perfect gas, and explain how the specific heat may be calculated without direct observation from the values of Joule's equivalent and of the velocity of sound.

January 21. 9—12.

iv. Prove that the potential energy of a string stretched with tension T on a smooth spherical surface of radius a, and slightly displaced from the position of equilibrium, is

$$V = \frac{aT}{2} \int \left\{ \left(\frac{d\theta}{d\phi} \right)^2 - \theta^2 \right\} d\phi,$$

where θ is the angular displacement in the lateral direction of the point whose longitude (measured round the string) is ϕ.

Thence investigate fully the small motion of a uniform string of given length, stretched on a smooth sphere between two fixed points, and show that the effect of the curvature is to diminish the square of the frequency of vibration by a quantity which is the same for all the possible modes of vibration.

vii. Investigate an approximate expression for the potential of a body of any form at a distant point P in terms of the distance of P from the centre of gravity G, and the moment of inertia of the body about the line GP.

Three equal particles rest on a smooth spherical surface of large radius described about G as centre, and are rigidly connected by a framework without mass in such a manner that the three sides of the spherical triangle formed by joining the particles are all quadrants. Prove that to the above order of approximation there is equilibrium in any position.

viii. Explain generally and briefly on what circumstances it depends whether a tide in frictionless fluid is in the same phase as the forces which generate it, or in the opposite phase.

Calculate the tidal motion of heavy liquid contained in a square vessel of uniform depth, due to a small horizontal disturbing force acting uniformly throughout the mass, whose magnitude is constant, and whose direction revolves uniformly in the horizontal plane.

How could the forces here imagined be realized experimentally?

January 21. $1\frac{1}{2}$—4.

iii. Investigate the figure of equilibrium of a revolving fluid covering a symmetrical centrobaric core, supposing the density of the fluid to be so small that the mutual attractions of its parts may be neglected; and find what ratio of equatorial and polar diameters corresponds to a vanishing apparent gravity at the equator.

v. Investigate the relation which must subsist between the pressure and volume of a fluid in order that plane waves of sound of any amplitude may be propagated with type unchanged.

Supposing the pressure to vary inversely as the volume, find what impressed forces, parallel to the direction of propagation, would be necessary in order to counteract the tendency to alteration of type.

vii. Investigate the transverse vibration of a uniform circular membrane, whose boundary is fixed, due to the action of a pressure, uniform over the area of the membrane, but varying with the time according to the simple harmonic law.

viii. Show that, according to Fresnel's theory of double refraction, the velocity of propagation of a plane wave whose direction-cosines estimated with reference to the principal axes of a biaxal crystal are l, m, n, is given by

$$\frac{l^2}{v^2 - a^2} + \frac{m^2}{v^2 - b^2} + \frac{n^2}{v^2 - c^2} = 0.$$

Supposing that for wave-normals lying in each principal plane one of the values of v is constant, prove that a wave-surface of the fourth degree can be no other than Fresnel's surface.

x. Show that if in a uniform mass conducting electricity the potential be zero over the plane xy, its value at neighbouring points out of that plane will be

$$z\chi - \frac{z^3}{1 \cdot 2 \cdot 3} \nabla^2 \chi + \frac{z^5}{\lfloor 5} \nabla^4 \chi - \&c.$$

where χ is a function of x and y, and $\nabla^2 \equiv d^2/dx^2 + d^2/dy^2$.

If a neighbouring equipotential surface, $z = z_1$, coincide with a plane parallel to xy, except over a certain finite region where there is a slight deviation, the conductivity between the equipotential surfaces $z = z_1$, and $z = 0$, is expressed approximately by

$$\iint \left\{ 1 + \frac{1}{3} \left(\frac{dz_1}{dx} \right)^2 + \frac{1}{3} \left(\frac{dz_1}{dy} \right)^2 \right\} \frac{dx\,dy}{z_1},$$

the area of integration including the whole of the above-mentioned region.

42.

ON THE RESISTANCE OF FLUIDS.

[*Phil. Mag.* II. pp. 430—441, 1876.]

THERE is no part of hydrodynamics more perplexing to the student than that which treats of the resistance of fluids. According to one school of writers, a body exposed to a stream of perfect fluid would experience no resultant force at all, any augmentation of pressure on its face due to the stream being compensated by equal and opposite pressures on its rear. And indeed it is a rigorous consequence of the usual hypotheses of perfect fluidity and of the continuity of the motion, that the resultant of the fluid pressures reduces to a *couple* tending to turn the broader face of the body towards the stream. On the other hand, it is well known that in practice an obstacle does experience a force tending to carry it down stream, and of magnitude too great to be the direct effect of friction; while in many of the treatises calculations of resistances are given leading to results depending on the inertia of the fluid without any reference to friction.

It was Helmholtz who first pointed out that there is nothing in the nature of a perfect fluid to forbid a finite slipping between contiguous layers, and that the possibility of such an occurrence is not taken into account in the common mathematical theory, which makes the fluid flow according to the same laws as determine the motion of electricity in uniform conductors. Moreover the electrical law of flow (as it may be called for brevity) would make the velocity infinite at every sharp edge encountered by the fluid; and this would require a negative pressure of infinite magnitude. It is no answer to this objection that a mathematically sharp edge is an impossibility, inasmuch as the electrical law of flow would require negative pressure in cases where the edge is not perfectly sharp, as may be readily proved from the theory of the simple circular vortex, in which the velocity varies inversely as the distance from the axis*.

---

* [1899. However, there is nothing in the constitution of a liquid to forbid negative pressures, even of considerable amount.]

The application of these ideas to the problem of the resistance of a stream to a plane lamina immersed transversely amounts to a justification of the older theory as at least approximately correct. Behind the lamina the fluid is at rest under a pressure equal to that which prevails at a distance, the region of rest being bounded by a surface of separation or discontinuity which joins the lamina tangentially, and is determined mathematically by the condition of constant pressure. On the anterior surface of the lamina there is an augmentation of pressure corresponding to the loss of velocity.

The relation between the velocity and pressure in a steady stream of incompressible fluid may be obtained immediately by considering the transference of energy along an imaginary tube bounded by stream-lines. In consequence of the steadiness of the motion, there must be the same amount of energy transferred in a given time across any one section of the tube as across any other. Now if p and v be the pressure and velocity respectively at any point, and ρ be the density of the fluid, the energy corresponding to the passage of the unit of volume is $p + \frac{1}{2}\rho v^2$, of which the first term represents potential, and the second kinetic energy; and thus $p + \frac{1}{2}\rho v^2$ must retain the same value at all points of the same stream-line. It is further true, though not required for our present purpose or to be proved so simply, that $p + \frac{1}{2}\rho v^2$ retains a constant value not merely on the same stream-line, but also when we pass from one stream-line to another, provided that the fluid flows throughout the region considered in accordance with the electrical law.

If u be the velocity of the stream, the increment of pressure due to the loss of velocity is $\frac{1}{2}\rho u^2 - \frac{1}{2}\rho v^2$, and can never exceed $\frac{1}{2}\rho u^2$, which value corresponds to a place of rest where the whole of the energy, originally kinetic, has become potential. The old theory of resistances went on the assumption that the velocity of the stream was destroyed over the whole of the anterior face of the lamina, and therefore led to the conclusion that the resistance amounted to $\frac{1}{2}\rho u^2$ for each unit of area exposed. It is evident at once that this is an overestimate, since it is only near the middle of the anterior face that the fluid is approximately at rest; towards the edge of the lamina the fluid moves outwards with no inconsiderable velocity, and at the edge itself retains the full velocity of the original stream. Nevertheless the amount of error involved in the theory referred to is not great, as appears from the result of Kirchhoff's calculation of the case of two dimensions, from which it follows that the resistance per unit of area is $\dfrac{\pi}{4 + \pi}\, \rho u^2$ instead of $\frac{1}{2}\rho u^2$.

It is worthy of notice that by a slight modification of the conditions of the problem the estimate $\frac{1}{2}\rho u^2$ may be made accurate. For this purpose the lamina is replaced by the bottom of a box-shaped vessel, whose sides project in the direction from which the stream is flowing, and are sufficiently extended

to cause approximate quiescence over the whole of the bottom (fig. 1). In the absence of friction, the sides themselves do not contribute anything to the resistance. It appears from this argument that the increase of resistance due to concavity can never exceed a very moderate value.

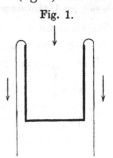

Fig. 1.

Although not very closely connected with the prin-cipal subject of this communication, it may be well to state the corresponding result in the case of a compressible fluid such as air. If p_0 be the normal pressure in the stream, a the velocity of sound corresponding to the general temperature, γ the ratio of the two specific heats, $\frac{1}{2}\rho u^2$ is replaced by

$$ p_0 \left\{ \left(1 + \frac{\gamma - 1}{2} \frac{u^2}{a^2} \right)^{\frac{\gamma}{\gamma - 1}} - 1 \right\}, $$

which gives the resistance per unit of area. The compression is supposed (as in the theory of sound) to take place without loss of heat; and the numerical value of γ is 1·408.

When u is small in comparison with a, the resistance follows the same law as if the fluid were incompressible; but in the case of greater velocities the resistance increases more rapidly. The resistance to a meteor moving at speeds comparable with 20 miles per second must be enormous, as also the rise of temperature due to compression of the air. In fact it seems quite unnecessary to appeal to friction in order to explain the phenomena of light and heat attending the entrance of a meteor into the earth's atmosphere.

But although the old theory of resistance was not very wide of the mark in its application to the case of a lamina against which a stream impinges directly, the same cannot be said of the way in which the influence of obliquity was estimated. It was argued that inasmuch as a lamina moved edgeways through still fluid would create no disturbance (in the absence of friction), such an edgeways motion would produce no alteration in the resist-ance due to a stream perpendicular to the plane of the lamina; and from this it would follow that when a lamina is exposed to an oblique stream, the resistance experienced would be that calculated from the same formula as before, on the understanding that u now represents the *perpendicular com-ponent* of the actual velocity of the stream. Or if the actual velocity of the stream be V, and α denote the angle between the direction of the stream and the lamina, the resistance would be per unit of area

$$ \tfrac{1}{2}\rho V^2 \sin^2 \alpha. \quad \dots\dots\dots\dots\dots\dots\dots\dots\dots\dots(1) $$

This force acts of course perpendicularly to the plane of the lamina; the component down the current is

$$ \tfrac{1}{2}\rho V^2 \sin^3 \alpha. \quad \dots\dots\dots\dots\dots\dots\dots\dots\dots\dots(2) $$

The argument by which this result is obtained, however, is quite worth-less; and the law of the squares of the sines expressed in (1) is known to practical men to be very wide of the mark, especially for small values of α. The resistance at high obliquities is much greater than (1) would make it, being more nearly in proportion to the first power of sin α than to the square.

As a proof that an edgeways motion of an elongated body through water is not without influence on the force necessary to move it with a given speed broadways, Mr Froude says*, "Thus when a vessel was working to windward, immediately after she had tacked and before she had gathered headway, it was plainly visible, and it was known to every sailor, that her leeway was much more rapid than after she had begun to gather headway. The more rapid her headway became, the slower became the lee-drift, not merely relatively slower, but absolutely slower."

"Again, anyone might obtain conclusive proof of the existence of this increase of pressure occasioned by the introduction of the edgeways com-ponent of motion, who would try the following simple experiment. Let him stand in a boat moving through the water, and, taking an oar in his hand, let him dip the blade vertically into the water alongside the boat, presenting its face normally to the line of the boat's motion, holding the plane steady in that position, and let him estimate the pressure of the water on the blade by the muscular effort required to overcome it. When he has consciously appre-ciated this, let him begin to sway the blade edgeways like a pendulum, and he will at once experience a very sensible increase of pressure. And if the edgeways sweep thus assigned to the blade is considerable and is performed rapidly, the greatness of the increase in the pressure will be astonishing until its true meaning has been realized. Utilizing this proposition, many boatmen, when rowing a heavy boat with narrow-bladed oars, were in the habit of alternately raising and lowering the hand with a reciprocating motion, so as to give an oscillatory dip to the blade during each stroke, and thus obtained an equally vigorous reaction from the water with a greatly reduced slip or sternward motion of the blade."

It is not difficult to see that in the case of obliquity we have to do with the whole velocity of the current, and not merely with the resolved part. Behind the lamina there must be a region of dead water bounded by a surface of discontinuity, within which the pressure is the same as if there were no obstacle. On the front face of the lamina there must be an augmen-tation of pressure, vanishing at the edges and increasing inwards to a maximum at the point where the stream divides. At this point the pressure is $\frac{1}{2}\rho V^2$, corresponding to the loss of the *whole* velocity of the stream. It is true that the maximum pressure prevails over only an infinitely small fraction

---

* *Proceedings of the Society of Civil Engineers*, vol. XXXII., in a discussion on a paper by Sir F. Knowles on the Screw Propeller.

of the area; but the same may be said even when the incidence of the stream is perpendicular.

The exact solution of the problem in the case of two dimensions, which covers almost all the points of practical interest, can be obtained by the analytical method of Helmholtz and Kirchhoff*. If an elongated blade be held vertically in a horizontal stream, so that the angle between the plane of the blade and the stream is α, the mean pressure is

$$\frac{\pi \sin \alpha}{4 + \pi \sin \alpha} \rho V^2, \quad \dots\dots\dots\dots\dots\dots\dots\dots\dots(3)$$

varying, when α is small, as $\sin \alpha$, and not as $\sin^2 \alpha$. The proof will be found at the end of the present paper.

The fact that the resistance to the broadways motion of a lamina through still fluid can be increased enormously by the superposition of an edgeways motion is of great interest. For example, it will be found to be of vital importance in the problem of artificial flight.

According to the old theory the component of resistance transverse to the stream varied as $\sin^2 \alpha \cos \alpha$, and attained its maximum for $\alpha = 55°$ nearly. The substitution of expression (3) for $\sin^2 \alpha$ will materially modify the angle at which the transverse force is greatest. The quantity to be made a maximum is

$$\frac{\sin \alpha \cos \alpha}{4 + \pi \sin \alpha};$$

and the value of α for which the maximum is attained is $\alpha = 39°$ nearly, being considerably less than according to the old theory, on account of the increased value of the normal pressure at high obliquities.

The pressure, whose mean amount is given in (3), is far from symmetrically distributed over the breadth of the blade, as might be anticipated from the fact that the region of maximum pressure, where the stream divides, is evidently nearer to the anterior or up-stream edge. If the breadth of the blade be called l, the distance (x) of the centre of pressure, reckoned from the middle, is

$$x = \frac{3}{4} \cdot \frac{l \cos \alpha}{4 + \pi \sin \alpha}. \quad \dots\dots\dots\dots\dots\dots\dots\dots(4)$$

If the blade be pivoted so as to be free to turn about an axis parallel to its edges, (4) gives the position of the axis corresponding to any angle of

* Formulæ (3) and (4) were given at the Glasgow Meeting of the British Association. I was then only acquainted with Kirchhoff's *Vorlesungen über mathematische Physik*, and was not aware that the case of an oblique stream had been considered by him (*Crelle*, Bd. LXX. 1869). However, Kirchhoff has not calculated the forces; so that the formulæ are new. [1899. They appear to have been given a little earlier by Thiesen. See *Phil. Mag.* v. p. 320, 1878.]

inclination α. If $\alpha = 90°$, $x = 0$, as is evident from symmetry. As α diminishes, the corresponding value of x increases and reaches a maximum, viz. $\frac{3}{16} l$, when $\alpha = 0$. The axis then divides the breadth of the blade in the ratio 11 : 5.

These results may be stated in another form as follows. If the axis of suspension divide the width in a more extreme ratio than 11 : 5, there is but one position of stable equilibrium, that namely in which the blade is parallel to the stream with the narrower portion directed upwards. If the axis be situated exactly at the point which divides the width in the ratio 11 : 5, this position becomes neutral, in the sense that for small displacements the force of restitution is of the second order, but the equilibrium is really stable. When the axis is still nearer the centre of figure, the position parallel to the stream becomes unstable, and is replaced by two inclined positions given by (4), making with the stream equal angles, which increase from zero to a right angle as the axis moves in towards the centre. With the centre line itself for axis, the lamina can only remain at rest when transverse to the stream, though of course with either face turned upwards.

The fact, rather paradoxical to the uninitiated, that a blade free to turn about its centre line sets itself transversely, may be easily proved by experiment. For this purpose it is sufficient to take a piece of thin brass plate shaped as in the figure (fig. 2), and mount it with its points bearing in two

Fig. 2.

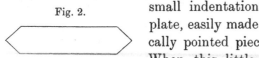

small indentations in a U-shaped strip of thicker plate, easily made by striking the strip with a conically pointed piece of steel driven by the hammer. When this little apparatus is moved through the water, the moveable piece at once sets itself across the direction of motion. The same result may be observed when the apparatus is exposed to the wind; but in this case an unexpected phenomenon often masks the stability of the transverse position. It is found that when the plate is set rotating, the force of the wind will maintain or accelerate the motion. This effect might be supposed to be due to a want of symmetry, were it not that the rotation occurs in either direction. It is evidently connected with the disturbance of the fluid due to the motion of rotation, and is not covered by the calculation leading to formula (4), which refers to the forces experienced when the blade is *at rest* in any position.

I am not aware of any experimental measurements with which (4) could be compared; but the result that the equilibrium parallel to the stream is indifferent when the axis is situated in the position defined by the ratio 11 : 5, is in agreement with the construction of balanced rudders, of which the front part is usually made of about one-half the width of the hinder part.

The accompanying Table contains some numerical examples of the general formulæ. The first column gives the angle between the lamina and stream, the second the value of $\sin^2 \alpha$, to which, on the old theory, the resistance should be proportional; the third column is derived from some experiments by Vince on water, published in the *Philosophical Transactions* for 1798. The quantity directly measured by Vince was the resolved part of the resistance in the direction of the stream, from which the tabulated number is

α	$\sin^2 \alpha$	Vince	$\dfrac{\sin \alpha \,(4+\pi)}{4+\pi \sin \alpha}$	$\dfrac{3}{4}\,\dfrac{\cos \alpha}{4+\pi \sin \alpha}$	
90	1·0000	1·000	1·0000	·0000	·5000
70	·8830	·974	·9652	·0369	·2676
50	·5868	·873	·8537	·0752	·0981
30	·2500	·663	·6411	·1166	·0173
20	·1170	·458	·4814	·1389	·0040
10	·0302	·278	·2728	·1625	·0004

derived by division by $\sin \alpha$. The fourth column represents the law of resistance according to the formula now proposed, a factor being introduced so as to make the maximum value unity. The fifth column gives the distance between the centre of pressure and the middle line of the blade, expressed as a fraction of the total width. The sixth column is the value of

$$\frac{2\,(1 - 2\cos \alpha + \cos^3 \alpha) + \alpha \sin \alpha}{4 + \pi \sin \alpha},$$

which is the distance from the anterior edge of the point where the stream divides, and where accordingly the pressure attains its greatest value. It will be seen that, as might be expected, this distance becomes small at moderate obliquities.

The result of Vince's experiments agrees with theory remarkably well; and the contrast with $\sin^2 \alpha$ is especially worthy of note. The experiments were made with a whirling machine, and appear to have been carefully conducted; but they were on too small a scale to be quite satisfactory. The subject might now be resumed with advantage*.

From theory it would appear that any part of the region of dead water behind the lamina might be filled up with solid matter without in any way disturbing the motion or altering the resistance; but in practice with actual fluids this statement must not be taken without qualification. If the boundary of the solid approach too nearly the natural position of the surface of separation, the intervening fluid appears to be sucked out until the lines of flow follow the surface of the obstacle. This is the state of things aimed at, and approximately attained, in well-designed ships, round which the

* [1899. The reader will hardly need to be reminded of Langley's experiments (*Smithsonian Contributions to Knowledge*, 1891). A notice of them in *Nature* (vol. XLV. p. 108, 1891) may also be referred to.]

water flows nearly according to the electrical law. The resistance is then of an entirely altered character, and depends only upon the friction against the skin.

It was observed by Sir William Thomson at Glasgow, that motions involving a surface of separation are unstable. This is no doubt the case, and is true even of a parallel jet moving with uniform velocity. If from any cause a slight swelling occurs at any point of the surface, an *increase* of pressure ensues tending not to correct but to augment the irregularity. I had occasion myself to refer to a case of this kind in a paper on Waves, published in the *Philosophical Magazine* for April, 1876 [Art. XXXVIII.]. But it may be doubted whether the calculations of resistance are materially affected by this circumstance, as the pressures experienced must be nearly independent of what happens at some distance in the rear of the obstacle, where the instability would first begin to manifest itself.

The formulæ proposed in the present paper are also liable to a certain amount of modification from friction which it would be difficult to estimate beforehand, but which cannot be very considerable, if the experiments of Vince are to be at all relied on.

In the following analysis ϕ and ψ are the potential and stream functions, $z = x + iy$, $\omega = \phi + i\psi$; and it is known that the general conditions of fluid motion in two dimensions are satisfied by taking z as an arbitrary function of ω. If

$$\frac{dz}{d\omega} = \zeta = \rho \left(\cos \theta + i \sin \theta \right), \quad \text{......................(A)}$$

Kirchhoff shows that ζ represents the velocity of the stream at any point, with the exception that its modulus ρ is proportional to the reciprocal of the velocity instead of to the velocity itself. If the general velocity of the stream be unity, the condition to be satisfied along a surface of separation bounding a region of dead water is $\rho = 1$. The value of ψ must of course also preserve a constant value along the same surface.

The form of ζ applicable to the present problem is

$$\zeta = \cos \alpha + \frac{1}{\sqrt{\omega}} + \sqrt{\left(\cos \alpha + \frac{1}{\sqrt{\omega}} \right)^2 - 1}. \quad \text{.................(B)}$$

When $\omega = \infty$, $\qquad\qquad\qquad \zeta = \cos \alpha - i \sin \alpha$.

The surface of separation corresponds to $\psi = 0$, for which value of ψ ω becomes real; and the point at which the stream divides corresponds to $\omega = 0$, for which $\zeta = \infty$. For $\psi = 0$ and real values of $\cos \alpha + 1/\sqrt{\omega}$ less than unity, $\rho = 1$. This portion therefore corresponds to the surface of separation, for which the pressure is constant. When $\cos \alpha + 1/\sqrt{\omega}$ is real and greater than unity, ζ is real, indicating that the direction of motion is parallel to the axis of x. This part corresponds to the anterior face of the lamina.

The augmentation of pressure at any point is represented by $\frac{1}{2}(1-\rho^{-2})$, if the density of the fluid be taken as unity; and thus the whole resistance is measured by the integral

$$\tfrac{1}{2}\int(1-\rho^{-2})\,dl,$$

if dl represents an element of the width of the lamina. Kirchhoff shows how to change the variable of integration from l to ω. The velocity of the fluid is $d\phi/dl$, or, since ψ is here zero, $d\omega/dl$. Thus, since ζ is real, $\pm \zeta = \rho = dl/d\omega$; and therefore the integral may be replaced by

$$\int \pm \frac{1}{2}\left(\zeta - \frac{1}{\zeta}\right)d\omega, \quad\dots\dots\dots\dots\dots\dots\dots(C)$$

in which all the elements are to be taken positive.

From the form of ζ in (B), it appears that

$$\frac{1}{2}\left(\zeta - \frac{1}{\zeta}\right) = \sqrt{\left(\cos\alpha + \frac{1}{\sqrt{\omega}}\right)^2 - 1.} \quad\dots\dots\dots\dots\dots(D)$$

The width of the lamina l is $\int\zeta\,d\omega$, where the limits of integration are such as make

$$\cos\alpha + 1/\sqrt{\omega} = \pm 1.$$

The integration may be effected by the introduction of a new variable β, where

$$\beta = \sin^2\alpha\,\sqrt{\omega} - \cos\alpha,$$

and the limits for β are ± 1. Thus

$$\int\left(\cos\alpha + \frac{1}{\sqrt{\omega}}\right)d\omega = \frac{\beta^2\cos\alpha}{\sin^4\alpha} + \frac{2\beta}{\sin^4\alpha} + \text{const.};$$

and therefore between the limits ± 1 we have

$$4 \div \sin^4\alpha.$$

The second part of ζ may be written $\sqrt{(1-\beta^2)} \div \sin\alpha\,\sqrt{\omega}$, giving the integral

$$\int \frac{d\omega}{\sin\alpha\,\sqrt{\omega}}\,\sqrt{(1-\beta^2)} = \frac{\beta\sqrt{(1-\beta^2)}}{\sin^3\alpha} + \frac{\sin^{-1}\beta}{\sin^3\alpha} + \text{const.}$$

Thus the complete value of z between the limits, or l, is

$$l = \frac{4}{\sin^4\alpha} + \frac{\pi}{\sin^3\alpha} = \frac{4 + \pi\sin\alpha}{\sin^4\alpha}. \quad\dots\dots\dots\dots\dots(E)$$

By (C) and (D) the whole pressure on the lamina is represented by the second part of l in (E), or $\pi \div \sin^3\alpha$; so that the mean pressure is

$$\frac{\pi}{\sin^3\alpha} \div \frac{4 + \pi\sin\alpha}{\sin^4\alpha} = \frac{\pi\sin\alpha}{4 + \pi\sin\alpha},$$

as was to be proved.

Again, the elementary moment of pressure about $z = 0$ is

$$\pm \tfrac{1}{2} \left(\zeta - \frac{1}{\zeta} \right) d\omega \cdot z = \frac{2 \sqrt{(1 - \beta^2)}}{\sin^3 \alpha} z \, d\beta.$$

Now if the arbitrary constant be taken suitably, the complete value of z is

$$z = \frac{\beta^2 \cos \alpha + 2\beta}{\sin^4 \alpha} + \frac{\beta \sqrt{(1 - \beta^2)} + \sin^{-1} \beta}{\sin^3 \alpha}.$$

The odd terms in z will contribute nothing to the integral; and therefore we may take for the moment of pressure about $z = 0$,

$$\int_{-1}^{+1} \frac{2 \sqrt{(1 - \beta^2)}}{\sin^3 \alpha} \cdot \frac{\beta^2 \cos \alpha}{\sin^4 \alpha} \, d\beta = \frac{\pi}{\sin^3 \alpha} \cdot \frac{\cos \alpha}{4 \sin^4 \alpha}.$$

In this result the first factor represents the total pressure, and therefore $\tfrac{1}{4} \cos \alpha / \sin^4 \alpha$ expresses the distance of the centre of pressure from the point $z = 0$. With the same origin the value of z for the middle of the lamina is $\cos \alpha / \sin^4 \alpha$; and thus the displacement of the centre of pressure from the middle of the lamina is $-\tfrac{3}{4} \cos \alpha / \sin^4 \alpha$. This distance must now be expressed in terms of l or

$$(4 + \pi \sin \alpha) \div \sin^3 \alpha,$$

which gives as the final result,

$$-\frac{3}{4} \frac{\cos \alpha \cdot l}{4 + \pi \sin \alpha}.$$

The negative sign indicates that the centre of pressure is on the up-stream side of the middle point.

As to the form of the surface of separation, its intrinsic equation is given at once by the value of ζ in terms of ω. The real part of ζ is $\cos \theta$ (since $\rho = 1$), where θ is the angle between the tangent at any point and the plane of the lamina. Along the surface of separation ω is identical with ϕ, and $d\phi/ds = 1$. Thus if s be the length of the arc of either branch measured from the point where it joins the lamina, the intrinsic equation is

$$\cos \theta = \cos \alpha \pm \frac{1}{\sqrt{(s + c)}},$$

and the constant is to be determined by the condition that $s = 0$ when $\cos \theta = \pm 1$. Since $\cos \theta = dx/ds$, the relation between x and s is readily obtained on integration; but the relation between y and s is more complicated.

In the case of perpendicular incidence $\cos \alpha = 0$, $c = 1$, so that

$$\cos \theta = \pm \frac{1}{\sqrt{(s + 1)}},$$

giving on integration

$$x = 2 \sqrt{(s + 1)} + \text{const.}$$

It appears that the value of x does not approach a finite limit as s increases indefinitely.

43.

NOTES ON HYDRODYNAMICS.

[*Phil. Mag.* II. pp. 441—447, 1876.]

The Contracted Vein.

THE contraction of a jet of fluid in escaping from a higher to a lower pressure through a hole in a thin plate has been the subject of much controversy. Of late years it has been placed in a much clearer light by a direct application of the principle of momentum to the circumstances of the problem by Messrs Hanlon and Maxwell* among others.

For the sake of simplicity the liquid will be supposed to be unacted upon by gravity, and to be expelled from the vessel by the force of compressed air through a hole of area σ in a thin plane plate forming part of the sides of the vessel. After passing the hole the jet contracts, and at a little distance assumes the form of a cylindrical bar of reduced area σ'. The ratio $\sigma' : \sigma$ is called the coefficient of contraction.

The velocity acquired by the fluid in escaping from the pressure p is determined, in the absence of friction, by the principle of energy alone. If the density of the fluid be unity, and the acquired velocity v,

$$v^2 = 2p. \quad\dots\dots\dots\dots\dots\dots\dots\dots\dots\dots\dots\dots(1)$$

The product of v, as given by (1), and σ is sometimes, though very improperly, called the theoretical discharge; and it differs from the true discharge for two reasons. In the first place, the velocity of the fluid is not equal to v over the whole of the area of the orifice. At the edge, where the jet is free, the velocity is indeed v; but in the interior of the jet the pressure is above atmosphere, and therefore the velocity is less than v. And, secondly, it is evident that the quantity of fluid passing the orifice depends, not upon the

---

* *Proceedings of the Mathematical Society*, November 11, 1869.

whole velocity with which the fluid may be moving at any point, but upon the resolved part of this velocity in a direction perpendicular to the plane of the orifice. Thus it is only in the middle of the jet that the whole velocity is efficient; near the edge the motion is tangential; and consequently this part contributes but little to the discharge. It is certain that the discharge will be considerably *less* than $v\sigma$, or, which is the same thing, that the jet must undergo considerable contraction before the liquid composing it can move in parallel lines with uniform velocity v.

Since the actual discharge is $\sigma'v$, the quantity of momentum passing away with the jet in unit time is $\sigma'v^2$, and the force generating this momentum is that necessary to hold the vessel at rest. If the whole of the interior surface of the vessel were subject to the pressure p, this force would have no existence. On account of the orifice the equilibrium of internal pressures is disturbed and a force $p\sigma$ is uncompensated. But this is not all. Not only is the pressure that would have acted over the area of the orifice wanting, but there is also a relief of pressure on the surface surrounding the orifice corresponding to the velocity with which the fluid there moves. The uncompensated force tending to produce recoil may therefore be represented by $(\sigma + \delta\sigma)p$, where $\delta\sigma$ is a small positive quantity; and if the vessel is to remain at rest, a force of this magnitude must be applied to it acting in the direction in which the jet escapes. Thus

$$(\sigma + \delta\sigma)p = \sigma'v^2; \quad\text{.............................(2)}$$

and therefore, by (1),

$$\sigma' = \tfrac{1}{2}(\sigma + \delta\sigma), \quad\text{..........................(3)}$$

expressing that the coefficient of contraction is greater than $\tfrac{1}{2}$.

In the absence of a mathematical solution of the problem it is impossible to estimate the magnitude of $\delta\sigma$ with any precision; but it is something to know from general principles that there must be a considerable contraction, and yet that the coefficient of contraction must exceed one-half. However, by a slight modification of the problem it is possible to get rid of the uncertainty arising from the unknown magnitude of $\delta\sigma$. Suppose the hole in a thin plate to be replaced by a thin parallel tube projecting into the interior of the vessel. If the tube be long enough, the sides of the vessel are sufficiently removed from the region of rapid flow to allow of the pressure acting upon them being treated as constant, while the relief of pressure on the sides of the *tube* does not add anything to the forces tending to produce momentum in the jet. Under these circumstances, if σ be the area of the section of the tube and σ' the area of the section of the jet after contraction, $\sigma' = \tfrac{1}{2}\sigma$, or the coefficient of contraction is one-half exactly. The rigorous mathematical solution of this problem, so far as relates to the case of motion in two dimensions, has been given by Helmholtz (*Phil. Mag.*, November, 1868); and the conclusion that the width of the emergent stream is ultimately one-half that of the channel follows from his analysis*.

This problem throws some light on the formation of a surface of discontinuity. If the electrical law of flow held good so that the tube were filled, twice as much momentum as before would have to be generated, and the extra momentum would have its origin in the infinite negative pressure which, according to that law, must prevail over the extreme edge of the tube. In the absence of forces capable of generating the extra momentum the tube could not flow full.

A generalization of the problem just considered may be effected by replacing the vessel, whose dimensions were supposed to be indefinitely great, by a cylinder of finite section σ'' (fig. 1), in which the fluid moves with finite velocity v''. If v' and σ' be the ultimate velocity and section of the escaping jet, the equation of continuity gives

Fig. 1.

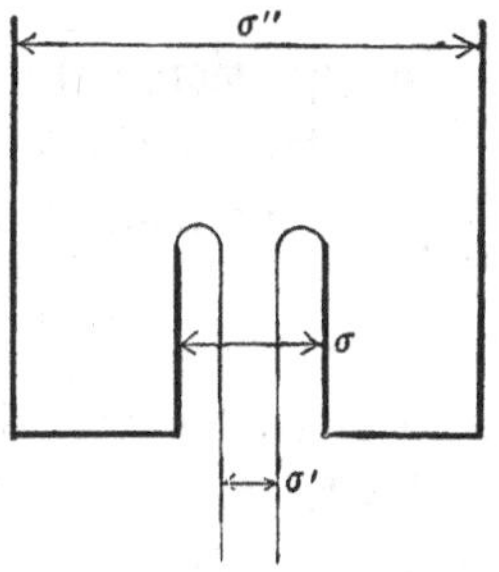

$$v' \sigma' = v'' \sigma''. \quad\quad\quad (4)$$

By the principle of energy,

$$p = \tfrac{1}{2}(v'^2 - v''^2); \quad\quad\quad (5)$$

and by the principle of momentum, if σ be the area of the tube,

$$p\sigma - \tfrac{1}{2}v''^2(\sigma'' - \sigma) = \sigma'v'^2 - \sigma''v''^2. \quad\quad\quad (6)\dagger$$

From these equations we obtain

$$(\sigma'' - \sigma')^2 = \sigma''(\sigma'' - \sigma). \quad\quad\quad (7)$$

The problem of the contracted vein for a hole in a thin plate has been solved mathematically by Kirchhoff‡ for the case of motion in two dimensions. As this solution is very little known, and many points of interest are passed over by Kirchhoff himself, a short account of it accompanied by a few remarks and calculations may not be out of place.

With the notation explained in the previous paper, the form of ζ proper to this problem is

$$\zeta = e^{-\omega} + \sqrt{(e^{-2\omega} - 1)}. \quad\quad\quad (8)$$

---

* The application of the principle of momentum to the case of the introverted tube was original with myself, but, as I learned at Glasgow, had been made previously by Mr Froude. [1899. Subsequently I learned that it had been given in the last century by Borda.] In small-scale experiments the result is liable to be vitiated by adhesion to the side of the tube.

† [1899. In the original paper the term $-\tfrac{1}{2}v''^2(\sigma'' - \sigma)$, representing the increase of pressure on the bottom of the vessel due to the loss of velocity v'', was omitted. The error was pointed out and corrected in an important paper by Michell (Phil. Trans. vol. CLXXXI. p. 408, 1890). The final result (7) has been altered accordingly.]

‡ Crelle, vol. LXX. 1868, and Vorlesungen über mathematische Physik.

The values of ψ corresponding to the boundaries of the jet are 0 and π; and the stream-line which passes symmetrically through the middle of the orifice is $\psi = \frac{1}{2}\pi$, for which value of ψ ζ is purely imaginary. For the stream-line $\psi = 0$,

$$\zeta = e^{-\phi} + \sqrt{(e^{-2\phi} - 1)}. \quad\dots\dots\dots\dots\dots\dots\dots(9)$$

When ϕ is negative in (9) ζ is wholly real and positive, so that this part of the stream-line is parallel to the axis of x, and answers to the bottom of the vessel up to the edge of the orifice. When ϕ is positive, ζ is complex, but its modulus is unity. This part therefore corresponds to the free boundary.

The width of the jet after contraction is π, since the velocity is unity; and the total flow between the stream-lines $\psi = 0$ and $\psi = \pi$ is measured by the difference of the values of ψ.

In equation (9) the real part of ζ (ϕ positive) is $\cos\theta$, where θ is the angle between the direction of motion at any point and the axis of x; so that the intrinsic equation to the boundary is

$$\cos\theta = dx/ds = e^{-s}, \quad\dots\dots\dots\dots\dots\dots\dots\dots(10)$$

no constant being added if s be measured from the edge of the orifice where $\cos\theta = 1$.

From (10), by integration,

$$x = 1 - e^{-s}, \quad\dots\dots\dots\dots\dots\dots\dots\dots\dots(11)$$

if the origin of x be taken at the edge of the orifice, where $s = 0$. This equation determines the width of the aperture. When $s = \infty$, $x = 1$, which corresponds to the abscissa of the boundary of the jet after contraction; and, as we have already seen, the width of the jet itself is π. Accordingly the whole width of the aperture is $2 + \pi$, and the coefficient of contraction $\pi : 2 + \pi$.

The numerical value of $\pi : 2 + \pi$ is ·611, agreeing very nearly with the coefficient of contraction found by observation.

From (10),

$$\frac{dy}{ds} = \sqrt{(1 - e^{-2s})},$$

whence

$$y = \sqrt{(1 - e^{-2s})} - \frac{1}{2}\log\frac{1 + \sqrt{(1 - e^{-2s})}}{1 - \sqrt{(1 - e^{-2s})}}, \quad\dots\dots\dots\dots(12)*$$

if the origin of y be taken at $s = 0$.

If we eliminate s between (11) and (12), we get as the equation of the curve in Cartesian coordinates,

$$y = \sqrt{(2x - x^2)} - \frac{1}{2}\log\frac{1 + \sqrt{(2x - x^2)}}{1 - \sqrt{(2x - x^2)}}, \quad\dots\dots\dots\dots(13)$$

* Equations (11) and (12) are given by Kirchhoff.

from which the following points are calculated:—

$x = \cdot 1$	$-y = \cdot 0313$	$x = \cdot 6$	$-y = \cdot 6494$
$x = \cdot 2$	$-y = \cdot 0932$	$x = \cdot 7$	$-y = \cdot 9203$
$x = \cdot 3$	$-y = \cdot 1815$	$x = \cdot 8$	$-y = 1 \cdot 3127$
$x = \cdot 4$	$-y = \cdot 2985$	$x = \cdot 9$	$-y = 1 \cdot 9915$
$x = \cdot 5$	$-y = \cdot 4509$	$x = 1 \cdot 0$	$-y = \infty$

By means of these points the curve (fig. 2), is constructed. From (10),

$$s = -\log \cos \theta ;$$

so that the radius of curvature is $\tan \theta$. The curvature is therefore infinite at the origin, and diminishes continually as s increases.

Fig. 2.

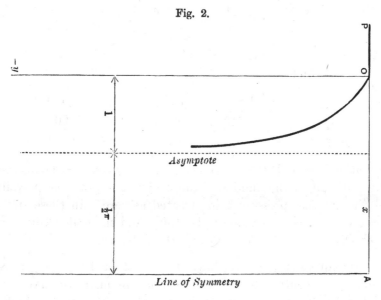

In discussions on the cause of the contraction of the jet doubts have been expressed as to the reality of the deficiency of velocity in the middle of the orifice ; and it may therefore be worth while to examine this point more closely. For this purpose it will be convenient to express z or $x + iy$ in terms of ζ.

From (8) we get

$$e^{-\omega} = \frac{1 + \zeta^2}{2\zeta} , \quad \ldots\ldots\ldots\ldots\ldots\ldots\ldots(14)$$

whence

$$\frac{d\omega}{d\zeta} = \frac{1}{\zeta}\frac{1 - \zeta^2}{1 + \zeta^2} . \quad \ldots\ldots\ldots\ldots\ldots\ldots(15)$$

Thus

$$z = \int \zeta d\omega = \int \zeta \frac{d\omega}{d\zeta} d\zeta = \int \frac{1 - \zeta^2}{1 + \zeta^2} d\zeta = 2 \tan^{-1} \zeta - \zeta + C.$$

In order to determine the value of the constant of integration, we may observe that ζ when real varies between $-\infty$ and -1, and between $+1$ and $+\infty$. The values ± 1 correspond to the edges of the orifice when $z = 0$ and $z = \pi + 2$. Hence $\tan^{-1}\zeta$ varies between $\frac{1}{4}\pi$ and $\frac{3}{4}\pi$, and $C = 1 - \frac{1}{2}\pi$. Accordingly

$$z = 2 \tan^{-1}\zeta - \zeta + 1 - \tfrac{1}{2}\pi. \quad\ldots\ldots\ldots\ldots\ldots\ldots\ldots(16)$$

If v be the velocity of the stream at any point y of the line of symmetry $x = \frac{1}{2}\pi + 1$, $\zeta = -i/v$; and therefore by (16) the relation between y and v is

$$iy = -2 \tan^{-1}\frac{i}{v} + \frac{i}{v} - \pi;$$

or, if $\tan^{-1}(i/v)$ be replaced by its logarithmic equivalent,

$$y = \frac{1}{v} - \log\frac{1+v}{1-v}. \quad\ldots\ldots\ldots\ldots\ldots\ldots(17)$$

A few pairs of corresponding values of y and v will give an idea of the relation expressed in (17).

$v = \frac{9}{10}$	$y = -1\cdot834$	$v = \frac{7}{12}$	$y = +\ \cdot3792$
$v = \frac{3}{4}$	$y = -\ \cdot613$	$v = \frac{1}{2}$	$y = +\ \cdot902$
$v = \frac{2}{3}$	$y = -\ \cdot1095$	$v = \frac{1}{4}$	$y = +3\cdot489$

By interpolation we find that, corresponding to $y = 0$, $v = \cdot6840$, $v^2 = \cdot420$. Hence the pressure in the middle of the orifice is 58 of that prevailing in the vessel, the external pressure being treated as zero. In these statements the ultimate velocity is understood to be unity, and the scale of linear magnitude is such that $2 + \pi$ represents the width of the orifice.

[1899. In equation (17) we have the relation between velocity and position along the central line. In a similar manner we may form the expression for the velocity in contact with the plate PO (fig. 2). Here

$$z = x, \qquad \zeta = 1/v,$$

and (16) gives

$$x = 2 \tan^{-1}\frac{1}{v} - \frac{1}{v} + 1 - \tfrac{1}{2}\pi.$$

From O to the left v varies from 1 to 0, $\tan^{-1}(1/v)$ from $\frac{1}{4}\pi$ to $\frac{1}{2}\pi$. When v is very small,

$$v^{-1} = -x + 1 + \tfrac{1}{2}\pi = AP.]$$

Meeting Streams.

The principle of momentum gives interesting information on the question of the mutual action of streams which come into collision. Suppose, for example, that the motion is in two dimensions, and that two equal streams moving with the same velocity meet at an angle 2α (fig. 3). After the collision the fluid resolves itself into two other streams of unequal width parallel to the line bisecting the angle 2α; and a question arises as to the relative magnitude of these streams. The ultimate velocity is of course in both parts the same as before the collision.

Fig. 3.

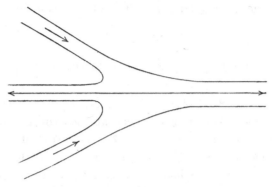

The widths of the original streams being unity, let us suppose that the width of that derived stream which is least diverted is x. The width of the other derived stream is then $2-x$; and the principle of conservation of momentum gives at once the relation

$$2 \cos \alpha = x - (2 - x),$$

whence $x = 1 + \cos \alpha$. The ratio of the two derived streams is

$$2 - x : x = \tan^2 \tfrac{1}{2}\alpha. \quad\dots\dots\dots\dots\dots\dots\dots\dots(18)$$

For example, if $\alpha = 60°$, the ratio of the derived streams is $1 : 3$. The effect of friction would be to make the ratio still more extreme.

Fig. 4.

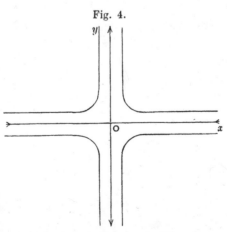

If we suppose the motion reversed, we obtain the solution of the problem of the direct impact of two streams of unequal widths which meet with equal velocities. The ratio of streams being known, (18) determines the angle of divergence.

In the case where α is a right angle the four streams are all equal, and the bounding surfaces are symmetrical with respect to the straight lines bisecting the streams (fig. 4). The exact solution of this case has been indicated by Kirchhoff. If θ be the angle between the tangent to the free surface at any point and the axis of x, and if s be the length of the arc measured from the middle point (where $\theta = \pm \frac{1}{4}\pi$), the intrinsic equation to the free surface is

$$\tan \theta = \pm e^{-s},$$

the scale of linear magnitude being such that the initial and final width of the streams is π.

Postscript, Nov. 14.—Although the accurate determination by experiment of the pressure in the interior of a stream is a matter of considerable difficulty, the theoretical value of the pressure in the middle of a jet issuing through a slit in a thin plate has been recently verified approximately by Mr A. Mallock in my laboratory. The fraction of the original head still remaining in the plane of the plate was found to be about ·59, which agrees very closely with the value obtained from theory, viz. ·58. For a circular hole the corresponding fraction is higher, about ·67 ; but for this there is at present no theory. The observations were made by introducing along the axis of the jet a fine glass tube in the side of which was a small hole, the interior of the tube being in connexion with a manometer.

44.

ON THE APPLICATION OF THE PRINCIPLE OF RECIPROCITY TO ACOUSTICS.

[*Proceedings of the Royal Society*, XXV. pp. 118—122, 1876.]

IN a memoir published some years ago by Helmholtz (*Crelle*, Bd. LVII.) it was proved that if a uniform frictionless gaseous medium be thrown into vibration by a simple source of sound of given period and intensity, the variation of pressure is the same at any point B when the source of sound is at A as it would have been at A had the source of sound been situated at B, and that this law is not interfered with by the presence of any number of fixed solid obstacles on which the sound may impinge.

A simple source of sound is a point at which the condition of continuity of the fluid is broken by an alternate introduction and abstraction of fluid, given in amount and periodic according to the harmonic law.

The reciprocal property is capable of generalization so as to apply to all acoustical systems whatever capable of vibrating about a configuration of equilibrium, as I proved in the *Proceedings of the Mathematical Society* for June 1873 [Art. XXI.], and is not lost even when the systems are subject to damping, provided that the frictional forces vary as the first power of the velocity, as must always be the case when the motion is small enough. Thus Helmholtz's theorem may be extended to the case when the medium is not uniform, and when the obstacles are of such a character that they share the vibration.

But although the principle of reciprocity appears to be firmly grounded on the theoretical side, instances are not uncommon in which a sound generated in the open air at a point A is heard at a distant point B, when an equal or even more powerful sound at B fails to make itself heard at A; and some phenomena of this kind are strongly insisted upon by Prof. Henry in opposition to Prof. Tyndall's views as to the importance of "acoustic clouds" in relation to the audibility of fog-signals. These observations were not, indeed, made with the simple sonorous sources of theory; but there is no reason to suppose that the result would have been different if simple sources could have been used.

In experiments having for their object the comparison of sounds heard under different circumstances there is one necessary precaution to which it may not be superfluous to allude, depending on the fact that the audibility of a particular sound depends not only upon the strength of that sound, but also upon the strength of other sounds which may be heard along with it. For example, a lady seated in a closed carriage and carrying on a conversation through an open window in a crowded thoroughfare will hear what is said to her far more easily than she can make herself heard in return; but this is no failure in the law of reciprocity.

The explanation of his observations given by Henry depends upon the peculiar action of wind, first explained by Prof. Stokes. According to this view a sound is ordinarily heard better with the wind than against it, in consequence of a curvature of the rays. With the wind a ray will generally be bent downwards, since the velocity of the air is generally greater overhead than at the surface, and therefore the upper part of the wave-front tends to gain on the lower. The ray which ultimately reaches the observer is one which started in some degree upwards from the source, and has the advantage of being out of the way of obstacles for the greater part of its course. Against the wind, on the other hand, the curvature of the rays is upwards, so that a would-be observer at a considerable distance is in danger of being left in a sound-shadow.

It is very important to remark that this effect depends, not upon the mere existence of a wind, but upon the velocity of the wind being greater overhead than below. A uniform translation of the entire atmosphere would be almost without effect. In particular cases it may happen that the velocity of the wind diminishes with height, and then sound is best transmitted *against* the wind. Prof. Henry shows that several anomalous phenomena relating to the audibility of signals may be explained by various suppositions as to the velocity of the wind at different heights. When the distances concerned are great, comparatively small curvatures of the ray may produce considerable results.

There is a further possible consequence of the action of wind (or variable temperature), which, so far as I know, has not hitherto been remarked. By making the velocity a suitable function of height it would be possible to secure an actual convergence of rays in a vertical plane upon a particular station. The atmosphere would then act like the lens of a lighthouse, and the intensity of sound might be altogether abnormal. This may perhaps be the explanation of the extraordinary distances at which guns have sometimes been heard.

The difference in the propagation of sound against and with the wind is no exception to the general law referred to at the beginning of this communication, for that law applies only to the vibrations of a system about a configuration of equilibrium. A motion of the medium is thus excluded. But the bending of the sound-ray due to a variable temperature, to which attention has been drawn by Prof. Reynolds, does not interfere with the application of the law.

An experiment has, however, been brought forward by Prof. Tyndall, in which there is an apparent failure of reciprocity not referable to any motion of the medium*. The source of sound is a very high-pitched reed mounted in a short tube and blown from a small bellows with which it is connected by rubber tubing. The variation of pressure at the second point is made apparent by means of the sensitive flame, which has been used by Prof. Tyndall with so much success on other occasions. Although the flame itself, when unexcited, is 18 to 24 inches high, it was proved by a subsidiary experiment that the root of the flame, where it issues from the burner, is the seat of sensitiveness. With this arrangement the effect of a cardboard or glass screen interposed between the reed and the flame was found to be different, according as the screen was close to the flame or close to the reed. In the former case the flame indicated the action of sound, but in the latter remained uninfluenced. Since the motion of the screen is plainly equivalent to an interchange of the reed and flame, there is to all appearance a failure in the law of reciprocity.

At first sight this experiment is difficult to reconcile with theoretical conclusions. It is true that the conditions under which reciprocity is to be expected are not very perfectly realized, since the flame ought not to be moved from one position to the other. Although the seat of sensitiveness may be limited to the root of the flame, the tall column of highly heated gas might not be without effect; and in fact it appeared to me possible that the response of the flame, when close to the screen, might be due to the conduction of sound downwards along it. Not feeling satisfied, however, with this explanation, I determined to repeat the experiment, and wrote to Prof. Tyndall, asking to be allowed to see the apparatus. In reply he very kindly proposed to arrange a repetition of the experiment at the Royal Institution for my benefit, an offer which I gladly accepted.

The effect itself was perfectly distinct, and, as it soon appeared, was not to be explained in the manner just suggested, since the response of the flame when close to the screen continued, even when the upper part

* *Proceedings of the Royal Institution*, January 1875; also Prof. Tyndall's work on *Sound*, 3rd edition.

of the heated column was protected from the direct action of the source by additional screens interposed. I was more than ever puzzled until Mr Cottrell showed me another experiment in which, I believe, the key of the difficulty is to be found.

When the axis of the tube containing the reed is directed towards the flame, situated at a moderate distance, there is a distinct and immediate response; but when the axis is turned away from the flame through a comparatively small angle, the effect ceases, although the distance is the same as before, and there are no obstacles interposed. If now a cardboard screen is held in the prolongation of the axis of the reed, and at such an angle as to reflect the vibrations in the direction of the flame, the effect is again produced with the same apparent force as at first.

These results prove conclusively that the reed does not behave as the simple source of theory, even approximately. When the screen is close (about 2 inches distant) the more powerful vibrations issuing along the axis of the instrument impinge directly upon the screen, are reflected back, and take no further part in the experiment. The only vibrations which have a chance of reaching the flame, after diffraction round the screen, are the comparatively feeble ones which issue nearly at right angles with the axis. On the other hand, when the screen is close to the flame, the efficient vibrations are those which issue at a small angle with the axis, and are therefore much more powerful. Under these circumstances it is not surprising that the flame is affected in the latter case and not in the former.

The concentration of sound in the direction of the axis is greater than would have been anticipated, and is to be explained by the very short wave-length corresponding to the pitch of the reed. If, as is not improbable, the overtones of the note given by the reed are the most efficient part of the sound, the wave-length will be still shorter and the concentration more easy to understand*.

The reciprocal theorem in its generalized form is not restricted to simple sources, from which (in the absence of obstacles) sound would issue alike in all directions; and the statement for *double sources* will throw light on the subject of this note. A double source may be thus defined:—Conceive two equal and opposite simple sources, situated at a short distance apart, to be acting simultaneously. By calling the two sources opposite, it is meant that they are to be at any moment in opposite phases. At a moderate distance the effects of the two sources are antagonistic and may be made to neutralize one another to any extent by diminishing the distance between the sources. If, however, at the same time that we diminish the interval, we augment the

---

* July 13. I have lately observed that the flame in question is extremely sensitive to one of Mr F. Galton's whistles, which gives notes near the limits of ordinary hearing.

intensity of the single sources, the effect may be kept constant. Pushing this idea to its limit, when the intensity becomes infinite and the interval vanishes, we arrive at the conception of a double source having an axis of symmetry coincident with the line joining the single sources of which it is composed. In an open space the effect of a double source is the same as that communicated to the air by the vibration of a solid sphere whose centre is situated at the double point and whose line of vibration coincides with the axis, and the intensity of sound in directions inclined to the axis varies as the square of the cosine of the obliquity.

The statement of the reciprocal theorem with respect to double sources is then as follows:—If there be equal double sources at two points A and B, having axes AP, BQ respectively, then the *velocity* of the medium at B resolved in the direction BQ due to the source at A is the same as the *velocity* at A resolved in the direction AP due to the source at B. If the waves observed at A and B are sensibly plane, and if the axes AP, BQ are equally inclined to the waves received, we may, in the above statement, replace "velocities" by "pressures," but not otherwise.

Suppose, now, that equal double sources face each other, so that the common axis is AB, and let us examine the effect of interposing a screen near to A. By the reciprocal theorem, whether there be a screen or not, the velocity at A in direction AB due to B is equal to the velocity at B in direction AB due to A. The waves received at B are approximately plane and perpendicular to AB, so that the relation between the velocity and pressure at B is that proper to a plane wave; but it is otherwise in the case of the sound received at A. Accordingly the reciprocal theorem does not lead us to expect an equality between the pressures at A and B, on which quantities the behaviour of the sensitive flames depends*. On the contrary, it would appear that the pressure at A corresponding to the given velocity along AB should be much greater than in the case of a plane wave, and then the relative advantage of the position A would be explained.

It will be seen that if the preceding arguments are correct, Prof. Tyndall's experiment does not bear out the conclusions that he has based upon it with respect to the observations of the French Commission at Villejuif and Montlhéry. No acoustic clouds could explain the failure of reciprocity then observed; and the more probable hypothesis that the effect was due to wind is not inconsistent with the observation that the air (at the surface) was moving in the direction against which the sound was best heard.

Further experiments on this subject are very desirable.

---

* [1899. See however *Phil. Mag.* vol. VII. p. 153, 1879, where it appears that the excitation of a flame is due, not to a variable *pressure*, but to transverse *motion* across the nozzle.]

45.

ON A PERMANENT DEFLECTION OF THE GALVANOMETER-NEEDLE UNDER THE INFLUENCE OF A RAPID SERIES OF EQUAL AND OPPOSITE INDUCED CURRENTS.

[*Philosophical Magazine*, III. pp. 43—46, 1877.]

To the Editors of the Philosophical Magazine and Journal.

GENTLEMEN,

The publication, in your December Number, of a memoir by Mr Chrystal on Bi- and Unilateral Galvanometer Deflection recalled to my mind some observations of a like character made some years ago by myself. I have lately succeeded in finding the manuscript of a communication with the above title read (literally) before the British Association at Norwich in 1868, which contains a short account of these observations. As the subject has acquired an additional interest in consequence of the investigations of Dr Schuster and Mr Chrystal, I shall be glad if you can find room for my paper, which has not been printed in full hitherto.

I am, Gentlemen,

Your obedient Servant,

RAYLEIGH.

The following paper contains a short account of some experiments which led to rather unexpected results, of which I can find no notice in the methodical treatises on Electricity, although they might seem to be in the way of any experimenter on induced currents. The arrangement of the first experiment was nearly the same as that described by Faraday in his original memoir on induction. Two thick copper wires were coiled together—

the circuit of one being completed by the battery and make-and-break apparatus, and that of the other by an ordinary astatic galvanometer of moderate sensitiveness. The make-and-break arrangement is a very rude one of my own construction, acting either by the dipping of needles into mercury, or by the intermittent contact of a spring with a toothed wheel. When the handle of the instrument is turned, there are generated in the second circuit, as is well known, a series of instantaneous currents which are alternately opposite in sign but whose magnitudes are equal, although that corresponding to the break of the battery-circuit is the most condensed. When, then, the instrument is worked with such rapidity that the interval between the currents is very small in comparison with the time of free oscillation of the needle, the latter might be expected to be sensibly unaffected. But so far was this from being the case, that although the swing of the needle produced by a single impulse was only a few degrees, yet under the influence of the series of equal and opposite currents it remained steady at 60 or 70 [degrees], and *that* on either side of the zero-point, which had in fact become a position of unstable equilibrium. Since it took place indifferently in either direction, the deflection cannot be ascribed to an inequality in the alternate currents, giving on the whole a balance in one direction such as, according to the experiments of Henry and Abria, might arise from imperfect contacts in the second circuit.

The first explanation which suggested itself to me was that, while no doubt the currents of the two series were strictly equal (numerically), the resulting impulses, or rather impulsive couples, acting on the needle might be slightly different owing to the change of the latter's position in reference to the coil in the small vibration which the series of currents must produce, however quickly they may follow one another, which would give one set an advantage over the other. Those currents would prevail which tend to increase the deviation of the needle; for they would have, as it were, the greatest purchase on it. To make this perfectly clear, suppose either the galvanometer to be turned round, or the direction of the magnetic force altered by permanent magnets, so that the position of equilibrium of the needle is now no longer zero, but say 20°, and then let the series of induced currents pass. There might appear at first sight to be two cases, according as the first current tends to diminish or increase the already existing deviation; but the result is the same in both, and I will take for the sake of illustration that in which the needle is first sent towards zero. When the second instantaneous current passes, it finds the needle nearer zero, and therefore acts upon it with greater force than did the first; and this process continues, so that if for the moment we imagine the needle to vibrate about 20°, there is an outstanding force tending to increase the deviation. As this is unbalanced, the equilibrium at 20° cannot be maintained, and the needle must move further from zero: instead of equilibrium, perhaps, I should say resultant equilibrium; for the rapid vibration of the needle just now referred to of course goes on in any case. I worked out the mathematical theory of this action fully for a tangent-galvanometer; and for the case, to which experiment is not limited, of an equal interval between consecutive instantaneous currents of opposite sorts. The most conspicuous result (which might, however, have been anticipated) was that the effect is independent of the rapidity with which the make-and-break apparatus works. As this was not at all what I had inferred from the experiment, I began to doubt whether I had hit upon the true cause of the phenomenon; and on more close examination of the mathematical result, it appeared that the needle could not remain permanently deflected from its position of equilibrium at zero, unless each instantaneous current were powerful enough to swing it right round when acting on it alone, although an already existing deviation would be always increased. I have already mentioned that the phenomenon was observed when the swing for a single current was only a few degrees, so that there is no doubt of the inadequacy of the foregoing explanation.

The real cause is, I believe, to be found in a deficiency in the hardness of the steel needles, rendering them to some extent capable of temporary magnetism when placed in a field of force. If this temporary magnetism alone be considered, the two sets of instantaneous currents conspire in their effects instead of opposing each other; for if a soft-iron needle be freely suspended in a uniform field of magnetic force, it has, as is known, four positions of equilibrium, of which those two are stable which would be positions of equilibrium (one stable and one unstable) for a magnetized steel bar. If while the needle is in equilibrium the direction of the magnetic force is reversed, no disturbance takes place, because the magnetism of the needle is at the same time reversed also. If such a needle be suspended in the coil of a galvanometer, the force with which a current acts upon it is independent of the direction and varies as the square of the current; or when there is a rapid series of varying but periodic currents, the deflecting force varies as the integral of the square of the current, and as the sine of twice the deviation from zero. The deflecting force would, according to this, be for a given position of the needle with reference to the coil (or deviation) proportional to the heating-power of the discontinuous current; but it must be remembered that the case is an ideal one, as no iron is perfectly soft or capable of at once assuming the magnetism due to the field of force in which it is placed. A remarkable illustration of this will be mentioned a little later.

In order to test the correctness of these views, I removed the steel needles from the galvanometer and replaced them by a single soft-iron needle, with which it was found that all the phenomena observed before were reproduced. Being anxious to submit the arrangement to a more severe test, I placed the galvanometer in a third circuit, so that it should be acted on by the currents induced by the induced currents of the second circuit, as in Henry's experiments. The effect was very marked, though for this it was necessary that the galvanometer should be turned round so that the position of equilibrium should be about 20° or 25°; in a tangent-galvanometer the most favourable position would be at 45°. Throughout these experiments the effect always increased with the velocity of the contact-breaker up to a certain point, about 100 per second, and then declined. The general increase is in accordance with the explanation here advanced, while the falling off might be owing to an imperfect action of the make-and-break machine when a certain velocity is reached. I am more inclined, however, to attribute it to a want of theoretical softness in the iron, which prevents it from taking the full magnetism when the alternation of currents is too rapid*. In support of this opinion I adduce one more experiment. Returning to the first arrangement, in which the galvanometer was placed in the second circuit, I arranged a third circuit in the neighbour-hood of the second or galvanometer circuit, whose ends could be joined or kept apart†. In the second case, of course, no effect is produced by the third circuit; but what will be the result of completing it? It is known that while the magnitude of each instantaneous current in the galvanometer circuit is unaffected, the duration of them is increased by the induction. This dilution, so to speak, of the induced currents diminishes their heating-power, which depends on the integral of the square of the current while it lasts, and would, if the iron were perfectly soft, diminish the deflecting force on the galvanometer-needle; but it was found, on the contrary, considerably to increase it. In fact, the induced currents are too condensed to produce their full action, passing away before the needle is properly magnetized.

It is too soon to say whether any use can be made of these results; but it is possible that such a soft-iron galvanometer might be available for measuring the currents produced by the new magneto-electric machines when the consecutive waves are opposite.

[1899. For an application of the soft-iron galvanometer here suggested to measure feeble alternating currents, see Bellati (*Venezia, Istit. Atti*, I. p. 563, 1883; *Wied. Beibl.* VII. p. 617). A similar instrument has been constructed by Giltay. More recently the author has applied the method to determine the angle of lag in the case of powerful alternating currents (*Phil. Mag.* XLIII. p. 343, 1897).]

* [1899. The influence of eddy currents in the iron would need to be considered here.]

† There were two similar coils, each containing two wires A_1, A_2, B_1, B_2. The battery-circuit included A_1 and the interrupter. The second circuit consisted of the wires A_2, B_1, and the galvanometer. B_2 gave the third circuit, which was closed or open, according as the ends of B_2 were in connexion or not.—*Note added* Dec. 1876.

46.

ACOUSTICAL OBSERVATIONS. I.

[*Philosophical Magazine*, III. pp. 456—464, 1877.]

Perception of the Direction of a Source of Sound.

IN a paper with the above title, communicated last year to the Musical Association and afterwards published in abstract in *Nature* [Art. XL.], I brought forward the fact that we are unable to distinguish whether a pure tone (obtained from a tuning-fork and air-resonator) is immediately in front of or immediately behind us—although with other sounds, and notably with the human voice, there is in general no difficulty. In order to make the experiment satisfactorily, it is necessary to provide two similar forks and resonators and to place the observer between them. At a given signal *both* forks are struck, but one of them only is held over its resonator. If this precaution be neglected, the noise attending the excitation of the fork vitiates the experiment. Subsequently to the reading of my paper, it occurred to me that if the ordinary view as to the functions of the two ears be correct, there must be other ambiguous cases besides those already experimented upon. To the right of the observer, and probably nearly in the line of the ears, there must be one direction in which the ratio of the intensity of sound as heard by the right ear to the intensity as heard by the left ear has a maximum value greater than unity. For sounds coming from directions in front of this the ratio of intensities has a less and less value, approaching unity as its limit when the sound is immediately in front. In like manner, for directions intermediate between the direction of maximum ratio and that immediately behind the observer, the ratio of intensities varies continuously between the same maximum value and unity. Accordingly, for every direction in front there must be a corresponding direction behind for which the ratio of intensities has the same value; and these two directions could not be distinguished in the case of a pure tone. The only directions as to which there would be no ambiguity are the direction of maximum ratio itself, and a corresponding direction of minimum ratio on the other side of the head.

The attitude of my mind with respect to this result was, I confess, one of considerable scepticism. A great number of miscellaneous experiments had been made with forks as well as with other sources of sound; and I thought that, if these ambiguities had existed, indications of them must have been perceived already. It was therefore with some curiosity that I took the first opportunity, last September, of submitting the matter to the test of experiment, the same forks (making 256 vibrations per second) being used as on previous occasions. The decision was soon given. An observer facing north, for example, made mistakes between forks bearing approximately north-east and south-east, though he could distinguish without a moment's hesitation forks bearing east and west. In all such experiments it is necessary that the observer keep his head perfectly still, a very slight motion being sufficient in many cases to give the information that was previously wanting.

A suggestion was made, in the discussion that followed the reading of my paper before the Musical Association, which I thought it proper to examine, though I had not much doubt as to the result. In order to meet the difficulty in the ordinary view as to the functions of the two ears arising out of the fact that a 256-fork seems to be heard nearly as well with the ear turned away as with the ear turned towards it, it was suggested that possibly the discrimination between forks right and left depended on something connected with the commencement of the sound. It might be supposed, for example, that we are able to recognize which ear is first affected. On trial, however, it appeared that the power of discrimination was not weakened, although the observer stopped his ears during the establishment of the sound.

When *one* ear is stopped, mistakes are made between forks right and left; but the direction of other sounds, such as those produced by clapping hands or by the voice, is often told much better than might have been expected. [1899. See *Phil. Mag.* vol. XIII. p. 343, 1882.]

The Head as an Obstacle to Sound.

The perfection of the shadow thrown by the head depends on the pitch of the sound. I have already mentioned that it appears to make but little difference in the audibility of a pure tone with a frequency of 256, whether the ear used be turned towards or from the source. But the case is very different with sounds of higher pitch, such as that of an ordinary whistle. The one that I employed was blown from a loaded gas-bag, and gave a very steady note of pitch f^{iv}. A hiss is also heard very badly with the averted ear. This observation may be made by first listening with both ears to a steady hiss on the right or left, and then closing one ear. It makes but little difference when the further ear is closed, but a great difference when the nearer ear is closed. A similar observation may be made on the sound of running water.

For the same reason a hiss or whisper, coming from a person whose face is averted, is badly heard. Under these circumstances even ordinary speech is difficult to understand, though the mere intensity of sound does not seem deficient.

Reflection of Sound.

In many cases sound-shadows appear much less perfect than theory would lead us to expect. The anomally is due in great measure, I believe, to an error of judgment, depending on the enormous range of intensity with which the ear is capable of dealing. The whistle of a locomotive is very loud at a distance of ten yards. At a mile off the intensity must be 30,000 times less; but the sound still appears rather loud, and would probably be audible under favourable circumstances even when enfeebled in the ratio of a million to one. For this reason it is not easy to obtain complete shadows; but another difficulty arises from the fact that there are generally obstacles capable of reflecting a more or less feeble sound into what might otherwise be a nearly complete shadow. An attempt to examine this point led me to a few simple experiments on the reflection of sound, which may be worth recording.

The principal obstacle throwing the shadow was the corner of a large house; and among the sources of sound tried were the human voice, tuning-forks, whistles steadily blown, and a small electric bell, of which the last (which was employed in Professor Reynolds' acoustical experiments) proved to be as convenient as any. The source was placed close to the south side of the house, at a distance of eight or ten yards from the south-west corner, while the observer took up a corresponding position on the west side. With these arrangements the sound-shadow was pretty good, though far from perfect. When, however, a flat reflector, such as a drawing-board of moderate dimensions, was held at the proper angle by an assistant placed at some distance outwards from the corner, the augmentation of sound was immense, and the hearer realized for the first time how very good the shadow really was.

A screen made by stretching a *Times* over a hoop about $2\frac{1}{2}$ feet in diameter gave apparently as good a reflection as the drawing-board; but when calico was substituted for the paper the reflecting-power was very feeble. By wetting the calico, however, it could be made to reflect very well. These results are in agreement with the striking experiments described by Professor Tyndall.

Audibility of Consonants.

I suppose it must have been noticed before now that the *s* sound is badly returned by an echo. Standing at a distance of about 150 yards from a large wall, I found that there was scarcely any response to even the most powerful hiss. *Sh* was heard a little better; *m*, *k*, *p*, *g* pretty well; *r* very well; *h* badly; *t* badly; *b* seemed half converted into *p* by the echo. The failure of the hiss seems to be the fault of the air rather than of the wall, for a powerful hiss heard directly at a distance of 200 yards had very little *s* left in it.

Interference of Sounds from two unisonant Tuning-forks.

In ordinary experiments on interference the sounds are only approximately in unison, and consequently the silences resulting from antagonism of the vibrations are of only momentary duration. I thought it of interest, therefore, to arrange an experiment in which the sounds should be pure tones, absolutely in unison, and should proceed from sources at a considerable distance apart. With the aid of electromagnetism the solution of the problem was comparatively easy. An intermittent electric current, obtained from a fork interrupter making 128 vibrations per second, excited by means of electromagnets two other forks, whose frequency was 256. These latter forks were placed at a distance of about ten yards apart, and were provided with suitably tuned resonators by which their sounds were reinforced. The pitch of both forks is necessarily identical, since the vibrations are *forced* by electromagnetic forces of absolutely the same period. The arrangement was successful; and with a battery-power of two Grove cells sounds of fair intensity were obtained. With one ear closed it was possible to define the places of silence with considerable accuracy, a motion of about an inch being sufficient to produce a marked revival of sound. At a point of silence, from which the line joining the forks subtended an angle of about 60°, the apparent striking up of one fork, when the other was stopped, had a very peculiar effect.

Symmetrical Bell.

I do not know whether it has ever been noticed that there ought to be no sound emitted along the axis of a symmetrical bell. It is easy to see that at any point of the axis any effect, whether condensation or rarefaction, which may be produced by one part of the surface of the bell must be neutralized by other parts, and that therefore on the whole there can be no variation of pressure during the vibration. The experiment may be made with a large glass bell (such as are used with air-pumps), set into vibration by friction with the wetted finger carried round the circumference. If the axis of the vibrating bell be turned exactly towards the observer, the sound is feeble as compared with that heard when the position of the bell is altered. The residual sound may be due to want of symmetry, or more probably to reflexion from the ground, which last cause of error it is almost impossible to get rid of.

Octave from Tuning-forks.

When a vibrating fork is held over an air-resonator in tune with itself, the sound emitted is very approximately a pure tone; but when the fork is placed in contact with a sounding-board, the octave may generally be perceived by a practised ear, and is often of remarkable loudness. By means of a resonator tuned to the octave the fact may be made apparent to any one. This result need not surprise us. By the construction of a fork the moving parts are carefully balanced, and the motion is approximately isolated. In the ideal tuning-fork, composed of equal masses moving to and fro in a straight line, the isolation would be complete, and there would be no tendency whatever to communicate motion to surrounding bodies. In an actual fork, however, even if the direction of motion of the masses were as nearly as possible perpendicular to the stalk, the necessary curvature of the paths would give rise to an unbalanced centrifugal force tending to set the sounding-board in vibration. The force thus arising is indeed of the second order, and might probably be neglected, were it not that the apparatus is especially suited to bring it into prominence.

In order to test the soundness of this view as to the origin of the octave, the following experiment was contrived. A 256 tuning-fork was screwed on to a resonance-box intended for a 512 tuning-fork, and therefore approximately in tune with the octave of the first fork. When a powerful vibration was excited by means of a bow, *the octave sound was predominant*, and but little could be heard of the proper tone of the fork. In order to place the two sounds on a more equal footing, a resonator, consisting of a bottle tuned by pouring water into it to a frequency of 256, was brought near the ends of the vibrating prongs. By adjusting the distance it was easy to arrange matters so that at the beginning of the vibration neither sound had a conspicuous advantage. But, as the amplitude of vibration diminished, the graver tone continually gained on its rival, and was left at last in complete possession of the field. The purity of the remaining sound could be tested at any time by the perfection of the silence obtained on removing the air-resonator. This arrangement may be recommended to any one who wishes to practise his ears in hearing octaves.

From the above experiment (in which, if desired, the ear may be replaced by König's manometric flames), it appears that the octave sound is to be attributed to a motion of the second order, which is rendered important by the peculiar isolation of the motion of the first order. The harmonic sounds heard when suitably tuned resonators are presented to the free ends of the prongs, though also dependent on orders of the motion higher than the first, have a somewhat different origin.

Influence of a Flange on the Correction for the Open End of a Pipe.

In theoretical investigations[*] as to the amount of the correction to the length of an open pipe due to the inertia of the external air, it has been usual, for the sake of facilitating the calculations, to suppose that the open end is provided with an infinite flange. Even with this simplification no exact solution of the problem has been obtained. It has been proved, however, that, provided the wave-length be sufficient in relation to the diameter of the pipe, the addition which must be supposed to be made to the length is very nearly equal to, though somewhat less than, $\cdot 8242\,R$, and is certainly greater than $\cdot 785\,R$[†], R being the radius of the pipe.

It is obvious that the removal of the flange would make a considerable difference, probably reducing the correction below the lower limit above mentioned. In the absence of any theoretical estimate, I thought it desirable to make an experimental determination of the effect of a flange, and ordered some years ago a pair of similar organ-pipes of circular section for the purpose. My idea was to tune the pipes to unison, and then to count the beats when the pitch of one of them was slightly lowered by the addition of a flange; but the experiment lay in abeyance until last winter. Instead of tuning the pipes to unison, I preferred simply to count the beats before and after the addition of the flange, which consisted of a large sheet of stiff millboard perforated with a hole sufficiently large to allow the passage of the pipe. In this way it appeared that the effect of the flange was to reduce the frequency by nearly $1\frac{1}{2}$ out of about 242. If we take the velocity of sound at 1123 feet per second, corresponding to 60° F., the calculated effective length of the pipe is about 28 inches, and the radius is 1 inch. Thus the correction to the length due to the flange is the same fraction of 28 inches that $1\frac{1}{2}$ is of 242, or is equal to about $\cdot 2R$. Combining this result with the theoretical estimate above referred to, we may conclude that the whole correction for an open end, when there is no flange, must be about $\cdot 6R$.

Mr Bosanquet, to whom I communicated the result at which I had arrived, informs me that he has since determined the correction for a flange as $\cdot 25R$.

---

[*] Helmholtz, *Crelle*, 1860. Also a memoir by myself "On Resonance," *Phil. Trans.* 1871. [Art. v.]

[†] See note to a paper "On the Approximate Solution of certain Problems relating to the Potential," *Math. Soc. Proc.* vol. VII. No. 93. [Art. XXXIX.]

The Pitch of Organ-pipes.

The whole correction to the length of an organ-pipe, necessary to make it agree with Bernoulli's theory, is considerably greater than any of those spoken of under the preceding heading. According to the rule of Cavaillé-Coll, the addition for an open pipe of circular section amounts to as much as $3\frac{1}{3}R$, whereas for a simple tube open at both ends it should be only about $1\cdot2R$. This discrepancy is, I believe, often attributed to a peculiar action of the stream of air by which the pipe is excited. Of course it is not to be denied that some disturbance arises from this source, as is proved by the dependence of the pitch on the strength of the wind; but the near agreement between theory and measurements by Wertheim and others, on the pitch of resonators caused to speak by a stream of air, has always seemed to me to prove that a comparatively small part only of the whole discrepancy is to be explained in this way. On the other hand, it is obvious that the "open" end at the base of the pipe is very much contracted, and that the correction thence arising may be several times as great as that applicable to the upper end, where the pipe retains its full section. I was therefore anxious to ascertain what was the proper note of an organ-pipe, regarded as a freely vibrating column of air, and thus to estimate in what proportion the two causes of disturbance contribute to the final result.

There are two methods by which the pitch of a resonator may be determined without the use of a stream of air. The simplest, and in many cases the most accurate, method consists merely in tapping the resonator with the finger or other hammer of suitable hardness, and estimating with the aid of a monochord the pitch of the sound so produced. In attempting, however, thus to determine the pitch of the organ-pipe, I found a difficulty arising from the uncertain character of the sound, and the results were by no means so accordant as I desired. Possibly an observer gifted with a more accurate ear than mine would have been more successful. The other method is one of which I have had a good deal of experience, and which I can generally rely upon to give results of moderate accuracy. It consists in putting the ear into communication with the interior of the resonator, and determining to what note of the scale the resonance is loudest. I have generally found it possible thus to fix the pitch of a resonator to within a quarter of a semitone. In the present case a small hole was made in the side of the pipe near the centre; and over the hole a short piece of tube was cemented, which could be put into communication with the ear by means of a rubber tube. In this way the effective length of the pipe was determined to be 28·7 inches, 4·7 inches more than the actual length. As a check upon this estimate, I closed the upper end of the pipe with a plate of wood and again determined the note of maximum resonance. The effective length of the pipe was now 29·1 inches, so that the correction due to want of openness at the lower end amounted to 5·1 inches. If we add ·6 as a correction for the upper end, we obtain as the corrected length of the pipe in its ordinary condition 29·7 inches. The difference between this and 28·7, obtained directly, is greater, I think, than can be ascribed to errors of experimenting, and is possibly connected with the excessive magnitude of the correction in relation to the wave-length of the sound. The actual note of the pipe, when blown in the ordinary way by a wind of pressure measured by 2½ inches of water, corresponded to an effective length of 28 inches, so that the blown note was actually *higher* in pitch than the note of maximum resonance. So far, therefore, from the depression of pitch in an organ-pipe below that calculated from the actual length, according to Bernoulli's theory, being principally due to the action of the wind, it would appear that in the absence of a peculiar action of the wind the depression would be even greater than it is. Too much stress, however, must not be laid on a single observation; and all I would maintain is, that by far the larger part of the depression of pitch is due to the insufficient openness of the lower end of the pipe.

[1899. For further observations upon this subject, see *Phil. Mag.* vol. XIII. p. 340, 1882.]

47.

ON PROGRESSIVE WAVES.

[*Proceedings of the London Mathematical Society*, IX. pp. 21—26, 1877.]

IT has often been remarked that, when a group of waves advances into still water, the velocity of the group is less than that of the individual waves of which it is composed; the waves appear to advance through the group, dying away as they approach its anterior limit. This phenomenon was, I believe, first explained by Stokes, who regarded the group as formed by the superposition of two infinite trains of waves, of equal amplitudes and of nearly equal wave-lengths, advancing in the same direction. My attention was called to the subject about two years since by Mr Froude, and the same explanation then occurred to me independently[*]. In my book on the *Theory of Sound*, 1877 (§ 191), I have considered the question more generally, and have shown that, if V be the velocity of propagation of any kind of waves whose wave-length is λ, and $k = 2\pi\lambda^{-1}$, then U, the velocity of a group composed of a great number of waves, and moving into an undisturbed part of the medium, is expressed by

$$U = \frac{d(kV)}{dk}, \quad \dots\dots\dots\dots\dots\dots\dots\dots(1)$$

---

[*] Another phenomenon, also mentioned to me by Mr Froude, admits of a similar explanation. A steam launch moving quickly through the water is accompanied by a peculiar system of diverging waves, of which the most striking feature is the obliquity of the line containing the greatest elevations of successive waves to the wave-fronts. This wave pattern may be explained by the superposition of two (or more) infinite trains of waves, of slightly differing wave-lengths, whose directions and velocities of propagation are so related in each case that there is no change of position relatively to the boat. The mode of composition will be best understood by drawing on paper two sets of parallel and equidistant lines, subject to the above condition, to represent the crests of the component trains. In the case of two trains of slightly different wave-lengths, it may be proved that the tangent of the angle between the line of maxima and the wave-fronts is half the tangent of the angle between the wave-fronts and the boat's course.

or, as we may also write it,

$$U : V = 1 + \frac{d \log V}{d \log k} . \quad \dots\dots\dots\dots\dots\dots\dots(2)$$

Thus, if $V \propto \lambda^n$, $\qquad\qquad U = (1 - n) V. \quad \dots\dots\dots\dots \dots\dots\dots(3)$

In fact, if the two infinite trains be represented by $\cos k (Vt - x)$ and $\cos k' (V't - x)$, their resultant is represented by

$$\cos k (Vt - x) + \cos k' (V't - x),$$

which is equal to

$$2 \cos \{ \tfrac{1}{2} (k'V' - kV) t - \tfrac{1}{2} (k' - k) x \} . \cos \{ \tfrac{1}{2} (k'V' + kV) t - \tfrac{1}{2} (k' + k) x \}.$$

If $k' - k$, $V' - V$ be small, we have a train of waves whose amplitude varies slowly from one point to another between the limits 0 and 2, forming a series of groups separated from one another by regions comparatively free from disturbance. The position at time t of the middle of that group, which was initially at the origin, is given by

$$(k'V' - kV) t - (k' - k) x = 0,$$

which shows that the velocity of the group is $(k'V' - kV) \div (k' - k)$. In the limit, when the number of waves in each group is indefinitely great, this result coincides with (1).

The following particular cases are worth notice, and are here tabulated for convenience of comparison :—

$V \propto \lambda$,	$U = 0$,	Reynolds' disconnected pendulums[*].
$V \propto \lambda^{\frac{1}{2}}$,	$U = \tfrac{1}{2}V$,	Deep-water gravity waves.
$V \propto \lambda^0$,	$U = V$,	Aërial waves, &c.
$V \propto \lambda^{-\frac{1}{2}}$,	$U = \tfrac{3}{2}V$,	Capillary water waves.
$V \propto \lambda^{-1}$,	$U = 2V$,	Flexural waves.

The capillary water waves are those whose wave-length is so small that the force of restitution due to capillarity largely exceeds that due to gravity. Their theory has been given by Thomson (*Phil. Mag.*, Nov. 1871). The flexural waves, for which $U = 2V$, are those corresponding to the bending of an elastic rod or plate (*Theory of Sound*, § 191).

In a paper read at the Plymouth meeting of the British Association (afterwards printed in *Nature*, Aug. 23, 1877), Prof. Osborne Reynolds gave a dynamical explanation of the fact that a group of deep-water waves advances with only half the rapidity of the individual waves. It appears that the energy propagated across any point, when a train of waves is

* [1899. For a further discussion of the case $U=0$, see *Phil. Mag.* vol. XLVI. p. 567, 1898.]

passing, is only one-half of the energy necessary to supply the waves which pass in the same time, so that, if the train of waves be limited, it is impossible that its front can be propagated with the full velocity of the waves, because this would imply the acquisition of more energy than can in fact be supplied. Prof. Reynolds did not contemplate the cases where *more* energy is propagated than corresponds to the waves passing in the same time; but his argument, applied conversely to the results already given, shows that such cases must exist. The ratio of the energy propagated to that of the passing waves is $U : V$; thus the energy propagated in the unit time is $U : V$ of that existing in a length V, or U times that existing in the unit length. Accordingly

Energy propagated in unit time : Energy contained (on an average) in unit length $= d\,(kV)/dk$, by (1).

As an example, I will take the case of small irrotational waves in water of finite depth l^*. If z be measured downwards from the surface, and the elevation (h) of the wave be denoted by

$$h = H \cos (nt - kx), \quad\dots\dots\dots\dots\dots\dots\dots(4)$$

in which $n = kV$, the corresponding velocity-potential (ϕ) is

$$\phi = - VH \frac{e^{k\,(z-l)} + e^{-k\,(z-l)}}{e^{kl} - e^{-kl}} \sin (nt - kx). \quad\dots\dots\dots\dots(5)$$

This value of ϕ satisfies the general differential equation for irrotational motion ($\nabla^2\phi = 0$), makes the vertical velocity $d\phi/dz$ zero when $z = l$, and $- dh/dt$ when $z = 0$. The velocity of propagation is given by

$$V^2 = \frac{g}{k} \frac{e^{kl} - e^{-kl}}{e^{kl} + e^{-kl}}. \quad\dots\dots\dots\dots\dots\dots\dots\dots(6)$$

We may now calculate the energy contained in a length x, which is supposed to include so great a number of waves that fractional parts may be left out of account.

For the potential energy we have

$$V_1 = g\rho \iint_0^h z\,dz\,dx = \tfrac{1}{2}g\rho \int h^2 dx = \tfrac{1}{4}g\rho H^2 . x. \quad\dots\dots\dots\dots(7)$$

For the kinetic energy,

$$T = \tfrac{1}{2}\rho \iint \left\{ \left(\frac{d\phi}{dx}\right)^2 + \left(\frac{d\phi}{dz}\right)^2 \right\} dx\,dz = \tfrac{1}{2}\rho \int \left(\phi \frac{d\phi}{dz} \right)_{z=0} dx = \tfrac{1}{4}g\rho H^2 . x, \quad\dots\dots(8)$$

by (1) and (6). If, in accordance with the argument advanced at the end of this paper, the equality of V_1 and T be assumed, the value of the velocity of

---

* Prof. Reynolds considers the trochoidal wave of Rankine and Froude, which involves molecular rotation.

propagation follows from the present expressions. The whole energy in the waves occupying a length x is therefore (for each unit of breadth)

$$V_1 + T = \tfrac{1}{2} g \rho H^2 . x, \quad \dots\dots\dots\dots\dots\dots(9)$$

H denoting the maximum elevation.

We have next to calculate the energy propagated in time t across a plane for which x is constant, or, in other words, the work (W) that must be done in order to sustain the motion of the plane (considered as a flexible lamina) in the face of the fluid pressures acting upon the front of it. The variable part of the pressure (δp) at depth z is given by

$$\delta p = -\rho \frac{d\phi}{dt} = -nVH \frac{e^{k(z-l)} + e^{-k(z-l)}}{e^{kl} - e^{-kl}} \cos(nt - kx),$$

while for the horizontal velocity

$$\frac{d\phi}{dx} = kVH \frac{e^{k(z-l)} + e^{-k(z-l)}}{e^{kl} - e^{-kl}} \cos(nt - kx);$$

so that

$$W = \iint \delta p \frac{d\phi}{dx} dz\, dt = \tfrac{1}{4} g \rho H^2 . Vt . \left[1 + \frac{4kl}{e^{2kl} - e^{-2kl}} \right], \quad \dots\dots(10)$$

on integration. From the value of V in (6) it may be proved that

$$\frac{d(kV)}{dk} = \tfrac{1}{2} V \left\{ 1 + \frac{1}{V^2} \frac{d(kV^2)}{dk} \right\} = \tfrac{1}{2} V \left\{ 1 + \frac{4kl}{e^{2kl} - e^{-2kl}} \right\};$$

and it is thus verified that the value of W for a unit time

$$= \frac{d(kV)}{dk} \times \text{energy in unit length.}$$

As an example of the direct calculation of U, we may take the case of waves moving under the joint influence of gravity and cohesion. It is proved by Thomson that

$$V^2 = g/k + T'k, \quad \dots\dots\dots\dots\dots\dots(11)$$

where T' is the cohesive tension. Hence

$$U = \tfrac{1}{2} V \left\{ 1 + \frac{1}{V^2} \frac{d(kV^2)}{dk} \right\} = \tfrac{1}{2} V \frac{g + 3k^2 T'}{g + k^2 T'}. \quad \dots\dots\dots(12)$$

When k is small, the surface tension is negligible, and then $U = \tfrac{1}{2} V$; but when, on the contrary, k is large, $U = \tfrac{3}{2} V$, as has already been stated. When $T'k^2 = g$, $U = V$. This corresponds to the minimum velocity of propagation investigated by Thomson.

Although the argument from interference groups seems satisfactory, an independent investigation is desirable of the relation between energy existing and energy propagated. For some time I was at a loss for a method applicable to all kinds of waves, not seeing in particular why the comparison of energies should introduce the consideration of a variation of wave-length. The following investigation, in which the increment of wave-length is *imaginary*, may perhaps be considered to meet the want.

Let us suppose that the motion of every part of the medium is resisted by a force of very small magnitude proportional to the mass and to the velocity of the part, the effect of which will be that waves generated at the origin gradually die away as x increases. The motion, which in the absence of friction would be represented by $\cos(nt - kx)$, under the influence of friction is represented by $e^{-\mu x} \cos(nt - kx)$, where μ is a small positive coefficient. In strictness the value of k is also altered by the friction; but the alteration is of the second order as regards the frictional forces, and may be omitted under the circumstances here supposed. The energy of the waves per unit length at any stage of degradation is proportional to the square of the amplitude, and thus the whole energy on the positive side of the origin is to the energy of so much of the waves at their greatest value, *i.e.*, at the origin, as would be contained in the unit of length, as

$$\int_0^x e^{-2\mu x}\, dx : 1,$$

or as $(2\mu)^{-1} : 1$. The energy transmitted through the origin in the unit time is the same as the energy dissipated; and, if the frictional force acting on the element of mass m be hmv, where v is the velocity of the element and h is constant, the energy dissipated in unit time is $h\Sigma mv^2$ or $2hT$, T being the kinetic energy. Thus, on the assumption that the kinetic energy is half the whole energy, we find that the energy transmitted in the unit time is to the greatest energy existing in the unit length as $h : 2\mu$. It remains to find the connection between h and μ.

For this purpose it will be convenient to regard $\cos(nt - kx)$ as the real part of $e^{int} e^{-ikx}$, and to inquire how k is affected, when n is given, by the introduction of friction. Now the effect of friction is represented in the differential equations of motion by the substitution of $d^2/dt^2 + h\, d/dt$ in place of d^2/dt^2, or, since the whole motion is proportional to e^{int}, by substituting $-n^2 + ihn$ for $-n^2$. Hence the introduction of friction corresponds to an alteration of n from n to $n - \frac{1}{2}ih$ (the square of h being neglected); and accordingly k is altered from k to $k - \frac{1}{2}ih\, dk/dn$. The solution thus becomes $e^{-\frac{1}{2}hx\, dk/dn} e^{i(nt - kx)}$, or, when the imaginary part is rejected, $e^{-\frac{1}{2}hx\, dk/dn} \cos(nt - kx)$; so that $\mu = \frac{1}{2}h\, dk/dn$, and $h : 2\mu = dn/dk$. The ratio of the energy transmitted in the unit time to the energy existing in the unit length is therefore expressed by dn/dk or $d(kV)/dk$, as was to be proved.

It has often been noticed, in particular cases of progressive waves, that the potential and kinetic energies are equal; but I do not call to mind any general treatment of the question. The theorem is not usually true for the individual parts of the medium*, but must be understood to refer either to an integral number of wave-lengths, or to a space so considerable that the outstanding fractional parts of waves may be left out of account. As an example well adapted to give insight into the question, I will take the case of a uniform stretched circular membrane (*Theory of Sound*, § 200) vibrating with a given number of nodal circles and diameters. The fundamental modes are not quite determinate in consequence of the symmetry, for any diameter may be made nodal. In order to get rid of this indeterminateness, we may suppose the membrane to carry a small load attached to it anywhere except on a nodal circle. There are then two definite fundamental modes, in one of which the load lies on a nodal diameter, thus producing no effect, and in the other midway between nodal diameters, where it produces a maximum effect (*Theory of Sound*, § 208). If vibrations of both modes are going on simultaneously, the potential and kinetic energies of the whole motion may be calculated by *simple addition* of those of the components. Let us now, supposing the load to diminish without limit, imagine that the vibrations are of equal amplitude and differ in phase by a quarter of a period. The result is a *progressive* wave, whose potential and kinetic energies are the sums of those of the stationary waves of which it is composed. For the first component we have $V_1 = E \cos^2 nt$, $T_1 = E \sin^2 nt$; and for the second component, $V_2 = E \sin^2 nt$, $T_2 = E \cos^2 nt$; so that $V_1 + V_2 = T_1 + T_2 = E$, or the potential and kinetic energies of the progressive wave are equal, being the same as the whole energy of either of the components. The method of proof here employed appears to be sufficiently general, though it is rather difficult to express it in language which is appropriate to all kinds of waves.

---

* Aërial waves are an important exception.

48.

ON THE AMPLITUDE OF SOUND-WAVES.

[*Proceedings of the Royal Society*, XXVI. pp. 248—249, 1877.]

SCARCELY any attempts have been made, so far as I am aware, to measure the actual amplitude of sound-bearing waves, and indeed the problem is one of considerable difficulty[*]. Even if the measurement could be effected, the result would have reference only to the waves actually experimented upon, and would be of no great value in the absence of some means of defining the intensity of the corresponding sound. It is bad policy, however, to despise quantitative estimates because they are rough ; and in the present case it is for many reasons desirable to have a general idea of the magnitudes of the quantities with which we have to deal. Now it is evident that a superior limit to the amplitude of waves giving an audible sound may be arrived at from a knowledge of the energy which must be expended in a given time in order to generate them, and of the extent of surface over which the waves so generated are spread at the time of hearing. An estimate founded on these data will necessarily be too high, both because sound-waves must suffer some dissipation in their progress, and also because a part, and in some cases a large part, of the energy expended never takes the form of sound-waves at all.

The source of sound in my experiment was a whistle, mounted on a Wolf's bottle, in connexion with which was a syphon manometer for the purpose of measuring the pressure of wind. This apparatus was inflated from the lungs through an india-rubber tube, and with a little practice there was no difficulty in maintaining a sufficiently constant blast of the requisite duration. The most suitable pressure was determined by pre-liminary trials, and was measured by a column of water $9\frac{1}{2}$ centimetres high.

---

[*] [1899. Reference should have been made to the work of Boltzmann and Toppler (*Pogg. Ann.* vol. CXLI. p. 321, 1870).]

The first point to be determined was the distance from the source to which the sound remained clearly audible. The experiment was tried in the middle of a fine still winter's day, and it was ascertained that the whistle was heard without effort at a distance of 820 metres. In order to guard against any effect of wind, the precaution was taken of repeating the observation with the direction of propagation reversed, but without any difference being observable.

The only remaining datum necessary for the calculation is the quantity of air which passes through the whistle in a given time. This was determined by a laboratory experiment. The india-rubber tube was put into connexion with the interior of a rather large bell-glass open at the bottom, and this was pressed gradually down into a large vessel of water in such a manner that the manometer indicated a steady pressure of $9\frac{1}{2}$ centimetres. The capacity of the bell-glass was 5200 cubic centimetres, and it was found that the supply of air was sufficient to last $26\frac{1}{2}$ seconds of time. The consumption of air was therefore 196 cubic centimetres per second.

In working out the result it will be most convenient to use consistently the c.g.s. system. On this system of measurement the pressure employed was $9\frac{1}{2} \times 981$ dynes per square centimetre, and therefore the work expended per second in generating the waves was $196 \times 9\frac{1}{2} \times 981$ ergs [or 1.8×10^6 ergs. It may be noted that a volt-ampère corresponds to 10^7 ergs per second]. Now the mechanical value of a series of progressive waves is the same as the kinetic energy of the whole mass of air concerned, supposed to be moving with the maximum velocity of vibration (v); so that, if S denote the area of the wave-front considered, a be the velocity of sound, and ρ be the density of air, the mechanical value of the waves passing in a unit of time is expressed by $\frac{1}{2}S \cdot a \cdot \rho \cdot v^2$, in which the numerical value of a is about 34100, and that of ρ about ·0013. In the present application S is the area of the surface of a hemisphere whose radius is 82000 centimetres; and thus, if the whole energy of the escaping air were converted into sound, and there were no dissipation on the way, the value of v at the distance of 82000 centimetres would be given by the equation

$$v^2 = \frac{2 \times 196 \times 9\frac{1}{2} \times 981}{2\pi (82000)^2 \times 34100 \times ·0013},$$

whence $\qquad\qquad v = ·0014$ centimetre per second.

[Also $\qquad\qquad s = v/a = 4·1 \times 10^{-8}$.]

This result does not require a knowledge of the pitch of the sound. If the period be τ, the relation between the maximum excursion x and the maximum velocity v is

$$x = \frac{v\tau}{2\pi}.$$

In the present case the note of the whistle was f^{iv}, with a frequency of about 2730. Hence

$$x = \frac{\cdot 0014}{2\pi \times 2730} = 10^{-8} \times 8\cdot 1,$$

or the amplitude of the aërial particles was less than a ten-millionth of a centimetre.

I am inclined to think that on a still night a sound of this pitch, whose amplitude is only a hundred-millionth of a centimetre, would still be audible.

[1899. Further investigations upon this subject will be found in *Phil. Mag.* vol. XXXVIII. pp. 365—370, 1894.]

49.

ABSOLUTE PITCH.

[*Nature*, XVII. pp. 12—14, 1877.]

AT the present time the question of absolute pitch is attracting attention in consequence of the discrepancy between König's scale and the numbers determined by Appunn's tonometer. This instrument is founded upon the same idea as Scheibler's fork tonometer, and consists of a series of sixty-five harmonium reeds, bridging over an entire octave, and so tuned that each reed gives with its immediate neighbours four beats per second. The application to determine absolute pitch, however, does not require precision of tuning, all that is necessary being to count with sufficient accuracy the number of beats per second between each pair of consecutive reeds. The sum of all these numbers gives the difference of frequencies of vibration between the first reed and its octave, which is, of course, the same as the frequency of the first reed itself.

The whole question of musical pitch has recently been discussed with great care by Mr Ellis, in a paper read before the Society of Arts (May 23, 1877). He finds by original observation with Appunn's instrument 258·4 as the actual frequency of a König's 256 fork, and Prof. Preyer, of Jena, has arrived at a similar result (258·2). On the other hand, Prof. Mayer in America, and Prof. Macleod in this country, using other methods, have obtained numbers not differing materially from König's. The discrepancy is so considerable that it cannot well be attributed to casual errors of experiment; it seems rather to point to some defect in principle in the method employed. Now it appears to me that there is such a theoretical defect in the reed tonometer, arising from a sensible mutual action of the reeds. The use of the instrument to determine absolute frequencies assumes that the pitch of each reed is the same, whether it be sounding with the reed above, or with the reed below; and the results arrived at would be vitiated by any mutual influence. In consequence of the ill-understood operation of the wind, it is difficult to predict the character of the mutual influence with certainty; but (*Theory of Sound*, §§ 112—115) there is reason to think that the sounds would repel one another, so that the frequency of the beats heard when both reeds are sounding, *exceeds* the difference of the frequencies of the reeds when sounding singly. However this may be, in view of the proximity of consecutive reeds and of the near approach to unison*, the assumption of complete independence could only be justified by actual observation, and this would be a matter of some delicacy. If the mutual influence be uniform over the octave it would require a difference of one beat per minute only to reconcile König's and Appunn's numbers.

As to the amount of the influence I am not in a position to speak with confidence, but I may mention an observation which seems to prove that it cannot be left out of account. If two sounds of nearly the same pitch are going on together, slow beats are heard as the result of the superposition of vibrations. Suppose now that a third sound supervenes whose pitch is such that it gives rapid beats with the other two. It is evident that these rapid beats will be subject to a cycle of changes whose frequency is the same as that of the slow beat of the first two sounds. For example, in the case of equal intensities of two sounds there is a moment of silence due to the superposition of equal and opposite vibrations, and at this moment a third sound would be heard alone and could not give rise to beats. The experiment may be made with tuning-forks, and the period of the cycle will be found to be sensibly the same whether it be determined from the slow beat of the two forks nearly in unison, or from the rattle caused by the simultaneous sounding of a third fork giving from four to ten beats per second with the other two. In the case of forks there is no fear of sensible mutual action, but if it were possible for the third sound to affect the pitch of one of the others the equality of the periods would be disturbed. The observation on Appunn's instrument was as follows:—The reeds numbered 0 and 64 being adjusted to an exact octave, it was found that the beats arising from the simultaneous sounding of reeds 0, 63, and 64 were by no means steady, but passed through a cycle of changes in a period no greater than about five seconds. In order to work with greater certainty a resonator of pitch corresponding to reed 64 was connected with the ear by a flexible tube and adjusted to such a position that the beats between reeds 0 and 64 (when put slightly out of tune) were as distinct as possible, indicating that the gravest tone of reed 64 and the octave over-tone of reed 0 were of equal intensity. By *flattening* reed 64 (which can be done very readily by partially cutting off the wind) the beats of the three sounds could be made nearly steady, and then when reed 63 was put out of operation, beats having

---

* It must not be forgotten that the vibration of the tongue involves a transference of the centre of inertia, so that there is a direct tendency to set the sounding-board into motion.

a 5 seconds' period were heard, indicating that reeds 0 and 64 were in tune no longer. It would appear, therefore, that when reed 63 sounds the pitch of reed 64 is raised, but in interpreting the experiment a difficulty arises from the amount of the disturbance being much in excess of what would be expected from the performance of the instrument when tested in other ways*.

I come now to an independent determination of absolute pitch, which it is the principal object of the present communication to describe. The method employed may be regarded as new, and it appears to be capable of giving excellent results.

The standard fork, whose frequency was to be measured, is one of König's, and is supposed to execute 128 complete vibrations in a second. When placed on its stand (which does not include a resonance box) and excited by a violin bow, it vibrates for a ·minute with intensity sufficient for the counting of beats. The problem is to compare the frequency of this fork with that of the pendulum of a clock keeping good time. In my experiments two clocks were employed, of which one had a pendulum making about $1\frac{1}{2}$ complete vibrations per second, and the other a so-called seconds' pendulum, making half a vibration per second. Contrary to expectation, the slower pendulum was found the more convenient in use, and the numerical results about to be given refer to it alone. The rate of the clock at the time of the experiments was determined by comparison with a watch that was keeping good time, but the difference was found to be too small to be worth considering. In what follows it will be supposed for the sake of simplicity of explanation that the vibrations of the pendulum really occupied two seconds of time exactly.

The remainder of the apparatus consists of an electrically maintained fork interrupter, with adjustable weights, making about $12\frac{1}{2}$ vibrations per second, and a dependent fork whose frequency is about 125. The current from a Grove cell is rendered intermittent by the interrupter, and, as in Helmholtz's vowel experiments, excites the vibrations of the second fork, whose period is as nearly as possible an exact submultiple of its own. When the apparatus is in steady operation, the sound emitted from a resonator associated with the higher fork has a frequency which is

---

* The value of my instrument has been greatly enhanced by the valuable assistance of Mr Ellis, who was good enough to count the entire series of beats, and to compare the pitch with that of the tuning-forks employed by him in previous investigations. Mr Ellis, however, is not responsible for the facts and opinions here expressed. It may be worth mentioning that the steadiness or unsteadiness of the beats heard when three consecutive reeds are sounding simultaneously is a convenient test of the equality of the consecutive intervals. The frequency of the cycle of the four a second beats is equal to the difference of the frequencies of either of the actual extreme notes and that which, in conjunction with the other two, would make the intervals exactly equal.

determined by that of the interrupter, and not by that of the higher fork itself; nevertheless, an accurate tuning is necessary in order to obtain vibrations of sufficient *intensity**. By counting the beats during a minute of time it is easy to compare the higher fork and the standard with the necessary accuracy, and all that remains is to compare the frequencies of the interrupter and of the pendulum. For this purpose the prongs of the interrupter are provided with small plates of tin so arranged as to afford an intermittent view of a small silvered bead carried by the pendulum and suitably illuminated. Under the actual circumstances of the experiment the bright point of light is visible in general in twenty-five positions, which would remain fixed, if the frequency of the interrupter were exactly twenty-five times that of the pendulum. In accordance, however, with a well-known principle, these twenty-five positions are not easily observed when the pendulum is simply looked at; for the motion then appears to be continuous. The difficulty thence arising is readily evaded by the interposition of a somewhat narrow vertical slit, through which only one of the twenty-five positions is visible. In practice it is not necessary to adjust the slit to any particular position, since a slight departure from exactness in the ratio of frequencies brings all the visible positions into the field of view in turn.

In making an experiment the interrupter is tuned, at first by sliding the weights and afterwards by soft wax, until the interval between successive appearances of the bright spot is sufficiently long to be conveniently observed. With a slow pendulum there is no difficulty in distinguishing in which direction the pendulum is vibrating at the moment when the spot appears on the slit, and it is best to attend only to those appearances which correspond to one direction of the pendulum's motion. This will be best understood by considering the case of a conical pendulum whose motion, really circular, appears to be rectilinear to an eye situated in the plane of motion. The restriction just spoken of then amounts to supposing the hinder half of the circular path to be invisible. On this understanding the interval between successive appearances is the time required by the fork to gain or to lose one complete vibration as compared with the pendulum. Whether the difference is a loss or a gain is easily determined in any particular case by observing whether the apparent motion of the spot across the slit (which should have a visible breadth) is in the same or in the opposite direction to that of the pendulum's motion.

In my experiment the interrupter *gained* one vibration on the clock in about eighty seconds, so that the frequency of the fork was a thousandth part greater than 12·5, viz. 12·51. The dependent fork gave the ninth

---

* This tuning is effected by prolonging as much as possible the period of the beat heard when the dependent fork starts from rest. This beat may be regarded as due to an interference of the forced and natural notes.

harmonic, with a frequency of 125·1. The beats between this fork and the standard (whose pitch was the higher) were 180 in sixty seconds, so that the frequency of the standard was as nearly as possible 128·1, agreeing very closely with König's scale. The error of the determination may amount to ·1, but could not, I think, exceed ·2.

I ought to add that the *approximate* determination of the frequency of the interrupter must be made independently, as the observation on the pendulum does not decide *which* multiple of $\frac{1}{2}$ nearly coincides with the frequency of the fork. Also the relation between the two auxiliary forks was *assumed*, and not determined; but as to this there can be no doubt, unless it be supposed that König's scale may be in error to the extent of a whole tone.

50.

ON MR VENN'S EXPLANATION OF A GAMBLING PARADOX.

[*Mind*, II. pp. 409—410, 1877.]

Two players, A and B, toss for pennies. A has the option of continuing or stopping the game at any moment as it suits him. Has he, in consequence of this option, any advantage over B?

From one point of view it would seem that A *has* an advantage; for, as the game proceeds, the balance of gains must pass backwards and forwards from one side to the other, and if A makes up his mind to continue until he has won (for example) 10, the time must come when he will have an opportunity of carrying off his gains. On the other hand, it seems obvious *à priori* that no combination of fair bets can be unfair, and that A's option is of no value to him, inasmuch as at any point it is a matter of perfect indifference to him whether he risks another penny or not.

In order to examine the matter more closely, let us suppose that A has originally 1000 pennies, and that he proposes to continue the game until he has won 10, and then to leave off. Under these circumstances, it is clear that in no case can B lose more than 10, whereas A, if unlucky, may lose his whole stock before he has an opportunity of carrying off B's. The case is in fact exactly the same as if B had originally only 10 pennies, and the agreement were to continue the game until either A or B was ruined. The problem thus presented was solved long ago (see Todhunter's *History of Probabilities*, p. 62); and the result, as might have been expected, is that the odds are exactly 100 : 1 that B will be ruined. But it does not follow from this that the arrangement is in any degree advantageous for A; for, if A loses, he loses a sum one hundred times as great as that which he gains from B in the other (and more probable) contingency. A like argument applies, however great the disproportion of capitals may be. If the sums risked, as well as the chances to which they are subjected, be taken into account, the compensation is complete.

Mr Venn, however, is of opinion that these considerations do not meet the difficulty. With respect to the argument that A will always win if he goes on long enough, he says—" It may be replied that we have no right to assume that the fortune of the player (A) will hold out in this way, for he may be ruined before his turn of luck comes. This . . . is quite true, but does not explain the difficulty. We have only to suppose the men to be playing on credit to remove the objection. There is no reason whatever why any money should pass between them until the affair is finally settled. All such transactions, really, must be carried on to some extent on credit, unless there is to be the trouble of perpetual payments backwards and forwards; and it is therefore perfectly legitimate to suppose a state of things in which no enquiry is made as to the solvency of either of the parties until the crisis agreed upon has been reached."—(*Logic of Chance*, 2nd edition, ch. XIV. p. 371.) And again a little further on, " A man might safely, for instance, continue to lay an *even* bet that he would get the single prize in a lottery of a thousand tickets, provided he thus doubled, or more than doubled, his stake each time, and unlimited credit was given."—*Ibid.* p. 373.

To me, on the contrary, it seems that the question is entirely altered by the introduction of indefinite credit. There is no object, of course, in insisting on perpetual payments, and a credit may properly be allowed to the extent of the *actual resources* of the parties; but the case is very different when insolvency is permitted. In order to make a comparison, let us suppose, in our previous example, that A has no fortune of his own but is allowed a credit of 1000. If he wins 10 from B without first losing 1000 himself, he retires a victor, and his actual poverty is not exposed. But how does the matter stand if the luck is against him, and he comes to the end of his credit before securing his prize? When called upon to pay at the termination of the transaction, he has no means of doing so, and thus B is defrauded of his 1000, which in the long run would otherwise compensate him for the more frequent losses of 10. The advantage which A possesses depends entirely, as it seems to me, on the credit which is allowed him, but to which he is not justly entitled, and is of exactly the same nature as that enjoyed by any man of straw, who is nevertheless allowed to trade. What would be thought of a beggar who proposed to toss Baron Rothschild for 1000 pound notes? and if the proposal were agreed to, would it be said that the beggar's advantage depended upon his power of arbitrarily calling for a stoppage when it suits him and refusing to permit it sooner, and not rather that the one-sided character of the agreement depended on the simple fact that one party could pay if he lost, while the other party could not?

51.

ON THE RELATION BETWEEN THE FUNCTIONS OF LAPLACE AND BESSEL.

[*Proceedings of the London Mathematical Society*, IX. pp. 61—64, 1878.]

IN § 783 of Thomson and Tait's *Natural Philosophy* (1867), a suggestion is made to examine the transition from formulæ dealing with Laplace's spherical functions to the corresponding formulæ proper to a plane. It is evident at once from this point of view that Bessel's functions are merely particular cases of Laplace's more general functions; but the fact seems to be very little known. Of two valuable works recently published on this subject*, one makes no mention of Bessel's functions, and the other states expressly that they are not connected with the main subject of the book; other mathematicians also, to whom I have mentioned the matter, have been unaware of the relation. Under these circumstances it may not be superfluous to point out briefly the correspondence of some of the formulæ.

The Bessel's function of zero order J_0 is the limiting form of Legendre's function $P_n(\mu)$, when n is indefinitely great, and $\mu\ (=\cos\theta)$ such that $n\sin\theta$ is finite, equal say to z. The simplest proof of this assertion is perhaps that obtained from Murphy's series for P_n. Thus (Todhunter, § 23),

$$P_n(\cos\theta) = \cos^{2n}\frac{\theta}{2}\left\{1 - \left(\frac{n}{1}\tan\frac{\theta}{2}\right)^2 + \left(\frac{n\,.\,n-1}{1\,.\,2}\tan^2\frac{\theta}{2}\right)^2 - \ \ldots\ldots\right\}\ldots\ldots(1)$$

The limit of $\cos^{2n}(\tfrac{1}{2}\theta)$ is easily proved to be unity, and thus ultimately

$$P_n(\cos\theta) = 1 - \frac{z^2}{2^2} + \frac{z^4}{2^2\,.\,4^2} - \frac{z^6}{2^2\,.\,4^2\,.\,6^2} + \ \ldots\ldots = J_0(z).\ \ \ldots\ldots(2)$$

---

* *An Elementary Treatise on Laplace's Functions, Lamé's Functions, and Bessel's Functions.* By I. Todhunter. 1875.

An Elementary Treatise on Spherical Harmonics and subjects connected with them. By the Rev. N. M. Ferrers. 1877.

Again (§ 47),

$$\pi P_n(\mu) = \int_0^\pi \{\mu + i\sqrt{(1 - \mu^2)}\cos\phi\}^n\, d\phi, \quad \ldots\ldots\ldots\ldots(3)$$

where $i = \sqrt{(-1)}$. In the limit μ is to be replaced by unity, and $\sqrt{(1 - \mu^2)}$ by zn^{-1}; the limit of $\{\mu + i\sqrt{(1 - \mu^2)}\cos\phi\}^n$ is $e^{iz\cos\phi}$, and thus we get ultimately

$$\pi P_n(\mu) = \int_0^\pi e^{iz\cos\phi}\, d\phi = \int_0^\pi \cos(z\cos\phi)\, d\phi = \pi J_0(z).$$

In like manner Bessel's functions of higher order are the limits of those Laplace's functions called by Todhunter Associated Functions, which, when multiplied by sines or cosines of multiples of the longitude (ω), satisfy Laplace's spherical surface equation,

$$\frac{d}{d\mu}\left\{(1 - \mu^2)\frac{dY_n}{d\mu}\right\} + \frac{1}{1 - \mu^2}\frac{d^2Y_n}{d\omega^2} + n(n + 1)Y_n = 0. \quad \ldots\ldots\ldots(4)$$

They may be expressed (§§ 99, 106) by

$$(1 - \mu^2)^{\frac{1}{2}m}\, \varpi(m, n, \cos\theta), \quad \ldots\ldots\ldots\ldots\ldots\ldots(5)$$

where

$$\varpi(m, n, \cos\theta) = \varpi(m, n, 1)\left\{\cos^{n-m}\theta - \frac{(n - m)(n - m - 1)}{4 \cdot (m + 1) \cdot 1}\cos^{n-m-2}\theta\sin^2\theta\right.$$

$$\left. + \frac{(n - m)\ldots(n - m - 3)}{4^2 \cdot (m + 1)(m + 2) \cdot 1 \cdot 2}\cos^{n-m-4}\theta\sin^4\theta - \ldots\ldots\right\}. \quad \ldots\ldots\ldots(6)^*$$

Now $(1 - \mu^2)^{\frac{1}{2}m}$ is proportional to z^m; and the function becomes

$$z^m\left\{1 - \frac{z^2}{2(2m + 2)} + \frac{z^4}{2 \cdot 4 \cdot (2m + 2)(2m + 4)} - \ldots\ldots\right\},$$

which is proportional to the Bessel's function of order m (§ 370).

The same conclusion may be arrived at from the expression for the associated functions as definite integrals. Thus (§ 150) c_m^n is an associated function†, where

$$c_m^n = \int_0^\pi \{\mu + i\sqrt{(1 - \mu^2)}\cos\phi\}^n\cos m\phi\, d\phi. \quad \ldots\ldots\ldots\ldots(7)$$

By the investigation in § 150, we have

$$c_m^n = \frac{(-1)^{\frac{1}{2}m} \cdot 1 \cdot 2 \cdot 3 \ldots n \cdot (1 - \mu^2)^{\frac{1}{2}m}}{1 \cdot 3 \cdot 5 \ldots (2m - 1) \cdot 1 \cdot 2 \cdot 3 \ldots (n - m)}$$

$$\times \int_0^\pi \{\mu + i\sqrt{(1 - \mu^2)}\cos\phi\}^{n-m}\sin^{2m}\phi\, d\phi, \quad \ldots\ldots\ldots\ldots(8)$$

---

* This series is not quite correctly given in § 106.

† In Heine's and Todhunter's works this integral is called J_n, a notation which will be seen to be inappropriate.

from which we find, as for the function of the first order, when $n = \infty$, $n \sqrt{(1 - \mu^2)} = z$,

$$c_m^\infty = (-1)^{\frac{1}{2}m} \pi J_m(z). \quad \dots\dots\dots\dots\dots(9)$$

It is known (§ 380) that the following relation subsists between three consecutive Bessel's functions:

$$\tfrac{1}{2} z \{J_{m-1}(z) + J_{m+1}(z)\} = m J_m(z); \quad \dots\dots\dots\dots(10)$$

and it is natural to suppose that a corresponding theorem holds for the general functions. By (7),

$$\tfrac{1}{2}(c_{m-1}^n - c_{m+1}^n) = \int_0^\pi \{\mu + i \sqrt{(1 - \mu^2)} \cos \phi\}^n \sin m\phi \sin \phi \, d\phi,$$

whence, integrating by parts, we find (m being integral)

$$\tfrac{1}{2}(c_{m-1}^n - c_{m+1}^n) = \frac{m}{i \sqrt{(1 - \mu^2)} \cdot (n + 1)} c_m^{n+1}; \quad \dots\dots\dots(11)$$

and this, as we see from (9), identifies itself with (10), when $n = \infty$.

Again, by (7),

$$\tfrac{1}{2}(c_{m-1}^n + c_{m+1}^n) = \int_0^\pi \{\mu + i \sqrt{(1 - \mu^2)} \cos \phi\}^n \cos m\phi \cos \phi \, d\phi; \dots\dots(12)$$

and thus

$$c_m^{n+1} = \mu c_m^n + \tfrac{1}{2} i \sqrt{1 - \mu^2} (c_{m-1}^n + c_{m+1}^n). \quad \dots\dots\dots\dots(13)$$

From (11) and (13) we may eliminate c_m^{n+1}; we thus obtain

$$2m \, c_m^n = i \tan \theta \{(n + 1 - m) c_{m-1}^n - (n + 1 + m) c_{m+1}^n\}, \quad \dots\dots\dots(14)$$

a relation given in a somewhat different form by Heine*.

Returning to (12), we find in the limit, by (9),

$$\tfrac{1}{2}(-1)^{\frac{1}{2}m} \pi \{-i J_{m-1}(z) + i J_{m+1}(z)\}$$

$$= \int_0^\pi e^{iz \cos \phi} \cos m\phi \cos \phi \, d\phi = -i \frac{d}{dz} \int_0^\pi e^{iz \cos \phi} \cos m\phi \, d\phi;$$

so that

$$\tfrac{1}{2}\{J_{m-1}(z) - J_{m+1}(z)\} = \frac{d}{dz} J_m(z), \quad \dots\dots\dots\dots(15)$$

a known theorem (§ 381).

The complete solution of the equation

$$\frac{d}{d\mu}\left\{(1 - \mu^2) \frac{d\psi_m}{d\mu}\right\} - \frac{m^2}{1 - \mu^2} \psi_m + n(n + 1) \psi_m = 0, \quad \dots\dots\dots(16)$$

obtained from (4) by assuming that ψ_m is proportional to $\cos m\omega$ or $\sin m\omega$, involves also another function called the associated function of the second kind. If we take

$$\psi_m = (1 - \mu^2)^{\frac{1}{2}m} \phi_m, \quad \dots\dots\dots\dots\dots(17)$$

* *Handbuch der Kugelfunctionen*, p. 146.

the equation for ϕ_m is

$$(1 - \mu^2)\frac{d^2\phi_m}{d\mu^2} - 2(m+1)\mu\frac{d\phi_m}{d\mu} + \left\{n(n+1) - m(m+1)\right\}\phi_m = 0. \quad\ldots..(18)$$

From this it may be readily proved that ϕ_m is of the same form as

$$\frac{d}{d\mu}\phi_{m-1},$$

so that

$$\phi_m = \left(\frac{d}{d\mu}\right)^m \phi_0, \quad\ldots\ldots\ldots\ldots\ldots\ldots\ldots\ldots\ldots(19)$$

in which, however, no connexion between the arbitrary constants is implied. In terms of ψ_m, (19) becomes

$$\psi_m = (1 - \mu^2)^{\frac{1}{2}m}\left(\frac{d}{d\mu}\right)^m \psi_0. \quad\ldots\ldots\ldots\ldots\ldots\ldots(20)$$

We have now to trace the transformation of (20), as n and the radius of the sphere become infinite, while $n\nu = n\sqrt{(1 - \mu^2)} = z$.   Since $\mu^2 + \nu^2 = 1$,

$$\psi_m = \nu^m\left(-\frac{\mu}{\nu}\frac{d}{d\nu}\right)^m \psi_0,$$

ψ_0 being regarded as a function of ν.

In the limit, μ (even though subject to differentiation) may be identified with unity, and thus we may take

$$\psi_m = -(2z)^m\left(\frac{d}{d.z^2}\right)^m \psi_0 \quad\ldots\ldots\ldots\ldots\ldots\ldots(21)$$

as the ultimate form of (20).   The function ψ_m is now the general solution of

$$\frac{d^2\psi_m}{dz^2} + \frac{1}{z}\frac{d\psi_m}{dz} + \left(1 - \frac{m^2}{z^2}\right)\psi_m = 0, \quad\ldots\ldots\ldots\ldots(22)$$

involving Bessel's functions of order m *of both kinds*, each function being affected with an arbitrary coefficient.   A theorem of precisely the same form holds good when ψ_m is limited to the functions of the first kind, for which the differential equations (16), (22) are satisfied at the pole ($\mu = 1$) as well as at all other points.   The equation corresponding to (21) may then be written (§ 390)

$$J_m(z) = (-2z)^m\left(\frac{d}{d.z^2}\right)^m J_0(z), \quad\ldots\ldots\ldots\ldots\ldots(23)$$

the constant multiplier left arbitrary by (21) being determined by a comparison of the leading terms.

Other comparisons relating to the functions of the second kind might also be made, some of which would appear to involve theorems not hitherto formulated.

52.

NOTE ON ACOUSTIC REPULSION.

[*Philosophical Magazine*, VI. pp. 270—271, 1878.]

PERHAPS the following explanation of the curious phenomenon of the repulsion of resonators observed by Dvořák and Mayer* may be of interest to the readers of the *Philosophical Magazine*.

The hydrodynamical equation of pressure for irrotational motion is (in the usual notation)

$$\varpi = \int \frac{dp}{\rho} = R - \frac{d\phi}{dt} - \tfrac{1}{2} U^2. \quad \dots\dots\dots\dots\dots(1)$$

If we suppose that there are no impressed forces, $R = 0$. Distinguishing the values of the quantities at two points of space by suffixes, we may write

$$\varpi_1 - \varpi_0 = \frac{d}{dt}(\phi_0 - \phi_1) + \tfrac{1}{2} U_0^2 - \tfrac{1}{2} U_1^2. \quad \dots\dots\dots\dots(2)$$

This equation holds good at every instant. Integrating it over a long range of time, we obtain as applicable to every case of fluid-motion in which the flow between the two points does not continually increase

$$\int \varpi_1 dt - \int \varpi_0 dt = \tfrac{1}{2} \int U_0^2 dt - \tfrac{1}{2} \int U_1^2 dt. \quad \dots\dots\dots\dots(3)$$

Let us now apply this equation to the case of a resonator excited by a source of sound nearly in unison with itself, taking the first point at a distance from the resonator, where neither the variation of pressure nor the velocity is sensible, and for the second a point in the interior of the cavity, where the velocity is negligible, but, on the other hand, the variation of pressure considerable. It follows that

$$\int (\varpi_1 - \varpi_0)\, dt = 0, \quad \dots\dots\dots\dots\dots\dots(4)$$

---

* *Phil. Mag.* September 1878, p. 225.

or that the mean value of ϖ in the interior is the same as at a distance outside.

The remainder of the investigation depends upon the relation between p and ρ. If the expansions and contractions are isothermal, $p = a^2\rho$, and $\varpi = a^2 \log p$. Thus

$$\int \log p_1 dt = \log p_0 . t ; \qquad \dots\dots\dots\dots\dots\dots(5)$$

or the mean logarithmic pressure in the interior is the same as the constant logarithmic pressure at a distance. Equation (5) may also be written

$$\int \log \left(1 + \frac{p_1 - p_0}{p_0}\right) dt = 0, \qquad \dots\dots\dots\dots\dots(6)$$

or

$$\int \left\{ \frac{p_1 - p_0}{p_0} - \tfrac{1}{2}\left(\frac{p_1 - p_0}{p_0}\right)^2 + \dots \right\} dt = 0 ; \qquad \dots\dots\dots\dots(7)$$

whence, if the changes of pressure be relatively small, we see that the mean value of $p_1 - p_0$ is positive, or, in other words, that the mean pressure inside the resonator is in *excess* of the atmospheric pressure.

If, as in practice, the expansions and contractions are adiabatic, $p \propto \rho^\gamma$, where $\gamma = 1\cdot4$, and (5) is replaced by

$$\int p_1^{(\gamma-1)/\gamma} dt = p_0^{(\gamma-1)/\gamma} . t \dots\dots\dots\dots\dots\dots\dots(8)$$

Thus, instead of (7),

$$\int \left\{ \left(1 + \frac{p_1 - p_0}{p_0}\right)^{\frac{\gamma-1}{\gamma}} - 1 \right\} dt = 0 ; \qquad \dots\dots\dots\dots(9)$$

whence, by the binomial theorem,

$$\int \frac{p_1 - p_0}{p_0} dt = \frac{1}{2\gamma} \int \left(\frac{p_1 - p_0}{p_0}\right)^2 dt \dots\dots\dots\dots(10)$$

approximately, showing that here again the mean pressure in the interior of the cavity exceeds the atmospheric pressure*.

Hence, on either supposition, the resonator tends to move as if impelled by a force acting normally over the area of its aperture and directed *inwards*.

* [1899. If in (10) we put $\gamma = 1$, we fall back upon (7).]

53.

ON THE IRREGULAR FLIGHT OF A TENNIS-BALL.

[*Messenger of Mathematics*, VII. pp. 14—16, 1877.]

IT is well known to tennis players that a rapidly rotating ball in moving through the air will often deviate considerably from the vertical plane. There is no difficulty in so projecting a ball against a vertical wall that after rebounding obliquely it shall come back in the air and strike the same wall again. It is sometimes supposed that this phenomena is to be explained as a sort of frictional rolling of the rotating ball on the air condensed in front of it, but the actual deviation is in the opposite direction to that which this explanation supposes. A ball projected horizontally and rotating about a vertical axis, deviates from the vertical plane, as if it were rolling on the air *behind* it. The true explanation was given in general terms many years ago by Prof. Magnus, in a paper "On the Deviation of Projectiles," published in the *Memoirs of the Berlin Academy*, 1852, and translated in Taylor's *Scientific Memoirs*, 1853, p. 210. Instead of supposing the ball to move through air which at a sufficient distance remains undisturbed, it is rather more convenient to transfer the motion to the air, so that a uniform stream impinges on a ball whose centre maintains its position in space—a change not affecting the relative motion on which alone the mutual forces can depend. Under these circumstances, if there be no rotation, the action of the stream, whether there be friction or not, can only give rise to a force in the direction of the stream, having no lateral component. But if the ball rotate, the friction between the solid surface and the adjacent air will generate a sort of whirlpool of rotating air, whose effect may be to modify the force due to the stream. If the rotation take place about an axis perpendicular to the stream, the superposition of the two motions gives rise on the one side to an augmented, and on the other to a diminished velocity, and consequently to a lateral force urging the ball towards that side on which the motions conspire.

The only weak place in this argument is in the last step, in which it is assumed that the pressure is greatest on the side where the velocity is least. The law that a diminished pressure accompanies an increased velocity is only generally true on the assumption that the fluid is frictionless and unacted on by external forces; whereas, in the present case, friction is the immediate cause of the whirlpool motion. The actual mode of generation of the lateral force will be perhaps better understood, if we suppose small vertical blades to project from the surface of the ball. On that side of the ball where the motion of the blades is up stream, their anterior faces are in part exposed to the pressure due to the augmented relative velocity, which pressure necessarily operates also on the contiguous spherical surface of the ball. On the other side the relative motion, and therefore also the lateral pressure, is less; and thus an uncompensated lateral force remains over.

The principal object of the present note is to propose and solve a problem which has sufficient relation to practice to be of interest, while its mathematical conditions are simple enough to allow of an exact solution being obtained. For this purpose I take the case of a cylinder round which a perfect fluid circulates without molecular rotation. At a great distance from the cylinder the fluid is supposed to move with uniform velocity, and the whole motion is in two dimensions. On these suppositions the stream function, on which the whole motion depends, is of the form

$$\psi = \alpha \left(1 - \frac{a^2}{r^2} \right) r \sin \theta + \beta \log r,$$

where r, θ are the polar coordinates of any point of the fluid, measured from the centre of the cylinder and the direction of the stream, as pole and initial line respectively, a is the radius of the cylinder, and α, β are constant coefficients proportional respectively to the velocity of the general current and the velocity of circulation round the cylinder. When $r = a$, ψ is constant, showing that the surface of the cylinder is a stream-line. The radial velocity at any point is given by

$$\frac{d\psi}{r \, d\theta} = \alpha \left(1 - \frac{a^2}{r^2} \right) \cos \theta \, ;$$

so that, when $r = \infty$ and $\theta = 0$, the radial velocity is α, which is therefore the general velocity of the stream.

At the surface of the cylinder there is no radial velocity, and the magnitude of the tangential velocity is given by

$$d\psi/dr = 2\alpha \sin \theta + \beta/a.$$

Hence, if p_0 be the pressure at a distance, and p the pressure at any point on the surface,

$$2 \, (p - p_0) = \alpha^2 - (2\alpha \sin \theta + \beta/a)^2,$$

the density of the fluid being taken as unity. Thus the lateral force

$$= \int_0^{2\pi} (p - p_0)\, a \sin \theta \, d\theta = - [2]\, \pi a \beta \, ;$$

it is therefore proportional both to the velocity of the motion of circulation, and also to the velocity with which the cylinder moves relatively to the fluid at a distance.

If the velocity of circulation depending on β be small, the character of the stream-lines differs but little from that given by

$$\psi = \alpha \, (1 - a^2/r^2) \, r \sin \theta,$$

corresponding to a simple stream; but when β attains a certain point of magnitude, the stream-lines in the neighbourhood of the cylinder become re-entrant.

Sir W. Thomson has proved that, if in an infinite mass of otherwise quiescent fluid there exist irrotational circulation round a moveable cylinder, the amount of the circulation cannot be changed by any forces applied to the cylinder*.   Hence, if the cylinder receive an impulse, it will afterwards move in a circle, and the direction of revolution will be [the same as] that of the circulation of the fluid.

It must not be forgotten that the motion of an actual fluid would differ materially from that supposed in the preceding calculation in consequence of the unwillingness of stream-lines to close in at the stern of an obstacle, but this circumstance would have more bearing on the force in the direction of motion than on the lateral component.

[1899.   The corrections introduced in square brackets were indicated by Greenhill in a paper (*Mess. of Math.* vol. IX. p. 113, 1880), where the subject is further developed.]

* Vortex Motion, *Edinburgh Transactions*, 1868.

54.

A SIMPLE PROOF OF A THEOREM RELATING TO THE POTENTIAL.

[*Messenger of Mathematics*, VII. p. 69, 1878.]

THE value of a function which satisfies Laplace's equation within a closed surface is determined by the values at the surface. If two surface distributions be superposed, the value at any internal point is the sum of those due to the two surface distributions considered separately. By means of this principle a very simple proof may be given of the known theorem that the value of the potential at the centre of a sphere is the mean of those distributed over the surface.

On account of the symmetry it is clear that the central value would not be affected by any rotation of the sphere, to which the surface values are supposed to be rigidly attached.

Thus, if we conceive the sphere to be turned into n different positions taken at random and the resulting surface distributions to be superposed, we obtain a new surface distribution, whose mean value is n times greater than before, determining a central value which is also n times greater than that due to the original distribution. When n is made infinite, the surface distribution becomes constant, in which case the central value is the same as the surface value. From this it follows that in the original state of things the central value was the mean of the surface values.

55.

THE EXPLANATION OF CERTAIN ACOUSTICAL PHENOMENA.

[*Roy. Inst. Proc.* VIII. pp. 536—542, 1878; *Nature*, XVIII. pp. 319—321, 1878.]

MUSICAL sounds have their origin in the vibrations of material systems. In many cases, *e.g.* the pianoforte, the vibrations are free, and are then necessarily of short duration. In other cases, *e.g.* organ pipes and instruments of the violin class, the vibrations are maintained, which can only happen when the vibrating body is in connexion with a source of energy, capable of compensating the loss caused by friction and generation of aerial waves. The theory of free vibrations is tolerably complete, but the explanations hitherto given of maintained vibrations are generally inadequate and in most cases altogether illusory.

In consequence of its connexion with a source of energy, a vibrating body is subject to certain forces, whose nature and effects are to be estimated. These forces are divisible into two groups. The first group operate upon the periodic time of the vibration, *i.e.* upon the pitch of the resulting note, and their effect may be in either direction. The second group of forces do not alter the pitch, but either encourage or discourage the vibration. In the first case only can the vibration be maintained; so that for the explanation of any maintained vibration, it is necessary to examine the character of the second group of forces sufficiently to discover whether their effect is favourable or unfavourable. In illustration of these remarks, the simple case of a common pendulum was considered. The effect of a small periodic horizontal impulse is in general both to alter the periodic time and the amplitude of vibration. If the impulse (supposed to be always in the same direction) acts when the pendulum passes through its lowest position, the force belongs to the second group. It leaves the periodic time unaltered, and encourages or discourages the vibration according as the direction of the pendulum's motion is the same or the opposite of that of the impulse. If, on the other hand, the impulse acts when the pendulum is at one or other of the limits of its swing, the effect is solely on the periodic time, and the vibration is neither encouraged nor discouraged. In order to encourage, *i.e.* practically in order to maintain, a vibration, it is necessary that the forces should not depend solely upon the position of the vibrating body. Thus, in the case of the pendulum, if a small impulse in a given direction acts upon it every time that it passes through its lowest position, the vibration is not maintained, the advantage gained as the pendulum makes a passage in the same direction as that in which the impulse acts being exactly neutralized on the return passage when the motion is in the opposite direction.

As an example of the application of these principles the maintenance of an electric tuning-fork was discussed. If the magnetic forces depended only upon the position of the fork, the vibration could not be maintained. It appears therefore that the explanations usually given do not touch the real point at all. The fact that the vibrations are maintained is a proof that the forces do not depend solely upon the position of the fork. The causes of deviation are two—the self-induction of the electric currents, and the adhesion of the mercury to the wire whose motion makes and breaks the contact. On both accounts the magnetic forces are more powerful in the latter than in the earlier part of the contact, although the position of the fork is the same; and it is on this *difference* that the possibility of maintenance depends. Of course the arrangement must be such that the retardation of force *encourages* the vibration, and the arrangement which in fact encourages the vibration would have had the opposite effect, if the nature of electric currents had been such that they were more powerful during the earlier than during the later stages of a contact.

In order to bring the subject within the limits of a lecture, one class of maintained vibrations was selected for discussion, that, namely, of which *heat* is the motive power. The best understood example of this kind of maintenance is that afforded by Trevelyan's bars, or rockers. A heated brass or copper bar, so shaped as to rock readily from one point of support to another, is laid upon a cold block of lead. The communication of heat through the point of support expands the lead lying immediately below in such a manner that the rocker receives a small impulse. During the interruption of the contact the communicated heat has time to disperse itself in some degree into the mass of lead, and it is not difficult to see that the impulse is of a kind to encourage the motion. But the most interesting vibrations of this class are those in which the vibrating body consists of a mass of air more or less completely confined.

If heat be periodically communicated to, and abstracted from, a mass of air vibrating (for example) in a cylinder bounded by a piston, the effect produced will depend upon the phase of the vibration at which the transfer of heat takes place. If heat be given to the air at the moment of greatest condensation, or taken from it at the moment of greatest rarefaction, the vibration is encouraged. On the other hand, if heat be given at the moment of greatest rarefaction, or abstracted at the moment of greatest condensation, the vibration is discouraged. The latter effect takes place of itself, when the rapidity of alternation is neither very great nor very small, in consequence of radiation; for when air is condensed it becomes hotter, and communicates heat to surrounding bodies. The two extreme cases are exceptional, though for different reasons. In the first, which corresponds to the suppositions of Laplace's theory of the propagation of sound, there is not sufficient time for a sensible transfer to be effected. In the second, the temperature remains nearly constant, and the loss of heat occurs during the *process* of condensation, and not when the condensation is effected. This case corresponds to Newton's theory of the velocity of sound. When the transfer of heat takes place at the moments of greatest condensation or of greatest rarefaction, the pitch is not affected.

If the air be at its normal density at the moment when the transfer of heat takes place, the vibration is neither encouraged nor discouraged, but the pitch is altered. Thus the pitch is *raised*, if heat be communicated to the air a quarter period *before* the phase of greatest condensation, and the pitch is *lowered* if the heat be communicated a quarter period *after* the phase of greatest condensation.

In general both kinds of effects are produced by a periodic transfer of heat. The pitch is altered, and the vibrations are either encouraged or discouraged. But there is no effect of the second kind if the air concerned be at a loop, *i.e.* a place where the density does not vary, nor if the communication of heat be the same at any stage of rarefaction as in the corresponding stage of condensation.

The first example of aerial vibrations maintained by heat was found in a phenomenon which has often been observed by glass-blowers, and was made the subject of a systematic investigation by Dr Sondhauss. When a bulb about three-quarters of an inch in diameter is blown at the end of a somewhat narrow tube, 5 or 6 inches in length, a sound is sometimes heard proceeding from the heated glass. It was proved by Sondhauss that a vibration of the glass itself is no essential part of the phenomenon, and the same observer was very successful in discovering the connexion between the *pitch* of the note and the dimensions of the apparatus. But no explanation (worthy of the name) of the production of sound has been given.

For the sake of simplicity, a simple tube, hot at the closed end and getting gradually cooler towards the open end, was first considered. At a quarter of a period *before* the phase of greatest condensation (which occurs almost simultaneously at all parts of the column) the air is moving inwards, *i.e.* towards the closed end, and therefore is passing from colder to hotter parts of the tube; but the heat received at this moment (of normal density) has no effect either in encouraging or discouraging the vibration. The same would be true of the entire operation of the heat, if the adjustment of temperature were instantaneous, so that there was never any sensible difference between the temperatures of the air and of the neighbouring parts of the tube. But in fact the adjustment of temperature takes *time*, and thus the temperature of the air deviates from that of the neighbouring parts of the tube, inclining towards the temperature of that part of the tube *from* which the air has just come. From this it follows that at the phase of greatest condensation heat is received by the air, and at the phase of greatest rarefaction is given up from it, and thus there is a tendency to maintain the vibrations. It must not be forgotten, however, that apart from transfer of heat altogether, the condensed air is hotter than the rarefied air, and that in order that the whole effect of heat may be on the side of encouragement, it is necessary that previous to condensation the air should pass not merely towards a hotter part of the tube, but towards a part of the tube which is hotter than the air will be when it arrives there. On this account a great range of temperature is necessary for the maintenance of vibration, and even with a great range the influence of the transfer of heat is necessarily unfavourable at the closed end where the motion is very small. This is probably the reason of the advantage of a bulb. It is obvious that if the *open* end of the tube were heated, the effect of the transfer of heat would be even more unfavourable than in the case of a temperature uniform throughout.

The sounds emitted by a jet of hydrogen, burning in an open tube, were noticed soon after the discovery of the gas, and have been the subject of several elaborate inquiries. The fact that the notes are substantially the same as those which may be elicited from the tube in other ways, *e.g.* by blowing, was announced by Chladni. Faraday proved that other gases were competent to take the place of hydrogen, though not without disadvantage. But it is to Sondhauss that we owe the most detailed examination of the circumstances under which the sound is produced. His experiments prove the importance of the part taken by the column of gas in the tube which supplies the jet. For example, sound cannot be obtained with a supply tube which is plugged with cotton in the neighbourhood of the jet, although no difference can be detected by the eye between the flame thus obtained and others which are competent to excite sound. When the supply tube is unobstructed, the sounds obtainable are limited as to pitch, often dividing themselves into detached groups. In the intervals between the groups no coaxing will induce a maintained sound, and it may be added that, for a part of the interval at any rate, the influence of the flame is inimical, so that a vibration started by a blow is damped more rapidly than if the jet were not ignited.

Partly in consequence of the peculiar behaviour of flames and partly for other reasons, the thorough explanation of these phenomena is a matter of some difficulty; but there can be no doubt that they fall under the head of vibrations maintained by heat, the heat being communicated periodically to the mass of air confined in the sounding tube at a place where, in the course of a vibration, the pressure varies. Although some authors have shown an inclination to lay stress upon the effects of the current of air passing through the tube, the sounds can readily be produced, not only when there is no through draught, but even when the flame is so situated that there is no sensible periodic motion of the air in its neighbourhood. In the course of the lecture a globe intended for burning phosphorus in oxygen gas was used as a resonator, and, when excited by a hydrogen flame well removed from the neck, gave a pure tone of about 95 vibrations per second.

In consequence of the variable pressure within the resonator, the issue of gas, and therefore the development of heat, varies during the vibration. The question is under what circumstances the variation is of the kind necessary for the maintenance of the vibration. If we were to suppose, as we might at first be inclined to do, that the issue of gas is greatest when the pressure in the resonator is least, and that the phase of greatest development of heat coincides with that of the greatest issue of gas, we should have the condition of things the most unfavourable of all to the persistence of the vibration. It is not difficult, however, to see that both suppositions are incorrect. In the supply tube (supposed to be unplugged, and of not too small bore) stationary, or approximately stationary, vibrations are excited, whose phase is either the same or the opposite of that of the vibration in the resonator. If the length of the supply tube from the burner to the open end in the gas-generating flask be less than a quarter of the wave-length in hydrogen of the actual vibration, the greatest issue of gas *precedes* by a quarter period the phase of greatest condensation; so that, if the development of heat is *retarded* somewhat in comparison with the issue of gas, a state of things exists *favourable* to the maintenance of the sound. Some such retardation is inevitable, because a jet of inflammable gas can burn only at the outside; but in many cases a still more potent cause may be found, in the fact that during the retreat of the gas in the supply tube small quantities of air may enter from the interior of the resonator, whose expulsion must be effected before the inflammable gas can again begin to escape.

If the length of the supply tube amounts to exactly one quarter of the wave-length, the stationary vibration within it will be of such a character that a node is formed at the burner, the variable part of the pressure just inside the burner being the same as in the interior of the resonator. Under these circumstances there is nothing to make the flow of gas, or the development of heat, variable, and therefore the vibration cannot be maintained. This particular case is free from some of the difficulties which attach themselves to the general problem, and the conclusion is in accordance with Sondhauss' observations.

When the supply tube is somewhat longer than a quarter of the wave, the motion of the gas is materially different from that first described. Instead of preceding, the greatest outward flow of gas *follows* at a quarter period interval the phase of greatest condensation, and therefore if the development of heat be somewhat retarded, the whole effect is unfavourable. This state of things continues to prevail, as the supply tube is lengthened, until the length of half a wave is reached, after which the motion again changes sign, so as to restore the possibility of maintenance. Although the size of the flame and its position in the tube (or neck of resonator) are not without influence, this sketch of the theory is sufficient to explain the fact, formulated by Dr Sondhauss, that the principal element in the question is the length of the supply tube.

The next example of the production of sound by heat, shown in the lecture, was a very interesting phenomenon discovered by Rijke. When a piece of fine metallic gauze, stretching across the lower part of a tube open at both ends and held vertically, is heated by a gas flame placed under it, a sound of considerable power, and lasting for several seconds, is observed almost immediately *after* the removal of the flame. Differing in this respect from the case of sonorous flames, the generation of sound was found by Rijke to be closely connected with the formation of a through draught, which impinges upon the heated gauze. In this form of the experiment the heat is soon abstracted, and then the sound ceases; but by keeping the gauze hot by the current from a powerful galvanic battery, Rijke was able to obtain the prolongation of the sound for an indefinite period. In any case from the point of view of the lecture the sound is to be regarded as a *maintained* sound.

In accordance with the general views already explained, we have to examine the character of the variable communication of heat from the gauze to the air. So far as the communication is affected directly by variations of pressure or density, the influence is unfavourable, inasmuch as the air will receive less heat from the gauze when its own temperature is raised by condensation. The maintenance depends upon the variable transfer of heat due to the varying *motions* of the air through the gauze, this motion being compounded of a uniform motion upwards with a motion, alternately upwards and downwards, due to the vibration. In the lower half of the tube these motions conspire a quarter period *before* the phase of greatest condensation, and oppose one another a quarter period after that phase. The rate of transfer of heat will depend mainly upon the temperature of the air in contact with the gauze, being greatest when that temperature is lowest. Perhaps the easiest way to trace the mode of action is to begin with the case of a simple vibration without a steady current. Under these circumstances the whole of the air which comes in contact with the metal, in the course of a complete period, becomes heated; and after this state of things is established, there is comparatively little further transfer of heat. The effect of superposing a small steady upwards current is now easily recognized. At the limit of the inwards motion, *i.e.* at the phase of greatest condensation, a small quantity of air comes into contact with the metal, which has not done so before, and is accordingly cool; and the heat communicated to this quantity of air acts in the most favourable manner for the maintenance of the vibration.

A quite different result ensues if the gauze be placed in the *upper* half of the tube. In this case the fresh air will come into the field at the moment of greatest rarefaction, when the communication of heat has an unfavourable instead of a favourable effect. The principal note of the tube therefore cannot be sounded.

A complementary phenomenon discovered by Bosscha and Riess may be explained upon the same principles. If a current of *hot* air impinge upon *cold* gauze, sound is produced; but in order to obtain the principal note of the tube the gauze must be in the upper, and not as before in the lower, half of the tube. An experiment due to Riess was shown in which the sound is maintained indefinitely. The upper part of a brass tube is kept cool by water contained in a tin vessel, through the bottom of which the tube passes. In this way the gauze remains comparatively cool, although exposed to the heat of a gas flame situated an inch or two below it. The experiment sometimes succeeds better when the draught is checked by a plate of wood placed somewhat closely over the top of the tube.

Both in Rijke's and Riess' experiments the variable transfer of heat depends upon the motion of vibration, while the effect of the transfer depends upon the variation of pressure. The gauze must therefore be placed where both effects are sensible, *i.e.* neither near a node nor near a loop. About a quarter of the length of the tube, from the lower or upper end, as the case may be, appears to be the most favourable position.

56.

UNIFORMITY OF ROTATION*.

[*Nature*, XVIII. p. 111, 1878.]

LORD RAYLEIGH exhibited and explained an arrangement which he has employed with advantage in certain acoustical experiments, in order to secure absolute uniformity in the rate of rotation of an axle. After referring to the mathematical principles involved in such a problem, he explained that the only hope of its solution lay in the employment of a vibratory movement, which by some suitable device must be converted into a motion of rotation. The axle whose motion it is required to maintain uniform is usually driven at an approximately uniform rate by means of a small horizontal water-wheel, or, in some cases, the electro-magnetic regulating apparatus presently described is sufficient by itself to supply the necessary power. At equal distances round the axle are arranged four soft iron armatures which successively come in front of the poles of a horse-shoe electro-magnet placed in the circuit of a four-cell Grove battery. The current is rendered intermittent by the following arrangement. Passing into the body of a tuning fork vibrating about 40 times per second, it leaves by a small platinum stud which is touched at each vibration of the fork; the current then traverses a second small electro-magnet between the prongs, and by this means the vibrations are maintained; passing to the magnet above referred to the current then returns to the battery. The velocity of the axle is such that it performs about one complete revolution for every four vibrations of the fork, and the

---

* [1899. It may be proper to mention that the report here reproduced was not written or corrected by me. I fully explained at the time the distinction between this apparatus, as controlling the *position* of the revolving piece, and an ordinary governor which regulates the *velocity*. An application to the determination of absolute pitch is given as an appendix to one of my papers upon the ohm (*Phil. Trans.* 1883, p. 295). It has sometimes appeared to me that the principle might be applied to the regulation of equatorial telescopes.]

exact adjustment is effected as follows. If the driving power be just sufficient to produce the desired speed, the armatures will be so attracted by the
magnet as to be exactly opposite to it at the middle of its period of magneti-
sation, and so long as this position is maintained the magnet will not, on the
whole, affect it. But if a disturbance occur in the driving power the
armature will be displaced from its former position and will be attracted by
the magnet until the error is compensated. Besides the armatures this axle
also carries, concentric with it, a hollow metallic ring filled with water, and
as this possesses a certain momentum in virtue of its rotation, it will act as a
drag tending to check the velocity in case it increases, and in a converse
manner when a diminution occurs. A blackened disc perforated with rings
of holes of various numbers also rotates with the axle, and by placing the eye
behind the ring of *four* holes and observing a prong of the fork it is easy to
ascertain whether the uniformity is maintained, since in that case the prong
will appear to remain stationary.

[1899. A similar apparatus was described somewhat earlier by La Cour
under the name of the *Phonic Wheel.*]

57.

ON THE DETERMINATION OF ABSOLUTE PITCH BY THE COMMON HARMONIUM*.

[*Nature*, XIX. pp. 275—276, 1879.]

THE methods described depend upon the principle that the absolute frequencies of vibration of two musical notes can be deduced from the *interval* between them, *i.e.* the *ratio* of their frequencies, and the number of beats which they occasion in a given time when sounded together. For example, if x and y denote the frequencies of two notes whose interval is an equal temperament major third, we know that $y = 1\cdot25992\,x$. At the same time the number of beats heard in a second, depending upon the deviation of the third from true intonation, is $4y - 5x$. In the case of the harmonium these beats are readily counted with the aid of a resonator tuned to the common over-tone, and thus are obtained two equations from which the absolute values of x and y may be found by the simplest arithmetic.

Of course, in practice, the truth of an equal temperament third could not be taken for granted, but the difficulty thence arising would be easily met by including in the counting all the three major thirds which together make up an octave. Suppose, for example, that the frequencies of c, e, $g\sharp$, c' are respectively x, y, z, $2x$, and that the beats per second between x and y are a, between y and z are b, and between z and $2x$ are c. Then

$$4y - 5x = a, \qquad 4z - 5y = b, \qquad 8x - 5z = c,$$

from which

$$x = \tfrac{1}{3}(25a + 20b + 16c), \qquad y = \tfrac{1}{3}(32a + 25b + 20c), \qquad z = \tfrac{1}{3}(40a + 32b + 25c).$$

In the above statements the octave c—c' is for simplicity supposed to be true. The actual error could be readily allowed for, if required; but in practice it is not necessary to use c' at all, inasmuch as the third set of beats can be counted equally well between $g\sharp$ and c.

* Abstract of a paper read before the Musical Association, December 2, 1878.

Although at first sight the method just sketched looks satisfactory, it is not practical in the case of the harmonium, in consequence of the pitch of the various notes not being sufficiently constant for the purpose, even when the blowing is carefully conducted with the aid of a pressure-gauge. A small variation in the absolute pitch of a chord when sounded under slightly varying pressures, would not be of much importance, but the slightest change of *interval* is fatal to the success of the method, and such a change actually occurs.

In order, therefore, to apply the fundamental principle with success, it is necessary to be able to check the accuracy of the interval which is supposed to be known, at the same time that the beats are being counted. If the interval be a major tone (9 : 8), its exactness is proved by the absence of beats between the ninth component of the lower, and the eighth component of the higher note, and a counting of the beats between the tenth component of the lower and the ninth of the higher note completes the necessary data for determining the absolute pitch.

The equal temperament whole tone (1·12246) is intermediate between the minor tone (1·11111) and the major tone (1·12500), but lies much nearer to the latter. Regarded as a disturbed major tone, it gives slow beats, and regarded as a disturbed minor tone it gives comparatively quick ones. Both sets of beats can be heard at the same time, and when counted give the means of calculating the absolute pitch of both notes. If x and y be the frequencies of the two notes, a and b the frequencies of the slow and quick beats respectively,

$$9x - 8y = a, \qquad 9y - 10x = b,$$

whence

$$x = 9a + 8b, \qquad y = 10a + 9b.$$

The application of this method in no way assumes the truth of the equal temperament whole tone, and in fact it is advantageous to flatten the interval somewhat by loading the upper reed with a minute fragment of soft wax, so as to make it lie more nearly midway between the major and the minor tone. In this way the rapidity of the quicker beats is diminished, which facilitates the counting.

It is impossible, of course, for the same observer to count both sets of beats, and the counting of even one set without the aid of resonators would present difficulties to most unpractised persons. Great assistance may be obtained by the choice of a suitable position. A room in which a pure tone is sounded is traversed by surfaces at which the intensity of sound is very much reduced in consequence of the superposition of vibrations reflected from the walls and ceiling. By choosing as the place of observation a position where the intensity of the beats which are not to be counted is a minimum, and with the aid of a resonator tuned to the pitch of the beats which are to be counted, the listener is able to work with ease and certainty.

The course of an experiment is then as follows:—The notes C and D are sounded, and at a given signal the listeners begin counting the beats whose pitch is about d'' and e'' respectively. At the expiration of a measured interval of time a second signal is given, and the number of both sets of beats is recorded.

In my experiments the interval of time was ten minutes (in one case eleven minutes), and the rapidity of the beats was about four a second. The listeners counted up to ten only, after each set of ten making a stroke with a pencil on a piece of paper. The number of strokes was afterwards counted, multiplied by ten, and added to the number which the listener was saying at the instant of the second signal. The following are the details of the actual observations :—

September 16, 1878.—Period of observation ten minutes. Numbers of beats 2392 and 2341.

$$a = \tfrac{2392}{600}, \; b = \tfrac{2341}{600}, \; \text{giving } x = \frac{9 \times 2392 + 8 \times 2341}{600} = 67{\cdot}09,$$

for the frequency of the lower note C.

September 17.—Period of observation ten minutes.

$$a = \tfrac{2423}{600}, \; b = \tfrac{2302}{600}, \; \text{giving } x = 67{\cdot}04.$$

September 18.—Period of observation ten minutes.

$$a = \tfrac{2476}{600}, \; b = \tfrac{2261}{600}, \; \text{giving } x = 67{\cdot}29.$$

September 19.—Period of observation eleven minutes.

$$a = \tfrac{2663}{600}, \; b = \tfrac{2547}{600}, \; \text{giving } x = 67{\cdot}19.$$

The discrepancies are hardly greater than may be attributed to errors in giving the signals, by which the intervals may have been unduly lengthened or shortened by about a second. On each day after the counting of the beats between C and D, the harmonium was compared with a König fork whose nominal frequency was 64. In order to obviate any objection arising from a mutual influence of the notes of the harmonium, *both* C and D were sounded at the same time as the fork. The beats between C and the fork were counted for about ninety seconds, during which time the fork was not bowed. In this way the pitch of the fork came out on the four days respectively as 64·06, 64·07, 64·17, 63·98, that is somewhat sharper than its nominal pitch, a result in agreement with that obtained by other methods.

The object of the experiments referred to was rather to prove the practicability of a method so unusually independent of special apparatus, than to obtain a result competing in point of accuracy with those of Prof. McLeod and other experimenters on this subject. Nevertheless it is believed that very accurate results might be obtained by the introduction of certain modifications. Ten minutes is near the limit of time over which beats can be conveniently counted by a single listener, but experiment proved that it is perfectly possible for one listener to relieve another without any break in the regularity of the counting. Even without an extension of time a more accurate result would be obtained if the listeners were able to fix the time for themselves, as they might do for example if they could conveniently observe the swinging of a clock pendulum. In this way the error in the time interval might be reduced to $\frac{1}{4}$ second, which would amount to but one part in 2400 in the case of a ten minutes' observation. In consequence, however, of the imperfect constancy of the pitch of the harmonium notes, even when the blower is assisted by a pressure-gauge, further attempts at accuracy would be useless unless the comparison with the fork were simultaneous with the other observations. In that case the result would be entirely independent of variations in the harmonium notes, and no difficulty would be experienced in carrying out the method excepting the necessity for more observers.

58.

ON THE INSTABILITY OF JETS.

[*Proceedings of the London Mathematical Society*, x. pp. 4—13, 1879.]

MANY, it may even be said most, of the still unexplained phenomena of Acoustics are connected with the instability of jets of fluid. For this instability there are two causes; the first is operative in the case of jets of heavy liquids, *e.g.*, water, projected into air (whose relative density is negligible), and has been investigated by Plateau in his admirable researches on the figures of a liquid mass withdrawn from the action of gravity. It consists in the operation of the capillary force, whose effect is to render the infinite cylinder an unstable form of equilibrium, and to favour its disintegration into detached masses whose aggregate surface is less than that of the cylinder. The other cause of instability, which is operative even when the jet and its environment are of the same material, is of a more dynamical character.

With respect to instability due to capillary force, the principal problem is the determination, as far as possible, of the mode of disintegration of an infinite cylinder, and in particular of the number of masses into which a given length of cylinder may be expected to distribute itself. It must, however, be observed that this problem is not so definite as Plateau seems to think it; the mode of falling away from unstable equilibrium necessarily depends upon the peculiarities of the small displacements to which a system is subjected, and without which the position of equilibrium, however unstable, could not be departed from. Nevertheless, in practice, the latitude is not very great, because some kinds of disturbance produce their effect much more rapidly than others. In fact, if the various disturbances be represented initially by $\alpha_1, \alpha_2, \alpha_3, \ldots$, and after a time t by $\alpha_1 e^{q_1 t}, \alpha_2 e^{q_2 t}, \alpha_3 e^{q_3 t}, \ldots$, the (positive) quantities q_1, q_2, q_3, &c., being in descending order of magnitude, it is easy to see that, when $\alpha_1, \alpha_2, \ldots$ are small enough, the first kind necessarily acquires the preponderance. For example, at time t the ratio of the second kind to the first is $(\alpha_2/\alpha_1)\, e^{-(q_1-q_2)t}$, which, independently of the value of $\alpha_2 : \alpha_1$, can be made as small as we please by taking t great enough. But, in order to allow the application of the analytical expressions for so extended a time, it is generally necessary to suppose the whole amount of disturbance to be originally extremely small *.

Let us, then, taking the axis of z along the axis of the cylinder, suppose that at time t the surface of the cylinder is of the form

$$r = a + \alpha \cos kz, \quad \dots\dots\dots\dots\dots\dots\dots(1)$$

where α is a small quantity variable with the time, and $k = 2\pi/\lambda$, λ being the *wave-length* of the original disturbance. The information that we require will be readily obtained by Lagrange's method, when we have calculated expressions for the potential and kinetic energies of the motion represented by (1).

The potential energy due to the capillary forces is a question merely of the surface of the liquid. If we denote the surface corresponding (on the average) to the unit length along the axis by σ, we readily find

$$\sigma = 2\pi a + \tfrac{1}{2}\pi a k^2 \alpha^2. \quad \dots\dots\dots\dots\dots\dots\dots(2)$$

In this, however, we have to substitute for a (which is not strictly constant) its value obtained from the condition that S, the volume enclosed per unit of length, is given. We have

$$S = \pi a^2 + \tfrac{1}{2}\pi \alpha^2, \quad \dots\dots\dots\dots\dots\dots\dots(3)$$

whence

$$a = \sqrt{\left(\frac{S}{\pi}\right)} \cdot \left(1 - \tfrac{1}{4}\frac{\pi \alpha^2}{S}\right). \quad \dots\dots\dots\dots\dots(4)$$

Using this in (2), we get with sufficient approximation

$$\sigma = 2\sqrt{(\pi S)} + \frac{\pi \alpha^2}{2a}\,(k^2 a^2 - 1); \quad \dots\dots\dots\dots\dots(5)$$

or, if σ_0 be the value of σ for the undisturbed condition,

$$\sigma - \sigma_0 = \frac{\pi \alpha^2}{2a}\,(k^2 a^2 - 1). \quad \dots\dots\dots\dots\dots(6)$$

From this we infer that, if $ka > 1$, the surface is greater after displace-. ment than before; so that, if $\lambda < 2\pi a$, the displacement is of such a character that with respect to it the system is stable. We are here concerned only

---

* Some of the theorems given in the Proceedings for June 1873 (*Theory of Sound*, §§ 88, 89), [Art. xxi.], for the period of vibrations about a configuration of stable equilibrium, are applicable, *mutatis mutandis*, to the times of falling away from unstable equilibrium when various types of displacement are considered. For example, the application of a constraint could never diminish the shortest time previously possible.

with values of ka less than unity. If T_1 denote the cohesive tension, the potential energy V reckoned per unit of length from the position of equilibrium is

$$V = - T_1 \frac{\pi a^2}{2a} (1 - k^2 a^2). \quad\dots\dots\dots\dots\dots\dots\dots(7)$$

We have now to calculate the kinetic energy of motion. It is easy to prove that the velocity potential is of the form

$$\phi = A J_0 (ikr) \cos kz, \quad\dots\dots\dots\dots\dots\dots\dots(8)$$

J_0 being the symbol of Bessel's functions of zero order, so that

$$J_0(x) = 1 - \frac{x^2}{2^2} + \frac{x^4}{2^2 \cdot 4^2} - \frac{x^6}{2^2 \cdot 4^2 \cdot 6^2} + \dots. \quad\dots\dots\dots\dots(9)$$

The coefficient A is to be determined from the consideration that the outwards normal velocity at the surface of the cylinder is equal to $\dot{a} \cos kz$. Hence

$$ik A J_0'(ika) = \dot{a}. \quad\dots\dots\dots\dots\dots\dots\dots(10)$$

Denoting the density by ρ, we have for the kinetic energy the expression

$$T = \tfrac{1}{2}\rho \int 2\pi a \cdot \phi \, \frac{d\phi}{dr}_{(=a)} \, dz\,;$$

or, if we reckon it in the same way as V per unit of length,

$$T = \tfrac{1}{2}\rho \pi a^2 \frac{J_0 (ika) \dot{a}^2}{ika \, J_0'(ika)}. \quad\dots\dots\dots\dots\dots\dots(11)$$

Thus, by Lagrange's method, if $\alpha \propto e^{qt}$,

$$q^2 = \frac{T_1}{\rho a^3} \frac{(1 - k^2 a^2) \cdot ika \cdot J_0'(ika)}{J_0(ika)}, \quad\dots\dots\dots\dots\dots(12)$$

which determines the law of falling away from equilibrium for a disturbance of wave-length λ. The solutions for the various values of λ and the corresponding energies are independent of one another; and thus, by Fourier's theorem, it is possible to express the condition of the system at time t, after the communication of any infinitely small disturbances symmetrical about the axis. But what we are most concerned with at present is the value of q^2 as a function of ka, and especially the determination of that value of ka for which q^2 is a maximum. That such a maximum must exist is evident *à priori*. Writing x for ka, we have to examine the values of

$$\frac{(1 - x^2) \cdot ix \cdot J_0'(ix)}{J_0(ix)} \quad\dots\dots\dots\dots\dots\dots(13)$$

Expanding in powers of x, we may write for (13),

$$\tfrac{1}{2}x^2(1-x^2)\left\{1 - \frac{x^2}{2^3} + \frac{x^4}{2^4 \cdot 3} - \frac{11x^6}{2^{10} \cdot 3} + \frac{19x^8}{2^{11} \cdot 3 \cdot 5} + \dots\right\}, \quad \dots\dots(14)$$

or

$$\tfrac{1}{2}\left\{x^2 - \tfrac{9}{8}x^4 + \frac{7}{2^4 \cdot 3}x^6 - \frac{25}{2^{10}}x^8 + \frac{91}{2^{11} \cdot 3 \cdot 5}x^{10} + \dots\right\} \dots\dots\dots(15)$$

Hence, to find the maximum, we obtain by differentiation

$$1 - \tfrac{9}{4}x^2 + \frac{7}{2^4}x^4 - \frac{100}{2^{10}}x^6 + \frac{91}{2^{11} \cdot 3}x^8 + \dots = 0. \quad \dots\dots\dots(16)$$

If the last two terms be neglected, the quadratic gives $x^2 = \cdot4914$. If this value be substituted in the small terms, the equation becomes

$$\cdot98928 - \tfrac{9}{4}x^2 + \tfrac{7}{16}x^4 = 0,$$

whence

$$x^2 = \cdot4858, \qquad [x = \cdot6970]. \quad \dots\dots\dots\dots\dots(17)$$

The corresponding value of λ is given by

$$\lambda = 4\cdot508 \times 2a, \quad \dots\dots\dots\dots\dots\dots\dots(18)$$

which gives accordingly the ratio of wave-length to diameter for the kind of disturbance which leads most rapidly to the disintegration of the cylindrical mass. The corresponding number obtained by Plateau from some experiments by Savart is 4·38; but this estimate involves a knowledge of the coefficient of contraction of a jet escaping through a small hole in a thin plate, and is probably liable to a greater error than its deviation from 4·51.

The following table exhibits the relationship between x^2 or k^2a^2 and the square root of expression (13) to which q is proportional:—

x^2		x^2	
·00	·0000	·40	·3382
·05	·1536	·50	·3432
·10	·2108	·60	·3344
·20	·2794	·80	·2701
·30	·3182	·90	·2015

In the cases just considered the cause of instability is statical, and the phenomena are independent of the general translatory motion of the jet; but the other kind of instability has its origin in this very translatory motion. In his work on the discontinuous movements of fluids, Helmholtz[*] remarks

* *Phil. Mag.* vol. xxxvi. 1868.

upon the instability of surfaces separating portions of fluid which move discontinuously, and Sir W. Thomson*, in treating of the influence of wind on waves in water, supposed frictionless, has shown under what conditions a level surface of water is rendered unstable. In the following investigations the method of Thomson's paper is applied to determine the law of falling away from unstable equilibrium in some of the simpler cases of a plane or cylindrical surface of separation.

Let us suppose that the equilibrium position of the surface of separation is represented by $z = 0$, and that on the positive and negative sides of it the velocities of the fluid are parallel to the axis of x, and of magnitudes V and V' respectively. In the absence of friction, the motion consequent upon any deformation of the surface of separation is determinate, in virtue of a well known hydrodynamical law. By Fourier's theorem, any displacement in two dimensions can be resolved into component displacements of the undulatory type, and the effect of any two undulatory displacements may be considered separately. We might, therefore, take as the initial equation of the surface of separation $h = H \cos kx$, in which h denotes the elevation at any point, λ the wave-length of the disturbance, and $k = 2\pi/\lambda$. But, as in almost all such cases, it is more convenient to use complex expressions, from which the imaginary parts are finally rejected. We will therefore assume

$$h = H e^{int} e^{ikx} ; \quad \dots\dots\dots\dots\dots\dots\dots(19)$$

and the principal question which we have to consider is the dependence of n upon k or λ.

For the velocity potential of the fluid on the positive side, we may take

$$\phi = A e^{int} e^{ikx} e^{-kz} + Vx, \dots\dots\dots\dots\dots\dots(20)$$

in which A is to be determined by equating the value of the normal velocity at the surface of separation with that obtained from (21). Thus (the positive direction of z being downwards)

$$-\frac{d\phi}{dz}(z = 0) = kA e^{int} e^{ikx} = \frac{dh}{dt} + V\frac{dh}{dx} = (in + ikV) H e^{int} e^{ikx},$$

whence

$$A = ik^{-1} (n + kV) H ; \dots\dots\dots\dots\dots\dots(22)$$

so that

$$\phi = ik^{-1} (n + kV) H e^{int} e^{ikx} e^{-kz} + Vx. \dots\dots\dots\dots(23)$$

Similarly, for the fluid on the negative side,

$$\phi' = -ik^{-1} (n + kV') H e^{int} e^{ikx} e^{kz} + V'x. \dots\dots\dots(24)$$

* *Phil. Mag.* Nov. 1871.

We have now to satisfy the condition of the equality of pressures. If σ denote the density, the hydrodynamical equation of pressure for the first fluid is

$$p = C - \sigma \frac{d\phi}{dt} - \tfrac{1}{2}\sigma U^2 ; \quad \dots\dots\dots\dots\dots(25)$$

and approximately, when $z = 0$,

$$\tfrac{1}{2}U^2 = \tfrac{1}{2}\left(\frac{d\phi}{dx}\right)^2 + \tfrac{1}{2}\left(\frac{d\phi}{dy}\right)^2 = \tfrac{1}{2}V^2 - V(n + kV)He^{int}e^{ikx}. \quad \dots\dots(26)$$

In like manner,

$$p' = C' - \sigma' \frac{d\phi'}{dt} - \tfrac{1}{2}\sigma'U'^2, \quad \dots\dots\dots\dots\dots(27)$$

where

$$\tfrac{1}{2}U'^2 = \tfrac{1}{2}V'^2 + V'(n + kV')He^{int}e^{ikx}. \quad \dots\dots\dots\dots(28)$$

Hence

$$\sigma(n + kV)^2 + \sigma'(n + kV')^2 = 0, \quad \dots\dots\dots\dots\dots(29)$$

which is the equation by which n and k are connected.

The simplest case of (29) occurs when $\sigma' = \sigma$, and $V' = -V$, so that the equilibrium motions of the portions of fluid are equal and opposite. We have then

$$n^2 + k^2V^2 = 0 ; \quad \dots\dots\dots\dots\dots\dots(30)$$

and for the elevations

$$h = He^{\mp kVt}\{\cos kx, \text{ or } \sin kx\}. \quad \dots\dots\dots\dots(31)$$

If initially $dh/dt = 0$, we get

$$h = H\cosh kVt\{\cos kx, \text{ or } \sin kx\} ; \quad \dots\dots\dots\dots(32)$$

indicating that the waves on the surface of separation are *stationary*, and increase in amplitude with the time according to the law of the hyperbolic cosine. By (31), Fourier's theorem allows us to express the consequences of arbitrary initial values of h and dh/dt, within the limits of time imposed by our methods of approximation.

Next, let us suppose that $\sigma' = \sigma$, $V' = 0$. We get from (29)

$$n = \tfrac{1}{2}kV(-1 \pm i), \quad \dots\dots\dots\dots\dots\dots(33)$$

whence

$$h = He^{\mp\frac{1}{2}kVt}e^{i(-\frac{1}{2}kVt + kx)},$$

of which the real part is

$$h = He^{\mp\frac{1}{2}kVt}\cos k(\tfrac{1}{2}Vt - x). \quad \dots\dots\dots\dots(34)$$

In (34) an arbitrary constant may of course be added to x. It appears that the waves travel in the same direction as the stream, and with *one-half* its velocity. In the case of the positive exponent, the rapidity with which the amplitude increases is very great.   Since $k = 2\pi/\lambda$, the amplitude is multiplied by e^π, or about 23, in the time occupied by the stream in passing over a distance λ.   If $\lambda = V\tau$, $e^{\frac{1}{2}kVt} = e^{\pi t/\tau}$, independent of V.

As a generalised form of (34), we may take

$$h = A \cosh(\tfrac{1}{2}kVt) \cos k(\tfrac{1}{2}Vt - x) + B \sinh(\tfrac{1}{2}kVt) \sin k(\tfrac{1}{2}Vt - x), \dots(35)$$

which gives, when $t = 0$, $h = A \cos kx$.

If $dh/dt = 0$ initially, $B = A$, by which the solution corresponding to a surface of separation initially displaced without velocity is determined.

If initially $h = 0$, and dh/dt is finite, we have, as the appropriate form,

$$h = B \sinh(\tfrac{1}{2}kVt) \cos k(\tfrac{1}{2}Vt - x). \dots(36)$$

Again, suppose that $\sigma' = \sigma$, $V' = V$.   In this case the roots of (29) are equal, but the general solution may be obtained by the usual method.   From (29) we have

$$n = \tfrac{1}{2}k[\pm i(V' - V) - (V' + V)]; \dots(39)$$

or, if we put $V' = V(1 + \alpha)$,

$$n = \tfrac{1}{2}kV[\pm i\alpha - (2 + \alpha)]. \dots(40)$$

The corresponding solution for h is

$$h = e^{ikx} e^{\frac{1}{2}i(2+\alpha)kVt}[A e^{\frac{1}{2}kVt \cdot \alpha} + B e^{-\frac{1}{2}kVt \cdot \alpha}], \dots(41)$$

where A and B are arbitrary constants.

Passing now to the limit when $\alpha = 0$, and taking new arbitrary constants, we get

$$h = e^{ikx} e^{-ikVt}[A + Bt]; \dots(42)$$

or, in real quantities,

$$h = [A + Bt][\cos k(Vt - x), \text{ or } \sin k(Vt - x)]. \dots(42)$$

If initially $h = \cos kx$, $dh/dt = 0$,

$$h = \cos k(Vt - x) + kVt \sin k(Vt - x). \dots(43)$$

The peculiarity of this case is that previous to the displacement there is no real surface of separation at all.   Its bearing upon the flapping of sails and flags will be evident.

The proportionality to λ/V of the time of falling away from equilibrium follows from the principle of dynamical similarity, as there is no linear element but λ.

When $V' = V$, the solution is the same, whether $\sigma' = \sigma$ or not.   For example, (31), (32), (33) are applicable when $\sigma' = 0$.

In general, the solution of (29) is

$$\frac{n}{k} = -\frac{\sigma V + \sigma' V' \pm i \sqrt{(\sigma \sigma')} \cdot (V - V')}{\sigma + \sigma'} \quad \dots\dots\dots\dots (44)$$

If $\sigma V + \sigma' V' = 0$, n is a pure imaginary, and the waves are stationary.

We will now suppose that the two portions of fluid are limited by rigid walls whose equations are respectively $z = l$, $z = -l'$. Then, corresponding to $h = H e^{int} e^{ikx}$, we get for the velocity potentials, in place of (23), (24),

$$\phi = ik^{-1} (n + kV) H \frac{\cosh k (z - l)}{\sinh kl} e^{int} e^{ikx} + Vx, \quad \dots\dots (45)$$

$$\phi' = - ik^{-1} (n + kV') H \frac{\cosh k (z + l')}{\sinh kl'} e^{int} e^{ikx} + V'x, \quad \dots\dots (46)$$

and, in place of (29),

$$\sigma (n + kV)^2 \coth kl + \sigma' (n + kV')^2 \coth kl' = 0. \quad \dots\dots\dots (47)$$

If $l' = l$, the result is the same as if l and l' were both infinite.

If l' be infinite, $\coth kl' = 1$; (47) may then be applied to a jet of width $2l$, symmetrical and symmetrically displaced with respect to the line $z = l$, and moving with velocity V in an infinite mass whose velocity is V'. If $V' = 0$, $\sigma' = \sigma$, so that the jet is of the same density as its stationary environment, (47) becomes

$$(n + kV)^2 \coth kl + n^2 = 0, \quad \dots\dots\dots\dots\dots (48)$$

of which the solution is

$$n = kV \frac{-1 \pm i \sqrt{(\tanh kl)}}{1 + \tanh kl}, \quad \dots\dots\dots\dots (49)$$

a generalisation of (33).

Thus

$$h = H e^{\pm \mu kVt} \cos k \left[\frac{Vt}{1 + \tanh kl} - x \right], \quad \dots\dots\dots (50)$$

where

$$\mu = \frac{\sqrt{(\tanh kl)}}{1 + \tanh kl}. \quad \dots\dots\dots\dots\dots\dots (51)$$

When kl is very small, we may take in place of (50)

$$h = H e^{\pm \sqrt{(kl)} kVt} \cos k (Vt - x). \quad \dots\dots\dots\dots (52)$$

We see, from (52), that when l is small the time of falling away from equilibrium is increased.

If the condition to be satisfied at $z = l$ be $\phi = 0$, in place of $d\phi/dz = 0$, the value of ϕ is

$$\phi = - ik^{-1} (n + kV) H \frac{\sinh k (z - l)}{\cosh kl} e^{int} e^{ikx} + Vx; \quad \dots\dots (53)$$

so that, if, as before, $d\phi'/dz = 0$ when $z = -l'$,

$$\sigma(n + kV)^2 \tanh kl + \sigma'(n + kV')^2 \coth kl' = 0. \quad\ldots\ldots\ldots\ldots(54)$$

If $l' = \infty$, $\sigma' = \sigma$, $V' = 0$,

$$(n + kV)^2 \tanh kl + n^2 = 0. \quad\ldots\ldots\ldots\ldots\ldots(55)$$

This is applicable to a jet of width $2l$, moving in still fluid with velocity V, and displaced in such a manner that the sinuosities of its two surfaces are parallel.

When kl is small, we have, approximately,

$$h = He^{\pm \sqrt{(kl)} \cdot kVt} \cos k(kl \cdot Vt - x). \quad\ldots\ldots\ldots\ldots(56)$$

By a combination of the solutions represented by (52), (56), we may determine the consequences of any displacements (in two dimensions) of the two surfaces of a thin jet moving with velocity V in still fluid of its own density.

These solutions may be extended to cases where the surface of separation is not plane, provided that the velocities of the fluids be constant (V, V') along it. Let us suppose that ϕ, ψ are the velocity-potential and stream functions for the steady motion of the first fluid, and that the surface of separation corresponds to $\psi = \psi_0$. At $\psi = \psi_1$ let there be a rigid barrier, which of course has no influence upon the steady motion. Then, if the elevation at any point s, measured along the surface of separation, be given by

$$h = He^{iks} e^{int},$$

the velocity-potential of the disturbed motion is

$$\phi + \delta\phi = \phi - ik^{-1} H(n + kV) \frac{\cosh kV^{-1}(\psi - \psi_1)}{\sinh kV^{-1}\psi_1} e^{int} e^{ik\phi/V}.$$

If l be the width of a uniform stream of velocity V, whose whole amount is equal to that of the stream between $\psi = \psi_0$ and $\psi = \psi_1$, and if dashed letters denote the corresponding quantities for the second fluid, we get finally for the equation in n

$$\sigma \coth kl (n + kV)^2 + \sigma' \coth kl' (n + kV')^2 = 0,$$

which is the same form as (47).

We will now pass to the consideration of cylindrical surfaces of separation, limiting ourselves for simplicity to the case of disturbances symmetrical about the axis (x). If h denote the increment of distance of any point on the surface from the axis, we may take, as before,

$$h = He^{int} e^{ikx}, \quad\ldots\ldots\ldots\ldots\ldots\ldots(57)$$

and the corresponding value of the velocity-potential for the fluid inside the cylinder is

$$\phi = A J_0 (ikr)\, e^{int}\, e^{ikx} + Vx, \quad \dots\dots\dots\dots\dots(58)$$

in which A is to be determined by the condition relating to the normal velocity at the surface ($r = a$). Thus

$$\phi = k^{-1} (n + kV)\, H \frac{J_0 (ikr)}{J_0'(ika)}\, e^{int}\, e^{ikx} + Vx. \quad \dots\dots\dots\dots(59)$$

For the motion of the fluid outside the cylinder, we have, in the first place, the general form

$$\phi' = \quad C\,(-r)^{-\frac{1}{2}} e^{+kr} \left\{ 1 - \frac{1}{1.(-8kr)} + \frac{1^2.3^2}{1.2.(-8kr)^2} - \dots \right\}$$

$$+ D\,(-r)^{-\frac{1}{2}} e^{-kr} \left\{ 1 + \frac{1}{1.(-8kr)} + \frac{1^2.3^2}{1.2.(-8kr)^2} + \dots \right\},$$

in which, however, by the condition at infinity, we are to put $C = 0$. Writing for brevity

$$(-r)^{-\frac{1}{2}} e^{-kr} \left\{ 1 + \frac{1}{1.(-8kr)} + \frac{1^2.3^2}{1.2.(-8kr)^2} + \dots \right\} = \chi(r), \dots\dots(60)$$

we have accordingly

$$\phi' = B\chi(r)\, e^{int}\, e^{ikx} + V'x; \quad \dots\dots\dots\dots\dots(61)$$

or, on determining the value of B,

$$\phi' = i (n + kV')\, H \frac{\chi(r)}{\chi'(a)}\, e^{int}\, e^{ikx} + V'x. \quad \dots\dots\dots\dots(62)$$

In the same manner as for plane surfaces, the condition of equality of pressures now gives

$$\sigma i k^{-1} (n + kV)^2 \frac{J_0 (ika)}{J_0'(ika)} + \sigma' (n + kV')^2 \frac{\chi(a)}{\chi'(a)} = 0, \quad \dots\dots\dots(63)$$

as the quadratic by which n is determined.

When ka is small, we may employ approximate values of the functions J_0 and χ.

Thus

$$\frac{J_0 (ika)}{J_0'(ika)} = - \frac{2}{ika}. \quad \dots\dots\dots\dots\dots\dots(64)$$

For χ we have ("Stokes on Pendulums," *Camb. Trans.*, Vol. IX.)

$$\chi = \left(\log (\tfrac{1}{2} kr) - \pi^{-\frac{1}{2}} \Gamma'(\tfrac{1}{2}) \right) \left(1 + \frac{k^2 r^2}{2^2} + \frac{k^4 r^4}{2^2.4^2} + \dots \right)$$

$$- \left(\frac{k^2 r^2}{2^2} S_1 + \frac{k^4 r^4}{2^2.4^2} S_2 + \dots \right), \quad \dots\dots\dots(65)$$

where
$$S_n = 1^{-1} + 2^{-1} + 3^{-1} + \dots + n^{-1}. \dots\dots\dots\dots(66)$$

When ka is small,
$$\chi(a) = \log(\tfrac{1}{8}ka) - \pi^{-\frac{1}{2}}\Gamma'(\tfrac{1}{2}), \qquad \chi'(a) = a^{-1},$$

so that
$$\frac{\chi(a)}{\chi'(a)} = a\{\log(\tfrac{1}{8}ka) - \pi^{-\frac{1}{2}}\Gamma'(\tfrac{1}{2})\}. \dots\dots\dots\dots(67)$$

If we suppose that $\sigma' = \sigma$, $V' = 0$, (63) becomes, for small values of ka,
$$(n + kV)^2 + k^2a^2\left\{\log\frac{8}{ka} + \pi^{-\frac{1}{2}}\Gamma'(\tfrac{1}{2})\right\}n^2 = 0. \dots\dots\dots\dots(68)$$

Writing μ^2 for
$$k^2a^2\left\{\log\frac{8}{ka} + \pi^{-\frac{1}{2}}\Gamma'(\tfrac{1}{2})\right\},$$

we get
$$n = kV\frac{-1 \pm i\mu}{1 + \mu^2}, \dots\dots\dots\dots\dots(69)$$

whence approximately
$$h = He^{\pm\mu kVt}\cos k(Vt - x). \dots\dots\dots\dots(70)$$

[1899. The subject of this paper is further considered in Arts. LX., LXVI. See also *Proc. Lond. Math. Soc.* XIX. p. 67, 1887; XXVII. p. 5, 1895; *Phil. Mag.* XXXIV. pp. 145, 177, 1892.]

59.

THE INFLUENCE OF ELECTRICITY ON COLLIDING WATER DROPS.

[*Proceedings of the Royal Society*, XXVIII. pp. 406—409, 1879.]

IT has been known for many years that electricity has an extraordinary influence upon the behaviour of fine jets of water ascending in a nearly vertical direction. In its normal state a jet resolves itself into drops, which even before passing the summit, and still more after passing it, are scattered through a considerable width. When a feebly electrified body is brought into its neighbourhood, the jet undergoes a remarkable transformation and appears to become coherent; but under more powerful electrical action the scattering becomes even greater than at first. The second effect is readily attributed to the mutual repulsion of the electrified drops, but the action of feeble electricity in producing apparent coherence has been a mystery hitherto.

It has been shown by Beetz that the coherence is apparent only, and that the place where the jet breaks into drops is not perceptibly shifted by the electricity. By screening various parts with metallic plates, Beetz further proved that, contrary to the opinion of earlier observers, the seat of sensitiveness is not at the root of the jet where it leaves the orifice, but at the place of resolution into drops. As in Sir W. Thomson's water-dropping apparatus for atmospheric electricity, the drops carry away with them an electric charge, which may be collected by receiving the water in an insulated vessel.

I have lately succeeded in proving that the normal scattering of a nearly vertical jet is due to the *rebound* of the drops when they come into collision with one another. Such collisions are inevitable in consequence of the different velocities acquired by the drops under the action of the capillary force, as they break away irregularly from the continuous portion of the jet.

Even when the resolution is regularised by the action of external vibrations of suitable frequency, as in the beautiful experiments of Savart and Plateau, the drops must still come into contact before they reach the summit of their parabolic path.   In the case of a continuous jet the 'equation of continuity" shows that as the jet loses velocity in ascending, it must increase in section. When the stream consists of drops following the same path in single file, no such increase of section is possible; and then the constancy of the total stream requires a gradual approximation of the drops, which in the case of a nearly vertical direction of motion cannot stop short of actual contact. Regular vibration has, however, the effect of postponing the collisions and consequent scattering of the drops, and in the case of a direction of motion less nearly vertical may prevent them altogether.

Under moderate electrical influence there is no material change in the resolution into drops, nor in the subsequent motion of the drops up to the moment of collision.   The difference begins here.   Instead of rebounding after collision, as the unelectrified drops of clean water generally or always do, the electrified drops *coalesce*, and thus the jet is no longer scattered about.   When the electrical influence is more powerful, the repulsion between the drops is sufficient to prevent actual contact, and then of course there is no opportunity for amalgamation.

These experiments may be repeated with extreme ease and with hardly any apparatus.   The diameter of the jet may be about $\frac{1}{20}$ inch [say 1 mm.], and may be obtained either from a hole in a thin plate or from a drawn-out glass tube.   I have generally employed a piece of glass tube fitted at the end with a perforated tin plate, and connected with a tap by india-rubber tubing.   The pressure may be such as to cause the jet to rise 18 or 24 inches [45 or 60 cm.], or even more.   A single passage of a rod of gutta-percha, or of sealing-wax, along the sleeve of the coat is sufficient to produce the effect. The seat of sensitiveness may be investigated by exciting the extreme tip only of a glass rod, which is then held in succession to the root of the jet and to the place of resolution into drops.   An effect is observed in the latter but not in the former position.   Care must be taken to use an electrification so feeble as to require close proximity for its operation; otherwise the discrimination of the positions will not be distinct.

The behaviour of the colliding drops becomes apparent under instantaneous illumination.   I have employed sparks from an inductorium, whose secondary terminals were connected with the coatings of a Leyden jar. The jet should be situated between the sparks and the eye, and the observation is facilitated by a piece of ground glass held a little beyond the jet, so as to diffuse the light; or the *shadow* of the jet may be received on the ground glass, which is then held as close as possible on the side towards the observer.

If the jet be supplied from an insulated vessel, the coalescence of colliding drops continues for a time after the removal of the influencing body. This is a consequence of the electrification of the vessel. If the electrified body be held for a time pretty close to the jet, and be then gradually withdrawn, a point may be found where the rebound of colliding drops is re-established. A small motion *to* or *from* the jet, or a discharge of the vessel by contact of the finger, again induces coalescence.

Although in these experiments the charges on the colliding drops are undoubtedly of the same name, it appeared to me very improbable that the result of contact of two equal drops, situated in the open, could be affected by any strictly equal electrifications. At the same time an opposite opinion makes the phenomena turn upon the very small *differences* of electrification due either to irregularities in the drops or to differences of situation, and is at first difficult of acceptance in view of the efficiency of such very feeble electric forces. Fortunately I am able to bring forward additional evidence bearing upon this point.

When two horizontal jets issue from neighbouring holes in a thin plate, they come into collision for a reason that I need not now stop to explain, and after contact they frequently rebound from one another without amalgamation. This observation, which I suppose must have been made before, allowed me to investigate the effect of a passage of electricity across two contiguous water surfaces. The jets that I employed were of about $\frac{1}{15}$ inch [$1\frac{1}{2}$ mm.] in diameter, and issued under a moderate pressure (5 or 6 inches [15 cm.]) from a large stoneware vessel. Below the place of rebound, but above that of resolution into drops, was placed a piece of insulated tin plate in connexion with a length of gutta-percha-covered wire. The source of electricity was a very feebly excited electrophorus, whose cover was brought into contact with the free end of the insulated wire. When both jets played upon the tin plate, the contact of the electrified cover had no effect in determining the union, but when only one jet washed the plate, union instantly followed the communication of electricity, and this notwithstanding that the jets were already in communication through the vessel. The quantity of electricity required is so small that the cover would act three or even four times without being re-charged, although no precautions were taken to insulate the reservoir.

In subsequent experiments the colliding jets, about $\frac{8}{100}$ inch [2 mm.] in diameter, issued horizontally from similar glass nozzles, formed by drawing out a piece of glass tubing and dividing it with a file at the narrowest part. One jet was supplied from the tap, and the other from the stoneware bottle placed upon an insulating stool. The sensitiveness to electricity was extraordinary. A piece of rubbed gutta-percha brought near the insulated bottle at once determined the coalescence of the jets. The influencing body being held still, it was possible to cause the jets again to rebound from one another, and then a small motion of the influencing body *to* or *from* the bottle again induced coalescence, but a *lateral* motion without effect. If an insulated wire be in connexion with the contents of the bottle, similar effects are produced when the electrified body is moved in the neighbourhood of the free end of the wire. With care it is possible to bring the electrified body into the neighbourhood of the free end of the wire so *slowly* that no effect is produced; a sudden movement of withdrawal will then usually determine the coalescence.

Hitherto statical electricity has been spoken of; but the electromotive force of even a single Grove cell is sufficient to produce these phenomena, though not with the same certainty. For this purpose one pole is connected through a contact key with the interior of the stoneware bottle, the other pole being to earth. If the fingers be slightly moistened, the body may be thrown into the circuit, apparently without diminution of effect. This perhaps ought not to surprise us, as in any case the electricity has to traverse several inches of a fine column of water. On the other hand, it appeared that most of the electromotive force of the Grove cell was necessary.

Further experiment showed that even the discharge of a condenser charged by a single Grove cell was sufficient to determine coalescence. Two condensers were used successively; one belonging to an inductorium by Ladd, the other made by Elliott Brothers, and marked "Capacity ½ Farad."* Sometimes even the "residual charge" sufficed.

It must be understood that coalescence of the jets would sometimes occur in a capricious manner, without the action of electricity or other apparent cause. I have reason to believe that some, at any rate, of these irregularities depended upon a want of cleanness in the water. The addition to the water of a very small quantity of soap makes the rebound of the jets impossible.

The last observation led me to examine the behaviour of a fine vertical jet of slightly soapy water; and I found, as I had expected, that *no scattering took place*†. Under these circumstances the approach of a moderately electrified body is without effect, but a more powerful influence scatters the drops as usual. The apparent coherence of a jet of water when the orifice is oiled was observed by Fuchs, and appears to have been always attributed to a diminution of adhesion between the jet and the walls of the orifice.

Some further details on this subject, and other investigations respecting the phenomena of jets, are reserved for another communication, which I hope

---

* [1899. The capacity is doubtless ½ *micro*-farad.]
† [1899. See however *Proc. Roy. Soc.* xxxiv. p. 130, 1882.]

soon to be able to present to the Royal Society [Art. LX.]; but I cannot close without indicating the probable application to meteorology of the facts already mentioned.   It is obvious that the formation of rain must depend very materially upon the consequences of encounters between cloud particles. If encounters do not lead to contacts, or if contacts result in rebounds, the particles remain of the same size as before; but, if the issue be coalescence, the bigger drops must rapidly increase in size and be precipitated as rain. Now, from what has appeared above we have every reason to suppose that the results of an encounter will be different according to the electrical condition of the particles, and we may thus anticipate an explanation of the remarkable but hitherto mysterious connexion between rain and electrical manifestations.

60.

ON THE CAPILLARY PHENOMENA OF JETS.

[*Proceedings of the Royal Society*, XXIX. pp. 71—97, 1879.]

WHEN water issues under high pressure from a circular orifice in a thin plate, a jet is formed whose section, though diminished in area, retains the circular form. But if the orifice be not circular, the section of the jet undergoes remarkable transformations, which were elaborately investigated by Bidone*, many years ago. The peculiarities of the orifice are exaggerated in the jet, but in an inverted manner. The following examples are taken from Bidone's memoir.

Fig. 1, orifice in the form of an ellipse (*A*), of which the major axis is horizontal, and 24 lines long; the minor axis is vertical, and 17 lines long. The head of water is 6 feet.

Fig. 1.

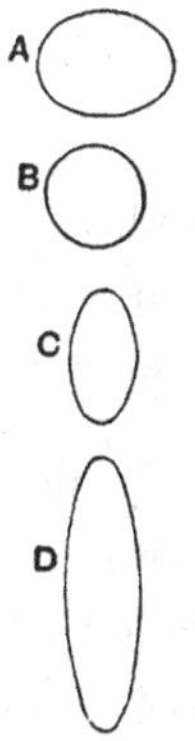

Near the orifice the sections of the vein are elliptical with major axis horizontal. The ellipticity gradually diminishes until at a distance of 30 lines from the orifice the section is circular. Beyond this point the vertical axis of the section increases, and the horizontal axis decreases, so that the vein reduces itself to a flat vertical sheet, very broad and thin. This sheet preserves its continuity to a distance of 6 feet from the orifice, where the vein is penetrated by air.

B represents the section at a distance of 30 lines from the orifice. It is a circle of 16 or 17 lines diameter.

C is the section at a distance of 6 inches from the orifice. It is an elliptical figure, whose major axis is 22 lines long and minor axis 14 lines long.

---

* *Expériences sur la Forme et sur la Direction des Veines et des Courans d'Eau lancés par diverses Ouvertures.* Par George Bidone.

D is the section at 24 inches from the orifice. It also is an elliptical figure, whose vertical axis is 45 lines long and horizontal axis about 12 lines long.

In fig. 2, the orifice (A) is an equilateral triangle, with sides 2 inches long. The head of water is 6 feet. The vein resolves itself into three flat sheets disposed symmetrically round the axis, the planes of the sheets being perpendicular to the sides of the orifice. These sheets are very thin, and retain their transparence and continuity to a distance of 42 inches, reckoned from the orifice. The sections represented by B, C, D, E are taken at distances from the orifice equal respectively to 1 inch, 6 inches, 12 inches, and 24 inches.

Fig. 2.

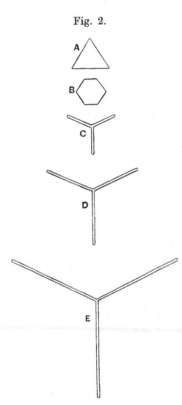

Similarly, a vein issuing from an orifice in the form of a regular polygon, of any number of sides, resolves itself into an equal number of thin sheets, whose planes are perpendicular to the sides of the polygon.

Bidone explains the formation of these sheets, in the main (as it appears to me), satisfactorily, by reference to simpler cases of meeting streams. Thus equal jets, moving in the same straight line with equal and opposite velocities, flatten themselves into a disk, situated in the perpendicular plane. If the axes of the jets intersect obliquely, a sheet is formed symmetrically in the plane perpendicular to that of the impinging jets. Those portions of a jet which proceed from the outlying parts of an unsymmetrical orifice are considered to behave, in some degree, like independent meeting streams.

In many cases, more especially when the orifices are small and the heads of water low, the extension of the sheets in directions perpendicular to the jet reaches a limit. Sections taken at greater distances from the orifice show a gradual shortening of the sheets, until a compact form is attained, similar to that of the first contraction. Beyond this point, if the jet retains its coherence, sheets are gradually thrown out again, but in directions bisecting the angles between the directions of the former sheets. These sheets may, in their turn, reach a limit of development, again contract, and so on. The forms assumed in the case of orifices of various shapes, including the rectangle, the equilateral triangle, and the square, have been carefully investigated and figured by Magnus*. Phenomena of this kind are of every-day occurrence, and may generally be observed whenever liquid falls from the lip of a moderately elevated vessel.

Admitting the substantial accuracy of Bidone's explanation of the formation and primary expansion of the sheets or excrescences, we have to inquire into the cause of the subsequent contraction. Bidone attributes it to the viscosity of the fluid, which may certainly be put out of the question. In Magnus's view the cause is "cohesion"; but he does not explain what is to be understood under this designation, and it is doubtful whether he had a clear idea upon the subject. The true explanation appears to have been first given by Buff†, who refers the phenomenon distinctly to the capillary force. Under the operation of this force the fluid behaves as if enclosed in an envelope of constant tension, and the recurrent form of the jet is due to vibrations of the fluid column about the circular figure of equilibrium, superposed upon the general progressive motion. Since the phase of vibration depends upon the time elapsed, it is always the same at the same point in space, and thus the motion is *steady* in the hydrodynamical sense, and the boundary of the jet is a fixed surface.

In so far as the vibrations may be considered to be isochronous, the distance between consecutive corresponding points of the recurrent figure, or, as it may be called, the *wave-length* of the figure, is directly proportional to the velocity of the jet, *i.e.*, to the *square root* of the head of water. This elongation of wave-length with increasing pressure was observed by Bidone and by Magnus, but no definite law was arrived at. As a jet falls under the action of gravity, its velocity increases, and thus an augmentation of wave-length might be expected; but, as will appear later, most of this augmentation is compensated by a change in the frequency of vibration due to the attenuation which is the necessary concomitant of the increased velocity. Consequently but little variation in the magnitudes of successive wave-lengths is to be noticed, even in the case of jets falling vertically with small initial velocity. In the following experiments the jets issued horizontally from orifices in thin plates, usually adapted to a large stoneware bottle which served as reservoir or cistern. The plates were of tin, soldered to the ends of short brass tubes rather more than an inch in diameter, by the aid of which they could be conveniently fitted to a tubulure in the lower part of the bottle. The pressure at any moment of the outflow could be measured by a water manometer read with a scale of millimetres. Some little uncertainty necessarily attended the determination of the zero point; it was usually taken to be the reading of the scale at which the

---

* "Hydraulische Untersuchungen." *Pogg. Ann.* vol. xcv. 1855.

† *Pogg. Ann.* vol. c. 1857.

jet ceased to clear itself from the plate on the running out of the water. At the beginning of an experiment, the orifice was plugged with a small roll of clean paper, and the bottle was filled from an india-rubber tube in connexion with a tap. After a sufficient time had elapsed for the water in the bottle to come sensibly to rest, the plug was withdrawn, and the observations were commenced. The jet is exceedingly sensitive to disturbances in the reservoir, and no arrangement hitherto tried for maintaining the level of the water has been successful. The measurements of wave-length (λ) were made with the aid of a pair of dividers adjusted so as to include one or more wave-lengths; and as nearly as possible at the same moment the manometer was read. The distance between the points of the dividers was afterwards taken from a scale of millimetres. The facility, and in some cases the success, of the operation of observing the wave-length depends very much upon the suitability of the illumination.

The first set of observations here given refers to a somewhat elongated orifice of rectangular form. The pressures and wave-lengths are measured in millimetres. The third column contains numbers proportional to the square roots of the pressures.

TABLE I.—November 11, 1878.

Pressure	Wave-length	$\sqrt{}$(Pressure)	Pressure	Wave-length	$\sqrt{}$(Pressure)
253	104	91	83	51	52
216	91	84	58	42	43
178	81	76	39	33	36
144	70	69	21	24	26
113*	61*	61*			

The agreement of the second and third columns is pretty good on the whole. Small discrepancies at the bottom of the table may be due to the uncertainty attaching to the zero point of pressure, and also to another cause which will be referred to later. At the higher pressures the observed wave-lengths have a marked tendency to increase more rapidly than the velocity of the jet. This result, which was confirmed by other observations, points to a departure from the law of isochronous vibration. Strict iso-chronism is only to be expected when vibrations are infinitely small, that is, in the present application when the section of the jet never deviates more than infinitesimally from the circular form. During the vibrations with which Table I. is concerned, however, the departures from circularity are very considerable, and there is no reason for supposing that such vibrations will be executed in exactly the same time as vibrations of infinitely small amplitude. Nevertheless, this consideration would not lead to an explanation of the discrepancies in Table I., unless it were the fact that the amplitude of vibration increased with the pressure under which the jet issues.

As a matter of observation the increase of amplitude is very apparent, and was noticed by Magnus. It is also a direct consequence of theory, inasmuch as the *lateral* velocities to which the vibrations are due vary in direct proportion to the longitudinal velocity of the jet. Consequently the amplitude varies approximately as the square root of the pressure, or as the wave-length. The amplitude here spoken of is measured, of course, by the *departure* from circularity, and not by the value of the maximum radius itself.

The law of the square root of the pressure thus applies only to small amplitudes, and unfortunately it is precisely these small amplitudes which it is difficult to experiment upon. Still it is possible to approach theoretical requirements more nearly than in the experiments of Table I.

The next set of measurements (Table II.) refer to an aperture in the form of an ellipse of moderate eccentricity. Two wave-lengths were included in the measurements; in other respects the arrangements were as before.

TABLE II.—November 12.

Pressure	Wave-length	√ (Pressure)	Pressure	Wave-length	√ (Pressure)
262	40½	40	69	20	20½
208	36½	35½	56	18	18½
182	34	33½	42	15	16
158	31	31	34	13½	14½
129	28½	28	27	12½	13
107*	25½*	25½*	21	10	11½
86	22½	23			

In this case the law is fully verified, the discrepancies being decidedly within the limits of experimental error.

On the other hand, the discrepancies may be exaggerated by the use of higher pressures. Table III. relates to the same orifice† as Table I. Instead of the stoneware bottle, a tall wooden box was used as reservoir.

TABLE III.—December 20.

Pressure	Wave-length	√ (Pressure)	Pressure	Wave-length	√ (Pressure)
757	200	155	189	79	77
672	184	145	154	70	70
587	171	136	123	62	62
497	152	125	107	58	58
442	141	118	89*	53*	53*
365	123	107	74	48	48
289	106	95½	61	44½	44
234	93	86			

† Its condition may have changed a little in the interval.

The wave-lengths at the high pressures very greatly exceed those calculated from the lower pressures according to the law applicable to small vibrations.

It is possible, however, to observe in cases where the amplitude is so small, that the discrepancies are moderate even at higher pressures than those recorded in Table III. The measurements in Table IV. are of a jet from an elliptical aperture of small eccentricity. The ratio of axes is about 5 : 6. The wooden box was used. Two wave-lengths were measured.

TABLE IV.—December 18.

Pressure	Wave-length	$\sqrt{}$ (Pressure)	Pressure	Wave-length	$\sqrt{}$ (Pressure)
1287	79†	83½	451	48¼	49¼
1195	82	80	371	45½	44¾
1117	79½	77½	290*	39½*	39½*
1023	73	74½	248	36½	36½
947	70½	71½	192	31½	32
852	66¾	68	158	28¾	29¼
770	64½	64½	133	26½	26¾
695	61¼	61	111	24¼	24½
620	58	58	94	21¾	22¼
532	54½	53½	85	21	21½

The following experiments relate to an orifice in the form of an equilateral triangle, with slightly rounded corners. The side measures about 3 millims. In this case the peculiarities of the contour are repeated *three* times in passing round the circumference. Two wave-lengths were measured.

TABLE V.—November 16.

Pressure	Wave-length	$\sqrt{}$ (Pressure)	Pressure	Wave-length	$\sqrt{}$ (Pressure)
215	36	35	66	19	19½
166	31½	31	43	14½	15½
127	27	27	27	11½	12½
92*	23*	23*			

Here again we observe the tendency of the wave-length to increase more rapidly than the square root of the pressure.

At higher pressures the difference is naturally still more marked. With the same aperture, and the wooden box as reservoir, the results were :—

† This is, doubtless, an error. At these high pressures the observation is difficult.

TABLE VI.—December 17.

Pressure	Wave-length	√ (Pressure)	Pressure	Wave-length	√ (Pressure)
1072	102	80·4	251	38½	38·9
992	94	77·5	213	34½	35·8
888	89	73·2		——	
827	86	70·7	189	33	33·7
762	81	67·8	163	31	31·3
702	77	65·0	140	28¼	29·1
619	70	61·1	111	24¾	25·9
539	66	57·0	90	22¾	23·3
468	59½	53·1	70	19¾	20·6
415	54½	50·0	57	17¼	18·5
337	47	44·6	45	16½	16·5
292	42	42·0			

The wave-lengths down to 34½ are immediate measurements; those below are deduced from measurements of two wave-lengths.

Similar experiments were made with jets from a *square* hole (side = 2 millims.), the peculiarities of which are repeated *four* times in passing round the circumference. Two wave-lengths were measured.

TABLE VII.—December 14.

Pressure	Wave-length	√ (Pressure)	Corrected	Pressure	Wave-length	√ (Pressure)	Corrected
447	32	30·2	29·9	167*	18½*	18·5*	18·0
377	29½	27·7	27·4	136	16½	16·6	16·1
312	27	25·2	24·9	107	14	14·8	14·2
269	24½	23·4	23·1	87	13	13·3	12·7
247	23	22·5	22·1	65	10¾	11·5	10·8
218	21½	21·1	20·7	47	8½	9·8	8·9
192	20	19·8	19·3				

The third column contains numbers proportional to the square roots of the pressures. In the fourth column a correction is introduced, the significance of which will be explained later.

The value of λ, other things being the same, depends upon the nature of the fluid. Thus methylated alcohol gave a wave-length about twice as great as tap water. This is a consequence of the smaller capillarity.

If a water jet be touched by a fragment of wood moistened with oil, the waves in front of the place of contact are considerably drawn out; but no sensible effect appears to be propagated up the stream.

If a jet of mercury discharging into dilute sulphuric acid be polarized by an electric current, the change in the capillary constant discovered by Lipmann shows itself by alterations in the length of the wave.

When the wave-length is considerable in comparison with the diameter of the jet, the vibrations about the circular form take place practically in two dimensions, and are easily calculated mathematically. The more general case, in which there is no limitation upon the magnitude of the diameter, involves the use of Bessel's functions. The investigation will be found in Appendix I. For the present we will confine ourselves to a statement of the results for vibrations in two dimensions.

Let us suppose that the polar equation of the section is

$$r = a_0 + a_n \cos n\theta, \quad \dots\dots\dots\dots\dots\dots\dots\dots(1)$$

so that the curve is an undulating one, repeating itself n times over the circumference. The mean radius is a_0; and, since the deviation from the circular form is small, a_n is a small quantity in comparison with a_0. The vibration is expressed by the variation of a_n as a harmonic function of the time. Thus if $a_n \propto \cos(pt - \epsilon)$, it may be proved that

$$p = \pi^{\frac{3}{4}} T^{\frac{1}{2}} \rho^{-\frac{1}{2}} A^{-\frac{3}{4}} \sqrt{(n^3 - n)}. \quad \dots\dots\dots\dots\dots\dots(2)$$

In this equation T is the superficial tension, ρ the density, A the area of the section (equal to πa_0^2), and the frequency of vibration is $p/2\pi$.

For a jet of given fluid and of given area, the frequency of vibration varies as $\sqrt{(n^3 - n)}$, or $\sqrt{\{(n-1) n (n+1)\}}$. The case of $n = 1$ corresponds to a displacement of the jet as a whole, without alteration in the *form* of the boundary. Accordingly there is no potential energy, and the frequency of vibration is zero. For $n = 2$ the boundary is elliptical, for $n = 3$ triangular with rounded corners, and so on. With most forms of orifice the jet is subject to more than one kind of vibration at the same time. Thus with a square orifice vibrations would occur corresponding to $n = 4$, $n = 8$, $n = 12$, &c. However, the higher modes of vibrations are quite subordinate, and may usually be neglected. The values of $\sqrt{(n^3 - n)}$ for various values of n are shown below.

n	$p/\sqrt{6}$	n	$p/\sqrt{6}$	n	$p/\sqrt{6}$
2	1·00	5	4·47	8	9·17
3	2·00	6	5·92	9	10·95
4	3·16	7	7·48	12	16·95

It appears that the frequency for $n = 3$ is just double that for $n = 2$, so that the wave-length for a triangular jet should be the half of that of an elliptical jet of equal area, the other circumstances being the same.

For a given fluid and mode of vibration (n), the frequency varies as $A^{-\frac{3}{4}}$, the thicker jet having the longer time of vibration. If v be the velocity of the jet, $\lambda = 2\pi v/p$. If the jet convey a given volume of fluid, $v \propto A^{-1}$, and thus $\lambda \propto A^{-\frac{1}{2}}$. Accordingly in the case of a jet falling vertically, the increase of λ due to velocity is in great measure compensated by the decrease due to diminishing area of section.

The law of variation of p for a given mode of vibration with the nature of the fluid and with the area of the section may be found by considerations of *dimensions*. T is a force divided by a line, so that its dimensions are 1 in mass, 0 in length, and -2 in time. The volume density ρ is of 1 dimension in mass, -3 in length, and 0 in time. A is of course of 2 dimensions in length, and 0 in mass and time. Thus the only combination of T, ρ, A, capable of representing a frequency, is $T^{\frac{1}{2}}\rho^{-\frac{1}{2}}A^{-\frac{3}{4}}$.

The above reasoning proceeds upon the assumption of the applicability of the law of isochronism. In the case of large vibrations, for which the law would not be true, we may still obtain a good deal of information by the method of dimensions. The *shape* of the orifice being given, let us inquire into the nature of the dependence of λ upon T, ρ, A, and P, the pressure under which the jet escapes. The dimensions of P, a force divided by an area, are 1 in mass, -1 in length, and -2 in time. Assume

$$\lambda \propto T^x \rho^y A^z P^u;$$

then by the method of dimensions we have the following relations among the exponents—

$$x + y + u = 0, \qquad -3y + 2z - u = 1, \qquad -2x - 2u = 0,$$

whence

$$u = -x, \qquad y = 0, \qquad z = \tfrac{1}{2}(1 - x).$$

Thus

$$\lambda \propto T^x A^{\frac{1}{2} - \frac{1}{2}x} P^{-x} \propto A^{\frac{1}{2}} \left(\frac{T}{PA^{\frac{1}{2}}} \right)^x.$$

The exponent x is undetermined; and since any number of terms with different values of x may occur together, all that we can infer is that λ is of the form

$$\lambda = A^{\frac{1}{2}} \cdot f \left(\frac{T}{PA^{\frac{1}{2}}} \right),$$

where f is an arbitrary function, or if we prefer it

$$\lambda = T^{-\frac{1}{2}} P^{\frac{1}{2}} A^{\frac{3}{4}} \cdot F \left(\frac{PA^{\frac{1}{2}}}{T} \right),$$

where F is equally arbitrary. Thus for a given liquid and shape of orifice, there is complete dynamical similarity if the pressure be taken inversely proportional to the linear dimension, and this whether the deviation from the circular form be great or small.

R. I.   25

In the case of water Quincke found $T = 81$ on the C.G.S. system of units. On the same system $\rho = 1$; and thus we get for the frequency of the gravest vibration ($n = 2$),

$$\frac{p}{2\pi} = 3\cdot51 a^{-\frac{1}{2}} = 8\cdot28 A^{-\frac{3}{4}}. \quad\dots\dots\dots\dots\dots\dots(3)$$

For a sectional area of one square centimetre, there are thus 8·28 vibrations per second. To obtain the pitch of middle C ($c' = 256$) we should require a diameter

$$2a = \left(\frac{3\cdot51}{2\cdot56}\right)^{\frac{2}{3}} = \cdot115,$$

or rather more than a millimetre.

For the general value of n, we have

$$\frac{p}{2\pi} = 1\cdot43 a^{-\frac{3}{2}} \sqrt{(n^3 - n)} = 3\cdot38 A^{-\frac{3}{4}} \sqrt{(n^3 - n)}. \quad\dots\dots\dots(4)$$

If h be the head of water to which the velocity of the jet is due,

$$\lambda = \frac{\sqrt{(2gh)} \cdot A^{\frac{3}{4}}}{3\cdot38 \sqrt{(n^3 - n)}}. \quad\dots\dots\dots\dots\dots\dots(5)$$

In applying this formula it must be remembered that A is the area of the section of the jet, and not the area of the aperture. We might indeed deduce the value of A from the area of the aperture by introduction of a coefficient of contraction (about ·62); but the area of the aperture itself is not very easily measured. It is much better to calculate A from an observation of the quantity of fluid (V), discharged under a measured head (h'), comparable in magnitude with that prevailing when λ is measured. Thus $A = V(2gh')^{-\frac{1}{2}}$. In the following calculations the C.G.S. system of units is employed.

In the case of the elliptical aperture of Table II., the value of A was found in this way to be ·0695. Hence at a head of 10·7 the wave-length should be

$$\lambda = \frac{\sqrt{(2g \times 10\cdot7)} \times (\cdot0695)^{\frac{3}{4}}}{3\cdot38 \times \sqrt{6}} = 2\cdot37,$$

the value of g being taken at 981. The corresponding observed value of λ is 2·55.

Again, in the case of the experiments recorded in Table IV., it was found that $A = \cdot0660$. Hence for $h = 29\cdot0$ the value of the wave-length should be given by

$$\lambda = \frac{\sqrt{(2g \times 29\cdot0)} \times (\cdot0660)^{\frac{3}{4}}}{3\cdot38 \times \sqrt{6}} = 3\cdot76.$$

The corresponding observed value is 3·95.

We will next take the triangular orifice of Table V. The value of A was found to be ·154. Hence for a head of 9·2 the value of λ, calculated à priori, is

$$\lambda = \frac{\sqrt{(2g \times 9·2)} \times (·154)^{\frac{3}{4}}}{3·38 \times \sqrt{24}} = 1·99,$$

as compared with 2·3 found by direct observation.

For the square orifice of Table VII., we have $A = ·153$. Hence, if $h = 16·7$,

$$\lambda = \frac{\sqrt{(2g \times 16·7)} \times (·153)^{\frac{3}{4}}}{3·38 \times \sqrt{60}} = 1·70,$$

as compared with 1·85 by observation.

It will be remarked that in every case the observed value of λ somewhat exceeds the calculated value. The discrepancies are to be attributed, not so much, I imagine, to errors of observation as to excessive amplitude of vibration, involving a departure from the frequency proper to infinitely small amplitudes. The closest agreement is in the case of Table IV., where the amplitude of vibration was smallest. It is also possible that the capillary tension actually operative in these experiments was somewhat less than that determined by Quincke for distilled water*.

When the pressures are small, the wave-lengths are no longer considerable in comparison with the diameter of the jet, and the vibrations cannot be supposed to take place sensibly in two dimensions. The frequency of vibration then becomes itself a function of the wave-length. This question is investigated mathematically in Appendix I. For the case of $n = 4$, it is proved that approximately

$$p^2 = \frac{60T}{\rho a^3} \left(1 + \frac{11\pi^2 a^2}{30\lambda^2}\right).$$

Hence for the aperture of Table VII.,

$$\lambda \propto \sqrt{h} \, (1 - ·088\lambda^{-2}),$$

λ being expressed in centimetres. The numbers in the fourth column of the table are calculated according to this formula.

On the other hand at high pressures the frequency becomes a function of the pressure. Since frequency is always an *even* function of amplitude, and in the present application, the square of the amplitude varies as h, the wave-length is given approximately by an expression of the form $\sqrt{h} \, (M + Nh)$, where M and N are constants. It appears from experiment, and might,

* [1899. As has been pointed out by Worthington (*Phil. Mag.* vol. xx. p. 66, 1885), the agreement with observation would be much improved by taking T at the now generally received value of about 72, in place of Quincke's 81.]

I think, have been expected, that frequency *diminishes* as amplitude increases, so that N is *positive*.

When the aperture has the form of an exact circle, and when the flow of fluid in its neighbourhood is unimpeded by obstacles, there is a perfect balance of lateral motions and pressures, and consequently nothing to render the jet in its future course unsymmetrical. Even in this case, however, the phenomena are profoundly modified by the operation of the capillary force. Far from retaining the cylindrical form unimpaired, the jet rapidly resolves itself in a more or less regular manner into detached masses. It has, in fact, been shown by Plateau*, both from theory and experiment, that in consequence of surface-tension the cylinder is an unstable form of equilibrium, when its length exceeds its circumference.

The circumstances attending the resolution of a cylindrical jet into drops have been admirably examined and described by Savart†, and for the most part explained with great sagacity by Plateau. There are, however, a few points which appear not to have been adequately treated hitherto; and in order to explain myself more effectually I propose to pass in review the leading features of Plateau's theory, imparting, where I am able, additional precision.

Let us conceive, then, an infinitely long circular cylinder of liquid, at rest‡, and inquire under what circumstances it is stable, or unstable, for small displacements, symmetrical about the axis of figure.

Whatever the deformation of the originally straight boundary of the axial section may be, it can be resolved by Fourier's theorem into deformations of the harmonic type. These component deformations are in general infinite in number, of every wave-length, and of arbitrary phase; but in the first stages of the motion, with which alone we are at present concerned, each produces its effect independently of every other, and may be considered by itself. Suppose, therefore, that the equation of the boundary is

$$r = a + \alpha \cos kz, \quad \dots\dots\dots\dots\dots\dots\dots\dots\dots\dots(6)$$

where α is a small quantity, the axis of z being that of symmetry. The wave-length of the disturbance may be called λ, and is connected with k by the equation $k = 2\pi/\lambda$. The capillary tension endeavours to contract the surface of the fluid; so that the stability, or instability, of the cylindrical form of equilibrium depends upon whether the surface (enclosing a given

---

* *Statique Expérimentale et Théorique des Liquides soumis aux seules Forces Moléculaires*, Paris, 1873.

† " Mémoire sur la Constitution des Veines Liquides lancées par des Orifices Circulaires en mince paroi." *Ann. d. Chim.* t. LIII. 1833.

‡ A motion common to every part of the fluid is necessarily without influence upon the stability, and may therefore be left out of account for convenience of conception and expression.

volume) be greater or less respectively after the displacement than before. It has been proved by Plateau (see also Appendix I.) that the surface is greater than before displacement if $ka > 1$, that is, if $\lambda < 2\pi a$; but less, if $ka < 1$, or $\lambda > 2\pi a$. Accordingly, the equilibrium is stable, if λ be less than the circumference; but unstable, if λ be greater than the circumference of the cylinder. Disturbances of the former kind, like those considered in the earlier part of this paper, lead to *vibrations* of harmonic type, whose amplitudes always remain small; but disturbances, whose wave-length exceeds the circumference, result in a greater and greater departure from the cylindrical figure. The analytical expression for the motion in the latter case involves exponential terms, one of which (except in case of a particular relation between the initial displacements and velocities) increases rapidly, being equally multiplied in equal times. The coefficient (q) of the time in the exponential term (e^{qt}) may be considered to measure the degree of dynamical instability; its reciprocal $1/q$ is the time in which the disturbance is multiplied in the ratio $1 : e$.

The degree of instability, as measured by q, is not to be determined from statical considerations only; otherwise there would be no limit to the increasing efficiency of the longer wave-lengths. The joint operation of superficial tension and *inertia* in fixing the wave-length of maximum instability was, I believe, first considered in a communication to the Mathematical Society*, on the "Instability of Jets." It appears that the value of q may be expressed in the form

$$q = \sqrt{\left(\frac{T}{\rho a^3}\right)} . F(ka), \quad\quad\quad\quad\dots\dots\dots\dots\dots(7)$$

where, as before, T is the superficial tension, ρ the density, and F is given by the following table:—

k^2a^2	$F(ka)$	k^2a^2	$F(ka)$
·00	·0000	·40	·3382
·05	·1536	·50	·3432
·10	·2108	·60	·3344
·20	·2794	·80	·2701
·30	·3182	·90	·2015

The greatest value of F thus corresponds, not to a zero value of k^2a^2, but approximately to $k^2a^2 = ·4858$, or to $\lambda = 4·508 \times 2a$. Hence the maximum instability occurs when the wave-length of disturbance is about half as great again as that at which instability first commences.

* *Math. Soc. Proc.*, November, 1878. [Art. LVIII.] See also Appendix I.

Taking for water, in C.G.S. units, $T = 81$, $\rho = 1$, we get for the case of maximum instability,

$$q^{-1} = \frac{a^{\frac{3}{2}}}{\sqrt{(81)} \times \cdot 343} = \cdot 115\, d^{\frac{3}{2}}, \dots\dots\dots\dots\dots(8)$$

if d be the diameter of the cylinder. Thus, if $d = 1$, $q^{-1} = \cdot 115$; or for a diameter of one centimetre the disturbance is multiplied 2·7 times in about one-ninth of a second. If the disturbance be multiplied 1000 fold in time t, $qt = 3 \log_e 10 = 6\cdot 9$, so that $t = \cdot 79\, d^{\frac{3}{2}}$. For example, if the diameter be one millimetre, the disturbance is multiplied 1000 fold in about one-fortieth of a second. In view of these estimates the rapid disintegration of a fine jet of water will not cause surprise.

The relative importance of two harmonic disturbances depends upon their initial magnitudes, and upon the rates at which they grow. When the initial values are very small, the latter consideration is much the more important; for, if the disturbances be represented by $\alpha_1 e^{q_1 t}$, $\alpha_2 e^{q_2 t}$, in which q_1 exceeds q_2, their ratio is $(\alpha_2 / \alpha_1) e^{-(q_1 - q_2)t}$; and this ratio decreases without limit with the time, whatever be the initial (finite) ratio $\alpha_2 : \alpha_1$. If the initial disturbances are small enough, that one is ultimately preponderant, for which the measure of instability is greatest. The smaller the causes by which the original equilibrium is upset, the more will the cylindrical mass tend to divide itself regularly into portions whose length is equal to 4·5 times the diameter. But a disturbance of less favourable wave-length may gain the preponderance in case its magnitude be sufficient to produce disintegration in a less time than that required by the other disturbances present.

The application of these results to actual jets presents no great difficulty. The disturbances, by which equilibrium is upset, are impressed upon the fluid as it leaves the aperture, and the continuous portion of the jet represents the distance travelled during the time necessary to produce disintegration. Thus the length of the continuous portion necessarily depends upon the character of the disturbances in respect of amplitude and wave-length. It may be increased considerably, as Savart showed, by a suitable isolation of the reservoir from tremors, whether due to external sources or to the impact of the jet itself in the vessel placed to receive it. Nevertheless it does not appear to be possible to carry the prolongation very far. Whether the residuary disturbances are of external origin, or are due to friction, or to some peculiarity of the fluid motion within the reservoir, has not been satisfactorily determined. On this point Plateau's explanations are not very clear, and he sometimes expresses himself as if the time of disintegration depended only upon the capillary tension, without reference to initial disturbances at all.

Two laws were formulated by Savart with respect to the length of the continuous portion of the jet, and have been to a certain extent explained by Plateau. For a given fluid and a given orifice the length is approximately proportional to the square root of the head. This follows at once from theory, if it can be assumed that the disturbances remain always of the same character, so that the *time* of disintegration is constant*. When the head is given, Savart found the length to be proportional to the diameter of the orifice. From (8) it appears that the time in which a disturbance is multiplied in a given ratio varies, not as d, but as $d^{\frac{3}{2}}$. Again, when the fluid is changed, the time varies as $\rho^{\frac{1}{2}} T^{-\frac{1}{2}}$. But it may be doubted, I think, whether the length of the continuous portion obeys any very simple laws, even when external disturbances are avoided as far as possible.

When the circumstances of the experiment are such that the reservoir is influenced by the shocks due to the impact of the jet, the disintegration usually establishes itself with complete regularity, and is attended by a musical note (Savart). The impact of the regular series of drops which is at any moment striking the sink (or vessel receiving the water), determines the rupture into similar drops of the portion of the jet at the same moment passing the orifice. The pitch of the note, though not absolutely definite, cannot differ much from that which corresponds to the division of the jet into wave-lengths of maximum instability; and, in fact, Savart found that the frequency was directly as the square root of the head, inversely as the diameter of the orifice, and independent of the nature of the fluid—laws which follow immediately from Plateau's theory.

From the pitch of the note due to a jet of given diameter, and issuing under a given head, the wave-length of the nascent divisions can be at once deduced. Reasoning from some observations of Savart, Plateau finds in this way 4·38 as the ratio of the length of a division to the diameter of the jet. The diameter of the orifice was 3 millims., from which that of the jet is deduced by the introduction of the coefficient ·8. Now that the length of a division has been estimated *à priori*, it is perhaps preferable to reverse Plateau's calculation, and to exhibit the frequency of vibration in terms of the other data of the problem. Thus

$$\text{frequency} = \frac{\sqrt{(2gh)}}{4\cdot508\,d}. \qquad\dots\dots\dots\dots\dots\dots(9)$$

But the most certain method of obtaining complete regularity of resolution is to bring the reservoir under the influence of an external vibrator, whose pitch is approximately the same as that proper to the jet. Magnus†

---

* For the sake of simplicity, I neglect the action of gravity upon the jet when formed. The question has been further discussed by Plateau.

† *Pogg. Ann.* Bd. cvi. 1859.

employed a Neef's hammer, attached to the wooden frame which supported
the reservoir.   Perhaps an electrically maintained tuning-fork is still better.
Magnus showed that the most important part of the effect is due to the
forced vibration of that side of the vessel which contains the orifice, and that
but little of it is propagated through the air.   With respect to the limits of
pitch, Savart found that the note might be a fifth above, and more than an
octave below, that proper to the jet.   According to theory, there would
be no well-defined lower limit; on the other side, the external vibration
cannot be efficient if it tends to produce divisions whose length is less than
the circumference of the jet.   This would give for the interval defining the
upper limit $\pi : 4\cdot508$, which is very nearly a fifth.   In the case of Plateau's
numbers ($\pi : 4\cdot38$) the discrepancy is a little greater.

The detached masses into which a jet is resolved do not at once assume
and retain a spherical form, but execute a series of vibrations, being
alternately compressed and elongated in the direction of the axis of symmetry.   When the resolution is effected in a perfectly periodic manner,
each drop is in the same phase of its vibration as it passes through a given
point of space; and thence arises the remarkable appearance of alternate
swellings and contractions described by Savart.   The interval from one
swelling to the next is the space described by the drop during one complete
vibration, and is therefore (as Plateau shows) proportional *cæteris paribus* to
the square root of the head.

The time of vibration is of course itself a function of the nature of the
fluid and of the size of the drop.   By the method of dimensions alone it may
be seen that the time of infinitely small vibrations varies directly as the
square root of the mass of the sphere and inversely as the square root
of the capillary tension; and in Appendix II. it is proved that its expression is

$$\tau = \sqrt{\left(\frac{3\pi\rho V}{8T}\right)}, \quad \ldots\ldots\ldots\ldots\ldots\ldots\ldots\ldots(10)$$

V being the volume of the vibrating mass.

In an experiment arranged to determine the time of vibration, a stream
of $19\cdot7$ cub. centims. per second was broken up under the action of a fork
making 128 vibrations per second.   Neglecting the mass of the small
spherules (of which more will be said presently), we get for the mass
of each sphere $19\cdot7 \div 128$, or $\cdot154$ grm.; and thence by (10), taking as
before $T = 81$,

$$\tau = \cdot0473 \text{ second.}$$

The distance between the first and second swellings was by measurement
$16\cdot5$ centims.   The level of the contraction midway between the two
swellings was $36\cdot8$ centims. below the surface of the liquid in the reservoir,

corresponding to a velocity of [269] centims. per second. These data give for the time of vibration,

$$\tau = 16\cdot5 \div [269] = \cdot0612 \text{ second.}$$

The discrepancy between the two values of τ, which is greater than I had expected, is doubtless due in part to excessive amplitude, rendering the vibration slower than that calculated for infinitely small amplitudes*.

A rough estimate of the degree of flattening to be expected at the first swelling may be arrived at by calculating the eccentricity of the *oblatum*, which has the same volume and *surface* as those appertaining to the portion of fluid in question when forming part of the undisturbed cylinder. In the case of the most natural mode of resolution, the volume of a drop is $9\pi a^3$, and its surface is $18\pi a^2$. The eccentricity of the *oblatum* which has this volume and this surface is ·944, corresponding to a ratio of principal axes equal to about 1 : 3.

In consequence of the rapidity of the motion some optical device is necessary to render apparent the phenomena attending the disintegration of a jet. Magnus employed a rotating mirror, and also a rotating disk from which a fine slit was cut out. The readiest method of obtaining instantaneous illumination is the electric spark, but with this Magnus was not successful. "The rounded masses of which the swellings consist reflect the light emanating from a point in such a manner that the eye sees only the single point of each, which is principally illuminated. Hence, when the stream is illuminated by the electric spark, the swellings appear like a string of pearls; but their form cannot be recognised, because the intensity of the light reflected from the remaining portions of the masses is too small to allow this, on account of the velocity with which the impression is lost†." The electric spark had, however, been used successfully for this purpose some years before by Buff‡, who observed the *shadow* of the jet on a white screen. Preferable to an opaque screen in my experience is a piece of ground glass, which allows the shadow to be examined from the further side. I have found also that the jet may be very well observed directly, if the illumination is properly managed. For this purpose it is necessary to place the jet between the source of light and the eye. The best effect is obtained when the light of the spark is somewhat diffused by being passed (for example) through a piece of ground glass.

The spark may be obtained from the secondary of an induction coil, whose terminals are in connexion with the coatings of a Leyden jar. By

* [1899. Experiments upon the vibrations of drops which fall singly from nozzles have been made by Lenard (*Wied. Ann.* vol. xxx. p. 209, 1887).]

† *Phil. Mag.* vol. xviii. 1859, p. 172.

‡ *Liebig's Ann.* vol. lxxviii. 1851.

adjustment of the contact breaker the series of sparks may be made to fit more or less perfectly with the formation of the drops. A still greater improvement may be effected by using an electrically maintained fork, which performs the double office of controlling the resolution of the jet and of interrupting the primary current of the induction coil. In this form the experiment is one of remarkable beauty. The jet, illuminated only in one phase of transformation, appears almost perfectly steady, and may be examined at leisure. The fork that I used had a frequency of 128, and communicated its vibration to the reservoir through the table on which both were placed without any special provision for the purpose. The only weak point in the arrangement was the rather feeble character of the sparks, depending probably upon the use of an induction coil too large for the rate of intermittence. A change in the phase under observation could be effected by pressing slightly upon the reservoir, whereby the vibration communicated was rendered more or less intense.

The jet issued horizontally from an orifice of about half a centimetre in diameter, and almost immediately assumed a rippled outline. The gradually increasing amplitude of the disturbance, the formation of the elongated ligament, and the subsequent transformation of the ligament into a spherule, could be examined with ease. In consequence of the transformation being in a more advanced stage at the forward than at the hinder end, the ligament remains for a moment connected with the mass behind, when it has freed itself from the mass in front, and thus the resulting spherule acquires a backwards relative velocity, which of necessity leads to a collision. Under ordinary circumstances the spherule rebounds, and may be thus reflected backwards and forwards several times between the adjacent masses. But if the jet be subject to moderate electrical influence, the spherule amalgamates with a larger mass at the first opportunity*. Magnus showed that the stream of spherules may be diverted into another path by the attraction of a powerfully electrified rod, held a little below the place of resolution.

Very interesting modifications of these phenomena are observed when a jet from an orifice in a thin plate† is directed obliquely upwards. In this case drops which break away with different velocities are carried under the action of gravity into different paths; and thus under ordinary circumstances a jet is apparently resolved into a "sheaf," or bundle of jets all lying in one vertical plane. Under the action of a vibrator of suitable periodic time the resolution is regularised; and then each drop, breaking away under like conditions, is projected with the same velocity, and therefore follows the

---

* *Proc. Roy. Soc.* March 13, 1879. On the Influence of Electricity on Colliding Water Drops. [Art. LIX.]

† Tyndall has shown that a pinhole gas burner may also be used with advantage.

same path. The apparent gathering together of the sheaf into a fine and well-defined stream is an effect of singular beauty.

In certain cases where the tremor to which the jet is subjected is compound, the single path is replaced by two, three, or even more paths, which the drops follow in a regular cycle. The explanation has been given with remarkable insight by Plateau. If for example besides the principal disturbance, which determines the size of the drops, there be another of twice the period, it is clear that the alternate drops break away under different conditions and therefore with different velocities. Complete periodicity is only attained after the passage of a *pair* of drops; and thus the odd series of drops pursues one path, and the even series another. All I propose at present is to bring forward a few facts connected with the influence of electricity, which are not mentioned in my former communication. To it, however, I must refer the reader for further explanations. The literature of the subject is given very fully in Plateau's second volume.

When the jet is projected upwards at a moderate obliquity, the sheaf is (as Savart describes it) confined to a vertical plane. Under these circumstances, there are few or no collisions, as the drops have room to clear one another, and moderate electrical influence is without effect. At a higher obliquity the drops begin to be scattered out of the vertical plane, which is a sign that collisions are taking place. Moderate electrical influence will now reduce the scattering again to the vertical plane, by causing the coalescence of drops which come into contact. When the projection is nearly vertical, the whole scattering is due to collisions, and is destroyed by electricity. If the resolution into drops is regularised by vibrations of suitable frequency, the principal drops follow the same path, and unless the projection is nearly vertical, there are no collisions, as explained in my former paper. It sometimes happens that the spherules are projected laterally in a distinct stream, making a considerable angle with the main stream. This is the result of collisions between the spherules and the principal drops. I believe that the former are often reflected backwards and forwards several times, until at last they escape laterally. Occasionally the principal drops themselves collide in a regular manner, and ultimately escape in a double stream. In all cases the behaviour under electrical influence is a criterion of the occurrence of collisions. The principal phenomena are easily observed directly, with the aid of instantaneous illumination.

[1899. Further experiments upon jets are described in *Proc. Roy. Soc.* vol. xxxiv. pp. 130—145, 1882.]

APPENDIX I.

The subject of this appendix is the mathematical investigation of the motion of frictionless fluid under the action of capillary force, the configuration of the fluid differing infinitely little from that of equilibrium in the form of an infinite circular cylinder.

Taking the axis of the cylinder as axis of z, and polar co-ordinates r, θ in the perpendicular plane, we may express the form of the surface at any time t by the equation

$$r = a_0 + f(\theta, z), \quad\ldots\ldots\ldots\ldots\ldots\ldots\ldots\ldots\ldots(11)$$

in which $f(\theta, z)$ is always a small quantity. By Fourier's theorem, the arbitrary function f may be expanded in a series of terms of the type $a_n \cos n\theta \cos kz$; and, as we shall see in the course of the investigation, each of these terms may be considered independently of the others. The summation extends to all positive values of k, and to all positive integral values of n, zero included.

During the motion the quantity a_0 does not remain absolutely constant, and must be determined by the condition that the inclosed *volume* is invariable. Now for the surface

$$r = a_0 + a_n \cos n\theta \cos kz, \quad\ldots\ldots\ldots\ldots\ldots\ldots(12)$$

we find

$$\text{volume} = \tfrac{1}{2}\iint r^2 \, d\theta \, dz = \int (\pi a_0^2 + \tfrac{1}{2}\pi a_n^2 \cos^2 kz)\, dz = z\,(\pi a_0^2 + \tfrac{1}{4}\pi a_n^2);$$

so that, if a denote the radius of the section of the undisturbed cylinder,

$$\pi a^2 = \pi a_0^2 + \tfrac{1}{4}\pi a_n^2,$$

whence approximately

$$a_0 = a\,(1 - \tfrac{1}{8}a_n^2/a^2). \quad\ldots\ldots\ldots\ldots\ldots\ldots\ldots(13)$$

For the case $n = 0$, (13) is replaced by

$$a_0 = a\,(1 - \tfrac{1}{4}a_0^2/a^2). \quad\ldots\ldots\ldots\ldots\ldots\ldots(14)$$

We have now to calculate the area of the surface of (22), on which the potential energy of displacement depends. We have [approximately]

$$\text{Surface} = \iint\{1 + \tfrac{1}{2}(dr/dz)^2 + \tfrac{1}{2}r^{-2}(dr/d\theta)^2\}\, r\, d\theta\, dz$$

$$= \iint\{1 + \tfrac{1}{2}k^2 a_n^2 \cos^2 n\theta \sin^2 kz + \tfrac{1}{2}n^2 a_n^2 a^{-2} \sin^2 n\theta \cos^2 kz\}\, r\, d\theta\, dz$$

$$= z\,\{2\pi a_0 + \tfrac{1}{4}\pi k^2 a_n^2 a + \tfrac{1}{4}\pi n^2 a_n^2 a^{-1}\};$$

so that, if σ denote the surface corresponding on the average to the unit of length,

$$\sigma = 2\pi a + \tfrac{1}{4}\pi a^{-1}(k^2 a^2 + n^2 - 1)\, a_n^2, \quad\ldots\ldots\ldots\ldots(15)$$

the value of a_0 being substituted from (13).

The potential energy P, estimated per unit length, is therefore expressed by

$$P = \tfrac{1}{4}\pi a^{-1} T (k^2 a^2 + n^2 - 1)\, \alpha_n{}^2, \quad \ldots\ldots\ldots\ldots\ldots\ldots(16)^*$$

T being the superficial tension.

For the case $n = 0$, (16) is replaced by

$$P = \tfrac{1}{2}\pi a^{-1} T (k^2 a^2 - 1)\, \alpha_0{}^2. \quad \ldots\ldots\ldots\ldots\ldots\ldots(17)$$

From (16) it appears that, when n is unity or any greater integer, the value of P is positive, showing that, for all displacements of these kinds, the original equilibrium is stable. For the case of displacements symmetrical about the axis, we see from (17) that the equilibrium is stable or unstable according as ka is greater or less than unity, i.e., according as the wavelength $(2\pi/k)$ is less or greater than the circumference of the cylinder.

If the expression for r in (12) involve a number of terms with various values of n and k, the corresponding expression for P is found by simple addition of the expressions relating to the component terms, and contains only the squares (and not the products) of the quantities α.

The velocity-potential (ϕ) of the motion of the fluid satisfies the equation

$$\frac{d^2\phi}{dr^2} + \frac{1}{r}\frac{d\phi}{dr} + \frac{1}{r^2}\frac{d^2\phi}{d\theta^2} + \frac{d^2\phi}{dz^2} = 0\,;$$

or, if in order to correspond with (12) we assume that the variable part is proportional to $\cos n\theta \cos kz$,

$$\frac{d^2\phi}{dr^2} + \frac{1}{r}\frac{d\phi}{dr} - \left(\frac{n^2}{r^2} + k^2\right)\phi = 0. \quad \ldots\ldots\ldots\ldots\ldots(18)$$

The solution of (18) under the condition that there is no introduction or abstraction of fluid along the axis of symmetry is—

$$\phi = \beta_n J_n (ikr) \cos n\theta \cos kz, \quad \ldots\ldots\ldots\ldots\ldots(19)$$

in which $i = \sqrt{(-1)}$, and J_n is the symbol of the Bessel's function of the nth order, so that

$$J_n(ikr) = \frac{(kr)^n}{2^n \Gamma(n+1)} \left\{ 1 + \frac{k^2 r^2}{2\,.\,2n+2} + \frac{k^4 r^4}{2\,.\,4\,.\,2n+2\,.\,2n+4} + \ldots \right\} \ldots\ldots(20)$$

The constant β_n is to be found from the condition that the radial velocity when $r = a$ coincides with that implied in (12). Thus

$$ik\,\beta_n J_n{}'(ika) = d\alpha_n/dt. \quad \ldots\ldots\ldots\ldots\ldots(21)$$

---

* [1899. If $k=0$, the right-hand member of (16), corresponding to $r=a_0+a_n \cos n\theta$, needs to be doubled.]

The kinetic energy of the motion is, by Green's theorem,

$$\tfrac{1}{2}\rho \iint \left[\phi \frac{d\phi}{dr} \right]_{r=a} a\, d\theta\, dz = \tfrac{1}{4}\pi\rho z \,.\, ika \,.\, J_n(ika)\, J_n'(ika)\,.\,\beta_n^2 \,;$$

so that, by (21), if K denote the kinetic energy per unit length,

$$K = \tfrac{1}{4}\pi\rho a^2\, \frac{J_n(ika)}{ika\,.\, J_n'(ika)} \left(\frac{da_n}{dt} \right)^2 . \quad\dots\dots\dots\dots\dots(22)$$

When $n = 0$, we must take, instead of (22),

$$K = \tfrac{1}{2}\pi\rho a^2\, \frac{J_0(ika)}{ika\,.\, J_0'(ika)} \left(\frac{da_0}{dt} \right)^2 . \quad\dots\dots\dots\dots\dots(23)$$

The most general value of K is to be found by simple summation, with respect to n and k, from the particular values expressed in (22) and (23). Since the expressions for P and K involve only the squares, and not the products, of the quantities a, da/dt, it follows that the motions represented by (12) take place in perfect independence of one another.

For the free motion we get by Lagrange's method from (16), (22),

$$\frac{d^2 a_n}{dt^2} + \frac{T}{\rho a^3}\, \frac{ika\,.\, J_n'(ika)}{J_n(ika)}\, (n^2 + k^2 a^2 - 1)\, a_n = 0, \dots\dots\dots(24)$$

which applies without change to the case $n = 0$. Thus, if $a_n \propto \cos(pt - \epsilon)$,

$$p^2 = \frac{T}{\rho a^3}\, \frac{ika\, J_n'(ika)}{J_n(ika)}\, (n^2 + k^2 a^2 - 1), \quad\dots\dots\dots\dots(25)$$

giving the frequency of vibration in the cases of stability. If $n = 0$, and $ka < 1$, the solution changes its form. If we suppose that $a_n \propto e^{\pm qt}$,

$$q^2 = \frac{T}{\rho a^3}\, \frac{ika\, J_0'(ika)}{J_0(ika)}\, (1 - k^2 a^2). \quad\dots\dots\dots\dots\dots(26)$$

From this the table in the text was calculated.

When n is greater than unity, the values of p^2 in (25) are usually in practical cases nearly the same as if ka were zero, or the motion took place in two dimensions. We may therefore advantageously introduce into (25) the supposition that ka is small. In this way we get

$$p^2 = n\,(n^2 - 1 + k^2 a^2)\, \frac{T}{\rho a^3}\, \left[1 + \frac{k^2 a^2}{n\,.\,2n + 2} + \dots \right], \quad\dots\dots\dots(27)$$

or, if ka be neglected altogether,

$$p^2 = (n^3 - n)\, \frac{T}{\rho a^3}, \quad\dots\dots\dots\dots\dots\dots\dots(28)$$

which agree with the formulæ used in the text. When $n = 1$, there is no force of restitution for the case of a displacement in two dimensions.

Combining in the usual way two stationary vibrations, whose phases differ by a quarter of a period, we find as the expression of a progressive wave,

$$r = a_0 + \gamma_n \cos n\theta \cos kz \cos pt + \gamma_n \cos n\theta \sin kz \sin pt$$

$$= a_0 + \gamma_n \cos n\theta \cos (pt - kz). \quad \dots\dots\dots\dots\dots\dots\dots(29)$$

For the application to a jet the progressive wave must be reduced to steady motion by the superposition of a common velocity (v) equal and opposite to that of the wave's propagation. The solution then becomes

$$r = a_0 + \gamma_n \cos n\theta \cos kz, \quad \dots\dots\dots\dots\dots(30)$$

in which γ_n is an absolute constant. The corresponding velocity-potential is

$$\phi = -vz + \frac{p\gamma_n J_n(ikr)}{ik J_n'(ika)} \sin kz \cos n\theta. \quad \dots\dots\dots\dots(31)$$

It is instructive to verify these results by the formulæ applicable to steady motion. The resultant velocity q at any point is approximately equal to $d\phi/dz$; and

$$\frac{d\phi}{dz} = -v + \frac{pk\gamma_n J_n(ikr)}{ik J_n'(ika)} \cos kz \cos n\theta.$$

At the surface we have approximately $r = a$, and

$$\tfrac{1}{2}q^2 = \tfrac{1}{2}v^2 - pkv \frac{\gamma_n J_n(ika)}{ik J_n'(ika)} \cos kz \cos n\theta.$$

Thus by the hydrodynamical equation of pressure, with use of (25), since $v = p/k$,

$$\text{Pressure} = \text{const.} + \gamma_n a^{-2} T (n^2 - 1 + k^2 a^2) \cos kz \cos n\theta. \quad \dots\dots(32)$$

The pressure due to superficial tension is $T(R_1^{-1} + R_2^{-1})$, if R_1, R_2, are the radii of curvature in planes parallel and perpendicular to the axis; and from (30)

$$- R_2^{-1} = d^2r/dz^2 = - k^2\gamma_n \cos n\theta \cos kz,$$

$$R_1^{-1} = r^{-1} + d^2r^{-1}/d\theta^2 = a^{-1} + \gamma_n a^{-2}(n^2 - 1) \cos n\theta \cos kz;$$

so that

$$\text{Pressure} = \text{const.} + \gamma_n a^{-2}(n^2 - 1 + k^2 a^2) \cos n\theta \cos kz.$$

Thus the pressure due to velocity is exactly balanced by the capillary force, and the surface condition of equilibrium is satisfied.

APPENDIX II.

We will now investigate in the same manner the vibrations of a liquid mass about a *spherical* figure, confining ourselves for brevity to modes of vibration symmetrical about an axis, which is sufficient for the application in the text. These modes require for their expression only Legendre's functions P_n; the more general problem, involving Laplace's functions, may be treated in the same way, and leads to the same results.

The radius r may be expanded at any time t in the series

$$r = a_0 + a_1 P_1(\mu) + \ldots\ldots\ldots + a_n P_n(\mu) +, \ldots\ldots\ldots\ldots(33)$$

where a_1, a_2 are small quantities relatively to a_0, and μ (according to the usual notation) represents the cosine of the colatitude (θ).

For the volume included within the surface (33) we have

$$V = \tfrac{2}{3}\pi \int_{-1}^{+1} r^3 d\mu = \tfrac{4}{3}\pi a_0^3 [1 + 3\Sigma (2n + 1)^{-1} a_n^2/a_0^2],$$

approximately. If a be the radius of the sphere of equilibrium,

$$a^3 = a_0^3 [1 + 3\Sigma (2n + 1)^{-1} a_n^2/a^2]. \quad \ldots\ldots\ldots\ldots\ldots(34)$$

We have now to calculate the area of the surface S.

$$S = 2\pi \int r \sin\theta \sqrt{\left\{r^2 + \left(\frac{dr}{d\theta}\right)^2\right\}} \, d\theta = 2\pi \int \left\{r^2 + \tfrac{1}{2}\left(\frac{dr}{d\theta}\right)^2\right\} \sin\theta \, d\theta.$$

For the first part

$$\int_{-1}^{+1} r^2 d\mu = 2a_0^2 + 2\Sigma (2n + 1)^{-1} a_n^2.$$

For the second part

$$\tfrac{1}{2}\int\left(\frac{dr}{d\theta}\right)^2 \sin\theta \, d\theta = \tfrac{1}{2}\int_{-1}^{+1}(1 - \mu^2)\left[\Sigma a_n \frac{dP_n}{d\mu}\right]^2 d\mu.$$

The value of the quantity on the right-hand side may be found with the aid of the formula*

$$\int_{-1}^{+1}(1 - \mu^2)\frac{dP_m}{d\mu}\frac{dP_n}{d\mu} \, d\mu = n(n + 1)\int_{-1}^{+1} P_m P_n d\mu.$$

Thus

$$\tfrac{1}{2}\int_{-1}^{+1}\left(\frac{dr}{d\theta}\right)^2 \sin\theta \, d\theta = \tfrac{1}{2}\int_{-1}^{+1}(1 - \mu^2)\Sigma a_n^2\left(\frac{dP_n}{d\mu}\right)^2 d\mu$$

$$= \tfrac{1}{2}\Sigma n(n + 1) a_n^2 \int_{-1}^{+1} P_n^2 d\mu = \Sigma n(n + 1)(2n + 1)^{-1} a_n^2.$$

---

* Todhunter's *Laplace's Functions*, § 62.

Accordingly

$$S = 4\pi a_0^2 + 2\pi \Sigma (2n+1)^{-1}(n^2+n+2) a_n^2;$$

or, since by (34)

$$a_0^2 = a^2 - 2\Sigma(2n+1)^{-1}a_n^2,$$

$$S = 4\pi a^2 + 2\pi \Sigma (n-1)(n+2)(2n+1)^{-1}a_n^2. \ldots\ldots\ldots(35)$$

If T be the cohesive tension, the potential energy is

$$P = 2\pi T \Sigma (n-1)(n+2)(2n+1)^{-1}a_n^2. \ldots\ldots\ldots(36)$$

We have now to calculate the kinetic energy of the motion. The velocity-potential ϕ may be expanded in the series

$$\phi = \beta_0 + \beta_1 r P_1(\mu) + \ldots\ldots + \beta_n r^n P_n(\mu) + \ldots; \ldots\ldots\ldots(37)$$

and thus for K we get

$$K = \tfrac{1}{2}\rho \iint \phi \frac{d\phi}{dr} dS = \tfrac{1}{2}\rho . 2\pi a^2 \int_{-1}^{+1} \phi \frac{d\phi}{dr} d\mu = \tfrac{1}{2}\rho . 4\pi a^2 . \Sigma (2n+1)^{-1} n a^{2n-1} \beta_n^2.$$

But by comparison of the value of $d\phi/dr$ from (37) with (33), we find $n a^{n-1}\beta_n = da_n/dt$; and thus

$$K = 2\pi \rho a^3 \Sigma (2n+1)^{-1} n^{-1} (da_n/dt)^2. \ldots\ldots\ldots\ldots(38)$$

Since the products of the quantities a_n and da_n/dt do not occur in the expressions for P and K, the motions represented by the various terms occur independently of one another. The equation for a_n is by Lagrange's method

$$\frac{d^2 a_n}{dt^2} + n(n-1)(n+2)\frac{T}{\rho a^3} a_n = 0; \ldots\ldots\ldots\ldots(39)$$

so that, if $a_n \propto \cos(pt+\epsilon)$,

$$p^2 = n(n-1)(n+2)\frac{T}{\rho a^3}. \ldots\ldots\ldots\ldots(40)$$

The periodic time τ given in the text (equation (10)) follows from (40) by putting $\tau = 2\pi p^{-1}$, $n = 2$, $V = \tfrac{4}{3}\pi a^3$.

To find the radius of the sphere of water which vibrates seconds, put $p = 2\pi$, $T = 81$, $\rho = 1$, $n = 2$. Thus $a = 2\cdot54$ centims., or one inch almost exactly. [1899. But a better value for T would be 74.]

61.

ACOUSTICAL OBSERVATIONS. II.

[*Phil. Mag.* VII. pp. 149—162, 1879.]

Pure Tones from Sounding Flames.

THE best approximation to a pure tone is doubtless that given by a fork held over a suitably tuned air resonator; but unless the vibrations are maintained, the sound is of but short duration, and varies in intensity throughout. On the other hand the introduction of an electro-magnetic maintenance (as in Helmholtz's vowel experiments) somewhat complicates the apparatus. For many purposes extreme purity and constancy of pitch are not important; and thus an arrangement which shall be simple and easy to manage, even though less perfect in its operation than a tuning-fork, is still a desideratum.

During the last year I have often used with good effect air resonators whose vibrations were maintained in a well-known manner by hydrogen flames. In the common form of the experiment an open cylindrical tube is employed as resonator, and gives a sound, usually of a highly compound character. In order to obtain a pure tone, it is only necessary to replace the tube by a resonator of different form, such as a rather wide-mouthed bottle or jar; but a difficulty then arises from the progressive deterioration of the limited quantity of air included. A better result is obtained from a tube with a central expansion, such as a bulbous paraffin-lamp chimney, which allows of a through draught, and yet departs sufficiently from the cylindrical form to give a pure tone. For ready speech, it is sometimes necessary to restrict the lower aperture, *e.g.* by a bored disk of wood attached with wax. Another plan which answers very well is to block the middle of a cylindrical tube by a loosely fitting plug. The tubes that I used are of cast iron, and were plugged by rectangular pieces of wood provided with springs of brass wire to keep them in position. The length of the plug may be about two diameters of the tube; the length of the tube itself should be about twelve diameters. In all cases the best result requires that the tubes through which the hydrogen is supplied be of suitable length, and be provided with suitable burners. These may be made of glass, and are easily adjusted by trial.

For ordinary purposes a common hydrogen-bottle is sufficient; but the note is rather more steady when the hydrogen is supplied from a gas-holder. In this way I have obtained pure tones, giving with tuning-forks pretty steady beats of more than two seconds' period. When the intensities are nearly equal, the phase of approximate silence is very well marked.

Points of silence near a wall from which a pure tone is reflected.

On this subject there are two papers by N. Savart[*], who advances views very difficult of acceptance. A criticism of some of Savart's positions was published soon after by Seebeck; but the question does not appear to have been thoroughly cleared up.

One source of confusion is imperfect recognition of the fact that the positions of the silences depend upon the nature of the apparatus used for the investigation. In the case of the ear a silence requires that there be no variation of pressure at the open end of the ear-passage, whether it be in its natural state, or prolonged by a tube fitted into the external ear. The addition of a small cone or resonator will not affect the truth of this statement. Thus, if the influence of the head and body of the observer acting as simple obstacles be put out of account (as may fairly be done when a tube is used), the silences occur at distances from the wall which are *odd* multiples of the quarter wave-length[†]. On the other hand, if a membrane simply stretched over a hoop and held parallel to the wall be used as the indicator, the positions of zero disturbance are at distances from the wall equal to *even* multiples of the quarter wave-length.

. In the theory of organ-pipes the places of zero velocity and of maximum pressure-variation are usually called *nodes*; and the places of zero pressure-variation and of maximum velocity are called *loops*. If we retain this nomenclature, we may say that silences as investigated by the ear occur at loops, and that the maximum sound is found at nodes; but in Savart's papers the silences are identified with *nodes*. Moreover the difference is not one of words merely; for Savart considers that (apart from the effects of obstacles) the silences are to be found at distances from the wall which are *even* multiples of the quarter wave-length. A large part of his work is thus an

---

[*] *Ann. d. Chim.* vol. LXXI. 1839, vol. XIV. 1845.

[†] The waves are here supposed complete. Savart's "ondes" are only half as long.

endeavour to bring the facts into accordance with a mistaken theoretical view.

When the median plane is parallel to the wall, the obstruction presented by the head displaces considerably the positions of the silences. In his first paper Savart proposes to add 27 mm. to the measured distances between the external ear nearer to the wall and the wall itself, in order to take account of the interval between the external ear and the sentient apparatus. In the case of the ear further from the wall a similar distance is to be subtracted. I am at a loss to understand how the situation of the sentient apparatus can be supposed to be an element in the question at all. Everything must surely depend upon whether there is or is not a variation of pressure at the outer end of the ear-passage. In the second paper Savart takes (as it appears to me) a further step in the wrong direction. He states that the positions of the silences are the same, whether they be observed with the ear nearer to the wall, or with the ear further from it, and draws the conclusion that the part of the head with which we have to deal is that situated in the median plane midway between the ears. Having already added 27 mm. to his measurements (in the case of the ear nearer to the wall), to take account of the distance between the external ear and the labyrinth, he now adds 50 mm. more. By this artificial treatment the distances of the silences from the wall are made to agree with the series of even multiples of the quarter wave-length, though considerable anomalies remain unexplained.

There can be no doubt, I imagine, that Savart's theoretical views are quite erroneous, and that what has to be explained by the action of the head as an obstacle is the displacement of the silences from the loops, and not from the nodes. An exact theoretical investigation of this subject is of course out of the question; but some information bearing upon it may be obtained from a calculation given in my *Theory of Sound*, § 328, relating to the character of the obstruction to sound presented by rigid spheres. It appears that if a source of sound be situated at the surface of a sphere whose circumference is moderate in comparison with the wave-length, the *phase* (which is the element on which the phenomena under consideration principally depend) at a distance is approximately the same as if the source were moved outward from the surface through a distance equal to half the radius, and the sphere were removed altogether. By the theorem of reciprocity, § 294, it follows that in the case of reflection of plane waves there is a silence at the point on the surface of the sphere nearest the wall when, not this point itself, but another further from the centre by half the radius, is distant from the wall by an odd multiple of the quarter wave-length, provided that the distance between the sphere and wall be not too small a multiple of the radius. Instead therefore of adding with Savart 27 mm., or 77 mm., to the observed distances in the expectation of so arriving at even multiples of the quarter wave-length, we ought rather to subtract some such distance as 50 mm. in the expectation of arriving at odd multiples of the same quantity.

The following are some of Savart's results given in the first paper:—

Designations des divers points.	Distances des points à la paroi. mètres.
Paroi.	0·000
1^r ventre.	·148
1^r nœud.	·373
2e ventre.	·716
2e nœud.	1·000
3e ventre.	1·358
3e nœud.	1·615
4e ventre.	1·997
4e nœud.	2·275

If we subtract 27 + 50 (= 77) from Savart's numbers for nodes we get

$$·296, \qquad ·923, \qquad 1·538, \qquad 2·198,$$

corresponding to

$$\tfrac{1}{2}(·619), \qquad \tfrac{3}{2}(·619), \qquad \tfrac{5}{2}(·619), \qquad \tfrac{7}{2}(·619),$$

or to

$$·309, \qquad ·927, \qquad 1·546, \qquad 2·163;$$

·619 being the value of the half-wave (onde). The "ventre" between the wall and the first node does not belong to the regular system at all.

From a theoretical point of view, it appeared to me highly improbable that the silences for the two ears should occur in the same position of the head, except perhaps in the case of a particular wave-length equal to about three diameters; and laboratory experiments with steadily maintained tones had made me familiar with the phenomenon of sounds apparently transferring themselves from one ear to the other when the head is moved; but I thought it desirable to try a few experiments in the open air especially directed to the examination of this point.

The source of sound was a lamp-chimney and hydrogen-flame, as described above, of pitch e' flat, so that the quarter wave-length was about eleven inches. The apparatus was placed at distances varying from about 18 to 50 feet in front of a tolerably flat wall; and the observer, with one ear stopped, investigated the positions of the silences, holding the middle plane of his head parallel to the wall. Although the positions of the silences were not very well defined, presumably in consequence of unequal amplitudes of the superposed vibrations, the most inattentive observer could not fail to notice that it was necessary to move the head considerably in order to pass from a place where the sound was a minimum for one ear to a place where it was a minimum for the other ear. We may therefore conclude that Savart's statement is not generally true, and that the views which he has founded upon it have no sufficient claim upon our acceptance.

When the median plane of the head is perpendicular to the wall, the silences are observed at distances which are *odd* multiples of the quarter wave-length, agreeably with theory.

Sensitive Flames.

The beautiful phenomenon of sensitive flames is now familiar to students of acoustics; but its rationale is by no means understood. An important contribution to the facts, from which some day a theory will doubtless arise, is contained in the observation of Prof. Tyndall as to the "seat of sensitiveness." My present purpose is to bring forward another fact which also will probably be found important. It may be thus stated. Under the action of stationary sonorous waves a flame is excited at *loops* and *not* at *nodes*.

The source of sound was a little contrivance on the principle of the birdcall, blown from a well-regulated bellows. Probably a very high organ-pipe or whistle might be employed; but it is necessary to use a nearly pure tone; and the pitch must be high, or the flame will be not affected sufficiently. At a distance of a few feet the sound was reflected perpendicularly from a large board. The flame itself was that called by Tyndall the vowel flame, issuing from a pin-hole steatite burner fed from a gas-holder with gas at high pressure (9 or 10 inches [25 cm.]).

The observations were made by moving the burner to and fro in front of the board, until the positions were found in which the flame was least disturbed by the sound; and it will be seen from the results that these positions were very well defined. The distance between the board and the orifice of the burner was then taken with a pair of dividers, and measured on a scale of millimetres. Two observers, J and E, adjusted the flame independently of one another. The following are the results obtained:—

First position	J. 16½, 16½ .... mean 16½ E. 16, 16 .... ,, 16	16¼	16·25
Second position ...	J. 31, 31½ .... ,, 31¼ E. 32½, 31, 32 ... ,, 31¾	31½	15·7
Third position ...	J. 47, 47½ .... ,, 47¼ E. 46½, 47, 45½... ,, 46¼	46¾	15·6
Fourth position ...	J. 62, 62½ .... ,, 62¼ E. 64, 60½, 62½... ,, 62¼	62¼	15·6
Fifth position	J. 78½, 78½ ...... ,, 78½	78½	15·5

The second column contains the individual measurements by the two observers; the third shows the mean of all the results for the same position. The numbers in the fourth column are the results of dividing those of the third column by 1, 2, 3, 4, 5, respectively, and agree very well together, proving that the positions coincide with nodes. If the positions had coincided with loops, the numbers of the third column would have been in the ratios $1 : 3 : 5 : 7 : 9$. The wave-length of the sound was thus $31\cdot2$ mm., corresponding to pitch $f^{\text{vi}}\sharp$.

A few observations were made at the same time on the positions of the silences, as estimated by the ear listening through a tube. As was to be expected, they coincided with the loops, bisecting the intervals given by the flame. When the flame was in a position of minimum effect, and the free end of the tube was held close to the burner at an equal distance from the reflecting wall, the sound heard was a maximum, and diminished when the end of the tube was displaced a little in either direction. It may therefore be taken as established that the flame is affected where the ear would not be affected, and *vice versâ*.

Aerial Vibrations of very Low Pitch maintained by Flames.

In a lecture "On the Explanation of certain Acoustical Phenomena*," I showed the production of a pure tone of about 95 vibrations per second from a glass resonator and a hydrogen-flame. With a larger resonator of the same kind—a globe with a short neck, intended for showing the combustion of phosphorus in oxygen, the pitch is 64 vibrations per second. I have lately made some further experiments, with the view of finding whether there is any obstacle to the maintenance by flames of vibrations of still lower frequency. The resonator, whose natural pitch is 64, was fitted with a pasteboard tube 2 inches in diameter and 14 inches long. In this condition its calculated frequency† is about 25; and it was found that vibrations could be maintained by a hydrogen-flame, or even by a flame of common gas. The supply-tube should be of considerable length; and the orifice must not be much contracted. Although the intensity of vibration was such as to make it a matter of difficulty to keep the flame alight, scarcely anything could be heard. I saw no reason to doubt that still slower vibrations might be maintained by flames.

In illustration of the mechanics of this subject, an apparatus was contrived, in which by the aid of electricity a periodic communication of heat to a limited mass of air could be effected. By means of a perforated cork one leg of a U-tube containing mercury was fitted air-tight to the neck of an

---

* *Proceedings of the Royal Institution*, March 15, 1875. *Nature*, vol. xviii. p. 319. [Art. lv.]
† *Theory of Sound*, vol. ii. § 307 (8). [1 inch = 2·54 cm.]

inverted bottle of about 200 cub. centims. capacity. The diameter of the column of mercury was about 1 centim., and the length of the column about 25 centims. The combination constituted a resonator, differing from an ordinary air resonator by the substitution of mercury for air in the channel joining the interior of the vessel with the external atmosphere. Inside the bottle was a spiral of fine platinum wire, at one end in communication through the cork with one pole of a battery of two or three small Grove cells. The other end of the platinum spiral was connected with a copper wire, which terminated in the U-tube near the equilibrium position of the mercury surface. The second pole of the battery was in permanent connexion with the outer extremity of the mercury column. As the mercury vibrates, the circuit is periodically completed and broken. The current passes, and the platinum wire glows, when the mercury rises in the leg connected with the bottle. Thus the communication of heat occurs when the air in the interior is *condensed* by the vibration, which is the necessary condition for maintenance, as is explained in the lecture referred to.

Rijke's Notes on a large scale.

The production of sound in tubes by heated gauze was discovered by Rijke*, and is perhaps the most interesting of all the cases in which vibrations are maintained by heat. The probable explanation is given in the Royal-Institution lecture. It is surprising that the phenomenon is not more generally known, as on a large scale the effect is extremely striking. I have employed a cast-iron pipe 5 feet long and $4\frac{3}{4}$ inches in diameter, hung over a table from a beam in the roof of my laboratory. The gauze (iron wire) is of about 32 meshes to the linear inch, and may advantageously be used in two thicknesses. It should be moulded with a hammer on a circular wooden block of somewhat smaller diameter than that of the pipe, and will then retain its position in the pipe by friction. When it is desired to produce the sound, the gauze caps are pushed up the pipe to a distance of about a foot, and a gas-flame from a large rose-burner is adjusted underneath, at such a level as to heat the gauze to a bright red heat. For this purpose the vertical tube of the lamp should be prolonged, if necessary, by an additional length of brass tubing. In making the adjustment a more convenient view of the interior of the pipe is obtained with the aid of a small piece of looking-glass held obliquely underneath. Sometimes a sound is excited by the flame itself independently of the gauze. This should be avoided if possible, as it impedes the due heating of the gauze. When a good red heat is attained the flame is suddenly removed, either by withdrawing the lamp or by stopping the supply of gas. In about a second the sound begins, and presently rises to such intensity as to shake the room, after which it gradually dies away. The whole duration of the sound may be about 10 seconds.

* *Pogg. Ann.* cvii. 339, 1859.

Mutual Influence of Organ-Pipes nearly in unison.

The easiest way of approaching the consideration of this subject is to take the case of an open or stopped pipe, divided into two similar parts by a rigid barrier along its middle plane. In the absence of the barrier, the vibrations of the two halves under the action of the wind are in the same phase; and at first sight there appears to be no reason why this state of things should be disturbed by the barrier. Nevertheless it is well known to physicists that the two halves do in fact take opposite vibrations, with the result that the sound in the external air at a distance from the compound pipe is a small fraction only of that due to either half acting alone. In the pipe itself the vibration is *more*, and not less, intense on account of the barrier. It is true that at the very beginning of the sound, when the wind first comes on, the vibrations in the two halves are similar, as is evidenced by the greater loudness; but the opposition of phase is rapidly established, usually in a fraction of a second of time. As a system with two degrees of freedom, the compound pipe is capable of two distinct modes of vibration, in one of which the vibrations of the component pipes are in the same phase, and in the other in opposite phases. Why the action of the wind should maintain the latter mode of vibration to the exclusion of the former has not hitherto been explained; but the fact remains that that mode of vibration, which depends for its possibility upon the barrier, is chosen in preference to the other mode, which is not dependent upon the barrier, and in the absence of the barrier is the one necessarily adopted.

The two possible modes of vibration have, as in almost all such cases, two distinct periods of vibration, the difference depending upon the behaviour of the air just outside the open ends. In consequence of the inertia of the external air at an open end, the effective length of a pipe exceeds its actual length by about six-tenths of the radius. The increment of effective length is therefore greater in the case of the compound column of air when its parts vibrate in the same phase, than it would be for either of the parts if removed from the influence of the other. On the other hand, when the vibrations are in opposite phases, the increment must be much less, one component pipe absorbing the air discharged from the other. Accordingly one note of the compound pipe is graver, and the other, which is the one actually sounded, is more acute, than the natural notes of the component pipes when supposed to act independently of one another.

In order to show this effect it is not necessary that the two pipes be similar, or even of exactly the same pitch. If two pipes in approximate unison be placed so that their open ends are contiguous, a mutual influence is exerted, which is usually sufficient to prevent the production of beats. The examples about to be given will show that the unison need not be exact; but the greater the deviation from unison, the more intense is the residual sound. Beyond the limit of the admissible departure from unison, beats ensue; but at first they are irregular, and liable to be disturbed by very slight causes, such as draughts of air. According to theory, the frequency of the beats ought to be a little greater than the difference of the frequencies of the notes given by the pipes independently; but I have not been able to detect the difference experimentally. It would therefore seem that over most of the range for which the mutual influence is sensible and regular, it is sufficiently powerful to prevent more than one note being sounded.

In the experiments that I have tried, the pipes were blown from a bellows provided with a special regulator, and the pitches of the various notes were determined by counting the beats for 20 seconds between them and a somewhat sharper note on a harmonium. Sometimes the blown ends of the pipes were near together, and sometimes (in the case of open pipes) the unblown ends; but during the course of an experiment the positions of the pipes were not altered. In order to prevent a pipe speaking, I placed some cotton-wool over the wind-way, and sometimes inserted a stopper; so that the pitch of the pipe as a resonator was entirely altered. The following are the details of some of the observations:—

I. Sept. 23. Open metal pipes about 2 feet long, one of them provided with an adjustable paper slider for modifying the pitch. Blown ends near one another; unblown ends distant.

Beats per second with harmonium-note.

One pipe alone.	Other pipe alone.	Both pipes together.
4·5, 4·5	5·0, 4·8	3·2, 3·1 ;

so that the note given by both pipes together is decidedly sharper than those of the separate pipes.

II. Sept. 23. Same pipes as in I. Unblown ends near one another; blown ends distant.

Beats per second with harmonium-note.

One pipe alone.	Other pipe alone.	Both pipes together.
4·8	5·1	3·8

III. Sept. 25. Same pipes placed parallel to one another at a distance of about 14 inches.

Beats per second with harmonium-note.

One pipe alone.	Other pipe alone.	Both pipes together.
5·15, 5·20	5·30, 5·45	5·00, 5·15

The note of both pipes together is somewhat higher than the notes of the single pipes.

IV. Sept. 26. Same pipes. Unblown ends near; blown ends distant.

Beats per second.

One pipe alone.	Other pipe alone.	Both pipes together.
5·80, 5·85	7·15, 7·50, 7·45	5·35, 5·50, 5·45

V. Sept. 26. Two bottles, tuned with water to about g, were blown by wind issuing from flattened tubes connected with the bellows by lengths of india-rubber tubing. When the bottles were sufficiently removed from one another, the mutual influence was very small, being insufficient to prevent the formation of slow and pretty steady beats of about four seconds' period. This experiment shows that the mutual influence depends upon the proximity of the open ends of the pipes, and not upon any effects propagated through the supply-pipes leading from a common bellows.

Some further remarks on this subject will be found in a paper read before the Musical Association, Dec. 2, 1878. Reference may also be made to some allied experiments by Gripon*, with which I have only lately become acquainted. They appear scarcely to extend to the case with which I have principally occupied myself, namely that in which both pipes are blown. M. Gripon had, however, anticipated me in the experimental determination of the effect of a flange in modifying the correction for an open end† of a pipe.

Kettledrums.

The theory of the vibrations of uniform and uniformly stretched flexible circular membranes, vibrating *in vacuo*, has been known for many years‡. In practice deviations from such theoretical results are to be expected, if only in consequence of the reaction of the air, which must operate with considerable force on a vibrating body exposing so large a surface in proportion to its mass. In the case of kettledrums, the problem is further complicated by the action of the shell, which limits the motion of the air on one side of the membrane.

From the fact§ that kettledrums are struck, not in the centre, but at a point about midway between the centre and edge, we may infer that the vibrations which it is desired to excite are not of the symmetrical class. I find, indeed, that the sound undergoes little, if any, change when the central point is touched by the finger. Putting therefore the symmetrical vibrations

---

* *Ann. d. Chim.* vol. III. p. 371, 1874.

† *Phil. Mag.* June, 1877. [Art. XLVI. p. 319.]

‡ *Theory of Sound*, ch. IX.

§ De Pontigny. *Proceedings of the Musical Association*, Feb. 1876.

out of account, we have to consider the parts played by vibrations of the following modes :—(1) that with one nodal diameter and no nodal circle; (2) that with two nodal diameters and no nodal circle; (3) that with three nodal diameters and no nodal circle; (4) that with one nodal diameter and one nodal circle, &c. The investigation proved to be of greater difficulty than I had expected, partly in consequence of the short duration of the sounds. Better ears than mine are liable to be puzzled in attempting to analyse compound sounds of such complication and irregularity. The following results, however, are believed to be trustworthy.

The principal tone corresponds to mode (1); the tone corresponding to (2) is about a *fifth* higher; that of mode (3) is about a major seventh above the principal tone; the tone of mode (4) is a little higher again, forming an imperfect octave with the principal tone. For the corresponding modes of a uniform membrane vibrating *in vacuo*, the theoretical intervals are those represented by the ratios 1·34, 1·66, 1·83, or about a fourth, a major sixth, and an interval nearly midway between a major and a minor seventh, respectively.

In experimenting on this subject it is important to bear in mind that the system of tones is really *double*, and that its components coincide only on the supposition of perfect symmetry. In practice the requirement of symmetry is difficult to attain even approximately; and thus it is that beats are generally heard, arising from the superposition of vibrations of nearly equal frequency. For the purpose of identifying the various modes, the want of symmetry is rather advantageous than otherwise. In the case of the gravest mode, I fastened with cement a small load (a halfpenny) to a point of the membrane situated about halfway between its centre and edge. In this way the two gravest tones fell asunder to about a semitone, one of them (the graver) being excited alone by a blow anywhere along the diameter through the load, the other alone by a blow anywhere along the perpendicular diameter. With the aid of a resonator tuned to the pitch of the *subordinate* tone, the nodal diameters of the two modes (1) may be fixed with great precision by the absence of beats. With a resonator tuned to a pitch midway between those of the two tones, the beats are most distinct when the blow is delivered at a point near the middle of one of the four quadrants formed by the two nodal diameters; but the position necessary for the most distinct beats varies with the pitch of the resonator, and also with the situation of the observer. It may be remarked that, provided the deviation from symmetry be moderate, the same vibrations (except as to phase) are excited, whether a blow be delivered at any point, or at the other point on the same diameter equally distant from the centre; and vibrations excited by striking one point are damped by touching the other. The other modes with nodal diameters only were identified in a similar way. The mode (4) with a nodal circle is known by the cessation of sound at a particular point when various places along a radius are tried; on either side of this point the sound revives.

The drum that I examined is of about 25 inches diameter; and the form of the shell is nearly hemispherical. During the experiments the pitch of the principal tone was about 120 vibrations per second. The vibrations were excited by a small wooden hammer, such as is used for harmonicons, the head being covered with cotton-wool tied on with string. For the graver tones the thickness of the cotton-wool may with advantage be greater than for higher tones.

I am not in a position to decide the question as to the function of the shell; but I think it at least doubtful whether it introduces any really advantageous modification into the relations of the component tones. It is possible that its advantage lies rather in obstructing the flow that would otherwise take place round the edge of the membrane. It must be remembered that the sounds due to the various parts of a vibrating membrane interfere greatly. In the case of a membrane simply stretched upon a hoop, and vibrating away from all obstacles, no sound at all would be heard at points in the prolongation of its plane. And even when there is a shell, no sound would be heard at points on the axis of symmetry, at least if the symmetrical vibrations may be left out of account.

The Æolian Harp.

So far as I am aware, it has always been assumed by writers who refer to this subject that the vibrations of the string are in the plane parallel to the direction of the wind; and, indeed, the action of the wind in maintaining the motion is usually explained as the result of friction, and as analogous to the action of a violin-bow. It is more than a year since I made some experiments with the view of testing a suspicion of the incorrectness of this view; and I then arrived at the conclusion that the vibrations are in fact executed in the plane perpendicular to the direction of the wind. I suppose for simplicity that the length of the string is perpendicular to the direction of the wind, as is usually the case in practice. Recently I have repeated these experiments in an improved form, and with confirmatory results.

The best draught is that obtained from a chimney. In my later experiments a fireplace was fitted with a structure of wood and paper, which could prevent all access of air to the chimney, except through an elongated horizontal aperture in the front (vertical) wall. The length of the aperture was 26 inches, and the width 4 inches; and along its middle a gut string was stretched over bridges. The strength of the draught could be regulated by slightly withdrawing the framework from the fireplace, so as to allow the passage of air to the chimney otherwise than through the slit.

A fine point of light was obtained from a fragment of a silvered bead attached to the string with wax, and illuminated by a suitably placed candle, and was observed in the direction of the length of the string through an extemporized telescope. In this way there could be no mistake as to the actual plane of vibration, or uncertainty as to the direction of the wind over the string. The path of the point of light was seen to be nearly rectilinear and *vertical*, showing that the vibration is *across* the wind. Sometimes the path was sensibly elliptic with the major axis vertical.

When a string is stretched across the slit at the bottom of a slightly open window, there is usually some difficulty in determining the actual direction of the wind where it plays upon the string. On a still night, and with a regular fire, the sound is sometimes steady for a long time, but it is wonderfully sensitive to the slightest changes in the draught. On one occasion it was found impossible to open a distant door so slightly as not to stop the sound, which would revive in a few seconds after the door was closed again. A piece of paper no larger than the hand thrown upon the fire (which was burning without flame) altered the draught sufficiently to stop the sound until the heated air due to its combustion had passed up the chimney. It is the irregularity, and not, as has been asserted, the insufficient intensity, of the wind which prevents the satisfactory performance of the harp in the open air.

62.

INVESTIGATIONS IN OPTICS, WITH SPECIAL REFERENCE TO THE SPECTROSCOPE.

[*Phil. Mag.* VIII. pp. 261—274, 403—411, 477—486, 1879; IX. pp. 40—55, 1880.]

§ 1. *Resolving, or Separating, Power of Optical Instruments.*

ACCORDING to the principles of common optics, there is no limit to resolving-power, nor any reason why an object, sufficiently well lighted, should be better seen with a large telescope than with a small one. In order to explain the peculiar advantage of large instruments, it is necessary to discard what may be looked upon as the fundamental principle of common optics, viz. the assumed infinitesimal character of the wave-length of light. It is probably for this reason that the subject of the present section is so little understood outside the circles of practical astronomers and mathematical physicists.

It is a simple consequence of Huyghens's principle, that the direction of a beam of limited width is to a certain extent indefinite. Consider the case of parallel light incident perpendicularly upon an infinite screen, in which is cut a circular aperture. According to the principle, the various points of the aperture may be regarded as secondary sources emitting synchronous vibrations. In the direction of original propagation the secondary vibrations are all in the same phase, and hence the intensity is as great as possible. In other directions the intensity is less; but there will be no sensible discrepancy of phase, and therefore no sensible diminution of intensity, until the obliquity is such that the (greatest) projection of the diameter of the aperture upon the direction in question amounts to a sensible fraction of the wave-length of the light. So long as the extreme difference of phase is less than a quarter of a period, the resultant cannot differ much from the maximum;

and thus there is little to choose between directions making with the principal direction less angles than that expressed in circular measure by dividing the quarter wave-length by the diameter of the aperture. Direct antagonism of phase commences when the projection amounts to half a wave-length. When the projection is twice as great, the phases range over a complete period, and it might be supposed at first sight that the secondary waves would neutralize one another. In consequence, however, of the preponderance of the middle parts of the aperture, complete neutralization does not occur until a higher obliquity is reached.

This indefiniteness of direction is sometimes said to be due to "diffraction" by the edge of the aperture—a mode of expression which I think misleading. From the point of view of the wave-theory, it is not the indefiniteness that requires explanation, but rather the smallness of its amount.

If the circular beam be received upon a perfect lens, an image is formed in the focal plane, in which *directions* are represented by *points*. The image accordingly consists of a central disk of light, surrounded by luminous rings of rapidly diminishing brightness. It was under this form that the problem was originally investigated by Airy[*]. The angular radius θ of the central disk is given by

$$\theta = 1\cdot2197\,\frac{\lambda}{2R}, \quad\dots\dots\dots\dots\dots\dots\dots\dots\dots(1)$$

in which λ represents the wave-length of light, and $2R$ the (diameter of the) aperture.

In estimating theoretically the resolving-power of a telescope on a double star, we have to consider the illumination of the field due to the superposition of the two independent images. If the angular interval between the components of the star were equal to 2θ, the central disks would be just in contact. Under these conditions there can be no doubt that the star would appear to be fairly resolved, since the brightness of the external ring-systems is too small to produce any material confusion, unless indeed the components are of very unequal magnitude.

The diminution of star-disks with increasing aperture was observed by W. Herschel; and in 1823 Fraunhofer formulated the law of inverse proportionality. In investigations extending over a long series of years, the advantage of a large aperture in separating the components of close double stars was fully examined by Dawes[†]. In a few instances it happened that a small companion was obscured by the first bright luminous ring in the image of a powerful neighbour. A diminution of aperture had then the effect of

[*] *Camb. Phil. Trans.* 1834.
[†] *Mem. Astron. Soc.* vol. xxxv.

bringing the smaller star into a more favourable position for detection; but in general the advantage of increased aperture was very apparent even when attended by considerable aberration.

The resolving-power of telescopes was investigated also by Foucault*, who employed a scale of equal bright and dark alternate parts: it was found to be proportional to the aperture and independent of the focal length. In telescopes of the best construction the performance is not sensibly prejudiced by outstanding aberration, and the limit imposed by the finiteness of the waves of light is practically reached. Verdet† has compared Foucault's results with theory, and has drawn the conclusion that the radius of the visible part of the image of a luminous point was nearly equal to the half of the radius of the first dark ring.

Near the margin of the theoretical central disk the illumination is relatively very small, and consequently the observed diameter of a star-disk is sensibly less than that indicated in equation (1), how much less depending in some measure upon the brightness of the star. That bright stars give larger disks than faint stars is well known to practical observers.

With a high power, say 100 for each inch [2·54 cm.] of aperture, the sharpness of an image given by a telescope is *necessarily* deteriorated, the apparent breadth of a point of light being at least $8\frac{1}{2}$ minutes. In this case the effective aperture of the eye is $\frac{1}{100}$ inch. In his paper on the limit of microscopic vision‡, Helmholtz has shown that the aperture of the eye cannot be much contracted without impairing definition—from which it follows that the limit of the resolving-power of telescopes is attained with a very moderate magnification, probably about 20 for each inch in the aperture of the object-glass or mirror.

We have seen that a certain width of beam is necessary to obtain a given resolving-power; but it does not follow that the whole of an available area of aperture ought to be used in order to get the best result. As the obliquity to the principal direction increases, the first antagonism of phase which sets in is between secondary waves issuing from marginal parts of the aperture; and thus the operation of the central parts is to retard the formation of the first dark ring. This unfavourable influence of the central rays upon resolving-power was well known to Herschel, who was in the habit of blocking them off by a cardboard stop. The image due to an annular aperture was calculated by Airy; and his results showed the contraction of the central disk and the augmented brightness of the surrounding rings§. More recently

* *Ann. de l'Observ. de Paris*, t. v. 1858.
† *Leçons d'Optique Physique*, t. i. p. 309.
‡ *Pogg. Ann.* Jubelband, 1874.
§ See also *Astron. Month. Notices*, xxxiii. 1872. [Art. xix.]

this subject has been ably treated by M. Ch. André*, who has especially considered the case in which the diameter of the central stop is half the full aperture How far it would be advantageous to carry the operation of blocking out the central rays would doubtless depend upon the nature of the object under examination. Near the limit of the power of an instrument a variety of stops ought to be tried. Possibly the best rays to block out are those not quite at the centre (see § 2).

The fact that the action of the central rays may be disadvantageous shows that in the case of full aperture the best effect is not necessarily obtained when all the secondary waves arrive in the same phase at the focal point. If by a retardation of half a wave-length the phase of any particular ray is reversed, the result is of the same character as if that ray were stopped. Hence an exactly parabolic figure is not certainly the best for mirrors.

The character of the image of a luminous *line* cannot be immediately deduced from that of a luminous point. It has, however, been investigated by M. André, who finds that the first minimum of illumination occurs at a somewhat lower obliquity than in the case of a point. A double *line* is therefore probably more easily resolvable than a double *point;* but the difference is not great. In the case of a line the minima are not absolute zeros of illumination.

§ 2. *Rectangular Sections.*

The diffraction phenomena presented by beams of rectangular section are simpler in theory than when the section is circular; and they have a practical application in the spectroscope, when the beam is limited by prisms or gratings rather than by the object-glasses of the telescopes.

Supposing, for convenience, that the sides of the rectangle are horizontal and vertical, let the horizontal aperture be a and the vertical aperture be b. As in § 1, there will be no direct antagonism among the phases of the secondary waves issuing in an oblique horizontal direction, until the obliquity is such that the projection of the horizontal aperture a is equal to $\frac{1}{2}\lambda$. At an obliquity twice as great the phases range over a complete period; and, *since all parts of the horizontal aperture have an equal importance*, there is in this direction a complete absence of illumination. In like manner, a zero of illumination occurs in every horizontal direction upon which the projection of a amounts to an exact multiple of λ.

The complete solution of the present problem, applicable to all oblique directions, is given in Airy's *Tracts*, 4th edition, p. 316, and in Verdet's *Leçons*, t. I. p. 265. If the focal length of the lens which receives the beam

* "Étude de la Diffraction dans les Instruments d'Optique," *Ann. de l'École Norm.* v. 1876.

be f, the illumination I^2 at a point in the focal plane whose horizontal and vertical coordinates (measured from the focal point) are ξ, η, is given by

$$I^2 = \frac{a^2 b^2}{\lambda^2 f^2} \frac{\sin^2 (\pi a \xi/\lambda f)}{\pi^2 a^2 \xi^2/\lambda^2 f^2} \frac{\sin^2 (\pi b \eta/\lambda f)}{\pi^2 b^2 \eta^2/\lambda^2 f^2}, \quad \ldots\ldots\ldots\ldots(1)$$

the intensity of the incident light being unity. The image is traversed by *straight* vertical and horizontal lines of darkness, whose equations are respectively

$$\sin (\pi a \xi/\lambda f) = 0, \qquad \sin (\pi b \eta/\lambda f) = 0. \ldots\ldots\ldots\ldots\ldots(2)$$

The calculation of the image due to a luminous line (of uniform intensity) is facilitated in the present case by the fact that the law of distribution of brightness, as one coordinate varies, is independent of the value of the other coordinate. Thus the distribution of brightness in the image of a vertical line is given by

$$\int_{-\infty}^{+\infty} I^2 d\eta = \frac{a^2 b}{\lambda f} \frac{\sin^2 (\pi a \xi/\lambda f)}{\pi^2 a^2 \xi^2/\lambda^2 f^2}, \quad \ldots\ldots\ldots\ldots\ldots(3)$$

the same law as obtains for a luminous point when horizontal directions are alone considered. It follows from (3) that in the spectroscope* *the definition is independent of the vertical aperture.*

In order to obtain a more precise idea of the character of the image of a luminous line, we must study the march of the function $u^{-2} \sin^2 u$. The roots occur when u is any multiple of π, except zero. The maximum value of the function is unity, and occurs when $u = 0$. Other maxima of rapidly diminishing magnitude occur in positions not far removed from those lying midway between the roots. The image thus consists of a central band of half width corresponding to $u = \pi$, accompanied by lateral bands of width π, and of rapidly diminishing brightness. The accompanying Table and diagram (fig. 1) will give a sufficient idea of the distribution of brightness for our purpose.

TABLE I.

u	$u^{-2} \sin^2 u$	u	$u^{-2} \sin^2 u$
0	1·0000	π	·0000
$\frac{1}{6}\pi$	·9119	$\frac{5}{4}\pi$	·0324
$\frac{1}{4}\pi$	·8106	$\frac{4}{3}\pi$	·0427
$\frac{1}{3}\pi$	·6839	$\frac{3}{2}\pi$	·0450
$\frac{1}{2}\pi$	·4053	$\frac{7}{4}\pi$	·0165
$\frac{2}{3}\pi$	·1710	2π	·0000
$\frac{3}{4}\pi$	·0901	$\frac{5}{2}\pi$	·0162
$\frac{5}{6}\pi$	·0365	3π	·0000

* [1899. Where the edges of the prisms are vertical.]

The curve $ABCD$ represents the values of $u^{-2}\sin^2 u$ from $u = 0$ to $u = 3\pi$. The part corresponding to negative values of u is similar, OA being a line of symmetry.

Fig. 1.

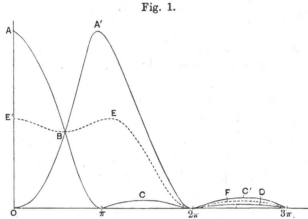

Let us now consider the distribution of brightness in the image of a double line whose components are of equal strength and at such an angular interval that the central line in the image of one coincides with the first zero of brightness in the image of the other. In fig. 1 the curve of brightness for one component is $ABCD$, and for the other $OA'C'$; and the curve representing half the combined brightnesses is $E'BEF$. The brightness (corresponding to B) midway between the two central points A, A' is ·8106 of the brightness at the central points themselves. We may consider this to be about the limit of closeness at which there could be any decided appearance of resolution. The obliquity corresponding to $u = \pi$ is such that the phases of the secondary waves range over a complete period, *i.e.* such that the projection of the horizontal aperture upon this direction is one wave-length. We conclude that *a double line cannot be fairly resolved unless its components subtend an angle exceeding that subtended by the wave-length of light at a distance equal to the horizontal aperture*[*].

This rule is convenient on account of its simplicity; and it is sufficiently accurate in view of the necessary uncertainty as to what exactly is meant by resolution. Perhaps in practice somewhat more favourable conditions are necessary to secure a resolution that would be thought satisfactory.

If the angular interval between the components of the double line be half as great again as that supposed above, the brightness in the middle is ·1802, viz. $(2 \times ·0901)$, as against 1·0450, viz. $(1 + ·0450)$, at the central line. Such a falling off in the middle must be more than sufficient for resolution. If the angle subtended by the components of the double line be twice that

---

[*] In the spectroscope the angular width of the slit should not exceed a moderate fraction of the angle defined in the text, if full resolving-power be wanted.

subtended by the wave-length at a distance equal to the horizontal aperture, the central bands are just clear of one another, and there is a line of absolute blackness in the middle of the combined images.

On the supposition that a certain horizontal aperture is available, a question (similar to that considered in § 1) arises, as to whether the whole of it ought to be used in order to obtain the highest possible resolving-power. From fig. 1 we see that our object must be to depress the curve $ABCD$ at the point B. Now the phase of the resultant is that of the waves coming from the centre; and at the obliquity corresponding to B the phases of the secondary waves range over half a period. It is not difficult to see that the removal of some of the central waves will depress the intensity-curve at B, not only absolutely, but relatively to the depression produced at A. In order to illustrate this question, I have calculated the illumination in the various directions on the supposition that one-sixth of the horizontal aperture is blocked off by a central screen. In this case the amplitude is represented by the function f, where

$$f = u^{-1} \{\sin u - \sin (\tfrac{1}{6}u)\}, \dots\dots\dots\dots\dots\dots\dots(4)$$

and, as usual, the intensity is represented by f^2.

TABLE II.

u	f	$f^2 \div f_0^2$	u	f	$f^2 \div f_0^2$
0	$+ \cdot 8333$	$1 \cdot 0000$	$\tfrac{7}{4}\pi$	$- \cdot 2729$	$\cdot 1072$
$\tfrac{1}{4}\pi$	$\cdot 7342$	$\cdot 7763$	2π	$\cdot 1378$	$\cdot 0274$
$\tfrac{1}{2}\pi$	$\cdot 4717$	$\cdot 3205$	$\tfrac{9}{4}\pi$	$- \cdot 0307$	$\cdot 0014$
$\tfrac{3}{4}\pi$	$+ \cdot 1377$	$\cdot 0273$	$\tfrac{12}{5}\pi$	$\cdot 0000$	$\cdot 0000$
$\tfrac{6}{7}\pi$	$\cdot 0000$	$\cdot 0000$	$\tfrac{5}{2}\pi$	$+ \cdot 0043$	$\cdot 0000$
π	$- \cdot 1592$	$\cdot 0365$	$\tfrac{18}{7}\pi$	$\cdot 0000$	$\cdot 0000$
$\tfrac{5}{4}\pi$	$\cdot 3351$	$\cdot 1617$	$\tfrac{11}{4}\pi$	$- \cdot 0329$	$\cdot 0016$
$\tfrac{3}{2}\pi$	$\cdot 3622$	$\cdot 1889$	3π	$\cdot 1061$	$\cdot 0162$

The third and sixth columns show the intensity in various directions relatively to the intensity in the principal direction ($u = 0$); and the curve $ABCD$ (fig. 2) exhibits the same results to the eye. A comparison with Table I. shows that a considerable advantage has been gained, the relative illumination at B being reduced from $\cdot 4053$ to $\cdot 3205$. On the other hand, the augmented brightness of the first lateral band (towards C) may be unfavourable to good definition. The second bright lateral band (towards D) is nearly obliterated. The curve $E'BEF$ represents the resultant illumination due to a double line whose components are of the same strength, and at the same angular interval as before. The relatively much more decided drop at B indicates a considerable improvement in resolving-power, at least on a double line of this degree of closeness.

Fig. 2.

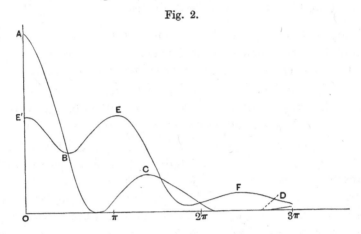

The increased importance of the first lateral band is a necessary consequence of the stoppage of the central rays; for in this direction the resultant has a phase *opposite* to that of the rays stopped. The defect may be avoided in great measure by blocking out rays somewhat removed from the centre on the two sides, and allowing the central rays themselves to pass. As an example, I have taken the case in which the two parts stopped have each a width of one-eighth of the whole aperture, with centres situated at the points of trisection (fig. 3).

Fig. 3.

The function f suitable to this case is readily proved to be

$$f = u^{-1} \{\sin u - 2 \sin (\tfrac{1}{8}u) \cos (\tfrac{1}{3}u)\}. \quad \ldots\ldots\ldots\ldots\ldots(5)$$

The values of f and $f^2 \div f_0^2$ are given in Table III.; and the intensity-curve $ABCD$ is shown in fig. 4.

TABLE III.

u	f	$f^2 \div f_0^2$	u	f	$f^2 \div f_0^2$
0	+ ·75	1·0000	$\frac{3}{2}\pi$	− ·2122	·0801
$\frac{1}{4}\pi$	·6594	·7727	$\frac{7}{4}\pi$	− ·0689	·0084
$\frac{1}{2}\pi$	·4215	·3158	2π	+ ·1125	·0225
$\frac{3}{4}\pi$	+ ·1259	·0282	$\frac{5}{2}\pi$	·2189	·0852
π	− ·1218	·0264	3π	·1960	·0684
$\frac{5}{4}\pi$	·2422	·1043			

The depression at *B* is even greater than in fig. 2, while the rise at *C* is much less. Probably this arrangement is about as efficient as any.

Fig. 4.

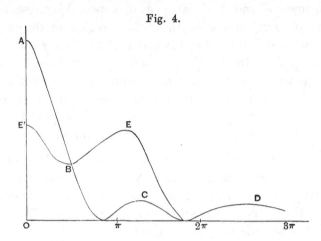

I have endeavoured to test these conclusions experimentally with the spectroscope, using the double soda-line. The horizontal aperture of a single prism instrument was narrowed by gradually advancing cardboard screens until there was scarcely any appearance of resolution. The interior rays were blocked out with vertical wires or needles, adjusted until they occupied the desired positions when seen through the telescope with eye-piece removed. With the arrangements either of fig. 2 or of fig. 4 a very decided improvement on the full aperture was observed; but there was no distinct difference between these two arrangements themselves. Indeed, no such difference was to be expected, since the brightness of the first lateral band has no bad effect on the combined images, as appears from the curve *E'BEF* (fig. 2). Under other circumstances the influence of the bright lateral band might be more unfavourable.

In powerful spectroscopes the beam is often rendered unsymmetrical in brightness by absorption. In such cases an improvement would probably be effected by stopping some of the rays on the preponderating side, for which purpose a sloping screen might be used giving a variable *vertical* aperture. It should be noticed, however, that it is only when the vertical aperture is constant that the image of a luminous line is immediately deducible from that of a luminous point.

§ 3. *Optical Power of Spectroscopes.*

As the power of a telescope is measured by the closeness of the double stars which it can resolve, so the power of a spectroscope ought to be measured by the closeness of the closest double lines in the spectrum which it is competent to resolve. In this sense it is possible for one instrument to be more powerful than a second in one part of the spectrum, while in another part of the spectrum the second instrument is more powerful than the first. The most striking cases of this inversion occur when one instrument is a diffraction-spectroscope and the other a dispersion-spectroscope. If the instruments are of equal power in the yellow region of the spectrum, the former will be the more powerful in the red, and the latter will be the more powerful in the green. In the present section I suppose the material and the workmanship to be perfect, and omit from consideration the effects of unsymmetrical absorption. Loss of light by reflection or by uniform absorption has no effect on resolving-power. Afterwards I propose to examine the effect of some of the errors most likely to occur in practice.

So far as relates to the diffraction-spectroscope, the problem of the present section was solved in the *Philosophical Magazine* for March, 1874 [Art. xxx. p. 216]. I there showed that if n denote the number of lines on a grating and m the order of the spectrum observed, a double line of wave-lengths λ and $\lambda + \delta\lambda$ will be just resolved (according to the standard of resolution defined in the previous section), provided

$$\frac{\delta\lambda}{\lambda} = \frac{1}{mn} , \quad\dots\dots\dots\dots\dots\dots\dots\dots\dots\dots\dots(1)$$

which shows that the resolving-power varies directly as m and n. When the ruling is very close, m is always small (not exceeding 3 or 4); and even when a considerable number of spectra are formed, the use of an order higher than the third or fourth is often inconvenient in consequence of the overlapping. But if the difficulty of ruling a grating may be measured by the total number of lines (n), it would seem that the intervals ought not to be so small as to preclude the convenient use of at least the third and fourth spectra.

In the case of the soda double line the difference of wave-lengths is a very little more than $\frac{1}{1000}$; so that, according to (1), about 1000 lines are necessary for resolution in the first spectrum. By experiment I found 1130*.

"Since a grating resolves in proportion to the total number of its grooves, it might be supposed that the defining-power depends on different principles in the case of gratings and prisms; but the distinction is not fundamental. The limit to definition arises in both cases from the impossibility of representing a line of light otherwise than by a band of finite though narrow width, the width in both cases depending on the horizontal aperture (for a given λ). If a grating and a prism have the same horizontal aperture and dispersion, they will have equal resolving-powers on the spectrum."

---

* In my former paper this number is given as 1200. On reference to my notebook, I find that I then took the full width of the grating as an English inch. The 3000 lines cover a *Paris* inch, whence the above correction. From the nature of the case, however, the experiment does not admit of much accuracy.

At the time the above paragraph was written, I was under the impression that the dispersion in a prismatic instrument depended on so many variable elements that no simple theory of its resolving-power was to be expected. Last autumn, while engaged upon some experiments with prisms, I was much struck with the inferiority of their spectra in comparison with those which I was in the habit of obtaining from gratings, and was led to calculate the resolving-power. I then found that the theory of the resolving-power of prisms is almost as simple as that of gratings.

Let $A_0 B_0$ (fig. 5) be a plane wave-surface of the light before it falls upon the prisms, AB the corresponding wave-surface for a particular part of the spectrum after the light has passed the prism or after it has passed the eye-piece of the observing-telescope. The path of a ray from the wave-surface $A_0 B_0$ to A or B is determined by the condition that the optical distance, represented by $\int \mu ds$, is a minimum; and as

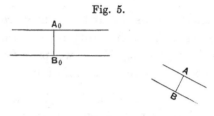

Fig. 5.

AB is by supposition a wave-surface, this optical distance is the same for both points. Thus

$$\int \mu ds \ (\text{for } A) = \int \mu ds \ (\text{for } B). \quad \dots \dots \dots \dots \dots \dots (2)$$

We have now to consider the behaviour of light belonging to a neighbouring part of the spectrum. The path of a ray from the wave-surface $A_0 B_0$ to A is changed; but in virtue of the minimum property the change may be neglected in calculating the optical distance, as it influences the result by quantities of the second order only in the change of refrangibility. Accordingly the optical distance from $A_0 B_0$ to A is represented by $\int (\mu + \delta \mu) \, ds$, the integration being along the path $A_0 \dots A$; and, similarly, the optical distance between $A_0 B_0$ and B is represented by $\int (\mu + \delta \mu) \, ds$, where the integration is along the path $B_0 \dots B$. In virtue of (2) the *difference* of the optical distances is

$$\int \delta \mu ds \ (\text{along } B_0 \dots B) - \int \delta \mu ds \ (\text{along } A_0 \dots A). \quad \dots \dots \dots (3)$$

The new wave-surface is formed in such a position that the optical distance is constant; and therefore the *dispersion*, or the angle through which the wave-surface is turned by the change in refrangibility, is found simply by dividing (3) by the distance AB. If, as in common flint-glass spectroscopes, there is only one dispersing substance, $\int \delta \mu ds = \delta \mu . s$, where s is simply the thickness traversed by the ray. If we call the width of the emergent beam a, the dispersion is represented by $\delta \mu (s_2 - s_1)/a$, s_1 and s_2 being the thicknesses traversed by the extreme rays. In a properly constructed instrument s_1 is negligible, and s_2 is the aggregate thickness of the prisms at their thick ends, which we will call t; so that the dispersion (θ) is given by

$$\theta = \frac{t\delta\mu}{a} . \quad\dots\dots\dots\dots\dots\dots\dots\dots\dots\dots(4)$$

By § 2 the condition of resolution of a double line whose components subtend an angle θ is that θ must exceed λ/a. Hence from (4), in order that a double line may be resolved whose components have indices μ and $\mu + \delta\mu$, it is necessary that t should exceed the value given by the following equation,

$$t = \frac{\lambda}{\delta\mu}, \quad\dots\dots\dots\dots\dots\dots\dots\dots\dots\dots(5)$$

which expresses that the relative retardation of the extreme rays due to the change of refrangibility is the same (λ) as that incurred without a change of refrangibility when we pass from the principal direction to that corresponding to the first minimum of illumination.

That the resolving-power of a prismatic spectroscope of given dispersive material is proportional to the total thickness used, without regard to the number, angles, or setting of the prisms, is a most important, perhaps the most important, proposition in connexion with this subject. Hitherto in descriptions of spectroscopes far too much stress has been laid upon the amount of dispersion produced by the prisms; but this element by itself tells nothing as to the power of an instrument. It is well known that by a sufficiently close approach to a grazing emergence the dispersion of a prism of given thickness may be increased without limit; but there is no corresponding gain in resolving-power. So far as resolving-power is concerned, it is a matter of indifference whether dispersion be effected by the prisms or by the telescope. Two things only are necessary:—first, to use a thickness exceeding that prescribed by (5); secondly, to narrow the beam until it can be received by the pupil of the eye, or rather, since with full aperture the eye is not a perfect instrument, until its width is not more than one-third or one-fourth of the diameter of the pupil.

The value of expression (3) on which resolving-power depends is readily calculable in all cases of practical interest. For a compound prism of flint and crown, $\delta\mu . t$ is replaced by

$$\delta\mu . t - \delta\mu' . t', \quad\dots\dots\dots\dots\dots\dots\dots\dots\dots\dots(6)$$

where t and t' denote the respective thicknesses traversed, and $\delta\mu$, $\delta\mu'$ the corresponding variations of refractive index.

The relation between $\delta\mu$ and $\delta\lambda$ may generally be obtained with sufficient approximation from Cauchy's formula

$$\mu = A + B\lambda^{-2}. \quad\dots\dots\dots\dots\dots\dots\dots\dots\dots(7)$$

Thus $$\delta\mu = -2B\lambda^{-3}\delta\lambda. \quad\quad\quad\quad (8)$$

The value of B varies of course according to the material of the prisms. As an example I will take Chance's "extra-dense flint." The indices for C and the more refrangible D are*

$$\mu_D = 1\cdot650388, \quad\quad\quad \mu_C = 1\cdot644866;$$

so that

$$\mu_D - \mu_C = \cdot005522. \quad\quad\quad\quad (9)$$

Also

$$\lambda_D = 5\cdot889 \times 10^{-5}, \quad\quad\quad \lambda_C = 6\cdot562 \times 10^{-5},$$

the unit of length being the centimetre; whence by (7),

$$B = \cdot984 \times 10^{-10}. \quad\quad\quad\quad (10)$$

Thus by (5) and (8),

$$t = \frac{\lambda^4}{2B\delta\lambda} = \frac{10^{10}\lambda^4}{1\cdot968\,\delta\lambda}. \quad\quad\quad\quad (11)$$

For the soda-line,

$$\lambda = 5\cdot889 \times 10^{-5}, \quad\quad\quad \delta\lambda = \cdot006 \times 10^{-5};$$

and thus the thickness necessary to resolve this line is given by

$$t = 1\cdot02 \text{ centimetres.} \quad\quad\quad\quad (12)$$

The number of times the power of a spectroscope exceeds that necessary to resolve the soda-lines might conveniently be taken as its practical measure. We learn from (12) that, according to this definition, the power of an instrument with simple prisms of "extra-dense glass" is expressed approximately by the number of centimetres of available thickness.

In order to confirm this theory, I have made some observations on the thickness necessary to resolve the soda-lines. The prism was of extra-dense glass of refractive index very nearly agreeing with that above specified, and had a refracting angle of 60°. Along one face sliding screens of cardboard were adapted, by which the horizontal aperture could be adjusted until, in the judgment of the observer, the line was barely resolved. A soda-flame was generally used, though similar observations have been made upon the D lines of the solar spectrum. When the adjustment was complete, the aperture along the face of the prism was measured, and gave at once the equivalent thickness, *i.e.* the *difference* of thicknesses traversed by the extreme rays, since the prism was in the position of minimum deviation. Care, of course, was taken that no ordinary optical imperfections of the apparatus interfered with the experiment.

One observer, familiar with astronomical work, fixed the point of resolution when the thickness amounted to from 1·00 to 1·20 centimetre.

* Hopkinson, *Proc. Roy. Soc.* June 1877.

I was myself less easily satisfied, requiring from 1·35 to 1·40 centimetre. But even with a less thickness than 1 centimetre, it was evident that the object under examination was not a single line. With the same prism I found the thickness necessary to resolve $b_3 b_4$ in the solar spectrum to be about 2·5 centimetres. According to (11), the thickness required for $b_3 b_4$ should be 2·2 times that required for $D_1 D_2$. Probably something depends upon the relative intensities of the component lines.

From (1) and (11) we see that if a diffraction and a dispersion instrument have equal resolving-powers,

$$t = \frac{mn\lambda^3}{2B}; \dots\dots\dots\dots\dots\dots(13)$$

so that the power of a dispersion instrument relatively to that of a diffraction instrument varies inversely as the third power of the wave-length.

For the kind of glass considered in (10), and for the region of the D lines,

$$t = 1\cdot037\, \frac{mn}{1000}. \dots\dots\dots\dots\dots(14)$$

To find what thickness is necessary to rival the fourth spectrum of a grating of 3000 lines, we have merely to put $m = 4$, $n = 3000$; so that the necessary thickness is about $12\frac{1}{2}$ centimetres—a result which abundantly explains the observations which led me to calculate the power of prisms.

§ 4. *Influence of Aberration.*

In the investigations of § 2 the wave-surface was considered to be plane, or (after passing through a condensing lens) spherical. As all optical instruments are liable to aberration, it is important to inquire what effects are produced thereby upon the intensity-curves, and especially to ascertain at what point a sensible deterioration of definition ensues. The only work bearing upon the present subject with which I am acquainted is Sir G. Airy's investigation " of the intensity of light in the neighbourhood of a caustic * ";; but the problem considered by him relates to an *unlimited* beam.

Considering in the first place the case of a beam of rectangular section, let us suppose that the aberration, or error of phase, is the same in all vertical lines, so that the actual wave-surface is cylindrical. With origin at the centre and axis of x horizontal, the aberration may be expressed in the form

$$cx^3 + fx^4 + \dots\dots \dots\dots\dots\dots\dots(1)$$

---

* *Camb. Phil. Trans.* vol. VI. 1838.

No terms appear in x or x^2: the first would be equivalent to a general turning of the beam; and the second would imply imperfect focussing of the central parts. In many cases the circumstances are symmetrical with respect to the centre; and then the first term which occurs is that containing x^4. But in general, since the whole error of linear retardation which we shall contemplate is exceedingly small in comparison with other linear magnitudes concerned in the problem, the term in x^3 is by far the more important, and those that follow may be neglected.

As in the case of no aberration (treated in § 2), the distribution of brightness in the image of a point is similar along every vertical line in the focal plane; and therefore the image of a vertical line follows the same law of brightness as applies in the case of a point to positions situated along the axis of ξ. The phase of the resultant at any point ξ is by symmetry the same as that of the secondary wave issuing from the centre $(x = 0)$; and thus the amplitude of the resultant is proportional to

$$\int_0^{+\frac{1}{2}a} \cos 2\pi \left(\frac{x\xi}{\lambda f} + cx^3\right) dx. \quad\ldots\ldots\ldots\ldots\ldots\ldots\ldots(2)$$

In Sir G. Airy's problem the upper limit of the integral (2) is infinite. Fortunately for my purpose the method of calculation employed by him is that of quadratures, and the intermediate results are recorded (p. 402) in sufficient detail. In order to bring (2) into conformity with Airy's notation, we must take

$$2\pi cx^3 = \tfrac{1}{2}\pi\omega^3, \qquad 2\pi x\xi/\lambda f = -\tfrac{1}{2}m\pi\omega; \quad\ldots\ldots\ldots\ldots(3)$$

we thus obtain

$$(4c)^{-\frac{1}{3}} \int_0^{\frac{1}{2}a \cdot \sqrt[3]{4c}} \cos \tfrac{1}{2}\pi (\omega^3 - m\omega)\, d\omega, \quad\ldots\ldots\ldots\ldots\ldots(4)$$

in which the upper limit of the integral is the cube root of the extreme aberration expressed in quarter-periods. For example, the upper limit is unity when the phase at one extremity is a quarter-period in advance, and that at the other extremity a quarter-period in the rear, of the phase at the centre.

The influence of aberration may be considered in two ways. We may suppose the aperture (a) constant, and inquire into the effect of an increasing aberration (c); or we may take a given value of c (i.e. a given wave-surface), and examine the effect of a varying aperture. To the latter comparison Airy's results are more immediately applicable. The following Table, easily derived from that given by him, exhibits the values of $\int \cos \tfrac{1}{2}\pi (\omega^3 - m\omega)\, d\omega$, between the limits specified in the headings of columns 2, 3, 4, 5. The results are applicable at once to the comparison of the amplitude-curves corresponding to various apertures, since the relation of m to ξ in (3) is independent of a. To obtain intensities, it would be necessary to square the numbers given in the Table.

Value of m	From 0 to 1·00	From 0 to 1·26	From 0 to 1·44	From 0 to ∞
− 4·0	+ ·0929	− ·0692	+ ·0588	+ ·0030
− 3·6	·0783	− ·0467	+ ·0197	·0062
− 3·2	+ ·0343	+ ·0142	− ·0309	·0124
− 2·8	− ·0203	·0849	− ·0461	·0239
− 2·4	·0563	·1320	+ ·0018	·0444
− 2·0	− ·0430	·1399	·1009	·0791
− 1·6	+ ·0411	·1263	·2095	·1346
− 1·2	·1997	·1377	·2906	·2184
− 0·8	·4140	·2266	·3462	·3362
− 0·4	·6449	·4185	·4211	·4886
0·0	·8422	·6873	·5672	·6653
+ 0·4	·9570	·9538	·7898	·8404
+ 0·8	·9559	1·1120	1·0157	·9701
+ 1·2	·8307	1·0748	1·1141	·9979
+ 1·6	·6024	·8170	·9681	·8705
+ 2·0	·3161	+ ·3952	+ ·5569	·5649
+ 2·4	+ ·0303	− ·0679	− ·0060	+ ·1172
+ 2·8	− ·1988	·4290	·5110	− ·3624
+ 3·2	·3309	·5826	·7545	·7087
+ 3·6	·3521	·5028	·6485	·7652
+ 4·0	− ·2761	− ·2525	− ·2725	− ·4745

The second column relates to the case where the aperture is such that the aberration between the extremities and the centre is one quarter of a period, or (which is the same thing) where the wave-surface at the extremities deviates by a quarter wave-length from the tangent plane at the central line of inflection. It will be seen that the position of maximum illumination deviates sensibly from the centre ($m = 0$, $\xi = 0$). This is no more than might have been expected, since the plane which most nearly coincides with the actual wave-surface is inclined to the central tangent plane. The third column relates to an aperture about a fourth part larger, for which the extreme aberration is $(1·26)^3$ quarter-periods or nearly *half* a period, and the fourth column relates to an aperture such that the extreme aberration amounts to about three-quarters of a period.

From columns 2 and 3 we see that an increase of aperture up to that corresponding to an extreme aberration of half a period has no ill effect upon the central band, but it increases unduly the intensity of the first lateral band at $m = + 3·2$. Indeed the principal objection to much greater apertures is this augmented importance of the lateral band. The practical conclusion is that the best results will be obtained with an aperture giving an extreme aberration of from a quarter to half a period, and that with a [still further] increased aperture aberration is not so much a direct cause of deterioration as an obstacle to the attainment of that improved definition which should accompany the increase of aperture.

We will now suppose the aperture given, and examine the effect of increasing aberration. In applying the tabular results we must have regard to the factor $(4c)^{-\frac{1}{3}}$, which occurs in (4), and we must take account of the variation of the relative scale of m and ξ in passing from one curve to another $(\xi \propto m \sqrt[3]{4c})$. The results for three cases are expressed graphically by the curves in fig. 6. The first, which rises highest, represents the *intensity* at the various points of the focal plane when there is no aberration—the same as in fig. 1. The second and third curves represent

Fig. 6.

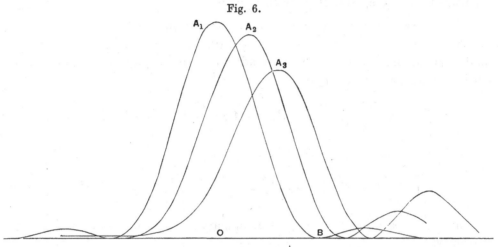

the intensities when the extreme aberrations are a quarter-period and half a period respectively. The total areas of these curves are the same, since the whole quantity of light in the beam is independent of the aberration; and this area is equal to that of a rectangle whose height is the maximum ordinate OA of the first curve, and width the distance OB between O and the first position of zero intensity. It appears that aberration begins to be distinctly mischievous when it amounts to about a quarter-period, *i.e.* when the wave-surface deviates at each end by a quarter wave-length from the true plane. The most marked effect is the increased importance of the lateral band on one side, and the approximate obliteration of the lateral band on the other side.

When the aberration is symmetrical about the centre of the beam, the term in x^3 vanishes, and the whole effect is of higher order. In general the term in x^4 will preponderate; and thus the problem for a symmetrical beam resolves itself into the investigation of aberration varying as x^4. In one respect the problem is simpler than the preceding, on account of the symmetry of the intensity-curves; but in another it is more complicated, since the phase of the resultant does not correspond with that of the central element. The intensity is represented by

$$\left[\int_{-\frac{1}{2}a}^{+\frac{1}{2}a} \cos 2\pi \left(\frac{x\xi}{\lambda f} + f x^4 \right) dx \right]^2 + \left[\int_{-\frac{1}{2}a}^{+\frac{1}{2}a} \sin 2\pi \left(\frac{x\xi}{\lambda f} + f x^4 \right) dx \right]^2, \quad .....(5)$$

and it requires for its calculation two integrations. These could be effected by quadratures; but the results would perhaps scarcely repay the labour, especially as the practical question differs somewhat from that here proposed. The intensity-curve derived from (5) represents the actual state of things on the supposition that the focussing adopted is that proper to a very small aperture; whereas in practice the aberration would be in some degree compensated for by a change of focus, as it is obvious that the real wave-surface, being curved only in one direction, could be more accurately identified with a sphere than with a plane.

Some idea of the effect of aberration may be obtained from a calculation of the intensity at the central point ($\xi = 0$), where it reaches a maximum; and this can be effected without quadratures by the aid of a series. In this case we have instead of (5),

$$4 \left[\int_0^{\frac{1}{2}a} \cos\left(2\pi f x^4\right) dx \right]^2 + 4 \left[\int_0^{\frac{1}{2}a} \sin\left(2\pi f x^4\right) dx \right]^2. \quad \ldots\ldots\ldots(6)$$

Now by integration by parts it can be proved that

$$\int_0^x e^{ihx^4} dx = e^{ihx^4} \left\{ x - 4ih\frac{x^5}{5} + \frac{(4ih)^2}{5}\frac{x^9}{9} - \frac{(4ih)^3}{5.9}\frac{x^{13}}{13} + \ldots \right\};$$

whence by separation of real and imaginary parts, and putting x equal to unity,

$$\int_0^1 \cos\left(hx^4\right) dx = \cos h \left\{ 1 - \frac{(4h)^2}{5.9} + \frac{(4h)^4}{5.9.13.17} - \ldots \right\}$$

$$+ \sin h \left\{ \frac{4h}{5} - \frac{(4h)^3}{5.9.13} + \frac{(4h)^5}{5.9.13.17.21} - \ldots \right\}, \ldots\ldots(7)$$

$$\int_0^1 \sin\left(hx^4\right) dx = \sin h \left\{ 1 - \frac{(4h)^2}{5.9} + \frac{(4h)^4}{5.9.13.17} - \ldots \right\}$$

$$- \cos h \left\{ \frac{4h}{5} - \frac{(4h)^3}{5.9.13} + \frac{(4h)^5}{5.9.13.17.21} - \ldots \right\}. \ldots\ldots(8)$$

Calculating from these series I find

$$\int_0^1 \cos\left(\tfrac{1}{4}\pi x^4\right) dx = \frac{1\cdot36704}{\sqrt{2}}, \qquad \int_0^1 \sin\left(\tfrac{1}{4}\pi x^4\right) dx = \frac{\cdot21352}{\sqrt{2}},$$

$$\left[\int_0^1 \cos\left(\tfrac{1}{4}\pi x^4\right) dx \right]^2 + \left[\int_0^1 \sin\left(\tfrac{1}{4}\pi x^4\right) dx \right]^2 = \cdot9576.$$

Again,

$$\int_0^1 \cos\left(\tfrac{1}{2}\pi x^4\right) dx = \cdot87704, \qquad \int_0^1 \sin\left(\tfrac{1}{2}\pi x^4\right) dx = \cdot26812,$$

$$\left[\int_0^1 \cos\left(\tfrac{1}{2}\pi x^4\right) dx \right]^2 + \left[\int_0^1 \sin\left(\tfrac{1}{2}\pi x^4\right) dx \right]^2 = \cdot84109.$$

Again,

$$\int_0^1 \cos(\pi x^4)\, dx = \cdot 64357, \qquad\qquad \int_0^1 \sin(\pi x^4)\, dx = \cdot 33363,$$

$$\left[\int_0^1 \cos(\pi x^4)\, dx\right]^2 + \left[\int_0^1 \sin(\pi x^4)\, dx\right]^2 = \cdot 52549.$$

Thus an extreme aberration of one-eighth of a period reduces the intensity at the central point from unity, corresponding to no aberration, to ·9576. With an aberration of one quarter of a period the intensity is ·84109; and with an aberration of half a period the intensity is reduced to ·52549. We must remember, however, that these numbers will be sensibly raised if a readjustment of focus be admitted.

In most optical instruments other than spectroscopes the section of the beam is circular, and there is symmetry about an axis. The calculation of the intensity-curves as affected by aberration could be performed by quadratures from tables of Bessel's functions; but, as in the case last considered, the results are liable to a modification in practice from re-adjustment of focus. For the central point we may obtain what we require from a series.

The intensity may be represented by

$$\left[2\int_0^1 \cos(hr^4)\, r\, dr\right]^2 + \left[2\int_0^1 \sin(hr^4)\, r\, dr\right]^2,$$

the scale being such that the intensity is unity in the case of no aberration ($h = 0$). As before, we find

$$2\int_0^1 e^{ihr^4} r\, dr = e^{ih}\left\{1 - \frac{4ih}{6} + \frac{(4ih)^2}{6 \cdot 10} - \frac{(4ih)^3}{6 \cdot 10 \cdot 14} + \ldots\right\},$$

whence

$$2\int_0^1 \cos(hr^4)\, r\, dr = \cos h\left\{1 - \frac{(4h)^2}{6 \cdot 10} + \frac{(4h)^4}{6 \cdot 10 \cdot 14 \cdot 18} - \ldots\right\}$$

$$+ \sin h\left\{\frac{4h}{6} - \frac{(4h)^3}{6 \cdot 10 \cdot 14} + \ldots\right\}, \ldots\ldots\ldots\ldots\ldots(9)$$

$$2\int_0^1 \sin(hr^4)\, r\, dr = \sin h\left\{1 - \frac{(4h)^2}{6 \cdot 10} + \frac{(4h)^4}{6 \cdot 10 \cdot 14 \cdot 18} - \ldots\right\}$$

$$- \cos h\left\{\frac{4h}{6} - \frac{(4h)^3}{6 \cdot 10 \cdot 14} + \ldots\right\}. \ldots\ldots\ldots\ldots\ldots(10)$$

Thus, when $h = \frac{1}{4}\pi$,

$$2\int_0^1 \cos(\tfrac{1}{4}\pi r^4)\, r\, dr = \frac{1 \cdot 32945}{\sqrt{2}}, \qquad 2\int_0^1 \sin(\tfrac{1}{4}\pi r^4)\, r\, dr = \frac{\cdot 35424}{\sqrt{2}},$$

$$\left[2\int_0^1 \cos(\tfrac{1}{4}\pi r^4)\, r\, dr\right]^2 + \left[2\int_0^1 \sin(\tfrac{1}{4}\pi r^4)\, r\, dr\right]^2 = \cdot 9464.$$

Again, when $h = \frac{1}{2}\pi$,

$$2\int_0^1 \cos\left(\tfrac{1}{2}\pi r^4\right) r\,dr = \cdot 77989, \qquad 2\int_0^1 \sin\left(\tfrac{1}{2}\pi r^4\right) r\,dr = \cdot 43828,$$

$$\left[2\int_0^1 \cos\left(\tfrac{1}{2}\pi r^4\right) r\,dr\right]^2 + \left[2\int_0^1 \sin\left(\tfrac{1}{2}\pi r^4\right) r\,dr\right]^2 = \cdot 8003.$$

Again, when $h = \pi$,

$$2\int_0^1 \cos\left(\pi r^4\right) r\,dr = \cdot 3740, \qquad 2\int_0^1 \sin\left(\pi r^4\right) r\,dr = \cdot 5048,$$

$$\left[2\int_0^1 \cos\left(\pi r^4\right) r\,dr\right]^2 + \left[2\int_0^1 \sin\left(\pi r^4\right) r\,dr\right]^2 = \cdot 3947.$$

Hence in this case, as in the preceding, we may consider that aberration begins to be decidedly prejudicial when the wave-surface deviates from its proper place by about a quarter of a wave-length.

As an application of this result, let us investigate what amount of temperature disturbance in the tube of a telescope may be expected to impair definition. According to the experiments of Biot and Arago, the refractive index μ for air at temperature t° C. and at atmospheric pressure is given by

$$\mu - 1 = \frac{\cdot 00029}{1 + \cdot 0037 t}.$$

If we take the freezing-point as standard temperature,

$$\delta\mu = -1\cdot 1 t \times 10^{-6}. \quad\dots\dots\dots\dots\dots\dots\dots(11)$$

Thus, supposing that the irregularity of temperature t extends through a length l, and produces a retardation of a quarter of a wave-length,

$$\tfrac{1}{4}\lambda = 1\cdot 1 lt \times 10^{-6};$$

or, if we take $\lambda = 5\cdot 3 \times 10^{-5}$,

$$lt = 12, \quad\dots\dots\dots\dots\dots\dots\dots\dots\dots\dots(12)$$

the unit of length being the centimetre.

We may infer that, in the case of a telescope-tube 12 centimetres long, a stratum of air heated one degree Cent., lying along the top of the tube and occupying a moderate fraction of the whole volume, would produce a not insensible effect. If the change of temperature progressed uniformly from one side of the tube to the other, the result would be a lateral displacement of the image without loss of definition; but in general both effects would be observable. In longer tubes a similar disturbance would be caused by a proportionally less difference of temperature.

In the ordinary investigations of the aberration of optical instruments attention is usually given to a quantity called the *longitudinal* aberration, which is the distance between the geometrical focus and the point at which the extreme ray meets the axis. In order to adapt these calculations to our purpose, it is necessary to establish the connexion between longitudinal aberration and the deviation of the actual surface of the converging waves from a truly spherical surface having its centre at the geometrical focus.

If the axis of symmetry be taken as that of z, and the tangent plane to the wave-surface as plane of xy, we have as the equation of the ideal wave-surface,

$$(z - f)^2 + x^2 + y^2 = f^2,$$

f being the distance of the focus from the origin; or if we limit our attention to the plane $y = 0$,

$$z = f - \sqrt{(f^2 - x^2)} = \tfrac{1}{2}\frac{x^2}{f} + \tfrac{1}{8}\frac{x^4}{f^3}. \quad\quad\quad\quad (13)$$

The actual wave-surface, having at the origin the same curvature, is represented by

$$z' = \tfrac{1}{2}\frac{x^2}{f} + \kappa\frac{x^4}{f^3}, \quad\quad\quad\quad\quad\quad (14)$$

where κ is a constant depending upon the amount of aberration. The distance (h) between the surfaces is given by

$$h = z - z' = (\tfrac{1}{8} - \kappa)\frac{x^4}{f^3}. \quad\quad\quad\quad\quad (15)$$

The equation to the normal to (14) at the point (z', x) is

$$\frac{\zeta - z'}{-1} = \frac{\xi - x}{x/f + 4\kappa x^3/f^3};$$

so that when $\xi = 0$,

$$\zeta = z' + \frac{f}{1 + 4\kappa x^2/f^2} = f + \tfrac{1}{2}\frac{x^2}{f}(1 - 8\kappa) + \dots$$

If the longitudinal aberration be called δf,

$$\delta f = \zeta - f = \tfrac{1}{2}\frac{x^2}{f}(1 - 8\kappa). \quad\quad\quad\quad (16)$$

Thus by (15) and (16),

$$\frac{h}{\delta f} = \frac{x^2}{4f^2} = \tfrac{1}{4}\alpha^2, \quad\quad\quad\quad\quad\quad (17)$$

where α denotes the angular semi-aperture. Taking the greatest admissible value of h as equal to $\tfrac{1}{4}\lambda$, we shall see that δf must not exceed the value given by

$$\delta f = \lambda \alpha^{-2}. \quad\quad\quad\quad\quad\quad\quad (18)$$

As a practical example, we may take the case of a single lens of glass collecting parallel rays to a focus. With the most favourable curvatures the longitudinal aberration is about $f\alpha^2$; so that α^4 must not exceed λ/f. For a lens of 3 feet [91 cm.] focus, this condition is satisfied if the aperture do not exceed 2 inches [5·1 cm.]. In spectroscopic work the chromatic aberration of single lenses does not come into play, and there is nothing to forbid their employment if the above-mentioned restriction be observed. I have been in the habit of using a plano-convex lens of plate-glass, the curved side being turned towards the parallel light, and have found its performance quite satisfactory. The fact that with a given focal length the extreme error of phase varies as the fourth power of the aperture is quite in accordance with practical experience; for it is well known that the difficulty of making object-glasses for telescopes increases very rapidly with the angular aperture.

When parallel rays fall directly upon a spherical mirror, the longitudinal aberration is only ·one-eighth as great as for the most favourably shaped lens of equal focal length and aperture. Hence a spherical mirror of 3 feet focus might have an aperture of $2\frac{1}{2}$ inches, and the image would not suffer materially from aberration.

§ 5. *On the Accuracy required in Optical Surfaces.*

Foucault, in the memoir already referred to, was, I believe, the first to show that the errors of optical surfaces should not exceed a moderate fraction of the wave-length of light. In the case of perpendicular reflection from mirrors, the results of § 4 lead to the conclusion that no considerable area of the surface should deviate from truth by more than one-eighth of the wave-length. For a glass surface refracting at nearly perpendicular incidence the admissible error is about four times as great. It will be understood, of course, that the errors of one surface in an optical train may compensate for those of another, all that is necessary being that the resultant error of retardation rise nowhere to importance.

In the case of oblique reflection at an angle ϕ, the error of retardation due to an elevation BD (fig. 7) is

$$QQ' - QS = BD \sec \phi \,(1 - \cos SQQ') = BD \sec \phi \,(1 + \cos 2\phi) = 2BD \cos \phi,$$

from which it follows that an error of given magnitude in the figure of

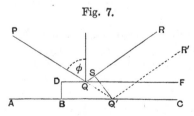

Fig. 7.

a surface is less important in oblique than in perpendicular reflection. At first sight this result appears to be contradicted by experience; for it is well known to practical opticians that it is more difficult to secure a satisfactory performance when reflection is oblique. The discrepancy is explained in great measure when we take into account the kind of error to which surfaces are most liable. No important deviation from a symmetrical form is to be feared; but a surface intended to be plane may easily assume a slight general convexity or concavity. Now in direct reflection, a small curvature is readily and almost completely compensated by a small motion of the eyepiece giving a change of focus; but the compensation obtainable in this way is much less perfect when the reflection is oblique. In the first case the family of surfaces approximating to a plane, which will answer the purpose, coincides with the family of surfaces most likely to be produced; in the second case the family of ellipsoidal or hyperbolic surfaces capable (with suitable focus) of giving good definition contains only one symmetrical member—the perfect plane. In order to test experimentally the correctness of the theoretical result, it would be necessary to retain the focus suitable to the true surface, and not to allow a readjustment by which its errors may be in greater or less degree compensated.

A further difficulty, not touched by the preceding considerations, still remains to be mentioned. In the ordinary method of testing plane surfaces by measuring the change of focus required when a distant point is viewed through a telescope, first directly, and then after reflection in the surface, the test is found to be more delicate as the reflection is more oblique. The explanation of the apparent inconsistency will be best understood by a calculation of the focal length of mirrors, founded directly upon the principles of the wave theory. Let ACB (fig. 8) be an arc of a (parabolic) mirror, which reflects parallel rays GA, HD, KB to a focus F. $AD = y$, $DF = f$, $CD = t$. In calculating the retardations of the various rays, we will take as standard the phase at F of a ray coincident with HD, reflected at D (as by a plane mirror ADB) instead of at C, so that the actual retardation at F of the central ray HCF is $2t$. The retardation of the extreme ray GAF is $AF - FD$, or $\sqrt{(f^2 + y^2)} - f$. Since F is by supposition an optical focus, the phases of all the rays must be the same, and thus

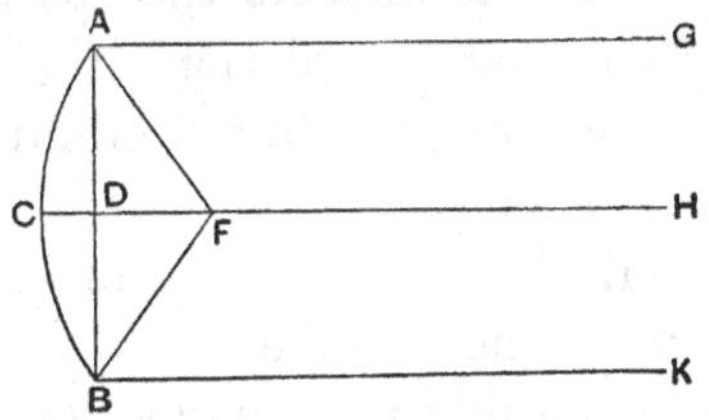

$$\sqrt{(f^2 + y^2)} - f = 2t. \quad\ldots\ldots(1)$$

If the aperture ($2y$) be small in proportion to the focal length,

$$\sqrt{(f^2 + y^2)} - f = \tfrac{1}{2}y^2/f \text{ approximately,}$$

so that

$$f = \frac{y^2}{4t}. \quad\ldots\ldots(2)$$

In the limit it is a matter of indifference whether f be measured from D or from C. If r be the radius of curvature at C,

$$t = \tfrac{1}{2}y^2/r,$$

Fig. 8.

and
$$f = \tfrac{1}{2}r, \quad \dots\dots\dots\dots\dots\dots(3)$$
the usual formula.

When the incidence is oblique, there are different foci in the primary and the secondary planes. Considering first the case of the primary plane, let ACB (fig. 9) be the mirror, and F the focus. $AD = y$, $CD = t$, $HCD = \phi$. AL is the course which the ray GA would take if reflected by the plane surface ADB. C is that point of the mirror at which the tangent is parallel to AB. The retardation at F of the ray GAF is $AF - AL$; and the retardation of the ray HCF due to the curvature of the mirror is $2t \cos \phi$. These retardations must be equal; and thus

Fig. 9.

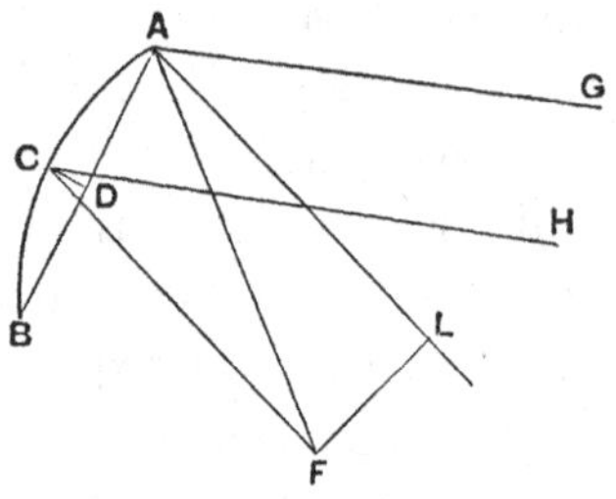

$$AF - AL = \sqrt{(AL^2 + FL^2)} - AL = 2t \cos \phi.$$

When the aperture is small, $\sqrt{(AL^2 + FL^2)} - AL$ is approximately $\tfrac{1}{2}FL^2/AL$; ultimately AL may be identified with f_1, the focal length in the primary plane, and FL may be identified with $y \cos \phi$, so that

$$\frac{y^2 \cos^2 \phi}{2f_1} = 2t \cos \phi, \quad \dots\dots\dots\dots\dots\dots(4)$$

or
$$f_1 = \frac{y^2 \cos \phi}{4t}. \quad \dots\dots\dots\dots\dots\dots(5)$$

Thus it appears that, so far as the primary focal length is concerned, the diminished retardation of the central ray due to obliquity is outweighed by the corresponding diminution of effective aperture (FL); but although in consequence of obliquity a greater change of focus (estimated from that required for the plane surface) is necessary in order to get the best result, still, if no change of focus be admitted, the error due to curvature is less sensible in oblique than in direct reflection.

The preceding discussion assumes that the same extent of surface is used in all cases. In testing planes by reflection it often happens that a *greater* extent of surface is used in the case of obliquity, the field being limited by the object-glass of the telescope rather than by the reflecting plane. Under such circumstances the loss of definition (with focus un-altered) due to curvature is aggravated by obliquity.

In the secondary focal plane there is no diminution of effective aperture due to obliquity. Instead of (4) we have

$$\frac{y^2}{2f_2} = 2t \cos \phi, \quad \dots\dots\dots\dots\dots\dots(6)$$

or
$$f_2 = \frac{y^2}{4t \cos \phi}. \quad \dots\dots\dots\dots\dots\dots(7)$$

In this case the favourable effect of obliquity shows itself directly in the increased value of f_2.

Fig. 10.

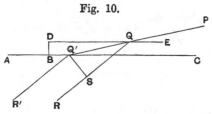

We will now consider the effect of errors in a refracting surface. The error of retardation due to an elevation BD is

$$\mu QS - QQ' = QQ' \{\mu \cos(\phi - \phi') - 1\}$$

$$= BD \frac{\mu \cos\phi \cos\phi' + \mu \sin\phi \sin\phi' - 1}{\cos\phi}.$$

$$= BD \frac{\mu \cos\phi \cos\phi' - (1 - \sin^2\phi)}{\cos\phi} = BD(\mu \cos\phi' - \cos\phi),$$

since

$$\mu \sin\phi' = \sin\phi.$$

As a function of obliquity $\mu \cos\phi' - \cos\phi$ is least $(\mu - 1)$ when the obliquity is zero; it is greatest $\sqrt{(\mu^2 - 1)}$ when the obliquity is 90°. Thus the retardation for a given error of elevation *increases* somewhat with the obliquity, being in the case of glass about twice as great at a grazing as at a perpendicular incidence.

Before concluding this section, it may be worth while to point out how the principles of the wave theory may be applied directly to calculate the focal length of lenses. The relative retardations of the rays DAF and ECF (fig. 11) are evidently $AF - CF$ and $(\mu - 1)t$, if t denote the thickness of the lens at the centre. Thus, if $AC = y$, $FC = f$,

$$(\mu - 1)t = \sqrt{(f^2 + y^2)} - f = \tfrac{1}{2}y^2/f \dots\dots\dots\dots(8)$$

approximately. For glass $\mu - 1 = \tfrac{1}{2}$ nearly; so that the old rule*, that "in glass lenses the half-breadth is a mean proportional between the thickness and the focal length," is more scientific than the usual formula in terms of the radii of curvature. If the lens do not terminate in a sharp edge, we may take as the

Fig. 11.

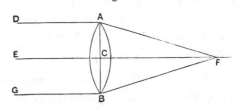

effective thickness the difference of the thicknesses at the centre and at the edge.

* Coddington's *Optics*, p. 96.

For an oblique central pencil, the focal lengths in the two principal planes may be obtained as in the case of the mirror. They take the form

$$\frac{y^2 \cos^2 \phi}{2f_1} = (\mu \cos \phi' - \cos \phi)\, t, \quad\text{.......................(9)}$$

$$\frac{y^2}{2f_2} = (\mu \cos \phi' - \cos \phi)\, t, \quad\text{.....................(10)}$$

in which, if we please, we may substitute for t its value in terms of the radii of curvature, viz.

$$\tfrac{1}{2} y^2 \left(\frac{1}{r} + \frac{1}{s} \right).$$

§ 6.   *The Aberration of Oblique Pencils.*

In treatises on geometrical optics it is usual to calculate the aberrations of mirrors and lenses for direct pencils, but in the case of oblique pencils to rest satisfied with determining the primary and secondary focal lengths. For most purposes indeed astigmatism is a worse defect than aberration, so that in the presence of the former it is not worth while to consider the latter; but in this respect the spectroscope is an exception, and the completion of its theory requires the consideration of the aberration of oblique pencils.

The reason of this peculiarity is not difficult to see. When a luminous point is observed through an optical instrument affected with astigmatism, there are three notable representations of it to be obtained by varying the focus. At the primary and secondary foci the point is represented by perpendicular lines of small width, and at a particular intermediate position by a circle of light, called the circle of least confusion. In most cases the last representation would be the best; but if the object under examination be itself a uniformly luminous line parallel to one or other of the focal lines, the best result will evidently be obtained at the corresponding focus. Under these circumstances the image is not prejudiced by the astigmatism, and its perfection depends upon the amount of aberration. In the case of a properly adjusted spectroscope the slit is parallel to the edges of the prisms, and the spectrum is seen with best definition at the *primary* focus.

The aberration that we have now to consider is of lower order than that which affects symmetrical pencils, and therefore, when it occurs, is presumably of greater importance. Before calculating its amount in particular cases, it will be convenient to consider the general character of the effects produced by it. The axis of the pencil being taken as axis of z, let the equation of the wave-surface, to which all rays are normal, be

$$z = \frac{x^2}{2\rho} + \frac{y^2}{2\rho'} + \alpha x^3 + \beta x^2 y + \gamma x y^2 + \delta y^3 + \dots \quad \dots\dots\dots\dots(1)$$

The principal focal lengths are ρ and ρ'. In the case of symmetry ρ and ρ' are equal, and the coefficients of the terms of the third order vanish. The aberration then depends upon terms of the fourth order; and even these are made to vanish in the formulæ for the object-glasses of telescopes by the selection of suitable curvatures. If this be effected, the outstanding aberration will be of the sixth order; whereas in the case of unsymmetrical pencils, even if we should succeed in destroying the terms of the third order, there will still remain an aberration of the fourth order. It follows that every effort should be made to retain symmetry about the axis; but in the case of the spectroscope this is usually impossible. If we could secure a perfect parallelism of the incident light, and perfectly flat faces for our prisms, we should indeed get rid of aberration, and at the same time render ourselves independent of the adjustments of the spectroscope; for it is evident that no repetition of refractions at plane surfaces, however situated, could disturb the original parallelism of the light. The fact that most large spectroscopes are more or less sensitive to maladjustment of the prisms proves either that the faces are not flat, or that it is difficult to obtain a sufficiently accurate adjustment of the collimator. We shall suppose that the faces of the prisms are surfaces of revolution, so that it is possible by proper adjustments to render everything symmetrical with respect to a plane bisecting at right angles the refracting edges. If ρ be the primary focal length, this plane is that represented by $y = 0$, and the equation of the wave-surface reduces to

$$z = \frac{x^2}{2\rho} + \frac{y^2}{2\rho'} + \alpha x^3 + \gamma x y^2, \quad \dots\dots\dots\dots\dots(2)$$

terms of higher order being omitted.

The constants α and γ in (2) may be interpreted in terms of the differential coefficients of the principal radii of curvature. By the usual formula, the radius of curvature at the point x of the intersection of (2) with the plane $y = 0$ is approximately $\rho(1 - 6\alpha\rho x)$. Since $y = 0$ is a principal plane throughout, this radius of curvature is a principal radius of the surface; so that, denoting it by ρ, we have

$$\alpha = -\frac{1}{6\rho^2}\frac{d\rho}{dx}. \quad \dots\dots\dots\dots\dots\dots(3)$$

In the neighbourhood of the origin the approximate value of the product of the principal curvatures is

$$\frac{1}{\rho\rho'} + \frac{6\alpha x}{\rho'} + \frac{2\gamma x}{\rho}.$$

Thus

$$\delta\left(\frac{1}{\rho\rho'}\right) = -\frac{\delta\rho}{\rho^2\rho'} - \frac{\delta\rho'}{\rho'^2\rho} = \frac{6\alpha x}{\rho'} + \frac{2\gamma x}{\rho};$$

whence by (3),

$$\gamma = -\frac{1}{2\rho'^2}\frac{d\rho'}{dx}. \quad\quad\quad\quad\quad\quad (4)$$

The equation of the normal at the point x, y, z is

$$z - \zeta = \frac{\xi - x}{\rho^{-1}x + 3\alpha x^2 + \gamma y^2} = \frac{\eta - y}{\rho'^{-1}y + 2\gamma xy}; \quad\quad\quad (5)$$

and its intersection with the plane $\zeta = \rho$ occurs at the point determined approximately by

$$\xi = -\rho(3\alpha x^2 + \gamma y^2), \quad\quad \eta = \frac{\rho' - \rho}{\rho'} - 2\rho\gamma xy, \quad\quad\quad (6)$$

terms of the third order being omitted.

According to geometrical optics, the thickness of the image of a luminous line at the primary focus is determined by the extreme value of ξ; and for good definition in the spectroscope it is necessary to reduce this thickness as much as possible. One way of attaining this result would be to narrow the aperture; but, as we have seen in preceding sections, to narrow the horizontal aperture is really to throw away the peculiar advantage of large instruments. The same objection, however, does not apply to narrowing the *vertical* aperture; and in many spectroscopes a great improvement in definition may be thus secured. In general it is necessary that *both* γ and α be small. Since the value of ξ is independent of ρ', it would seem that in respect of definition there is no advantage in avoiding astigmatism.

We will now examine more closely the character of the image at the primary focus in the case of a pencil of circular section. Unless $\rho' = \rho$, the second term in the value of η may be neglected. The rays for which $x^2 + y^2 = r^2$ intersect the plane $\zeta = \rho$ in the parabola

$$\frac{\rho\rho'^2(3\alpha - \gamma)}{(\rho' - \rho)^2}\eta^2 - \xi = 3\alpha\rho r^2; \quad\quad\quad\quad (7)$$

and the various parabolas corresponding to different values of r differ from one another only in being shifted along the axis of ξ. To find out how much of the parabolic arcs are included, we observe that for any given value of r the value of η is greatest when $x = 0$. Hence the rays starting in the secondary plane give the remainder of the boundary of the image. Its equation, formed from (6) after putting $x = 0$, is

$$\eta^2 = -\frac{(\rho' - \rho)^2}{\rho\rho'^2\gamma}\xi, \quad\quad\quad\quad\quad (8)$$

and represents a parabola touching the axis of η at the origin. The whole of the image is included between this parabola and the parabola of form (7) corresponding to the maximum value of r.

The width of the image when $\eta = 0$ is $3\alpha\rho r^2$, and vanishes when $\alpha = 0$, i.e. when there is no aberration for rays in the primary plane. In this case the two parabolic boundaries coincide, and the image is reduced to a linear arc. If further $\gamma = 0$, this arc becomes *straight*, and then the image of a luminous line is perfect (to this order of approximation) at the primary focus. In general if $\gamma = 0$, the parabola (7) reduces to the straight line $\eta = 0$; that is to say, the rays which start in the secondary plane remain in that plane.

We will now consider the image formed at the secondary focus. Putting $\zeta = \rho'$ in (5), we obtain

$$\xi = \frac{\rho - \rho'}{\rho} x, \qquad \eta = -2\gamma\rho' xy. \quad \ldots\ldots\ldots\ldots\ldots(9)$$

If $\gamma = 0$, the secondary focal line is formed without aberration, but not otherwise. In general the curve traced out by the rays for which $x^2 + y^2 = r^2$, is

$$\left(\frac{\rho}{\rho - \rho'}\right)^2 \xi^2 + \frac{(\rho - \rho')^2 \eta^2}{4\gamma^2 \rho^2 \rho'^2 \xi^2} = r^2, \quad \ldots\ldots\ldots\ldots\ldots(10)$$

in the form of a figure of eight symmetrical with respect to both the axes. The rays starting either in the primary or in the secondary plane pass through the axis of ξ, the thickness of the image being due to the rays for which $x = y = r/\sqrt{2}$*.

This subject can be illustrated without difficulty by experiment. A radiant point is obtained by admitting sunlight into a darkened room through a lens of short focus placed in the window-shutter. A real image of the radiant is received upon a piece of ground glass and examined from behind. To render the light approximately homogeneous, a piece of red glass is employed. The following results relate to an equiconvex lens of 6 inches [15 cm.] aperture and about 3 feet [91 cm.] focus, on which the light falls obliquely.

As the screen is moved gradually back from the lens, the illuminated area diminishes. At a certain point it begins to double back upon itself, until at the primary focus the whole area is double. The light is seen to be very unequally distributed. At the edges, corresponding to the boundary of the lens, the illumination is feeble; while at the folded edge, corresponding to the central vertical line of the lens, a caustic is formed. On this account

---

* I have lately found that the aberration of unsymmetrical pencils was very generally treated by Sir W. Hamilton in his work on *Systems of Rays*. Even if I could have supposed Hamilton's results to be known to the reader, the investigation in the text would still be necessary, as my purpose is very different from his. In the general theory (with β and δ finite) there is no distinction between the primary and secondary images.

it would seem that curvature of the primary focal line is a worse fault than thickness for the purposes of the spectroscope.

The accompanying figures show the general character of the image at the primary focus under various circumstances. The thick line represents the folded and highly illuminated edge, the thin line the double edge corresponding to the margin of the lens. The quantities u, v_1, v_2 are the distances from the lens of the radiant point, the primary, and the secondary focus respectively, expressed in inches.

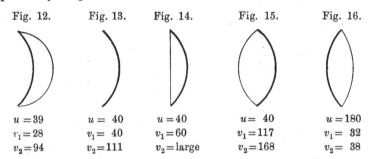

Fig. 12.	Fig. 13.	Fig. 14.	Fig. 15.	Fig. 16.
$u = 39$	$u = 40$	$u = 40$	$u = 40$	$u = 180$
$v_1 = 28$	$v_1 = 40$	$v_1 = 60$	$v_1 = 117$	$v_1 = 32$
$v_2 = 94$	$v_2 = 111$	$v_2 = $ large	$v_2 = 168$	$v_2 = 38$

In all these cases the line of intersection of the plane of the lens with the screen lies to the *right* of the diagram. It will be seen that the primary focal line is thin, though curved, when $v_1 = u$. This is true in general for an equiconvex lens, as may be shown from considerations of symmetry.

Experimenting on a plano-convex lens held at an obliquity of about 30°, I found that the focal lines were far better formed when the convex side was turned towards parallel rays than when the flat side was so turned. The theory, which I subsequently investigated, is given in the following section. I think that spherical lenses inclined at the most suitable obliquity might in many cases, perhaps in star-spectroscopes, replace cylindrical lenses. If it were necessary to cause no convergency at all in the secondary plane, a compensating concave lens, held perpendicularly, would be used.

§ 7. *Aberration of Lenses and Prisms.*

The following investigation refers to the aberration of rays in the primary plane. Let Q be a radiant point in air, from which rays fall upon the spherical surface of glass APB, of radius r. We require an approximate expression for the length of any ray QP referred to that of a standard ray QA.

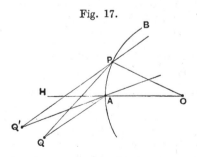

Fig. 17.

$$POA = \omega, \quad QAH = \phi,$$
$$QA = u. \quad PA = 2r \sin (\tfrac{1}{2}\omega).$$
$$QAP = \tfrac{1}{2}\pi + \phi + \tfrac{1}{2}\omega.$$
$$QP^2 = u^2 + 4r^2 \sin^2 \tfrac{1}{2}\omega + 4ru \sin \tfrac{1}{2}\omega \; \sin (\phi + \tfrac{1}{2}\omega),$$

so that as far as the cube of ω,

$$QP - u = r \sin \phi . \omega + \frac{r}{2} \left(\cos \phi + \frac{r \cos^2 \phi}{u} \right) \omega^2$$

$$- \frac{r \sin \phi}{2} \left(\frac{1}{3} + \frac{r \cos \phi}{u} + \frac{r^2 \cos^2 \phi}{u^2} \right) \omega^3. \quad \ldots\ldots\ldots\ldots(1)$$

Similarly if Q' be any other point, and $Q'A = u'$, $Q'AH = \phi'$,

$$Q'P - u' = r \sin \phi' . \omega + \frac{r}{2} \left(\cos \phi' + \frac{r \cos^2 \phi'}{u'} \right) \omega^2$$

$$- \frac{r \sin \phi'}{2} \left(\frac{1}{3} + \frac{r \cos \phi'}{u'} + \frac{r^2 \cos^2 \phi'}{u'^2} \right) \omega^3. \quad \ldots\ldots\ldots(2)$$

If QP, $Q'P$ be incident and refracted rays corresponding to $\omega = \theta$, the condition must be satisfied that $\delta(QP) = \mu \delta(Q'P)$ as ω passes from the value θ to $\theta + \delta\theta$. Thus

$$\sin \phi + \left(\cos \phi + \frac{r \cos^2 \phi}{u} \right) \theta - \frac{\sin \phi}{2} \left(1 + \frac{3r \cos \phi}{u} + \frac{3r^2 \cos^2 \phi}{u^2} \right) \theta^2$$

$$= \mu \sin \phi' + \mu \left(\cos \phi' + \frac{r \cos^2 \phi'}{u'} \right) \theta$$

$$- \mu \frac{\sin \phi'}{2} \left(1 + \frac{3r \cos \phi'}{u'} + \frac{3r^2 \cos^2 \phi'}{u'^2} \right) \theta^2. \quad \ldots\ldots\ldots\ldots(3)$$

If QA, $Q'A$ be also corresponding rays, $\sin \phi - \mu \sin \phi' = 0$, as appears by supposing $\theta = 0$ in (3), which then becomes

$$\left(\cos \phi + \frac{r \cos^2 \phi}{u} \right) \left(1 - \frac{3r\theta \sin \phi}{2u} \right) = \mu \left(\cos \phi' + \frac{r \cos^2 \phi'}{u'} \right) \left(1 - \frac{3r\theta \sin \phi'}{2u'} \right). \quad \ldots(4)$$

If we make $\theta = 0$ in (4),

$$\cos \phi + \frac{r \cos^2 \phi}{u} = \mu \left(\cos \phi' + \frac{r \cos^2 \phi'}{u'} \right), \quad \ldots\ldots\ldots\ldots(5)$$

the usual formula for the primary focus. For our present purpose θ is small, but it is not zero.

We will now apply the fundamental formula (4) to the case of a lens whose thickness can be neglected in comparison with the radii of curvature and the distances of the foci. The pencil is supposed to fall centrically, so that the angle of incidence at the second surface is equal to the angle of refraction at the first surface. The distance corresponding to PA is the same for both surfaces, and will be denoted by y. Thus, for the first refraction,

$$\left(\frac{\cos\phi}{r}+\frac{\cos^2\phi}{u}\right)\left(1-\frac{3y\sin\phi}{2u}\right)=\mu\left(\frac{\cos\phi'}{r}+\frac{\cos^2\phi'}{u'}\right)\left(1-\frac{3y\sin\phi'}{2u'}\right). \quad\ldots(6)$$

For the second refraction we have to interchange ϕ and ϕ', writing s for r, u' for u, μ^{-1} for μ, and v for u'. Thus

$$\mu\left(\frac{\cos\phi'}{s}+\frac{\cos^2\phi'}{u'}\right)\left(1-\frac{3y\sin\phi'}{2u'}\right)=\left(\frac{\cos\phi}{s}+\frac{\cos^2\phi}{v}\right)\left(1-\frac{3y\sin\phi}{2v}\right). \quad\ldots(7)$$

By addition of (6) and (7), and writing for brevity c for $\cos\phi$ and c' for $\cos\phi'$, we get

$$\left(\frac{c}{r}+\frac{c^2}{u}\right)\left(1-\frac{3y\sin\phi}{2u}\right)-\left(\frac{c}{s}+\frac{c^2}{v}\right)\left(1-\frac{3y\sin\phi}{2v}\right)$$

$$=\mu\left(\frac{c'}{r}+\frac{c'^2}{u'}\right)\left(1-\frac{3y\sin\phi'}{2u'}\right)-\mu\left(\frac{c'}{s}+\frac{c'^2}{u'}\right)\left(1-\frac{3y\sin\phi}{2u'}\right)$$

$$=\mu c'\left(\frac{1}{r}-\frac{1}{s}\right)-\frac{3y\sin\phi}{2u'}\left(\frac{c'}{r}-\frac{c'}{s}\right)$$

or

$$\frac{c^2}{u}-\frac{c^2}{v}=(\mu c'-c)\left(\frac{1}{r}-\frac{1}{s}\right)$$

$$-\frac{3y\sin\phi}{2}\left\{\frac{c'}{u'}\left(\frac{1}{r}-\frac{1}{s}\right)-\frac{1}{u}\left(\frac{c}{r}+\frac{c^2}{u}\right)+\frac{1}{v}\left(\frac{c}{s}+\frac{c^2}{v}\right)\right\}. \quad\ldots\ldots(8)$$

The quantity within brackets in (8) may be written

$$\frac{c'}{u'}\left(\frac{1}{r}-\frac{1}{s}\right)-\frac{c}{u}\left(\frac{1}{r}-\frac{1}{s}\right)+\left(\frac{c^2}{v}-\frac{c^2}{u}\right)\left(\frac{1}{v}+\frac{1}{u}+\frac{1}{cs}\right).$$

On substitution for $c^2/v-c^2/u$ of its approximate value from (8), $(1/r-1/s)$ becomes a factor of the whole expression, and we get

$$\left(\frac{1}{r}-\frac{1}{s}\right)\left\{\frac{c'}{u'}-\frac{c}{u}-(\mu c'-c)\left(\frac{1}{v}+\frac{1}{u}+\frac{1}{cs}\right)\right\}.$$

Again, from (5) with sufficient approximation for our purpose,

$$\frac{c'}{u'}-\frac{c}{u}=\frac{c'}{u'}-\frac{\mu}{c}\left(\frac{c'}{r}+\frac{c'^2}{u'}\right)+\frac{1}{r}=-(\mu c'-c)\frac{c'}{cu'}-(\mu c'-c)\frac{1}{cr};$$

so that the bracket in (8) assumes the form

$$-(\mu c'-c)\left(\frac{1}{r}-\frac{1}{s}\right)\left\{\frac{c'}{cu'}+\frac{1}{v}+\frac{1}{u}+\frac{1}{cr}+\frac{1}{cs}\right\}. \quad\ldots\ldots\ldots(9)$$

From (6) and (7),

$$2\frac{\mu c'^2}{u'}=c^2\left(\frac{1}{u}+\frac{1}{v}\right)-(\mu c'-c)\left(\frac{1}{r}+\frac{1}{s}\right). \quad\ldots\ldots\ldots(10)$$

Using this in (9), we get

$$-\frac{\mu c' - c}{2\mu c c'}\left(\frac{1}{r} - \frac{1}{s}\right)\left\{c\,(2\mu c' + c)\left(\frac{1}{u} + \frac{1}{v}\right) + (\mu c' + c)\left(\frac{1}{r} + \frac{1}{s}\right)\right\};$$

so that

$$\frac{c^2}{u} - \frac{c^2}{v} = (\mu c' - c)\left(\frac{1}{r} - \frac{1}{s}\right)\left[1 + \frac{3y\sin\phi}{4cc'\mu}\left\{c\,(2\mu c' + c)\left(\frac{1}{u} + \frac{1}{v}\right)\right.\right.$$

$$\left.\left. + (\mu c' + c)\left(\frac{1}{r} + \frac{1}{s}\right)\right\}\right]. \quad\ldots\ldots\ldots\ldots\ldots\ldots(11)^*$$

From this we see at once that in the case of an equiconvex or equiconcave lens, for which $1/r + 1/s = 0$, the aberration vanishes if $1/u + 1/v = 0$, i.e. if the primary focus be at the same distance on one side of the lens as the radiant point is on the other.

For some purposes a more convenient expression may be obtained by substituting for v its approximate value. Writing μ' for $\mu c'/c$, we get

$$\frac{c^2}{u} - \frac{c^2}{v} = (\mu c' - c)\left(\frac{1}{r} - \frac{1}{s}\right)\left[1 + \frac{3y\sin\phi}{2\mu c'}\left\{\frac{(2\mu' + 1)\,c}{u} + \frac{\mu'^2}{s} + \frac{\mu' - \mu'^2 + 1}{r}\right\}\right].$$

$$\ldots\ldots\ldots\ldots\ldots\ldots(12)$$

If the incident rays be parallel, $u = \infty$, and the aberration vanishes when

$$\frac{s}{r} = -\frac{\mu'^2}{\mu' - \mu'^2 + 1} = -\frac{\mu^2\cos^2\phi'}{\mu\cos\phi'\cos\phi - \mu^2\cos^2\phi' + \cos^2\phi}.\quad\ldots\ldots(13)$$

If the second surface be plane ($s = \infty$), this condition is

$$\mu' - \mu'^2 + 1 = 0, \quad\text{or}\quad \mu' = \tfrac{1}{2}(1 + \sqrt{5}) = 1\cdot62.$$

For a refractive index $\mu = 1\cdot5$, the value of ϕ which makes $\mu' = 1\cdot62$ is about 29°. This is the obliquity at which a plano-convex lens of plate glass must be held in order to give a thin primary image. A more refractive lens must be held at a less obliquity. If the refractive index exceed $1\cdot62$, there is no position of the lens for which the image is free from aberration.

From the above formulæ there is no difficulty in calculating the aberration due to any combination of lenses. As an example I will take the case of two lenses [of similar glass and] of equal focal length, inclined at equal angles but, in opposite directions to the axis of the pencil. Denoting the radii of the second lens by r' and s', the final focal length v' is to be found

---

* [1899. The fraction $\dfrac{\mu c' - c}{(\mu - 1)\,c^2}$, representing the ratio in which the focal length is altered by obliquity, has the following approximate expression

$$1 + \frac{2\mu + 1}{2\mu}\sin^2\phi + \text{term in } \sin^4\phi.]$$

from (11) in combination with

$$\frac{c^2}{v} - \frac{c^2}{v'} = (\mu c' - c)\left(\frac{1}{r} - \frac{1}{s}\right)\left[1 - \frac{3y \sin \phi}{4\mu c' c}\left\{c\,(2\mu c' + c)\left(\frac{1}{v} + \frac{1}{v'}\right)\right.\right.$$
$$\left.\left. + (\mu c' + c)\left(\frac{1}{r'} + \frac{1}{s'}\right)\right\}\right].$$

By addition,

$$\frac{c^2}{u} - \frac{c^2}{v'} = (\mu c' - v)\left(\frac{1}{r} - \frac{1}{s}\right)\left[2 + \frac{3y \sin \phi}{4\mu c c'}\left\{c\,(2\mu c' + c)\left(\frac{1}{u} - \frac{1}{v'}\right)\right.\right.$$
$$\left.\left. + (\mu c' + c)\left(\frac{1}{r} + \frac{1}{s} - \frac{1}{r'} - \frac{1}{s'}\right)\right\}\right]. \quad \dots\dots\dots(14)$$

If the lenses be of the same curvatures and be similarly turned, $r' = r$, $s' = s$. The aberration then varies as $(1/u - 1/v')$, and cannot vanish.

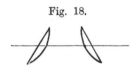

Fig. 18.

If the lenses be of the same curvature and be turned opposite ways (fig. 18), $r' = -s$, $s' = -r$. The aberration is proportional to

$$c\,(2\mu c' + c)\left(\frac{1}{u} - \frac{1}{v'}\right) + 2\,(\mu c' + c)\left(\frac{1}{r} + \frac{1}{s}\right);$$

or, on substitution of the approximate value of $1/u - 1/v'$,

$$4c\left\{\frac{\mu'^2}{r} + \frac{1 - \mu'^2 + \mu'}{s}\right\}, \quad \dots\dots\dots\dots\dots(15)$$

where, as before,

$$\mu' = \mu \cos \phi'/\cos \phi.$$

Thus the combination is aplanatic, independently of the value of u, if

$$\frac{s}{r} = -\frac{1 - \mu'^2 + \mu'}{\mu'^2}. \quad \dots\dots\dots\dots\dots(16)$$

By supposing the light to be parallel between the lenses, we obtain (13) as a particular case of (16).

We will next investigate the aberration in the case of a prism with flat faces on which divergent or convergent light falls at the angle of minimum deviation. From (4), with notation similar to that employed above,

$$\frac{c^2}{u}\left(1 - \frac{3y \sin \phi}{2u}\right) = \mu\,\frac{c'^2}{u'}\left(1 - \frac{3y \sin \phi'}{2u'}\right), \quad \dots\dots\dots(17)$$

$$\mu\,\frac{c'^2}{u' + l}\left(1 + \frac{3y' \sin \phi'}{2\,(u' + l)}\right) = \frac{c^2}{v}\left(1 + \frac{3y' \sin \phi}{2v}\right), \quad \dots\dots\dots(18)$$

and

$$y' : y = u' + l : u', \quad \dots\dots\dots\dots\dots(19)$$

l denoting the length of the path of the axis of the pencil within the prism.

If y be neglected,

$$\frac{c^2}{u} = \frac{\mu c'^2}{u'}, \qquad \frac{\mu c'^2}{u' + l} = \frac{c^2}{v}, \quad \dots\dots\dots\dots\dots(20)$$

whence

$$\frac{v}{u} = \frac{u' + l}{u'}; \quad \dots\dots\dots\dots\dots(21)$$

so that

$$\frac{y'}{v} = \frac{u' + l}{u'}\frac{y}{v} = \frac{y}{u}. \quad \dots\dots\dots\dots\dots(22)$$

Accordingly (18) becomes

$$\mu \frac{c'^2}{u' + l}\left(1 + \frac{3y \sin \phi}{2u'}\right) = \frac{c^2}{v}\left(1 + \frac{3y \sin \phi}{2u}\right). \quad \dots\dots\dots(23)$$

From (17) and (23), by multiplication,

$$\frac{\mu^2 c'^4}{u'(u' + l)} = \frac{c^4}{uv}, \quad \text{or} \quad v = \frac{c^4 u'(u' + l)}{\mu^2 c'^4 u}. \quad \dots\dots\dots(24)$$

Again, from (17),

$$u' = \frac{\mu c'^2 u}{c^2}\left\{1 + \frac{3y \sin \phi\,(\mu^2 c'^2 - c^2)}{2\mu^2 c'^2 u}\right\} = \frac{\mu c'^2 u}{c^2}\left\{1 + \frac{3y \sin \phi\,(\mu^2 - 1)}{2\mu^2 c'^2 u}\right\} \cdot \dots(25)$$

From (24) and (25),

$$v = u + \frac{c^2 l}{\mu c'^2} + \frac{3y \sin \phi\,(\mu^2 - 1)}{\mu^2 c'^2}\left(1 + \frac{c^2 l}{2\mu c'^2 u}\right). \quad \dots\dots\dots(26)$$

In most cases l/u is negligible in the small term, and the longitudinal aberration is simply

$$\delta v = \frac{3y \sin \phi\,(\mu^2 - 1)}{\mu^2 c'^2}. \quad \dots\dots\dots\dots\dots(27)$$

The linear width of the image is $\delta v \cdot \theta$, where θ denotes the angular semi-aperture; and the angular width of the image is $\delta v \theta / u$, or, since $y \cos \phi / u = \theta$,

$$\frac{3(\mu^2 - 1)\sin \phi \cdot \theta^2}{\mu^2 c c'^2}. \quad \dots\dots\dots\dots\dots(28)$$

To apply this result to the spectroscope, it is convenient to compare (28) with the angular deviation due to an alteration of refractive index from μ to $\mu + \delta\mu$. This is equal to

$$\frac{2\delta\mu \sin \phi'}{\cos \phi} = \frac{2\delta\mu \sin \phi}{\mu \cos \phi}. \quad \dots\dots\dots\dots\dots(29)$$

If (28) and (29) be equal,

$$\delta\mu = \frac{3\theta^2\,(\mu - \mu^{-1})}{2 \cos^2 \phi'}, \quad \dots\dots\dots\dots\dots(30)$$

which determines the degree of confusion of refrangibilities produced by the aberration.

For a glass prism, $\mu - \mu^{-1}$ differs very little from unity. If the refracting angle be 60°, $\cos^2 \phi' = \frac{3}{4}$, and thus

$$\delta\mu = 2\theta^2. \quad\dots\dots\dots\dots\dots\dots\dots\dots\dots\dots(31)$$

By (8), (10), § 4, the value of $\delta\mu$ for the soda-lines in the case of a prism of "extra-dense glass" is $5\cdot8 \times 10^{-5}$. Hence, if the image of one soda-line just touch the image of the other, $\theta = \cdot0054$, or the distance u must be about 100 times the width of the beam where it falls upon the prism. With small instruments, whose whole resolving-power is only a small multiple of that necessary to resolve the soda-lines, it is possible to work without a collimator by allowing a distance of 10 or 20 feet between the slit and the prisms; but in the case of powerful instruments a pretty accurate adjustment of the collimator is indispensable.

The next thing to consider is the effect of curvature in the faces of prisms. We shall neglect l, the length of path within the prism, as small in comparison with the other distances concerned; its influence is doubtless no more important than when the faces are flat. From (4),

$$\left(\frac{c}{r} + \frac{c^2}{u}\right)\left(1 - \frac{3y\sin\phi}{2u}\right) = \mu\left(\frac{c'}{r} + \frac{c'^2}{u'}\right)\left(1 - \frac{3y\sin\phi'}{2u'}\right), \quad\dots\dots(32)$$

$$\mu\left(\frac{c'}{s} + \frac{c'^2}{u'}\right)\left(1 + \frac{3y\sin\phi'}{2u'}\right) = \left(\frac{c}{s} + \frac{c^2}{v}\right)\left(1 + \frac{3y\sin\phi}{2v}\right), \quad\dots\dots(33)$$

giving by subtraction

$$\left(\frac{c}{r} + \frac{c^2}{u}\right)\left(1 - \frac{3y\sin\phi}{2u}\right) - \left(\frac{c}{s} + \frac{c^2}{v}\right)\left(1 + \frac{3y\sin\phi}{2v}\right)$$

$$= \mu c'\left(\frac{1}{r} - \frac{1}{s}\right) - \frac{3y\sin\phi}{2u'}\left(\frac{c'}{r} + \frac{c'}{s} + \frac{2c'^2}{u'}\right).$$

Hence

$$\frac{c^2}{u} - \frac{c^2}{v} = (\mu c' - c)\left(\frac{1}{r} - \frac{1}{s}\right) - \frac{3y\sin\phi}{2}\left\{\frac{1}{r}\left(\frac{c'}{u'} - \frac{c}{u}\right)\right.$$

$$\left. + \frac{1}{s}\left(\frac{c'}{u'} - \frac{c}{v}\right) + \frac{2c'^2}{u'^2} - \frac{c^2}{u^2} - \frac{c^2}{v^2}\right\}$$

$$= (\mu c' - c)\left(\frac{1}{r} - \frac{1}{s}\right) - \frac{3y\sin\phi}{2}\left\{\left(\frac{c'}{u'} - \frac{c}{u}\right)\left(\frac{1}{r} + \frac{c'}{u'} + \frac{c}{u}\right)\right.$$

$$\left. + \left(\frac{c'}{u'} - \frac{c}{v}\right)\left(\frac{1}{s} + \frac{c'}{u'} + \frac{c}{v}\right)\right\}. \quad\dots\dots\dots\dots\dots(34)$$

The condition that there shall be no aberration requires that the quantity within brackets vanish. In order to discuss it further it will be convenient to express u and v in terms of u'. By the original equations, y being neglected,

$$\frac{c}{u} = \frac{\mu c'^2}{cu'} + \frac{\mu c' - c}{cr} = \frac{\mu' c'}{u'} + \frac{\mu' - 1}{r}, \quad \dots\dots\dots\dots(35)$$

$$\frac{c}{v} = \frac{\mu c'^2}{cu'} + \frac{\mu c' - c}{cs} = \frac{\mu' c'}{u'} + \frac{\mu' - 1}{s}, \quad \dots\dots\dots\dots(36)$$

μ' being written as before for $\mu c'/c$.

By substitution of these values in (34) the condition becomes

$$\left(\frac{c'}{u'} + \frac{1}{r}\right)\left\{(\mu' + 1)\frac{c'}{u'} + \frac{\mu'}{r}\right\} + \left(\frac{c'}{u'} + \frac{1}{s}\right)\left\{(\mu' + 1)\frac{c'}{u'} + \frac{\mu'}{s}\right\} = 0. \quad \dots\dots(37)$$

This condition assumes a simplified form when $r = s$, i.e. when one face is as much convex as the other face is concave; it is satisfied either by

$$\frac{c'}{u'} + \frac{1}{r} = 0, \quad \dots\dots\dots\dots\dots\dots\dots\dots\dots(38)$$

or by

$$(\mu' + 1)\frac{c'}{u'} + \frac{\mu'}{r} = 0. \quad \dots\dots\dots\dots\dots\dots\dots\dots(39)$$

In the first case, by (35), $u = -cr$; so that if the first face be convex, the incident rays must be convergent. In the second case,

$$u = -c(\mu' + 1)r = -(\mu\cos\phi' + \cos\phi)r.$$

With general values of r and s, (37) may be written

$$2(\mu' + 1)\frac{c'^2}{u'^2} + (2\mu' + 1)\left(\frac{1}{r} + \frac{1}{s}\right)\frac{c'}{u'} + \mu'\left(\frac{1}{r^2} + \frac{1}{s^2}\right) = 0, \quad \dots\dots\dots(40)$$

which determines the values of u' (if any) for which the aberration vanishes. The roots of (40) are real and therefore aplanaticism is possible, if

$$\frac{2(2\mu' + 1)^2}{rs} - (4\mu'^2 + 4\mu' - 1)\left(\frac{1}{r^2} + \frac{1}{s^2}\right) \quad \dots\dots\dots\dots(41)$$

be positive. Unless both faces are flat, one must be convex and the other concave. The limits within which a suitable value of u' will secure aplanaticism are found by equating (41) to zero. The roots are reciprocals, and are given by

$$\frac{s}{r} = \frac{(2\mu' + 1)^2 \pm 4\sqrt{(\mu'^2 + \mu')}}{4\mu'^2 + 4\mu' - 1}. \quad \dots\dots\dots\dots\dots(42)$$

If the actual value of $s : r$ be nearer unity than the values determined by (42), the aberration may be destroyed by selecting for u' one of the roots of (40).

If, for example, $\mu = 1\cdot62$, and the refracting angle be 60°, $\mu' = 2\cdot392$,

$$\frac{s}{r} = 1\cdot43 \text{ or } (1\cdot43)^{-1}.$$

Unless, therefore, the curvatures are opposite and pretty nearly equal, no adjustment of the focus of the collimator can destroy the aberration.

In any optical instrument whatever, the aberration in the primary plane may be eliminated by sloping one of the lenses to a suitable angle ; but, as was shown in the preceding section, we have also in many cases to contend with the aberration of the rays in the secondary plane. The latter aberration is more difficult to calculate than the former. Among lenses, the only case that I have investigated is that of a plano-convex lens on the curved side of which parallel light falls. It appears that the aberration of the rays in the secondary plane vanishes when the following relation obtains between the refractive index and the obliquity:

$$\sin^2 \phi = \frac{3\mu^2 - \mu^4 - 1}{3 - \mu^2}. \quad\ldots\ldots\ldots\ldots\ldots\ldots\ldots(43)$$

The obliquity is zero when $\mu = 1\cdot62$—the same value of μ that gives no aberration in the primary plane for rays of small obliquity. Neither kind of aberration is important for a glass lens of this sort moderately sloped. If $\mu = 1\cdot5$, (43) gives $\phi = 73°$. If μ exceed $1\cdot62$, (43) cannot be satisfied.

Another case of considerable interest can be investigated more easily. It is that of a prism with flat sides, through which convergent or divergent light passes. The prism is supposed to be adjusted until the symmetrical (horizontal) plane contains the radiant point; but it need not be in the position of minimum deviation. The problem depends upon the same principles as are applied by Professor Stokes* to investigate the curvature of the image of a straight line when seen through a prism. If θ denote the altitude of a ray before it falls upon the prism, θ' the altitude of the ray within the prism, the horizontal projection of the ray follows a course which would be that of an actual ray if the refractive index were changed from μ to $\mu \cos \theta'/\cos \theta$. Since $\sin \theta' = \mu^{-1} \sin \theta$, and θ, θ' are supposed to be small, the virtual change in the refractive index is from μ to $\mu + \frac{1}{2}(\mu - \mu^{-1})\theta^2$. Thus, if $\delta\mu$ represent the range of refrangibilities confused in the resulting spectrum of an infinitely thin slit,

$$\delta\mu = \tfrac{1}{2}(\mu - \mu^{-1})\theta^2. \quad\ldots\ldots\ldots\ldots\ldots\ldots(44)$$

The factor $\mu - \mu^{-1}$ is equal to unity when $\mu = 1\cdot62$, and for glass will never differ much from unity. By comparison with (31), it would appear that in the case of a 60° prism in the position of minimum deviation, the admissible

* *Royal-Society Proceedings*, April 30, 1874.

vertical angular aperture is twice as great as the admissible horizontal angular aperture; but, on account of the variable distribution of light in the image, this conclusion probably requires modification in favour of the horizontal aperture.

If there is to be no confusion of the two soda-lines when seen with a prism of extra-dense glass one inch high, the distance of the radiant points must exceed four feet.

Equation (44) is applicable without change to a spectroscope of any number of properly adjusted prisms of similar material.

§ 8.   *The Design of Spectroscopes.*

The circumstances under which spectroscopes are used are so various that it is probably impossible to lay down any one construction as absolutely the best; but the principles of the foregoing sections allow us to impose certain limitations within which the choice of the designer must be confined. The first point to be considered is the resolving-power. This, in the case of prismatic instruments of one given material, carries with it the total thickness traversed; and the question is simply in what form it is most advantageous to employ this thickness. The other points to be attended to are principally illumination, quantity of material, ready applicability to various parts of the spectrum, simplicity and ease of construction.

To a certain extent the requirements of illumination and resolving-power are antagonistic. If, indeed, the eye were a perfect instrument, a beam of diameter equal to that of the pupil would present the full degree of brightness, and a resolving-power corresponding to the thickness employed. But, as was explained in § 3, in order to obtain the full value of the thickness it is necessary further to narrow the beam; and then illumination suffers. If full illumination be required, we must employ a thickness three or four times as great as that defined by (5), § 3.

Apart from the question of the area of the pupil occupied by the beam, illumination suffers from the effects of absorption and reflection. The first depends simply upon the thickness traversed, and is therefore an invariable quantity when the material and resolving-power are given. Some years since it was laid down by Pickering[*] that in spectroscopes composed of prisms of the same material and in the position of minimum deviation (which disperse equally and admit the same amount of light) the loss by absorption will be the same. In accordance with what has just been proved, we are now able to dispense with the restriction to minimum deviation.

[*] " On the Comparative Efficiency of Different Forms of the Spectroscope," *Phil. Mag.* xxxvi. p. 41, 1868.

In powerful spectroscopes the transparency of the material of which the prisms are made is a point of great importance. Some specimens even of well-made flint and crown glass examined by Christie* stopped as much as half the light in a thickness of 4 inches [10 cm.]. Such a degree of absorption renders the glass unsuited for instruments of more than moderate power. From measurements by Robinson and Grubb†, however, it would seem that absorption need not stand in the way of much more powerful instruments than any yet attempted. One specimen of Chance's glass was of such transparency that 111 inches [282 cm.] would be necessary to reduce the transmitted light in the ratio 2·7 to 1.

The loss of light by reflection depends upon the number of surfaces and upon the angles at which the rays are incident. It might be thought that a great multiplication of surfaces was necessarily very unfavourable to brightness; but, as has been pointed out by Pickering in the paper referred to, this difficulty may be overcome by using prisms of such angle that the reflected light is perfectly polarized. Under these circumstances *half* the light at least escapes reflection; and the necessary angles (64° for ordinary flint glass) are not otherwise objectionable. The least loss of light is incurred when the whole thickness is thrown into one prism of moderate angle; but the gain in brightness would rarely compensate for the other disadvantages of such a construction‡.

With regard to the material for prisms, the choice lies principally between various kinds of glass and fluids. It is not without difficulty that glass is prepared free from striæ and well annealed; but solid prisms have the great advantage that, when once good ones are obtained, there is no further trouble. In consequence probably of the practice of using in all cases a standard angle of 60°, an exaggerated idea is often entertained of the advantage of great density. According to Hopkinson, the differences of indices relative to the lines D and B for dense flint, extra-dense flint, and double-extra-dense flint are respectively ·0067, ·0075, and ·0091, which numbers are inversely as the equivalent thicknesses.

Of fluids, bisulphide of carbon has the merits of cheapness and very high dispersive power, the difference of indices for D and B being ·0126. When pure, it is also in a high degree transparent. On the other hand, its sensitiveness to variations of temperature is so great that 1° C. makes about nine times as great a change of refrangibility as a passage from D_1 to D_2. Great precautions are therefore required to prevent inequalities of temperature from destroying definition; and frequent shaking is generally necessary.

---

* "On the Magnifying-power of the Half-prism as a Means of obtaining great Dispersion, and on the general Theory of the Half-prism Spectroscope," *Proceedings of the Royal Society*, March 1, 1877.

† "Description of the Great Melbourne Telescope," *Phil. Trans.* 1869, p. 160.

‡ [1899. This arrangement, employed by Fraunhofer, has been revived for star spectroscopes.]

Some observers have thought that, apart from inequalities of temperature, ordinary bisulphide of carbon tends to arrange itself in strata of different refracting-power; but this does not seem to be established satisfactorily. In some recent experiments with a rotating stirrer, introduced with the view of promoting uniformity of temperature, I obtained evidence of a thin layer of moisture floating on the surface. Under the action of the stirrer this layer was broken up and the liquid rendered very irregular. A few lumps of chloride of calcium introduced into the liquid absorbed most of the layer; but an arrangement such that the free surface remained undisturbed would have been preferable.

Within the last few months Prof. Liveing* has proposed the use of a solution of iodide of mercury, which is considerably more dispersive than even bisulphide of carbon, the difference of indices between D and B amounting to ·017. This liquid is of a yellowish colour, and is hardly sufficiently transparent, even at the lower end of the spectrum, to make its use advisable in very powerful instruments. But for some purposes its great dispersion is an important recommendation. Using a single prism of 60° with an available thickness of $1\frac{1}{2}$ inch, I have obtained results which many pretentious instruments could not surpass.

Experience has not yet decided the question as to the relative advantages of large and small prisms. As generally used, large prisms have the disadvantages of requiring a greater quantity of material for a given thickness and of involving cumbrous and more expensive telescopes. The first might be avoided partially, and the second wholly, by the use of higher refracting angles, or (perhaps preferably) by the addition of half-prisms to the ends of

Fig. 19.

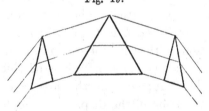

the train (fig. 19). That prisms may act as cylindrical telescopes was observed many years ago by Brewster†; and recently Christie has constructed half-prism spectroscopes in which this property is taken advantage of. In these instruments, however, the total thickness of glass is too small for high resolving-power.

In the arrangement of fig. 19 the rays from the collimator are received on *diminishing* half-prisms, by which the width of the beam is increased up to

---

* "On the Dispersion of a Solution of Mercuric Iodide," *Cambridge Proceedings*, May 19, 1879.

† *Optics* (London, 1853), p. 513.

the point suitable for the big prisms.   Afterwards the rays are concentrated
by *magnifying* half-prisms until the width of the beam is the same as at
first.   On this plan the larger prisms need be no *higher* than small ones;
and the quantity of glass corresponding to a given total thickness varies as
the first power, instead of as the square, of the linear dimension.

I have experimented with a pair of 60° prisms, 3 inches thick and only
¾ inch high.   The glass is of rather low density; so that when the position
of minimum deviation is adopted, the emergent beam is inconveniently wide.
With this material a larger angle would have been preferable; but much the
same result may be arrived at by turning the prisms a little, so as to increase
the angle of incidence on the first surface and the angle of emergence from
the last surface *.   In this way the incident and emergent beams are so far
narrowed that they can be received on small telescopes; and the combination
is very economical in comparison with one in which the position of minimum
deviation is adopted, involving, as it would, larger telescopes, and, unless a
loss of brightness were accepted, higher prisms.

Another point requiring attention in the discussion of the question of
large *versus* small prisms is the relative difficulty of securing the necessary
accuracy of surfaces in the two cases; but it can hardly be treated satisfactorily *à priori*.   It would, however, seem that small prisms have the advantage in this respect also.

So far I have supposed the prisms to be simple.   Of late years compound
prisms of flint and crown have come largely into fashion; and quite recently
M. Thollon has revived the use of compound prisms of bisulphide of carbon.
Mr Christie considers that compound prisms are, on the whole, to be preferred;
but I cannot help thinking that, except where direct vision is important,
their advantages have been overrated.   The difference of indices in the case
of crown glass for the rays D and B is ·0035, which is quite sufficiently high
to come into comparison with the corresponding numbers for flint and
bisulphide of carbon.   In a direct-vision prism the thickness of the crown is
greater than that of the flint; so that the loss of efficiency is even more
serious than a mere comparison of the dispersions would suggest.   See (6) § 4.

The principle of the compound prism is carried to its limit by employing
media of *equal* refracting-power for the part of the spectrum under examination.   For this purpose I chose bisulphide of carbon and flint glass.   With
Chance's "dense flint" the refractions are the same, and the difference of
dispersions relative to D and B is ·0056, *i.e.* about as great as the difference
for "double-extra-dense flint" and crown.   A dozen glass prisms of 90° were
cemented in a straight line on a strip of glass and immersed in bisulphide of
carbon.   The liquid is contained in a tube closed at its ends with glass
plates.   The "thickness" of each prism is 1 inch [2·54 cm.] and the height

---

* A pair of prisms thus arranged is called by Thollon a *couple*.

[parallel to the refracting edge] about $\frac{3}{4}$ inch, so that the total thickness is 12 inches [30·5 cm.]. The character of the glass is such that at ordinary temperatures the red rays pass without deviation. To observe other parts of the spectrum with advantage, it is necessary to mix a little ether [or benzole] with the bisulphide of carbon in order to lower its power. A similar result could be obtained by slightly warming the tube; but this course is not to be recommended.

The instrument, which was made by Hilger, gave excellent results at times, but showed the usual caprice of bisulphide of carbon. With the refracting edges vertical, the definition was usually good for twenty or thirty seconds after shaking up, but would often rapidly deteriorate afterwards. Although care was taken not to touch the tube with the hands, this effect was sometimes so persistent that I began to think I had evidence of a tendency to separate into distinct layers. When the edges of the prisms (and slit) were horizontal, the tube being also horizontal as before, the loss of definition after shaking still occurred, but could be remedied in great measure by a change of focus. Further experience, however, led me to attribute these effects to temperature-differences, caused perhaps by the ceiling of the laboratory being warmer than the floor. At any rate, they were greatly mitigated by wrapping round the tube strips of copper and numerous folds of cloth; and they could be produced with considerable persistency by touching the top of the naked tube for a few seconds with the hands. A difference of even $\frac{1}{100}$ of a degree Cent. between the upper and lower halves of the prisms might be expected to make itself apparent when the edges are vertical. The advantages of this construction are the elimination of reflection and the almost absolute immunity from defects due to errors in the separating surfaces; but they are rather dearly purchased. As might be expected, the best results as to definition are obtained when the tube is vertical; but such an arrangement is inconvenient, as it involves the aid of reflectors. Spectroscopes on this plan may perhaps be useful for special purposes; but the want of ready adaptability to different parts of the spectrum is a serious objection.

The general result of this discussion would seem to be in favour of a spectroscope with simple glass prisms of such angle that the reflected light is wholly polarized, the number of prisms being increased up to the point at which mechanical difficulties begin to interfere. With the aid of reflection, at least six prisms may be used twice over. When it becomes necessary to increase the size of the prisms, considerable economy may be effected by the introduction of half-prisms at the ends of the train, as already explained.

When the surfaces are not quite flat, or when the focus of the collimator is imperfectly adjusted, it becomes important to secure a rather exact perpendicularity between the edges of the prisms and a plane passing through their middle points; otherwise the linear image of a point may not be parallel to the length of the slit. Even when the plane of symmetry is obtained, there may remain considerable errors, due to *curvature* of the image, dependent upon the quantity denoted by γ in § 6. Much information as to the cause of bad definition may often be obtained by replacing the slit by a simple hole and examining the character of the resultant image. In many instruments [with refracting edges vertical] a great improvement is effected by narrowing the vertical aperture whenever the light will bear diminution. Such a result may be anticipated when with full aperture the top and bottom of the spectrum appear better defined than the central parts. The principal cause of error is probably a deviation of the extreme rays from horizontality in their passage through the prisms, due either to imperfect action of the collimator or to the curved faces of preceding prisms.

It is not easy to decide whether the highest resolving-power is more likely to be obtained by gratings or by prisms. Up to a certain point the resolving-power of gratings is principally a question of the *accuracy* with which the lines can be ruled. If the deviations of the lines from their proper places do not exceed (say) one-fourth of the interval between consecutive lines, the definition in the first spectrum will not be materially injured. To obtain corresponding results in the 2nd, 3rd, 4th,... spectrum, the errors must not exceed $\frac{1}{8}, \frac{1}{12}, \frac{1}{16},...$ respectively of the same interval[*]. Every effort should also be made to rule as great a number of lines as possible, even if it be necessary for this purpose to reduce their length.

I have lately had the opportunity of experimenting with two Rutherfurd gratings, one on glass, with nearly 12,000 lines (a present from Mr Rutherfurd himself), the other on speculum-metal, with nearly 14,000 lines (kindly lent me by Mr Spottiswoode). The lines are at the rate of 17,296 to the inch.

---

[*] [1899. This is a consequence of the general principle that errors of phase less than the quarter period do not seriously disturb definition. It has been suggested that the statement in the text is inconsistent with what was proved on p. 210, with regard to the performance of a grating composed of two halves, in each of which the ruling is accurate but is subject to an alteration of interval in passing from one half to the other. But the standard of accuracy now laid down holds good in any case, and the question regarding the performance (in the first spectrum) of the compound grating is simply whether the error of placing accumulates so far as to reach the amount of the quarter interval. Whether it will do so, or not, depends not only upon the assumed alteration of interval in the ratio 1000 : 1001, but also upon the total number of lines. The ideal grating with which the actual grating is to be compared must be supposed to have an interval equal to the mean of the two, and therefore differing from each by $\frac{1}{2000}$ part. In the middle of the grating the actual lines lie close to their ideal positions, but deviate increasingly on either side. The deviation attains the quarter interval at the 500th line, and thus if the whole grating include less than 1000 lines, its performance should not be much worse than if all the lines were accurately placed. The fact, proved on p. 210, that such a grating would fail to resolve the D-lines is quite in harmony, as will appear when it is remembered that an accurate grating of 1000 lines can only just do so. Though not important in the present connexion, it may be noted that a part of the error just discussed may be compensated by a change of focus. (*Enc. Brit.* "Wave Theory," p. 438, 1888.)]

Both these gratings give admirable results in the second spectrum, where, according to (14) § 4, the resolving-power in the orange exceeds that obtainable from 25 centims. of extra-dense flint. In the third spectrum the gain of resolving-power is still apparent, but illumination is rather deficient.

It is much to be wished that spectroscopists in possession of powerful instruments would compare their actual performances with those of which they are theoretically capable. A carefully arranged succession of tests of gradually increasing difficulty, like those applied to telescopes, would be of especial value. In my own observations I have usually attended principally to the definition of the fine lines bordering b_2 and to the double line b_3.

For experiments on dark heat, to which in some respects gratings are well adapted, resolving-power is secondary to illumination. In order to avoid confusion of spectra, it would be well to eliminate the second spectrum altogether, which could be effected by ruling equally strong lines at alternate intervals in the ratio of one to three.

[1899. A direct vision spectroscope, on the principle described on pp. 256, 257, has recently been constructed (see *Proc. Roy. Inst.* March 24, 1899) in which the linear dimension of the prisms is doubled. The "thickness" of each prism (along the hypothenuse) is 2 inches, and the total thickness for the 10 prisms employed is 20 inches. The edges of the prisms are horizontal and are disposed downwards; so that the upper part of a rectangular beam of light (which traverses the instrument once) passes through 20 inches of "dense flint" glass, while the lower part passes an equal thickness of bisulphide of carbon. When with the aid of reflectors the light is made to traverse the instrument *three* times, the thickness of glass (or CS_2) is 60 inches (152 cm.). Used in this way the spectroscope, when tested upon the lines E or b of Fraunhofer, performs as well as a good Rowland grating.]

63.

ON REFLECTION OF VIBRATIONS AT THE CONFINES OF TWO MEDIA BETWEEN WHICH THE TRANSITION IS GRADUAL.

[*Proceedings of the London Mathematical Society*, XI. pp. 51—56, 1880.]

MANY physicists, of whom may be especially mentioned Young, Fresnel, Poisson, Green, and Cauchy, have investigated the reflection of light or sound at the surface of separation of two uniform media of different mechanical properties. The transition from one medium to the other being treated as abrupt, the problem is of no great difficulty for the case of plane waves incident upon a plane surface of separation. It is of some interest to inquire what modifications would be caused by the substitution of a gradual for an abrupt transition, and the principal object of this paper is to give the details of one particular case, which admits of pretty simple treatment.

It is evident, from the first, that the transition may be supposed to be so gradual that no sensible reflection would ensue. No one would expect a ray of light to undergo reflection in passing through the earth's atmosphere as a consequence of the gradual change of density with elevation. At first sight, indeed, the case of so-called total reflection may appear to be an exception, as it is independent of the suddenness of transition; but this only shows that the phenomenon is inaccurately described by its usual title. It is, in strictness, a particular case of refraction, rather than of reflection, and must be so considered in theoretical work, although, no doubt, the name of total reflection will be retained whenever, as in constructing optical instruments, we have to deal with effects rather than with causes.

Admitting, then, that reflection proper is due to suddenness of transition, we have still to inquire what degree of suddenness is requisite. It is not difficult to see that the quantity with which the thickness of the transitional layer comes into comparison is the wave-length of the vibration; so that, when the thickness is a large multiple of the wave-length, there is little reflection, but when, on the other hand, the wave-length is a large multiple of the thickness, the reflection is sensibly as copious as if the transition were absolutely abrupt.   There is thus a considerable distinction, in practice, between the cases of sound and of light, the wave-length of the one being some million times greater than that of the other.   When sound is reflected in air from ordinary solids or liquids, the transition between the media may be treated as abrupt with abundant accuracy; but it is not so certain that a similar treatment is adequate to the case of light.   It is probable, however, that in the case of light passing from fluids to solids, or from one fluid to a second which does not mix with it—*e.g.*, from bisulphide of carbon to water—the phenomena are not materially influenced by in-sufficient suddenness of transition.   On the other hand, when the two fluids are miscible, it would not be easy to complete an experiment before the abruptness of transition is so far broken down by diffusion that the thickness of the transitional layer amounts to a multiple of $\frac{1}{40000}$ of an inch.

The problem of gradual transition includes, of course, that of a variable medium.   The particular case which I have taken is that of a stretched string, whose longitudinal density varies as the inverse square of the abscissa (measured along its length), vibrating transversely.   The same analysis is strictly applicable to other cases of vibrations, which I need not stop to specify, and the results will illustrate the general character of the phenomena to be expected in all cases.   If y, denoting the transverse displacement, be proportional to e^{ipt}, the equation which it must satisfy as a function of x, is

$$\frac{d^2y}{dx^2} + n^2 x^{-2} y = 0, \quad \dots\dots\dots\dots\dots\dots\dots(1)$$

where n^2 is some positive constant, of the nature of an abstract number*.

The solution of (1) is

$$y = A x^{\frac{1}{2}+im} + B x^{\frac{1}{2}-im}, \quad \dots\dots\dots\dots\dots\dots(2)$$

where

$$m^2 = n^2 - \tfrac{1}{4}. \quad \dots\dots\dots\dots\dots\dots\dots(3)$$

If m be real, that is, if $n > \frac{1}{2}$, we may obtain, by supposing $A = 0$, as a final solution in real quantities,

$$y = C x^{\frac{1}{2}} \cos(pt - m \log x + \epsilon), \quad \dots\dots\dots\dots\dots\dots(4)$$

which represents a positive progressive wave, in many respects similar to those propagated in uniform media.

Let us now suppose that, to the left of the point $x = x_1$, the variable medium is replaced by one of uniform constitution, such that there is

---

* *Theory of Sound*, § 141.

no discontinuity of density at the point of transition; and let us inquire, what reflection a positive progressive wave in the uniform medium will undergo on arrival at the variable medium. It will be sufficient to consider the case where m is real, that is, where the change of density is but moderately rapid.

By supposition, there is no negative wave in the variable medium, so that $A = 0$ in (2). Thus

$$y = Bx^{\frac{1}{2}-im}, \qquad \frac{dy}{dx} = (\tfrac{1}{2} - im)\, Bx^{-\frac{1}{2}-im};$$

and, when $x = x_1$,

$$\frac{dy}{y\,dx} = \frac{\tfrac{1}{2} - im}{x_1}. \quad\dots\dots\dots\dots\dots\dots\dots\dots(5)$$

The general solution for the uniform medium may be written

$$y = He^{-in(x-x_1)/x_1} + Ke^{+in(x-x_1)/x_1}, \quad\dots\dots\dots\dots(6)$$

from which, when $x = x_1$,

$$\frac{dy}{y\,dx} = -\frac{in}{x_1}\frac{H - K}{H + K}. \quad\dots\dots\dots\dots\dots\dots(7)$$

In equation (6), H represents the amplitude of the incident positive wave, and K the amplitude of the reflected negative wave. The condition to be satisfied at $x = x_1$ is expressed by equating the values of $dy/y\,dx$ given by (5) and (7). Thus

$$\frac{K}{H} = \frac{i(n - m) + \tfrac{1}{2}}{i(n + m) - \tfrac{1}{2}}, \quad\dots\dots\dots\dots\dots\dots(8)$$

which gives, in symbolical form, the ratio of the reflected to the incident vibration.

Having regard to (3), we may write (8) in the form

$$\frac{K}{H} = \frac{e^{i\gamma}}{2(n + m)}; \quad\dots\dots\dots\dots\dots\dots\dots(9)$$

so that the real amplitude of the reflected wave is $\tfrac{1}{2}(n + m)^{-1}$ of that of the incident. Thus, as was to be expected, when n and m are great, i.e., when the density changes slowly in the variable medium, there is but little reflection.

Passing on now to the more important problem, we will suppose that the variable medium extends only so far as the point $x = x_2$, beyond which the density retains uniformly its value at that point. A positive wave travelling at first in a uniform medium of density proportional to x_1^{-2}, passes at the point $x = x_1$ into a variable medium of density proportional to x^{-2}, and again, at the point $x = x_2$, into a uniform medium of density proportional to x_2^{-2}. The velocities of propagation are inversely proportional to the square roots of the densities, *i.e.*, vary as x, so that, if μ be the refractive index between the extreme media,

$$\mu = \frac{x_1}{x_2}. \quad \dots\dots\dots\dots\dots\dots\dots\dots\dots\dots(10)$$

The thickness (d) of the layer of transition is

$$d = x_2 - x_1. \quad \dots\dots\dots\dots\dots\dots\dots\dots\dots(11)$$

The wave-lengths in the two media are given by

$$\lambda_1 = 2\pi x_1/n, \qquad \lambda_2 = 2\pi x_2/n;$$

so that

$$n = \frac{2\pi d}{\lambda_2 - \lambda_1} = \frac{2\pi d}{(\mu^{-1} - 1)\,\lambda_1}. \quad \dots\dots\dots\dots\dots\dots\dots(12)$$

For the first medium we take, as before,

$$y = H e^{-in(x-x_1)/x_1} + K e^{+in(x-x_1)/x_1}, \quad \dots\dots\dots\dots\dots(6)$$

giving, when $x = x_1$,

$$\frac{dy}{y\,dx} = -\frac{in}{x_1}\frac{H - K}{H + K} = -\frac{in\theta}{x_1}, \quad \dots\dots\dots\dots(7)$$

if, for brevity, we write θ for $\dfrac{H - K}{H + K}$.

For the variable medium,

$$y = A x^{\frac{1}{2}+im} + B x^{\frac{1}{2}-im}, \quad \dots\dots\dots\dots\dots\dots(2)$$

giving, when $x = x_1$,

$$\frac{dy}{y\,dx} = x_1^{-1}\frac{(\frac{1}{2} + im)\,A x_1^{im} + (\frac{1}{2} - im)\,B x_1^{-im}}{A x_1^{im} + B x_1^{-im}}. \quad \dots\dots\dots\dots(13)$$

Hence the condition to be satisfied at $x = x_1$ gives

$$\tfrac{1}{2} + im\,\frac{A x_1^{im} - B x_1^{-im}}{A x_1^{im} + B x_1^{-im}} = -in\theta;$$

whence

$$\frac{A}{B} = x_1^{-2im}\frac{im - in\theta - \frac{1}{2}}{im + in\theta + \frac{1}{2}}. \quad \dots\dots\dots\dots\dots(14)$$

The condition to be satisfied at $x = x_2$ may be deduced from (14), by substituting x_2 for x_1, putting at the same time $\theta = 1$ in virtue of the supposition that in the second medium there is no negative wave. Hence, equating the two values of $A : B$, we get

$$x_1^{-2im}\frac{im - in\theta - \frac{1}{2}}{im + in\theta + \frac{1}{2}} = x_2^{-2im}\frac{im - in - \frac{1}{2}}{im + in + \frac{1}{2}}, \quad \dots\dots\dots\dots(15)$$

as the equation from which the reflected wave in the first medium is to be found. Having regard to (3), we get

$$\theta = \frac{H - K}{H + K} = \frac{m + n + \frac{1}{2}i + \mu^{2im}(m - n - \frac{1}{2}i)}{m + n - \frac{1}{2}i + \mu^{2im}(m - n + \frac{1}{2}i)};$$

so that

$$\frac{K}{H} = \frac{-i + \mu^{2im}i}{2(m + n) + 2\mu^{2im}(m - n)}. \quad\ldots\ldots\ldots\ldots\ldots\ldots(16)$$

This is the symbolical solution. To interpret it in real quantities, we must distinguish the cases of m real and m imaginary. If the transition be not too sudden, m is real, and (16) may be written

$$\frac{K}{H} = \frac{i}{2} \frac{-1 + \cos(2m \log \mu) + i \sin(2m \log \mu)}{m + n + (m - n) \cos(2m \log \mu) + i(m - n) \sin(2m \log \mu)}$$

Thus the expression for the ratio of the *intensities* of the reflected and incident waves is, after reduction,

$$\frac{\sin^2(m \log \mu)}{4m^2 + \sin^2(m \log \mu)}. \quad\ldots\ldots\ldots\ldots\ldots\ldots\ldots(17)$$

If m be imaginary, we may write $im = m'$; (16) then gives, for the ratio of intensities,

$$\frac{(\mu^{m'} - \mu^{-m'})^2}{(\mu^{m'} - \mu^{-m'})^2 + 16m'^2}; \quad\ldots\ldots\ldots\ldots\ldots\ldots(18)$$

or, if we introduce the notation of hyperbolic trigonometry,

$$\frac{\sinh^2(m' \log \mu)}{\sinh^2(m' \log \mu) + 4m'^2}. \quad\ldots\ldots\ldots\ldots\ldots\ldots(19)$$

For the critical value $m = 0$, we get, from (17) or (19),

$$\frac{(\log \mu)^2}{4 + (\log \mu)^2}. \quad\ldots\ldots\ldots\ldots\ldots\ldots\ldots(20)$$

These expressions allow us to trace of the effect of a more or less gradual transition between media of given index. If the transition be absolutely abrupt, $n = 0$, by (12); so that $m' = \frac{1}{2}$. In this case, (18) gives us Young's well-known formula

$$\left(\frac{\mu - 1}{\mu + 1}\right)^2. \quad\ldots\ldots\ldots\ldots\quad\ldots\ldots\ldots(21)$$

Since $x^{-1} \sinh x$ increases continually from $x = 0$, the ratio (19) increases continually from $m' = 0$ to $m' = \frac{1}{2}$, *i.e.*, diminishes continually from the case of sudden transition $m' = \frac{1}{2}$, when its value is (21), to the critical case $m' = 0$, when its value is (20), after which this form no longer holds good. When $m' = 0$, $n = \frac{1}{2}$, and, by (12), $d = (\lambda_1 - \lambda_2)/4\pi$.

When $n > \frac{1}{2}$, (17) is the appropriate form. We see from it that, as n increases, the reflection diminishes, until it vanishes, when $m \log \mu = \pi$, *i.e.*, when

$$n^2 = \tfrac{1}{4} + \frac{\pi^2}{(\log \mu)^2}. \quad\dots\dots\dots\dots\dots\dots(22)$$

With a still more gradual transition the reflection revives, reaches a maximum, again vanishes when $m \log \mu = 2\pi$, and so on.

As a numerical example I will take the case of $\mu = \frac{2}{3}$, applicable to the reflection of luminous waves in passing from glass to air. The value of $\log_e \frac{3}{2}$ is ·40546520. For the sudden transition, we have, from (21), ·040. When $m' = 0$, *i.e.*, when $d = (\lambda_2 - \lambda_1)/4\pi$, the reflection is, by (20), ·039478, but very little less. For moderate values of m', we may derive, from (18),

$$\frac{(\log \mu)^2}{(\log \mu)^2 + 4} \left\{ 1 + \frac{(\log \mu)^2}{(\log \mu)^2 + 4} \frac{4m'^2}{3} + \dots \right\}; \quad\dots\dots\dots\dots(23)$$

or, in the present example,

$$·039478 \left(1 + \frac{4m'^2}{3} \times ·039478 \right).$$

There is thus no appreciable variation of reflection from $m' = \frac{1}{2}$ to $m' = 0$. The reflection vanishes when

$$n^2 = \tfrac{1}{4} + \frac{\alpha^2 \pi^2}{(·4055)^2},$$

α being any integer. Even when $\alpha = 1$, the first term in the value of n^2 is small relatively to the second, and we may take, with sufficient approximation,

$$n = \frac{\alpha \pi}{·4055}; \text{ or, by (12)}, \ d = \frac{\alpha (\lambda_2 - \lambda_1)}{·811}.$$

The thickness of the layer of transition at which the reflection first vanishes is thus a little greater than the difference of the wave-lengths. The revivals of reflection which occur when the transition is still more gradual are not important. The approximate values are $(2/3\pi)^2$, $(2/5\pi)^2$, &c., the reflection with abrupt transition being taken as standard. Thus, at the first revival, the reflection is only about $\frac{1}{22}$ of its original value.

64.

ON THE MINIMUM ABERRATION OF A SINGLE LENS FOR PARALLEL RAYS.

[*Proceedings of the Cambridge Philosophical Society*, III. pp. 373—375, 1880.]

IT is well known that when the material of a lens is plate glass ($\mu = 1\cdot5$), the aberration is least when the lens is double convex, the radius of the anterior surface r being equal to $\frac{7}{12}$ of the focal length f, and that of the posterior surface $(-s)$ equal to $\frac{7}{2}f$. The residual aberration δf is then given by

$$\delta f = -\tfrac{15}{14} \frac{y^2}{f}, \qquad \dots\dots\dots\dots\dots\dots\dots\dots\dots\dots(1)$$

y being the semiaperture*.

In the older works on Optics the special supposition that $\mu = 1\cdot5$ is introduced at the beginning of the calculations, so that the results are not available for an examination of the effect of a varying refractive index; but it has been repeatedly asserted that lenses formed of diamonds or of other precious stones of high refracting power have an almost inappreciable aberration†. In Coddington's *Optics*, § 89, the minimum aberration for $\mu = 2$ is stated to be only $\frac{1}{16}y^2/f$, but no algebraical calculation is given.

The general expression of the aberration for parallel rays is‡

$$-\delta f : \frac{y^2}{f} = \frac{\mu-1}{2\mu^2} \left\{ \frac{1}{r^3} + \left(\frac{\mu+1}{f} - \frac{1}{s} \right) \left(\frac{1}{f} - \frac{1}{s} \right)^2 \right\} f^3, \qquad \dots\dots\dots\dots(2)$$

while r, s, and f are connected by

$$\frac{1}{f} = (\mu - 1) \left(\frac{1}{r} - \frac{1}{s} \right). \qquad \dots\dots\dots\dots\dots\dots\dots\dots(3)$$

---

* Parkinson's *Optics*, §§ 130, 131.
† E.g. *Optics. Encyclopædia Britannica*, 1842.
‡ Parkinson's *Optics*, § 129.

Writing for brevity R, S, F respectively for r^{-1}, s^{-1}, f^{-1}, and taking $G = \dfrac{\mu F}{\mu - 1}$, so that $-S = \dfrac{G}{\mu} - R$, we get

$$\frac{1}{r^3} + \left(\frac{\mu+1}{f} - \frac{1}{s}\right)\left(\frac{1}{f} - \frac{1}{s}\right)^2 = G\left\{R^2(\mu+2) - RG(2\mu+1) + \mu G^2\right\}$$

$$= G\left\{\sqrt{(\mu+2)}\cdot R - \frac{(2\mu+1)\,G}{2\sqrt{(\mu+2)}}\right\}^2 + \frac{\mu^3 F^3}{(\mu-1)^3}\frac{4\mu-1}{4(\mu+2)}\cdot \quad \text{.......(4)}$$

Since $\mu > 1$, both terms are of the same sign, and the aberration can never vanish. If f and y be given, the aberration is least when

$$\sqrt{(\mu+2)}\cdot R = \frac{2\mu+1}{2\sqrt{(\mu+2)}}\cdot G,$$

that is, when

$$r = \frac{2(\mu+2)(\mu-1)}{\mu(2\mu+1)}f. \quad\text{.........................(5)}$$

The corresponding value of s is

$$-s = \frac{2(\mu+2)(\mu-1)}{4+\mu-2\mu^2}f, \quad\text{.........................(6)}$$

so that

$$-s : r = \frac{\mu(2\mu+1)}{4+\mu-2\mu^2}, \quad\text{.........................(7)}$$

which agrees with the result of [Parkinson's *Optics*] § 130.

When this condition is satisfied, the second term of (4) gives for the minimum aberration

$$-\delta f : \frac{y^2}{f} = \frac{\mu(4\mu-1)}{8(\mu-1)^2(\mu+2)}, \quad\text{.....................(8)}$$

which is applicable to all values of μ.

If $\mu = 2$, (8) gives

$$-\delta f : \frac{y^2}{f} = \tfrac{7}{16},$$

not $\tfrac{1}{16}$, as stated by Coddington. The aberration tends indeed to become less as μ increases, but it remains considerable for all substances known in nature.

It seems to have been thought evident that great advantage would result from higher refracting power on account of its allowing the use of more moderate curvatures. It appears however from (5) and (6) that as μ increases, r and s do not tend to become infinite for the form of minimum aberration, but approach the finite value f.

65.

ACOUSTICAL OBSERVATIONS. III.

[*Philosophical Magazine*, IX. pp. 278—283, 1880.]

Intermittent Sounds.

IN the *Philosophical Magazine* for May 1875, Prof. A. M. Mayer describes some experiments on this subject, made by rotating a perforated cardboard disk between a resonator and a vibrating fork. "When the disk is stationary with one of its openings opposite the mouth of the resonator, it is evident that the ear will experience a simple sonorous sensation when a tuning-fork is brought near the mouth of the resonator. On revolving the perforated disk, two additional or secondary sounds appear—one slightly above, the other slightly below the pitch of the fork. An increasing velocity of rotation causes the two secondary sounds to diverge yet further from the note of the beating fork, until, on reaching a certain velocity, the two secondary sounds become separated from each other by a major sixth, while at the same moment a resultant sound appears, formed by the union of the sound of the fork with the upper and lower of the secondary sounds. This resultant is the lower second octave of the note given by the fork. On further increasing the velocity of rotation of the disk, the two secondary sounds and the resultant disappear, and the ear experiences only the sensation of the simple sound produced by the fork, whose beats at this stage of the experiment have blended into a smooth continuous sensation."

In concluding his paper Prof. Mayer calls attention "to the evident difference existing between the dynamic constitution of the sonorous waves belonging to beating pulses produced by the action of a perforated rotating disk on a continuous stream of sonorous vibrations, and those waves which cause beats, and which are formed by the joint action of sonorous vibrations differing in pitch. That these two kinds of beats are alike in their effects when following in the same rapidity I have assumed to be the fact in this paper."

At the time when Prof. Mayer's paper first appeared, I examined this question more closely; and some of my results were referred to in a discussion before the Musical Association. The difference between the two kinds of beats is considerable. If there are two vibrations of equal amplitude and slightly differing frequencies, represented by $\cos 2\pi n_1 t$ and $\cos 2\pi n_2 t$, the resultant may be expressed by

$$2 \cos \pi (n_1 - n_2) t . \cos \pi (n_1 + n_2) t,$$

and may be regarded as a vibration of frequency $\frac{1}{2}(n_1 + n_2)$, and of amplitude $2 \cos \pi (n_1 - n_2) t$. Hence, in passing through zero, the amplitude changes sign, which is equivalent to a *change of phase of* 180°, if the amplitude be regarded as always positive. This change of phase is readily detected by measurement in drawings traced by machines for compounding vibrations. If a force of the above character act upon a system whose natural frequency is $\frac{1}{2}(n_1 + n_2)$, the effect produced is comparatively small. If the system start from rest, the successive impulses cooperate at first, but after a time the later impulses begin to destroy the effect of former ones. The greatest response is given to forces of frequency n_1 and n_2, and not to a force of frequency $\frac{1}{2}(n_1 + n_2)$.

On the other hand, when a single vibration is rendered intermittent by the periodic interposition of an obstacle, there is no such change of phase in consecutive revivals. If a force of this character act upon an isochronous system, the effect is indeed less than if there were no intermittence; but as all the impulses operate in the same sense without any antagonism, the response is powerful. An intermittent vibration or force may be represented by

$$2 (1 + \cos 2\pi mt) \cos 2\pi nt,$$

in which n is the frequency of the vibration, and m the frequency of intermittence. The amplitude is always positive, and varies between the values 0 and 4. By ordinary trigonometrical transformation the above expression may be put into the form

$$2 \cos 2\pi nt + \cos 2\pi (n + m) t + \cos 2\pi (n - m) t;$$

which shows that the intermittent vibration is equivalent to *three* simple vibrations of frequencies n, $n + m$, and $n - m$. This is the explanation of the secondary sounds observed by Mayer. When m is equal to $\frac{1}{4}n$, $n + m : n - m = 5 : 3$, or the interval between the secondary sounds is a major sixth. The frequency of the resultant sound is m (that is, $\frac{1}{4}n$); and its pitch is two octaves below that of the original vibration.

In the *Edinburgh Proceedings* for June 1878, an experiment similar to Mayer's is described by Professors Crum Brown and Tait, and is explained in the above manner.

If the intensity of the intermittent sound rise more suddenly to its maximum, we may take $4\cos^4 \pi mt \cos 2\pi nt$; and this may be transformed into

$$\tfrac{3}{2}\cos 2\pi nt + \cos 2\pi (n+m) t + \cos 2\pi (n-m) t$$
$$+ \tfrac{1}{4}\cos 2\pi (n+2m) t + \tfrac{1}{4}\cos 2\pi (n-2m) t.$$

There are now *four* secondary sounds, the frequencies of the two new ones differing twice as much as before from that of the primary sound.

Other cases might be treated in the same way; but my object at present is to describe a modified form of the experiment which I planned some years ago, but first carried out last summer.

Fig. 1.

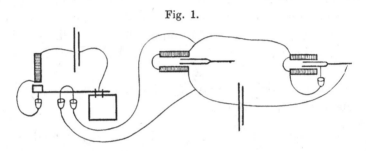

The desired forces were obtained electromagnetically. A fork interrupter of frequency 128 gave a periodic current, by the passage of which through an electromagnet a second fork of like pitch could be excited. The action of this current on the second fork could be rendered intermittent by short-circuiting the electromagnet. This was effected by another interrupter of frequency 4, worked by an *independent* current from a Smee cell. To excite the main current a Grove cell was employed. When the contact of the second interrupter was permanently broken, so that the main current passed continuously through the electromagnet, the fork was, of course, most powerfully affected when tuned to 128. Scarcely any response was observable when the pitch was changed to 124 or 132. But if the second interrupter was allowed to operate, so as to render the periodic current through the electromagnet intermittent, then the fork would respond powerfully when tuned to 124 or 132 as well as when tuned to 128, but not when tuned to intermediate pitches, such as 126 or 130.

The operation of the intermittence in producing a sensitiveness which would not otherwise exist, is easily understood. When a fork of frequency 124 starts from rest under the influence of a force of frequency 128, the impulses cooperate at first, but after $\tfrac{1}{8}$ of a second the new impulses begin to oppose the earlier ones. After $\frac{1}{4}$ of a second, another series of impulses begins whose effect agrees with that of the first, and so on. Thus if all these impulses are allowed to act, the resultant effect is trifling; but if every alternate series is stopped off, a large vibration accumulates.

A New Form of Siren.

Some years ago[*], I observed that a light pivoted blade is set into rapid rotation when exposed to wind. The phenomenon is of the same character as the rotation of a slip of paper falling freely in air, which was discussed a long while since by Prof. Maxwell[†]. In both cases the rotation may occur in *either direction*, proving that its cause is not to be looked for in any want of symmetry. But the view expressed by Maxwell does not appear to apply to the pivoted blade; and I think that the real explanation is yet to be discovered. At present, however, I am concerned merely with an application.

Fig. 2.

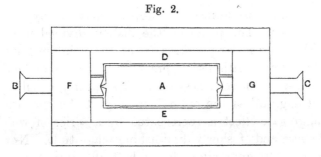

In fig. 2, A is a blade cut out of sheet brass and provided with sharp projecting points, which bear in hollows at the ends of the screws B, C. These screws pass through a small wooden frame FG, and are adjusted until the blade can turn freely but without perceptible shake. D and E are pieces of cardboard or sheet metal, fitting pretty closely to the blade when in the same plane with them, so that in this position of the blade the passage through the frame is almost closed. As the blade turns, it acts the part of a revolving stopcock.

In the summer of 1878 I made several sirens on this plan, which performed well. One of them is represented about full size in the figure. If the wind from the bellows is admitted symmetrically, they will revolve in either direction, and soon acquire sufficient speed to give a note of moderate pitch. The position of maximum obstruction is for small displacements a position of stable equilibrium. If a larger displacement is given, the vibration tends of itself to increase up to a certain point, or even to pass

---

* "On the Resistance of Fluids," *Phil. Mag.* Dec. 1876. [Art. XLII, p. 292.]
† *Cambridge and Dublin Math. Journal*, 1854.

into continuous rotation; but the precise behaviour in this respect probably depends upon the details of construction.

The Acoustical Shadow of a Circular Disk.

In a well-known experiment, suggested by Poisson, a bright point is observed in the centre of the shadow of a circular disk on which waves of light are directly incident. It is some years since I first attempted to obtain the acoustical analogue of this beautiful phenomenon; but my efforts were without success until a few months since. The difficulties to be overcome are entirely different in optics and in acoustics, on account of the immense disproportion of wave-lengths. In the former case the disk must be small and accurately shaped, and the source of light must be of very small angular magnitude—in practice, an image of the sun formed by a distant lens of short focus. In the latter case the difficulty is to arrange the experiment on a scale that shall be adequate in comparison with the wave-length of the sound.

The best way of considering the subject theoretically is with the use of Huyghens's zones. The plane of the disk is divided into zones by its intersection with spheres whose centres are at the point under consideration, and whose radii form an arithmetical progression with common difference $\frac{1}{2}\lambda$. The vibrations due to these zones are at first nearly equal, but gradually diminish to nothing, unless the outer boundary of the aperture is circular; and thus the aggregate effect is represented by a series of which the terms are of opposite sign and of slowly diminishing magnitude. Now the sum of such a series is equal approximately to half its first term; so that the whole effect of the aperture outside the disk is independent of the disk's diameter—that is to say, is the same as if no obstacle at all were present.

This way of regarding the matter shows at once what degree of accuracy is required in the figure of the circumference, which must not sensibly encroach upon the first exterior zone. If x be the radius of a circle in the plane of the disk, b the distance of the point under consideration, and $r^2 = x^2 + b^2$, $dx = r\,dr/x$; so that if $dr = \frac{1}{2}\lambda$, $dx = r\lambda/2x$. If, therefore, x be the radius of the disk, the radial error should be a small fraction of $r\lambda/2x$.

In like manner, we may form an estimate of the *size* of the bright spot, a subject which has been treated analytically by Airy[*]. If the disk be moved laterally through the width of one zone, it is clear that the effect at the old point will be materially changed. Hence the diameter of the bright spot is comparable with $r\lambda/x$, and its apparent magnitude as seen from the disk is comparable with λ/x. For the full success of the experiment, the apparent magnitude of the luminous source should be of the same order.

[*] *Phil. Mag.* May 1841.

When we pass to the analogous experiment in acoustics, it is of course impossible to retain any approximation to optical conditions. Instead of a ratio of $\lambda : x$, equal to, say, $\frac{1}{10000}$, we must be satisfied with some such value as $\frac{1}{20}$. In order to diminish λ as far as possible, it is advisable to use sounds of very high pitch, which have the additional advantage of readily exciting sensitive flames. I have found it best to work indoors, in which case a disk of 15 inches diameter is suitable; with a much larger disk and in an ordinary room there would hardly be sufficient free space on all sides. I have tried a considerable variety of sources of sound, including electric sparks, a small electric bell, and a Galton's whistle; but the best results were obtained with a bird-call and with a squeaky toy-reed.

On November 20 the source was a bird-call blown with a pressure of 4 inches [inch = 2·54 cm.] of water, and was placed about 20 inches from the 15 inches disk. The observation was made at a distance of 24 inches on the further side of the disk, and was successful both with the ear and with the sensitive flame. In the former case I employed a plate of wood bored with a hole about $\frac{1}{4}$ inch in diameter, and held against the side of the head in such a position that the hole was opposite the ear-passage. The head was moved about until the position of maximum sound was determined, and was then withdrawn, leaving the plate *in situ*. In order to verify conveniently that the position of maximum sound was really at the centre of the shadow, a hole was bored through the centre of the disk, which could be closed with a cork during the adjustment of the ear-plate. When the adjustment was complete, the cork was removed; and then the eye placed behind the ear-plate would see the source of sound through the two holes. With a little practice the central point could be picked out almost as well by ear as by eye.

With a sensitive flame the observation was even easier. The most suitable is that from a pin-hole burner brought near the flaring-point by a gas-pressure of about 10 inches of water. To get the best result, the pressure must be carefully adjusted; and in order to avoid disturbance, it is advisable to move the source rather than the flame. When the place of maximum effect has been determined, the cork is removed from the central hole of the disk, and the gas is lowered. By looking just over the burner it is then easy to see whether or not the source of sound occupies the central position.

On November 24 the toy-reed was substituted for the bird-call, the disk and distances being the same as before. In the case of this source the experiment succeeded better with the flame than with the ear.

On a subsequent occasion a larger disk, of 30 inches diameter, was tried; but the results were not so good, probably in consequence of the increased effect of reflection from the floor and walls of the room.

66.

ON THE STABILITY, OR INSTABILITY, OF CERTAIN FLUID MOTIONS.

[*Proceedings of the London Mathematical Society*, XI. pp. 57—70, 1880.]

IN a former communication to the Society on the "Instability of Jets[*]," I applied a method due to Sir W. Thomson, to calculate the manner of falling away from equilibrium of jets bounded by one or more surfaces of discontinuity. Such interest as these investigations possessed was due principally to the possibility of applying their results to the explanation of certain acoustical phenomena relating to sensitive flames and smoke jets. But it soon appeared that in one important respect the calculations failed to correspond with the facts.

To fix the ideas, let us take the case of an originally plane surface of separation, on the two sides of which the fluid moves with equal and opposite constant velocities ($\pm V$). In equilibrium the elevation h, at every point x along the surface, is zero. It is proved that, if initially the surface be at rest in the form defined by $h = H \cos kx$, then, after a time t, its form is given by

$$h = H \cos kx \, \cosh kVt, \dots\dots\dots\dots\dots\dots\dots(1)$$

provided that, throughout the whole time contemplated, the disturbance is small. In the same sense as that in which the frequency of vibration measures the stability of a system vibrating about a configuration of stable equilibrium, so the coefficient kV of t, in equations such as (1), measures the instability of an unstable system; and we see, in the present case, that the instability increases without limit with k; that is to say, the shorter the wave-length of the sinuosities on the surface of separation, the more rapidly are they magnified.

The application of this result to sensitive jets would lead us to the conclusion that their sensitiveness increases indefinitely with pitch. It is

---

[*] *Proceedings*, vol. x. p. 4, Nov. 14, 1878. [Art. LVIII.]

true that, in the case of certain flames, the pitch of the most efficient sounds is very high, not far from the upper limit of human hearing; but there are other kinds of sensitive jets on which these high sounds are without effect, and which require for their excitation a moderate or even a grave pitch.

A probable explanation of the discrepancy readily suggests itself. The calculations are founded upon the supposition that the changes of velocity are discontinuous—a supposition which cannot possibly agree with reality. In consequence of fluid friction a surface of discontinuity, even if it could ever be formed, would instantaneously disappear, the transition from the one velocity to the other becoming more and more gradual, until the layer of transition attained a sensible width. When this width is comparable with the wave-length of the sinuosity, the solution for an abrupt transition ceases to be applicable, and we have no reason for supposing that the instability would increase for much shorter wave-lengths.

In the following investigations, I shall suppose that the motion is entirely in two dimensions, parallel (say) to the plane xy, so that (in the usual notation) w is zero, as well as the rotations ξ, η. The rotation ζ parallel to z is connected with the velocities u, v by the equation

$$\zeta = \tfrac{1}{2}\left(\frac{du}{dy} - \frac{dv}{dx}\right). \quad\dots\dots\dots\dots\dots\dots(2)$$

When the phenomena under consideration are such that the compressibility may be neglected, the condition that no fluid is anywhere introduced or abstracted, gives

$$\frac{du}{dx} + \frac{dv}{dy} = 0. \quad\dots\dots\dots\dots\dots\dots(3)$$

In the absence of friction, ζ remains constant for every particle of the fluid; otherwise, if ν be the kinematic viscosity, the general equation for ζ is

$$\frac{\partial\zeta}{\partial t} = \xi\frac{dw}{dx} + \eta\frac{dw}{dy} + \zeta\frac{dw}{dz} + \nu\nabla^2\zeta^*, \quad\dots\dots\dots\dots(4)$$

where

$$\partial/\partial t = d/dt + u\,d/dx + v\,d/dy + w\,d/dz, \quad\dots\dots\dots\dots(5)$$

and

$$\nabla^2 = d^2/dx^2 + d^2/dy^2 + d^2/dz^2. \quad\dots\dots\dots\dots(6)$$

For the proposed applications to motion in two dimensions, these equations reduce to

$$\partial\zeta/\partial t = \nu\nabla^2\zeta, \quad\dots\dots\dots\dots\dots\dots(7)$$

$$\partial/\partial t = d/dt + u\,d/dx + v\,d/dy, \quad\dots\dots\dots\dots\dots(8)$$

$$\nabla^2 = d^2/dx^2 + d^2/dy^2, \quad\dots\dots\dots\dots\dots(9)$$

while the two other equations similar to (4) are satisfied identically.

* Lamb's *Motion of Fluids*, p. 243.

In order to investigate the influence of friction on stratified motion, we may now suppose that v is zero, while u and ζ are functions of y only. Our equations then give simply

$$d\zeta/dt = \nu\, d^2\zeta/dy^2, \quad\dots\dots\dots\dots\dots\dots(10)$$

which shows that the rotation ζ is *conducted* according to precisely the same laws as heat. In the case of air at atmospheric pressure, the value of ν is, according to Maxwell's experiments, $\nu = {\cdot}16$*, not differing greatly from the number $({\cdot}22)$ corresponding to the conduction of temperature in *iron*.

The various solutions of (10), discovered by Fourier, are at once applicable to our present purpose. In the problem already referred to, of a surface of discontinuity $y = 0$, separating portions of fluid moving with different but originally constant velocities, the rotation is at first zero, except upon the surface itself, but it is rapidly diffused into the adjacent fluid. At time t its value at any point y is

$$\zeta = \int_{-\infty}^{+\infty} \zeta\,dy \cdot \frac{e^{-y^2/4\nu t}}{2\sqrt{(\pi\nu t)}}, \quad\dots\dots\dots\dots\dots(11)$$

and

$$\int_{-\infty}^{+\infty} \zeta\,dy = \tfrac{1}{2}\int \frac{du}{dy}\,dy = \tfrac{1}{2}(V_2 - V_1), \quad\dots\dots\dots(12)$$

if V_2, V_1 are the velocities on the positive and negative sides of the surface respectively. If $y^2 = 4\nu t$, the value of ζ is less than that to be found at $y = 0$, in the ratio $e:1$. Thus, after a time t, the thickness of the layer of transition $(2y)$ is comparable in magnitude with $1{\cdot}6\sqrt{t}$; for example, after one second it may be considered to be about $1\frac{1}{2}$ centimetres. In the case of water, the coefficient of conductivity is much less. It seems that $\nu = {\cdot}014$; so that, after one second, the layer is about half a centimetre thick.

The circumstances of a two-dimensional jet will be represented by supposing the velocity to be limited initially to an infinitely thin layer at $y = 0$. It is convenient here to use the velocity u itself instead of ζ. Since $\zeta = \tfrac{1}{2}du/dy$,

$$du/dt = \nu\, d^2u/dy^2, \quad\dots\dots\dots\dots\dots\dots(13)$$

and thus the solution is of the same form as before:

$$u = \int_{-\infty}^{+\infty} u\,dy \cdot \frac{e^{-y^2/4\nu t}}{2\sqrt{(\pi\nu t)}} \cdot \quad\dots\dots\dots\dots\dots(14)$$

We may conclude that, however thin a jet of air may be initially, its thickness after one second is comparable with $1\frac{1}{2}$ centimetres. A similar calculation may be made for the case of a linear jet, whose whole velocity is originally concentrated in one line.

There is, therefore, ample foundation for the opinion that the phenomena of sensitive jets may be greatly influenced by fluid friction, and deviate

---

* The centimetre and second being units.

materially from the results of calculations based upon the supposition of discontinuous changes of velocity. Under these circumstances, it becomes important to investigate the character of the equilibrium of stratified motion in cases more nearly approaching what is met with in practice. Fully to include the effects of friction, would immensely increase the difficulties of the problem. For the present, at least, we must treat the fluid as frictionless, and be satisfied if we can obtain solutions for laws of stratification which are free from discontinuity. For the undisturbed motion, the component velocity v is zero, and u is a function of y only. A curve, in which u is ordinate and y is abscissa, represents the law of stratification and may be called, for brevity, the velocity curve.

A class of problems which can be dealt with by fairly simple methods, is obtained by supposing the rotation ζ to be constant throughout layers of finite thickness, and only to change its value in passing a limited number of planes for which y is constant. In such cases, the velocity curve is composed of portions of straight lines which meet one another at finite angles. This state of things may be supposed to be slightly disturbed by bending the surfaces of transition, and the determination of the subsequent motion depends upon that of the form of these surfaces. For ζ retains its constant value throughout each layer unchanged in the absence of friction, and by a well-known theorem the whole motion depends upon ζ. We shall suppose that the functions deviate from their equilibrium values by quantities proportional to e^{ikx}, so that everything is periodic with respect to x in a distance λ equal to $2\pi/k$. By Fourier's theorem, the solution may be generalised sufficiently to cover the case of an arbitrary deformation of the surfaces. As functions of the time, the disturbances will be assumed to be proportional to e^{int}, where n may be either real or complex, and the character of the resulting motion is determined in great measure by the value of n, found by the solution of the problem.

By a theorem due to Helmholtz, the effect of any element dA rotating with angular velocity ζ, is to produce, at a point whose distance from the element is r, a transverse velocity q, such that

$$q = \frac{\zeta dA}{\pi r} . \quad\dots\dots\dots\dots\dots\dots\dots\dots\dots(15)$$

In the application of this result to the problems in hand, it will be convenient to regard the actual value of ζ at time t as made up of two parts, (i) its undisturbed value, (ii) the difference between its actual value and (i). The effect of (i) is to produce the undisturbed system of velocities, on which the small effect of (ii) is superposed; and the calculation of the latter effects evidently involves integrations which extend only over the infinitely small areas included between the disturbed and undisturbed surfaces of transition. Suppose that the equation of one of these surfaces, reckoned from its undisturbed position, is

$$\eta = He^{ik\xi}, \quad \dots\dots\dots\dots\dots\dots\dots(16)$$

in which H is not necessarily real.   Then $dA = \eta\, d\xi$, and if $\Delta\zeta$ be the excess of the value of ζ on the upper above that on the lower side of the surface, we get, by (15), at any point whose abscissa is x and distance from the surface is b,

$$q = -\frac{\Delta\zeta.\eta\, d\xi}{\pi r}, \quad \dots\dots\dots\dots\dots\dots(17)$$

where

$$r^2 = b^2 + (\xi - x)^2. \quad \dots\dots\dots\dots\dots(18)$$

The velocity q, given by (17), is perpendicular to r.   The next step, previous to integration, is to resolve it in the fixed directions of x and y.   The resolution is effected by introduction of the factors b/r, and $(\xi - x)/r$; and thus, for the whole effect of the surface under consideration,

$$u = -\frac{b\Delta\zeta}{\pi}\int_{-\infty}^{+\infty}\frac{\eta\, d\xi}{r^2}, \qquad v = -\frac{\Delta\zeta}{\pi}\int_{-\infty}^{+\infty}\frac{\eta\,(\xi - x)\, d\xi}{r^2};$$

or, by (16),

$$u = -\frac{bH\Delta\zeta}{\pi}\int_{-\infty}^{+\infty}\frac{e^{ik\xi}\, d\xi}{b^2 + (\xi - x)^2}, \quad \dots\dots\dots\dots\dots(19)$$

$$v = -\frac{H\Delta\zeta}{\pi}\int_{-\infty}^{+\infty}\frac{e^{ik\xi}\,(\xi - x)\, d\xi}{b^2 + (\xi - x)^2}. \quad \dots\dots\dots\dots(20)$$

The integrals are readily evaluated by the theorem

$$\int_{-\infty}^{+\infty}\frac{\cos\alpha x}{1 + x^2}\, dx = \int_{-\infty}^{+\infty}\frac{x\sin\alpha x}{1 + x^2}\, dx = \pi e^{-\alpha};$$

and we obtain

$$u = -H\Delta\zeta e^{ikx}\, e^{-kb}, \qquad v = -iH\Delta\zeta e^{ikx}\, e^{-kb}. \quad \dots\dots\dots\dots(21)$$

In the derivation of (21), η has been treated as infinitesimal in comparison with b, but in the sequel we shall require to apply the formula to points situated upon the surface itself.   The value of u would need more careful examination, but that of v is easily seen to be equally applicable when b is zero, since the neighbouring elements do not contribute sensibly to the value of the integral.   In fact, the value is the same on whichever side of the surface the point under consideration is situated, and b is in both cases to be taken positive.   Accordingly, when b is zero, we are to take simply

$$v = -iH\Delta\zeta e^{ikx}. \quad \dots\dots\dots\dots\dots(22)$$

We are now prepared to enter upon the consideration of special problems. As a first example, let us suppose that on the upper side of a layer of thickness b the undisturbed velocity u is equal to $+ V$, and on the lower side to $- V$, while inside the layer it changes uniformly.   Thus

$$\zeta = \tfrac{1}{2} du/dy = V/b \dots\dots\dots\dots\dots(23)$$

inside the layer, and outside the layer $\zeta = 0$.   In the disturbed motion, let the equations of the upper and lower surfaces be respectively, at time t,

$$\eta = H e^{int} e^{ikx}, \quad \eta' = H' e^{int} e^{ikx} ; \dots\dots\dots\dots(24)$$

then, by (21), (22), (23), the whole value of v for a point on the upper surface is

$$v = i b^{-1} V e^{int} e^{ikx} (H - H' e^{-kb}), \dots\dots\dots\dots(24)'$$

and for the lower surface

$$v = i b^{-1} V e^{int} e^{ikx} (H e^{-kb} - H'). \dots\dots\dots\dots(25)$$

From these values of v the position of the surfaces at time $t + dt$ may be calculated.   At time t, η corresponds to x; at time $t + dt$, $\eta + v dt$ corresponds to $x + u dt$, u being the whole component velocity parallel to x.   Thus, at time $t + dt$, corresponding to x, we have

$$\eta + v dt - \frac{d}{dx}(\eta + v dt) . u dt ;$$

or, on neglecting the squares of small quantities,

$$\eta + \left(v - V \frac{d\eta}{dx}\right) dt.$$

Now, from (24), $d\eta/dt = in\eta$; so that, equating the two values of $d\eta/dt$, we get, from (24),

$$in H = i b^{-1} V (H - H' e^{-kb}) - ik V H,$$

or

$$\left(\frac{nb}{V} - 1 + kb\right) H + e^{-kb} H' = 0. \dots\dots \dots\dots\dots(26)$$

In like manner, by considering the motion of the lower surface, we get

$$e^{-kb} H + \left(-\frac{nb}{V} - 1 + kb\right) H' = 0. \dots\dots\dots\dots(27)$$

By eliminating the ratio $H' : H$ between (26) and (27), we obtain, as the equation giving the admissible values of n,

$$n^2 = \frac{V^2}{b^2} \{(kb - 1)^2 - e^{-2kb}\}. \dots\dots\dots\dots\dots(28)$$

When kb is small, that is, when the wave-length is great in comparison with b, the case approximates to that of a sudden transition.   Thus

$$n^2 = \frac{V^2}{b^2} \{1 - 2kb + k^2 b^2 - (1 - 2kb + 2k^2 b^2 + \dots)\} = - k^2 V^2 \text{ approximately}$$

$$\dots\dots\dots\dots(29)$$

in agreement with equation (30) of my former paper. In this case the motion is unstable. On the other hand, when kb is great, we find, from (28),

$$n^2 = k^2 V^2; \quad \dots\dots\dots\dots\dots\dots\dots\dots\dots(30)$$

and, since the values of n are real, the motion is *stable*. It appears, therefore, that so far from the instability increasing indefinitely with diminishing wave-length, as when the transition is sudden, a diminution of wave-length below a certain value entails an instability which gradually decreases, and is finally exchanged for actual stability. The following table exhibits more in detail the progress of $b^2 n^2 / V^2$ as a function of kb:—

kb	$b^2 n^2 / V^2$	kb	$b^2 n^2 / V^2$
·2	− ·03032	1·0	− ·13534
·4	− ·08933	1·2	− ·05072
·6	− ·14120	1·3	+ ·01573
·8	− ·16190	2·0	+ ·98168

We see that the instability is greatest when $kb = ·8$ nearly, that is, when $\lambda = 8b$; and that the passage from instability to stability takes place when $kb = 1·3$ nearly, or when $\lambda = 5b$.

Corresponding with the two values of n, there are two ratios of $H' : H$ determined by (26) or (27), each of which gives a normal mode of disturbance, and by means of these normal modes the results of an arbitrary displacement of the two surfaces may be represented. It will be seen that for the stable disturbances the ratio $H' : H$ is real, indicating that the sinuosities of the two surfaces are at every moment in the same phase.

We may next take an example of a jet of thickness $2b$ moving in still fluid, supposing that the velocity in the middle of the jet is V, and that it falls uniformly to zero on either side. Taking the middle line as axis of x, we may write

$$u = V(1 \mp y/b), \quad \dots\dots\dots\dots\dots\dots\dots(31)$$

in which the − sign applies to the upper, and the + sign to the lower half of the jet. Thus

$$\zeta = \tfrac{1}{2} du/dy = \mp \tfrac{1}{2} V/b \quad \dots\dots\dots\dots\dots\dots(32)$$

within the jet, and outside the jet $\zeta = 0$. In this problem there are three surfaces to be considered. We will suppose the equation of the upper surface, for which $\Delta\zeta = V/2b$, to be $\eta = H e^{ikx} e^{int}$; that of the middle surface, for which $\Delta\zeta = - V/b$, to be $\eta' = H' e^{ikx} e^{int}$; and that of the lower surface, for which $\Delta\zeta = V/2b$, to be $\eta'' = H'' e^{ikx} e^{int}$.

From (22), the velocities v are to be calculated as before.   We find

$$v \text{ (upper surface)} = \frac{iV}{2b} e^{int} e^{ikx} \{-H + 2e^{-kb} H' - e^{-2kb} H''\},$$

$$v \text{ (middle surface)} = \frac{iV}{2b} e^{int} e^{ikx} \{-e^{-kb} H + 2H' - e^{-kb} H''\},$$

$$v \text{ (lower surface)} = \frac{iV}{2b} e^{int} e^{ikx} \{-e^{-2kb} H + 2e^{-kb} H' - H''\}.$$

For the upper and lower surfaces the horizontal velocity is zero, and for the middle surface it is V.   In the same manner as for (26), we thus obtain

$$\left. \begin{array}{r} mH - 2\gamma H' + \gamma^2 H'' = 0 \\ \gamma H + (m + 2kb - 3) H' + \gamma H'' = 0 \\ \gamma^2 H - 2\gamma H' + mH'' = 0 \end{array} \right\}, \quad \dots\dots\dots\dots(33)$$

in which γ is written for e^{-kb}, and m is written $2bV^{-1}n + 1$.   The elimination from (33) of $H : H' : H''$ gives the following cubic in m:—

$$m^3 + (2kb - 3) m^2 + \gamma^2 (4 - \gamma^2) m - \gamma^4 (1 + 2kb) = 0. \quad \dots\dots\dots(34)$$

By inspection of (33), we see that one of the normal disturbances is defined by $H' = 0$, $H + H'' = 0$, and that the corresponding value of m is γ^2.   It follows that $m - \gamma^2$ is a factor of the cubic expression in (34), and the remaining quadratic factor is readily obtained by division.   Thus (34) assumes the form

$$(m - \gamma^2) \{m^2 + (2kb - 3 + \gamma^2) m + \gamma^2 (1 + 2kb)\} = 0. \quad \dots\dots\dots(35)$$

For the symmetrical disturbance

$$n = -\frac{V}{2b} (1 - e^{-2kb}), \quad \dots\dots\dots\dots\dots\dots(36)$$

a *real* quantity, indicating that the motion is *stable* so far as this mode of disturbance is concerned.

The other two values of n are real, if

$$(2kb - 3 + \gamma^2)^2 - 4\gamma^2 (1 + 2kb) \quad \dots\dots\dots\dots\dots(37)$$

be positive, but not otherwise.   When kb is infinite, $\gamma = 0$, and (37) reduces to $4k^2b^2$, which is positive; so that the motion is stable when the wave-length is small in comparison with the thickness of the jet.   On the other hand, as may readily be proved by expanding γ in (37), the motion is unstable, when the wave-length is great in comparison with the thickness of the jet.   The values of (37) can be more easily computed when it is thrown into the form

$$(5 + 2kb - e^{-2kb})^2 - 16 (1 + 2kb). \quad \dots\dots\dots\dots(38)$$

Some corresponding values of (38) and $2kb$ are shown below :—

$2kb$	(38)	$2kb$	(38)
·5	$-$ ·054	2·5	$-$ ·975
1·0	$-$ ·279	3·0	$-$ ·794
1·5	$-$ ·59()	3·5	$-$ ·263
2·0	$-$ ·876	4·0	$+$ ·671

The wave-length of maximum instability is about $2\frac{1}{2}$ times the thickness $(2b)$ of the jet; while, for a [value of kb] about half as great again, or more, the motion becomes stable.

Although it is the fact, as I have found by experiment, that a sensitive jet breaks up by becoming sinuous as a whole, the result that a symmetrical mode of disturbance is stable, is special to the law of velocity assumed in the foregoing example. In order to illustrate this, I will state the results for the more general law of velocity obtained by supposing the maximum velocity V to extend through a layer of finite thickness b' in the middle of the jet. The rotation ζ is zero in this central layer; in the adjacent layers of thickness b, $\zeta = \mp V/2b$, as before. The equations of the four surfaces, in crossing which ζ changes its value, being represented by

$$\eta, \eta', \eta'', \eta''' = (H, H', H'', H''') e^{int} e^{ikx},$$

we may obtain four equations involving n and the three ratios $H : H' : H'' : H'''$. The elimination of these ratios gives a biquadratic in n, which, however, is easily split into two quadratics, one of which relates to symmetrical disturbances, for which $H + H''' = 0$, $H' + H'' = 0$; and the other to disturbances for which $H - H''' = 0$, $H' - H'' = 0$. The resulting equation in n is

$$\left(\frac{2bn}{V}\right)^2 + (\pm \gamma' \mp \gamma'\gamma^2 + 2kb)\frac{2bn}{V} \pm \gamma' - 1 + 2kb + \gamma^2(1 \mp \gamma' \mp 2kb\gamma') = 0, \dots(39)$$

γ' being written for $e^{-kb'}$. In (39) the upper signs correspond to the symmetrical displacements. The roots are real, and the disturbances are stable, if

$$(\pm \gamma' \mp \gamma'\gamma^2 + 2kb)^2 - 4[\pm \gamma' - 1 + 2kb + \gamma^2(1 \mp \gamma' \mp 2kb\gamma')]\dots\dots(40)$$

be positive.

In what follows, we will limit our attention to the symmetrical disturbances, that is, to the upper signs in (40), and to terms of orders not higher than the first in b'. The expression (40) may then be reduced to

$$(1 - \gamma^2 - 2kb)^2 + 2kb'(1 + \gamma^2)(1 - \gamma^2 - 2kb). \dots\dots\dots(41)$$

If kb be very small, this becomes

$$4k^4b^4 - 8kb' \cdot k^2b^2. \qquad \dots\dots\dots\dots\dots\dots(42)$$

If b' be zero, (42) is positive, and the disturbance is stable, as we found before; but if b and b' be of the same order of magnitude, and both very small compared to λ, it follows from (42) that the disturbance is unstable.

If in (39) we suppose that b is zero, we fall back upon the case of a jet of uniform velocity V and thickness b' moving in still fluid. The equation for n, after division by b^2, becomes

$$n^2 + (1 \pm \gamma') kV \cdot n + \tfrac{1}{2}(1 \pm \gamma') k^2 V^2 = 0,$$

or

$$(n + kV)^2 \frac{1 \pm \gamma'}{1 \mp \gamma'} + n^2 = 0. \qquad \dots\dots\dots\dots\dots(43)$$

In the notation of my former paper, $[\tfrac{1}{2}b' = l]$, so that

$$\frac{1+\gamma'}{1-\gamma'} = \coth kl, \qquad \frac{1-\gamma'}{1+\gamma'} = \tanh kl;$$

and the equations there numbered (48) and (55) agree with (43).

Another particular case of (39), comparable with previous results, is obtained by supposing b' to be infinite.

I now pass to the consideration of certain cases in which the moving layers are bounded by fixed walls, instead of by an unlimited expanse of stationary fluid. The effect of the walls may be imitated by the introduction of an unlimited number of similar layers, in the same way as the vibrations of a string fixed at two points are often deduced from the theory applicable to an unlimited string. The displacements of the surfaces at which ζ changes its value being taken equal and opposite in consecutive layers, the value of v, at the places occupied by the walls, is, by symmetry, zero; and thus the presence or absence of the actual walls is a matter of indifference.

Let us first suppose that the distribution of velocity within the layer is that given in (31), uniformly increasing from zero at the walls to a maximum V in the middle, the distance between the walls being $2b$. The actual surface of transition and its successive images make contributions to the value of v at the surface, which are alternately opposite in sign, and, as regards numerical value, form a geometrical progression with common ratio e^{-2kb} or γ^2. Thus, with the same notation as before, we have at the surface, from (21), (22),

$$v = \frac{iVH}{b} e^{int} e^{ikx} \{1 - 2(\gamma^2 - \gamma^4 + \gamma^6 - \dots)\} = \frac{iVH}{b} e^{int} e^{ikx} \cdot \frac{1-\gamma^2}{1+\gamma^2};$$

so that, as in previous problems,

$$n = -kV + \frac{V}{b}\frac{1-\gamma^2}{1+\gamma^2};$$

or, as it may also be written,

$$\frac{bn}{V} = \tanh kb - kb. \quad \dots\dots\dots\dots\dots\dots(44)$$

[1899.   Solutions for the case where there is a layer of finite width b' in the centre, throughout which the undisturbed velocity is V, were subsequently (*Proc. Math. Soc.* XIX. p. 67, 1887) found to be erroneous and are accordingly omitted.]

In these examples the velocity curves are those represented by figs. (1) and (2).   I have taken a further step in the direction of generalisation by calculating the motion for a velocity curve in the form of (3).   The criterion of stability is complicated in its expression, but it is not difficult to show that the motion is stable if the angle N be a projecting angle.   From these examples there seemed to be some reason for thinking that the motion would be stable, whenever the velocity curve was of one curvature throughout; and this led me to attack the question by a more general method, which I will now explain.

Let us suppose that the conditions of steady motion are satisfied by $u = U$, $v = V$, $\zeta = Z$; and let us trace the effects of superposing upon this motion a disturbance for which $u = \delta u$, $v = \delta v$, $\zeta = \delta \zeta$.   Both the original motion and the disturbance satisfy the equation of continuity (3).

Since, in the absence of friction, the rotation of every element remains unchanged, $\partial(Z + \delta\zeta)/\partial t = 0$, or

$$\frac{d}{dt}(Z + \delta\zeta) + (U + \delta u)\frac{d}{dx}(Z + \delta\zeta) + (V + \delta v)\frac{d}{dy}(Z + \delta\zeta) = 0.$$

This equation is satisfied by supposition, if δu, δv, $\delta\zeta = 0$.   If we omit the squares and products of the small quantities, it becomes

$$\frac{d\delta\zeta}{dt} + U\frac{d\delta\zeta}{dx} + V\frac{d\delta\zeta}{dy} + \delta u\frac{dZ}{dx} + \delta v\frac{dZ}{dy} = 0. \quad \dots\dots\dots(47)$$

If $V = 0$, and U be a function of y only, (47) reduces to

$$\frac{d\delta\zeta}{dt} + U\frac{d\delta\zeta}{dx} + \delta v\frac{dZ}{dy} = 0;$$

or, since in this case $Z = \frac{1}{2} dU/dy$,

$$\left(\frac{d}{dt} + U\frac{d}{dx}\right)\left(\frac{d\delta u}{dy} - \frac{d\delta v}{dx}\right) + \frac{d^2 U}{dy^2}\delta v = 0. \quad\ldots\ldots\ldots\ldots(48)$$

We now introduce the supposition that, as functions of x, δu and δv are proportional to e^{ikx}, so that, by (3),

$$ik\,\delta u + \frac{d\delta v}{dy} = 0. \quad\ldots\ldots\ldots\ldots\ldots\ldots\ldots(49)$$

We thus obtain, by elimination of δu,

$$\left(\frac{1}{ik}\frac{d}{dt} + U\right)\left(\frac{d^2\delta v}{dy^2} - k^2\delta v\right) - \frac{d^2 U}{dy^2}\delta v = 0. \quad\ldots\ldots\ldots\ldots(50)$$

If we further suppose that, as a function of t, δv is proportional to e^{int}, where n is a real or complex constant, we get

$$\left(\frac{n}{k} + U\right)\left(\frac{d^2\delta v}{dy^2} - k^2\delta v\right) - \frac{d^2 U}{dy^2}\delta v = 0. \quad\ldots\ldots\ldots\ldots(51)$$

On this equation the solution of the special problems already considered may be founded. If, throughout any layer, the rotation Z be constant, $d^2 U/dy^2 = 0$, and, wherever $n + kU$ is not equal to zero, (51) reduces to

$$\frac{d^2\delta v}{dy^2} - k^2\delta v = 0. \quad\ldots\ldots\ldots\ldots\ldots\ldots(52)$$

Equation (52) may, in fact, be easily established independently, on the assumption that the rotation throughout the layer is the same after disturbance as before. From (2),

$$\frac{d\zeta}{dx} = \frac{1}{2}\left(\frac{d^2 u}{dx\,dy} - \frac{d^2 v}{dx^2}\right) = -\frac{1}{2}\left(\frac{d^2 v}{dx^2} + \frac{d^2 v}{dy^2}\right),$$

by (3); so that, when ζ is constant, $\nabla^2 v = 0$. In like manner $\nabla^2 u = 0$. If $\delta v \propto e^{ikx}$, (52) follows immediately.

The solution of (52) is

$$\delta v = Ae^{ky} + Be^{-ky}, \quad\ldots\ldots\ldots\ldots\ldots\ldots(53)$$

where A and B are constants, not restricted to be real. For each layer of constant Z, a fresh solution with fresh arbitraries is to be taken, and the partial solutions are to be fitted together by means of the proper boundary conditions. The first of these conditions is evidently

$$\Delta\delta v = 0. \quad\ldots\ldots\ldots\ldots\ldots\ldots\ldots(54)$$

The second may be obtained by integrating (51) across the boundary. Thus

$$\left(\frac{n}{k} + U\right) . \Delta\left(\frac{d\delta v}{dy}\right) - \Delta\left(\frac{dU}{dy}\right) . \delta v = 0. \quad\ldots\ldots\ldots\ldots(55)$$

At a fixed wall $\delta v = 0$.

The reader may apply this method to the problem whose solution is expressed in (44).

In cases where $d^2U/dy^2 = 0$, the substitution of (52) for (51), or the corresponding supposition that ζ is unchanged by the disturbance, amounts to a limitation on the generality of the solution. Suppose, for example, that the motion takes place between two fixed walls, at each of which $\delta v = 0$. Under these circumstances (53) shows that $\delta v = 0$ throughout, or no disturbance is possible; and this is obviously true if no new rotation is introduced by the disturbance. In order to obtain a general solution, we must retain the factor $n + kU$ in (51). For any value of y which gives $n + kU = 0$, (52) need not be satisfied; and thus any value of $-kU$ is an admissible value of n, satisfying all the conditions of the problem.

I will now inquire, under what conditions (51) admits of a solution with a complex value of n; or, in other words, under what conditions the steady motion is unstable, assuming that, for two finite or infinite values of y, $\delta v = 0$. Let $n/k = p + iq$, $\delta v = \alpha + i\beta$, where p, q, α, β are real. Substituting in (51), we get

$$\frac{d^2\alpha}{dy^2} + i\frac{d^2\beta}{dy^2} = \left[k^2 + \frac{d^2U}{dy^2}\cdot\frac{p + U - iq}{(p + U)^2 + q^2}\right](\alpha + i\beta);$$

or, on equating to zero the real and imaginary parts,

$$\frac{d^2\alpha}{dy^2} = k^2\alpha + \frac{d^2U}{dy^2}\cdot\frac{(p + U)\alpha + q\beta}{(p + U)^2 + q^2}, \quad\dots\dots\dots\dots(56)$$

$$\frac{d^2\beta}{dy^2} = k^2\beta + \frac{d^2U}{dy^2}\cdot\frac{-q\alpha + (p + U)\beta}{(p + U)^2 + q^2}. \quad\dots\dots\dots(57)$$

Multiplying (56) by β, (57) by α, and subtracting, we get

$$\beta\frac{d^2\alpha}{dy^2} - \alpha\frac{d^2\beta}{dy^2} = \frac{d^2U}{dy^2}\cdot\frac{q(\alpha^2 + \beta^2)}{(p + U)^2 + q^2} = \frac{d}{dy}\left(\beta\frac{d\alpha}{dy} - \alpha\frac{d\beta}{dy}\right). \quad\dots\dots(58)$$

At the limits δv, and therefore both α and β, are, by hypothesis, zero. Hence, integrating (58) between the limits, we see that q must be zero, if d^2U/dy^2 be of one sign throughout the range of integration; so that, if the velocity curve is either wholly convex or wholly concave for the space between two limits at which $\delta v = 0$, the motion is thoroughly stable*. This result covers all the special problems of motion between walls previously investigated. Its application to jets, for which d^2U/dy^2 changes sign, leaves the question of stability or instability still open.

---

* More generally, the same conclusion follows if the ratio $\dfrac{d\,\delta v}{dy}$: δv has real values at both limits.

Another general result worth notice may be obtained from (51). Writing it in the form

$$\frac{d^2\delta v}{dy^2} = \left\{k^2 + \frac{d^2U/dy^2}{U + n/k}\right\}\delta v,$$

we see that, if n is real, δv cannot pass from one zero value to another zero value, unless d^2U/dy^2 and $n + kU$ be somewhere of contrary signs. Thus, if we suppose that U is positive and d^2U/dy^2 negative throughout, and that V is the greatest value of U, we find that $n + kV$ must be positive. For an example see the equation immediately preceding (44).

If the stream lines of the steady motion be concentric circles instead of parallel straight lines, the character of the problem is not greatly changed. It may be proved that, if the fluid move between two rigid concentric circular walls, the motion is stable, provided that in the steady motion the rotation either continually increases or continually decreases in passing outwards from the axis.

[1899. Further investigations upon this subject will be found in *Proc. Math. Soc.* vol. XIX. p. 68, 1887 ; vol. XXVII. p. 5, 1895.]

67.

ON THE RESOLVING-POWER OF TELESCOPES.

[*Philosophical Magazine*, x. pp. 116—119, 1880.]

ALTHOUGH I have recently treated of this subject in the *Philosophical Magazine**, its importance induces me to return to it in order to explain how easily it may be investigated in the laboratory. There can be no reason why the experiment about to be described should not be included in every course on physical optics.

The only work on this subject with which I am acquainted is that of Foucault†, who investigated the resolving-power of a telescope of 10 centimetres aperture on a distant scale illuminated by direct sunshine. In this form the experiment is troublesome and requires expensive apparatus—difficulties which are entirely obviated by the plan which I have followed of using a much smaller aperture.

The object, on which the resolving-power of the telescope is tested, is a grating of fine wires, constructed on the plan employed by Fraunhofer for diffraction-gratings. A stout brass wire or rod is bent into a horse-shoe, and its ends are screwed. On these screws fine wire is wound of diameter equal to about half the pitch, and secured with solder. The wires on one side being now cut away, we obtain a grating of considerable accuracy. A wire grating thus formed is preferable to a scale ruled on paper, and placed in front of a lamp it presents a very suitable subject for examination. The one that I employed has 50 wires to the inch [2·54 cm.], and for security is mounted in a frame between two plates of glass. For rough purposes a piece of common gauze with 30 or 40 meshes to the inch may be substituted with good effect.

---

* Oct., Nov., and Dec. 1879, Jan. 1880. [Art. LXII.]

† "Mémoires sur la construction des télescopes," *Annales de l'Observatoire*, t. v. ; also Verdet's *Leçons d'Optique physique*, t. I. p. 309.

For the sake of definiteness of wave-length the grating was backed by a soda-flame, though fair results are obtainable with a common paraffine-lamp. The telescope is a small instrument mounted on a stand, and provided with a cap by means of which various diaphragms can be conveniently fitted in front of the object-glass. The apertures in these diaphragms may be either circular or rectangular. In the latter case the length of the slit is placed parallel to the wires of the grating, and we have the advantage of greater illumination than with a circle of equal width. The observation consists in ascertaining the greatest distance at which the wires can be seen resolved. For this purpose the telescope, focused all the while, is gradually drawn back until in the judgment of the observer the periodic structure is no longer seen; and the distance between the grating and the diaphragm is then measured with a steel tape. The distance thus determined is more definite than might be expected, the differences in the case of various observers not usually amounting to more than 2 or 3 per cent.

Two slits were tried, half an inch long, and of widths ·107, ·196 inch respectively. These widths were measured by inserting a graduated wedge. It was found, however, that the graduations could not be trusted; so that the wedge was in fact used merely to convey the length to be measured to a pair of callipers reading to one-thousandth of an inch. The distances at which resolution just ceased were estimated respectively as 91·5 and 168·5 inches, corresponding to angular intervals between consecutive lines equal to $\frac{1}{4575}$ and $\frac{1}{8425}$. According to theory, the minimum angle is approximately equal to that subtended by the wave-length of light, λ, at a distance equal to the width of the slit, a. In the present case $\lambda = 5\cdot89 \times 10^{-5}$ centimetres, and $a = \cdot107 \times 2\cdot54$, or $\cdot196 \times 2\cdot54$ centimetres, so that

$$\lambda/a = \tfrac{1}{4615} \text{ or } \tfrac{1}{8413},$$

agreeing with the angles found by observation more closely than we should have any right to expect.

Besides these slits, four circular apertures were examined. Their diameters were measured under a magnifier on a glass scale divided to tenths of a millimetre, and were found to be ·172, ·315, ·48, ·63 centimetre respectively. In the case of the two smaller holes the illumination given by an ordinary soda-flame was hardly sufficient; but with the assistance of a jet of oxygen the observation could be made.

The following distances are the means of those found by two observers[*]— 51·5, 98, 149, 196 inches, corresponding to angular intervals $\frac{1}{2575}$, $\frac{1}{4900}$, $\frac{1}{7450}$, $\frac{1}{9800}$ respectively. If D represent the diameter of the aperture, the values of

---

[*] Mr Glazebrook and myself.

λ/D are respectively $\frac{1}{2920}$, $\frac{1}{5850}$, $\frac{1}{8150}$, $\frac{1}{10700}$. The ratios of the observed angular intervals to λ/D are thus

$$1\cdot13, \quad 1\cdot09, \quad 1\cdot09, \quad 1\cdot09.$$

That a circular aperture would be less effective than a slit of the same width might have been expected. Even in the case of a slit it is advantageous to stop some of the central, in order to increase the relative importance of the extreme, rays; and with a circular aperture the extreme rays are much worse represented than with a slit. From the above results it appears that, to have an equal resolving-power, the circular aperture must be about a tenth part wider than the slit.

Merely to show the dependence of resolving-power on aperture it is not necessary to use a telescope at all. It is sufficient to look at wire gauze backed by the sky, or by a flame, through a piece of blackened cardboard pierced by a needle and held close to the eye. By varying the distance the point is easily found at which resolution ceases; and the observation is as sharp as with a telescope. The function of the telescope is in fact to allow the use of a wider, and therefore more easily measurable, aperture.

An interesting modification of the experiment is obtained by using lights of various wave-lengths. For this purpose we may have recourse to coloured glasses; but the best results would doubtless require the rays of the spectrum.

68.

ON THE RESULTANT OF A LARGE NUMBER OF VIBRATIONS OF THE SAME PITCH AND OF ARBITRARY PHASE.

[*Philosophical Magazine*, x. pp. 73—78, 1880.]

VERDET[*], in an investigation upon this subject, has arrived at the conclusion that the resultant of n vibrations of unit amplitude and arbitrary phase approaches the definite value $\sqrt{n}$ when n is very great. It can be shown[†], however, that this conclusion is inaccurate, and that the resultant tends to no definite value, however great the number of components may be.

But there is a modified form of the question, which admits of a definite answer, and was perhaps vaguely before Verdet's mind. If we inquire what is the *average* intensity in a great number of cases, or, in the language of the theory of probabilities, what is the *expectation* of intensity in a single case of composition, we shall find that the result is that assigned by Verdet, namely n.

A simple but instructive variation of the problem may be obtained by supposing the possible phases limited to *two opposite* phases, in which case it is convenient to discard the idea of phase altogether, and to regard the amplitudes as at random positive or negative. If all the signs are the same, the resultant intensity is n^2; if, on the other hand, there are as many positive as negative, the result is zero. But although the intensity may range from 0 to n^2, the smaller values are much more *probable* than the greater; and to calculate the expectation of intensity, these different degrees of probability must be taken into account. By well-known rules the expression for the expectation is

$$\frac{1}{2^n}\left\{1 \cdot n^2 + n \cdot (n-2)^2 + \frac{n(n-1)}{1 \cdot 2}(n-4)^2 + \frac{n(n-1)(n-2)}{1 \cdot 2 \cdot 3}(n-6)^2 + \ldots\right\}.$$

* *Leçons d'Optique physique*, t. I. p. 297.
† *Math. Soc. Proc.* May 1871. [Art. VI.]

The value of the series, which is to be continued so long as the terms are finite, is simply n, as may be proved by comparison of coefficients of x^2 in the equivalent forms

$$(e^x + e^{-x})^n = 2^n (1 + \tfrac{1}{2}x^2 + \ldots)^n = e^{nx} + ne^{(n-2)x} + \frac{n(n-1)}{1 \cdot 2} e^{(n-4)x} + \ldots$$

The expectation of intensity is therefore n, and this whether n be great or small.

In the more general problem, where the phases are distributed at random over the complete period, the expression for the expectation of intensity is

$$\int_0^{2\pi} \int_0^{2\pi} \int_0^{2\pi} \ldots \frac{d\theta}{2\pi} \frac{d\theta'}{2\pi} \ldots [(\cos\theta + \cos\theta' + \ldots)^2 + (\sin\theta + \sin\theta' + \ldots)^2].$$

If we effect the integration with respect to θ, we get

$$\int_0^{2\pi} \int_0^{2\pi} \ldots \frac{d\theta'}{2\pi} \frac{d\theta''}{2\pi} \ldots [1 + (\cos\theta' + \cos\theta'' + \ldots)^2 + (\sin\theta' + \sin\theta'' + \ldots)^2].$$

Continuing the process by successive integrations with respect to θ', θ'', ..., we see that, as before, the expectation of intensity is n.

So far there is no difficulty ; but a complete investigation of this subject involves an estimate of the relative probabilities of resultants lying between assigned limits of magnitude. For example, we ought to be able to say what is the probability that the intensity due to a large number (n) of equal components is less than $\tfrac{1}{2}n$. It will be convenient to begin by taking the problem under the restriction that the phases are of two opposite kinds only. When this has been dealt with, we shall not find much difficulty in extending our investigation to phases entirely arbitrary.

By Bernoulli's theorem[*] we find that the probability that of n vibrations, which are at random positive or negative, the number of positive vibrations lies between

$$\tfrac{1}{2}n - \tau \sqrt{(\tfrac{1}{2}n)} \quad \text{and} \quad \tfrac{1}{2}n + \tau \sqrt{(\tfrac{1}{2}n)}$$

is, when n is great,

$$\frac{2}{\sqrt{\pi}} \int_0^\tau e^{-t^2} dt,$$

where $\tau = r\sqrt{(2n)}$, and r must not surpass $\sqrt{n}$ in order of magnitude. In the extreme cases the amplitude is $\pm 2\tau\sqrt{(\tfrac{1}{2}n)}$, and the intensity is $2\tau^2 n$. Thus, if we put $\tau = \tfrac{1}{2}$, we see that the chance of intensity less than $\tfrac{1}{2}n$ is

$$\frac{2}{\sqrt{\pi}} \int_0^{\tfrac{1}{2}} e^{-t^2} dt = \cdot 5205 ;$$

so that however great n may be, there is always more than an even chance that the intensity will be less than $\tfrac{1}{2}n$. This, of course, is inconsistent with any such tendency to close upon the value n as Verdet supposes.

---

[*] Todhunter's *History of the Theory of Probability*, § 993.

From the tables of the definite integral, given in De Morgan's *Differential Calculus*, p. 657, we may find the probabilities of intensities less than any assigned values. The probability of intensity less than $\frac{1}{2}n$ is ·2764.

Again, the chance that in a series n the number of positive vibrations lies between

$$\tfrac{1}{2}n + \tau \sqrt{(\tfrac{1}{2}n)} \quad \text{and} \quad \tfrac{1}{2}n + (\tau + \delta\tau)\sqrt{(\tfrac{1}{2}n)}$$

is

$$\frac{1}{\sqrt{\pi}} e^{-\tau^2}\, \delta\tau,$$

which expresses accordingly the chance of a positive amplitude lying between

$$2\tau \sqrt{(\tfrac{1}{2}n)} \quad \text{and} \quad 2(\tau + \delta\tau)\sqrt{(\tfrac{1}{2}n)}.$$

Let these limits be called x and $x + \delta x$, so that $\tau = x/\sqrt{(2n)}$; then the chance of amplitude between x and $x + \delta x$ is

$$\frac{1}{\sqrt{(2\pi n)}} e^{-x^2/2n}\, \delta x.$$

The expectation of intensity is expressed by

$$\frac{1}{\sqrt{(2\pi n)}} \int_{-\infty}^{+\infty} e^{-x^2/2n} x^2 dx = n,$$

as before.

It will be convenient in what follows to consider the vibrations to be represented by lines (of unit length) drawn from a fixed point O, the intersection of rectangular axes Ox and Oy.

If n of these lines be taken at random in the directions $\pm x$, the probability of resultants also along $\pm x$, and of various magnitudes, is given by preceding expressions. We will now suppose that $\frac{1}{2}n$ are distributed at random along $\pm x$, and $\frac{1}{2}n$ along $\pm y$, and inquire into the probabilities of the various resultants. The probability that the end of the representative line, or, as we may consider it, the representative *point*, lies in the rectangle $dx\,dy$ is evidently

$$\frac{1}{\pi n} e^{-\frac{x^2+y^2}{n}} dx\,dy.$$

Substituting polar coordinates r, θ and integrating with respect to θ, we see that the probability of the representative point of the resultant lying between the circles r and $r + dr$ is

$$\frac{2}{n} e^{-r^2/n}\, r\,dr.$$

This is therefore the probability of a resultant vibration with amplitude between the values r and $r + dr$. In this case there are n components distributed in four rectangular directions; and we have supposed that $\frac{1}{2}n$

exactly are distributed along $\pm x$, and $\frac{1}{2}n$ along $\pm y$. It is important to remove this restriction, and to show that the result is the same when the distribution is perfectly arbitrary in respect to all four directions.

In order to see this, let us suppose that $\frac{1}{2}n + m$ are distributed along $\pm x$ and $\frac{1}{2}n - m$ along $\pm y$, and inquire how far the result is influenced by the value of m. The chance of the representative point of the resultant lying in the rectangle $dx\,dy$ is now expressed by

$$\frac{1}{\pi \sqrt{(n^2 - 4m^2)}} e^{-\frac{x^2}{n+2m} - \frac{y^2}{n-2m}} dx\,dy$$

$$= \frac{1}{\pi \sqrt{(n^2 - 4m^2)}} e^{-\frac{n(x^2+y^2)+2m(y^2-x^2)}{n^2-4m^2}} dx\,dy$$

$$= \frac{1}{\pi \sqrt{(n^2 - 4m^2)}} e^{-\frac{nr^2}{n^2-4m^2}} e^{-\frac{2mr^2}{n^2-4m^2}\cos 2\theta} r\,dr\,d\theta.$$

Also

$$\int_0^{2\pi} e^{-\frac{2mr^2 \cos 2\theta}{n^2-4m^2}} d\theta = 2\pi \left\{ 1 + \frac{m^2 r^4}{(n^2 - 4m^2)^2} + \cdots \right\},$$

as we find on expanding the exponential and integrating. Thus the chance of the representative point lying between the circles r and $r + dr$ is

$$\frac{2r\,dr}{\sqrt{(n^2 - 4m^2)}} e^{-\frac{nr^2}{n^2-4m^2}} \left\{ 1 + \frac{m^2 r^4}{(n^2 - 4m^2)^2} + \cdots \right\}.$$

Now, if the distribution be entirely at random, all the values of m of which there is a finite probability are of order not higher than $\sqrt{n}$, n being treated as infinite. But if m be of this order, the above expression is the same as if m were zero; and thus it makes no difference whether the numbers of components along $\pm x$ and along $\pm y$ are limited to be equal or not. The previous result, viz.

$$\frac{2}{n} e^{-r^2/n} r\,dr,$$

is accordingly applicable to a thoroughly arbitrary distribution among the four rectangular directions.

The next point to notice is that the result is symmetrical, and independent of the direction of the axes, so long as they are rectangular, from which we may conclude that it has a still higher generality. If a total of n components, to be distributed along one set of rectangular axes, be divided into any number of groups, it makes no difference whether we first obtain the probabilities of various resultants of the groups separately and afterwards of the final resultant, or whether we regard the whole n as one group. But the resultant of each group is the same, notwithstanding a change in the system of rectangular axes; so that the probabilities of various resultants are unaltered, whether we suppose the whole number of components restricted to one set of rectangular axes or divided in any manner between any number of sets of axes. This last state of things, however, is equivalent to no restriction at all; and we thus arrive at the important conclusion that, if n unit vibrations of equal pitch and of arbitrary phases be compounded, the probability of a resultant intermediate in amplitude between r and $r + dr$ is

$$\frac{2}{n} e^{-r^2/n} r\, dr,$$

a similar result applying, of course, in the case of any other vector quantities.

The probability of a resultant of amplitude less than r is

$$\frac{2}{n} \int_0^r e^{-r^2/n} r\, dr = 1 - e^{-r^2/n};$$

or, which is the same thing, the probability of a resultant greater than r is

$$e^{-r^2/n}.$$

The following table gives the probabilities of intensities less than the fractions of n named in the first column. For example, the probability of intensity less than n is $\cdot 6321$.

·05	·0488	·80	·5506
·10	·0952	1·00	·6321
·20	·1813	1·50	·7768
·40	·3296	2·00	·8647
·60	·4512	3·00	·9502

It will be seen that, however great n may be, there is a reasonable chance of considerable relative fluctuations of intensity in consecutive trials.

The *average* intensity, expressed by

$$\frac{2}{n} \int_0^\infty e^{-r^2/n} . r^2 . r\, dr,$$

is, as we have seen already, equal to n.

If the amplitude of each component be α, instead of unity, as we have hitherto supposed for brevity, the probability of a resultant amplitude between r and $r + dr$ is

$$\frac{2}{n\alpha^2} e^{-r^2/n\alpha^2} r\, dr.$$

The result is therefore in all respects the same as if, for example, the amplitude of the components had been $\frac{1}{2}\alpha$ and their number equal to $4n$.

From this we see that the law is not altered, even if the components have different amplitudes, provided always that the whole number of each kind is very great; so that if there be n components of amplitude α, n' of amplitude β, and so on, the probability of a resultant between r and $r + dr$ is

$$\frac{2}{n\alpha^2 + n'\beta^2 + \ldots} e^{-\frac{r^2}{n\alpha^2 + n'\beta^2 + \ldots}} r\, dr.$$

The conclusion that the resultant of a large number of independent sounds is practically, and to a considerable extent, uncertain may appear paradoxical; but its truth, I imagine, cannot be disputed. Perhaps even the appearance of paradox will be removed if we remember that with two sounds of equal intensity the degree of uncertainty is far greater, as is evidenced in the familiar experiment with tuning-forks in approximate unison. That the beats should not be altogether obliterated by a multiplication of sources can hardly be thought surprising.

[1899. The problem of the present paper is treated by another method in *Theory of Sound*, 2nd ed. § 42*a*, 1894.]

69.

NOTE ON THE THEORY OF THE INDUCTION BALANCE.

[*British Association Report*, Swansea, pp. 472—473, 1880.]

THIS subject has been treated by Dr Lodge in the *Phil. Mag.* for February, 1880, who has arrived at several interesting results. The investigation may be considerably simplified by taking the case of pure tones, as is usual in acoustics. We may also suppose, for distinctness of conception, that the current in the primary circuit (x_1) is sensibly unaffected by the reaction of derived currents, though our results will be independent of this hypothesis.

If $x_1 x_2 \ldots\ldots$ be the currents, $R_1 R_2 \ldots\ldots$ the resistances, $M_{11} M_{22} M_{12} \ldots\ldots$ the coefficients of self-induction and of mutual induction, the equations for three circuits are

$$M_{22} \frac{dx_2}{dt} + M_{23} \frac{dx_3}{dt} + R_2 x_2 = - M_{12} \frac{dx_1}{dt},$$

$$M_{23} \frac{dx_2}{dt} + M_{33} \frac{dx_3}{dt} + R_3 x_3 = - M_{13} \frac{dx_1}{dt}.$$

We now assume that $x_1, x_2 \ldots\ldots$ are proportional to e^{int}, where $n/2\pi$ is the frequency of vibration. Thus

$$in (M_{22} x_2 + M_{23} x_3) + R_2 x_2 = - in\, M_{12} x_1,$$

$$in (M_{23} x_2 + M_{33} x_3) + R_3 x_3 = - in\, M_{13} x_1,$$

whence by elimination of x_3

$$x_2 \left\{ in\, M_{22} + R_2 + \frac{M^2_{23} n^2}{in\, M_{33} + R_3} \right\} = - in\, M_{12} x_1 - \frac{n^2 M_{13} M_{23} x_1}{in\, M_{33} + R_3}.$$

From this it appears that a want of balance depending on M_{12} cannot compensate for the action of the tertiary circuit, so as to produce silence in the secondary (telephone) circuit, unless R_3 be negligible in comparison with $n M_{33}$, that is unless the time-constant of the tertiary circuit be very great in comparison with the period of the vibration. Otherwise the effects are of different phases, and therefore incapable of balancing.

We will now introduce a fourth circuit, and suppose that the primary and secondary circuits are accurately conjugate, so that $M_{12} = 0$, and also that the mutual induction between the third and fourth circuits (M_{34}) may be neglected. Thus

$$in\,(M_{22}x_2 + M_{23}x_3 + M_{24}x_4) + R_2x_2 = 0,$$

$$in\,(M_{32}x_2 + M_{33}x_3) + R_3x_3 = -\,in\,M_{13}x_1,$$

$$in\,(M_{42}x_2 + M_{44}x_4) + R_4x_4 = -\,in\,M_{14}x_1,$$

whence

$$x_2\left(in\,M_{22} + R_2 + \frac{n^2\,M^2_{23}}{in\,M_{33} + R_3} + \frac{n^2\,M^2_{24}}{in\,M_{44} + R_4}\right)$$

$$= -\,n^2\,x_1\left(\frac{M_{13}\,M_{23}}{in\,M_{33} + R_3} + \frac{M_{14}\,M_{24}}{in\,M_{44} + R_4}\right).$$

Two conditions must be satisfied to secure a balance, since both the phases and the intensities of the separate effects must be the same. The first condition requires that the time-constants of the third and fourth circuits be equal, unless both be either very great or very small in comparison with the period. If this condition be satisfied, a balance may be obtained by shifting the circuits so as to bring $M_{13}\,M_{23}$ into equality with $M_{14}\,M_{24}$*.

For a coil of mean radius a, and radius of section equal to $a \div 3\cdot22$, the coefficient of self-induction (L) is† $12\pi n^2 a$, n being the number of turns. Also, if r be the specific resistance,

$$R = \frac{2\,(3\cdot22)^2\,n^2\,r}{a}\,.$$

For copper $r = 1640$, so that

$$\tau = L/R = a^2/1810 \text{ on the c.g.s. system.}$$

In the case of a *shilling* the time-constant can scarcely be so high as a ten-thousandth of a second, but periods smaller than this may be concerned when a microphone clock is employed.

For similar discs or coins the time-constant varies as a^2/r, a being the linear dimension and r the specific resistance. Equal coins cannot in general be balanced if the specific resistances are different. To obtain a balance, a^2 should vary as r. In this case

$$\frac{M_{13}\,M_{23}}{in\,M_{33} + R} \propto \frac{a^2}{r\,a^{-1}} \propto \frac{a^3}{r} \propto a,$$

---

* [1899. The particular statement applies only when $M_{33} = M_{44}$, $R_3 = R_4$.]

† Maxwell, *Electricity and Magnetism*, § 707.

on the supposition that the positions of the coins relatively to the primary and secondary coils are the same.

A perfect balance is not to be expected in general without two adjustments, though in some cases a fair approximation may be obtained with the sliding wedge employed by Hughes.

If the condition of equality of time-constants be satisfied, the remaining condition is independent of the value of n, so that a perfect balance for one pitch secures a perfect balance for all pitches. From this it follows that the results are not limited to simple tones, and that the two conditions are sufficient to secure a balance in all cases. It should be remembered, however, that this indifference to pitch does not apply to approximate balances, which may be satisfactory with one sound, but quite inadequate when another is substituted.

70.

ON A NEW ARRANGEMENT FOR SENSITIVE FLAMES.

[*Cambridge Phil. Soc. Proc.* IV. pp. 17—18, 1880.]

A JET of coal-gas from a pin-hole burner rises vertically in the interior of a cavity from which the air is excluded. It then passes into a brass tube a few inches long, and on reaching the top, burns in the open. The front wall of the cavity is formed of a flexible membrane of tissue paper, through which external sounds can reach the burner.

The principle is the same as that of Barry's flame described by Tyndall. In both cases the unignited part of the jet is the sensitive agent, and the flame is only an indicator. Barry's flame may be made very sensitive to sound, but it is open to the objection of liability to disturbance by the slightest draught. A few years since Mr Ridout proposed to enclose the jet in a tube air-tight at the bottom, and to ignite it only on arrival at the top of this tube. In this case however external vibrations have very imperfect access to the sensitive part of the jet, and when they reach it they are of the wrong quality, having but little motion transverse to the direction of the jet. The arrangement now exhibited combines very satisfactorily sensitiveness to sound and insensitiveness to wind, and it requires no higher pressure than that of ordinary gas-pipes. If the extreme of sensitiveness be aimed at, the gas pressure must be adjusted until the jet is on the point of flaring without sound.

The apparatus exhibited was made in Prof. Stuart's workshop. An adjustment for directing the jet exactly up the middle of the brass tube is found necessary, and some advantage is gained by contracting the tube somewhat at the place of ignition.

[1899. For further particulars of construction with a drawing, see *Proc. Roy. Inst.* June 1898.]

71.

THE PHOTOPHONE.

[*Nature*, XXIII. pp. 274—275, 1881.]

THE following calculation, made with the view of examining whether the remarkable phenomena recently discovered by Prof. Bell could be explained on recognised principles, may interest the readers of *Nature*. I refer to the *un-electrical* sounds produced by the simple impact of intermittent radiation upon thin plates of various substances.

It has been thought by some that in order that a body exposed to variable radiation may experience a sensible fluctuation of temperature its rate of cooling must be rapid. This however is a mistake. The variable radiation may be divided into two parts—a constant part, and a periodic part—and each of these acts independently. Under the influence of the constant part the temperature of the body will rise until the loss of heat by radiation and conduction balances the steady inflow; but this is not appreciable by the ear, and may for the present purpose be left out of account. The question is as to what is the effect of the periodic part of the whole radiation, that is, of a periodic communication and *abstraction* of heat which leaves the mean temperature unaltered. It is not difficult to see that if the radiating power of the body were sufficiently high, the resulting fluctuation of temperature would diminish to any extent, and that what is wanted in order to obtain a considerable fluctuation of temperature is a *slow* rate of cooling in consequence of radiation or convection.

If θ denote the temperature at time t, reckoned from the mean temperature as zero, q be the rate of cooling, $E \cos pt$ the measure of the heating effect of the incident radiation, the equation regulating the fluctuation of temperature is

$$\frac{d\theta}{dt} + q\theta = E \cos pt.$$

Thus

$$\theta = \frac{E \cos (pt + \epsilon)}{\sqrt{\{p^2 + q^2\}}} ,$$

showing that if p and E be given, θ varies most when $q = 0$.

Let us suppose now that intermittent sunlight falls upon a plate of solid matter. If the plate be transparent, or absorb only a small fraction of the radiation, little sonorous effect will be produced, not merely because the radiation transmitted is lost, but because the heating due to the remainder is nearly uniform throughout the substance. In order that the plate may bend, as great a difference of temperature as possible must be established between its sides, and for this purpose the radiation should be absorbed within a distance of the order of half the thickness of the plate. If the absorption be still more rapid, it would appear that the thickness of the plate may be diminished with advantage, unless heat conduction in the plate itself interferes. The numerical calculation relates to a plate of iron of thickness d. It is supposed that q is negligible in comparison with p, i.e. that no sensible gain or loss of heat occurs in the period of the inter-mittence, due to the fluctuations of temperature themselves.

If the posterior surface remains unextended the extension of the anterior surface corresponding to a curvature ρ^{-1} is d/ρ, and the average extension is $d/2\rho$. Let us inquire what degree of curvature will be produced by the absorption of sunlight during a time t, on the supposition that the absorption is distributed throughout the substance of the plate, so as to give the right proportional extension to every stratum.

If Ht denote the heat received in time t per unit area, c the specific heat of the material per unit volume, e the linear extension of the material per degree centigrade, then

$$\frac{1}{\rho} = \frac{2eHt}{cd^2} .$$

In the case of sunshine, which is said to be capable of melting 100 feet of ice per annum, we have approximately in c.g.s. measure

$$Ht = \cdot008 \, t.$$

Thus

$$\frac{1}{\rho} = \cdot016 \, \frac{et}{cd^2} .$$

For iron $e = \cdot000012$, $c = \cdot86$.

Thus if $t = \frac{1}{500}$ (of a second), $d = \cdot02$ cm.

$$1/\rho = 1\cdot12 \times 10^{-6}.$$

This estimate will apply roughly to a period of intermittence equal to $\frac{1}{250}$th of a second, i.e. to about the middle of the musical scale. If the plate be a disk of radius r, held at the circumference, the displacement at the centre will be $r^2/2\rho$, or $\cdot 56 r^2 \times 10^{-6}$. In the case of a diameter of 6 centimetres this becomes $5\cdot 0 \times 10^{-6}$.

Five millionths of a centimetre is certainly a small amplitude, but it is probable that the sound would be audible. In an experiment (made, it is true, at a higher pitch) I found sound audible whose amplitude was less than a ten-millionth of a centimetre*. We may conclude, I think, that there is at present no reason for discarding the obvious explanation that the sounds in question are due to the bending of the plates under unequal heating.

* *Proc. Roy. Soc.* 1877. [Art. xlviii. p. 328.]

72.

ON COPYING DIFFRACTION-GRATINGS, AND ON SOME PHENOMENA CONNECTED THEREWITH.

[*Philosophical Magazine*, XI. pp. 196—205, 1881.]

IN the *Phil. Mag.* for February and March 1874 [Art. XXX.] I gave an account of experiments in the photographic reproduction of gratings ruled with lines at a rate of 3000 and 6000 to the inch. Since that time I have had further experience, extending to more closely ruled gratings, and have examined more minutely certain points which I was then obliged to leave unexplained. The present communication is thus to be regarded as supplementary to the former.

Some years ago Prof. Quincke described an unphotographic process by which he had succeeded in copying engraved glass gratings. He began by depositing a thin coating of silver by the chemical method upon the face of the grating. The conducting layer thus obtained was then transferred to an electrolytic cell, and thickened by the deposit of copper, until stout enough to be detached from the glass substratum. In this way he prepared an accurate *cast* of the glass surface, faced with highly reflecting silver. Since the optical depth of the lines is increased some four times, these gratings usually give much brighter spectra than the glass originals.

Prof. Quincke was kind enough to send me some specimens of his work, giving extremely beautiful spectra. I found, however, that, though carefully preserved, these gratings deteriorated after a time, apparently either from insufficient thickness, or from imperfect adhesion, of the silver layer. In my own attempts I endeavoured to remedy this defect by not allowing the silver to dry before transference to the electrolytic cell, and by commencing the electric deposit with a *silver* instead of with a *copper* solution.

I did not, however, succeed in finding a thoroughly satisfactory plating-liquid. In the ordinary cyanide solution the silver was at once loosened from the glass. In other solutions the grating could be immersed with impunity, but the film began to strip as soon as the current passed. Using *acetate* of silver, however, I was able to obtain a certain degree of thickening. I also found advantage from commencing the deposit of copper with a *neutral* solution. After the layer had attained a moderate thickness, its edges were dipped in melted paraffine; and it was then transferred to the usual acid solution of copper. I did not find it necessary to take any precautions against too great an adhesion between the silver and the glass.

These copies are now four years old, and they do not seem to have deteriorated. A slight yellow tarnish, due probably to sulphur, can be removed with cyanide of potassium. There is, however, one defect which I have not been able to avoid. The silver surface is never sufficiently flat to bear much magnifying-power. Unless this difficulty can be overcome, the use of such gratings must be limited to cases where brilliancy, and not high defining-power, is the desirable quality. For most purposes the photographic method of reproduction is to be preferred as far easier and quicker. Among various processes of this kind, I am still inclined to give the preference to that in which collodio-chloride of silver is employed, with subsequent treatment with mercury. The only trouble that I have met with is the tendency of the soluble salts to crystallize in the film; but this can generally be avoided with a little judgment. As these photographs cannot well be varnished, some doubts might have been entertained as to their permanence; but I find that copies now more than seven years old are none the worse. For gratings to be subjected to rough treatment, the various albumen processes offer decided advantages.

In my former paper I stated my opinion that the photographic method of reproduction would be applicable to lines finer than any that I had then tried (6000 to inch). In the summer of 1879 an opportunity afforded itself of submitting the matter to the test of actual trial through the kindness of Mr Rutherfurd, who presented me with a beautiful glass grating containing nearly 12,000 lines, ruled at the rate of 17,280 to the inch. The copies, taken with suitable precautions to secure a good contact, were completely successful, so far as the spectrum of the first order is concerned. Indeed careful comparison showed no appreciable difference between the defining-power of the original and of the copies; and with respect to brightness some of the copies had the advantage. On a former occasion*
I have shown that the theoretical resolving-power in the orange region of the spectrum is equal to that obtainable from a prismatic spectroscope with $12\frac{1}{2}$ cm. of "extra dense flint"; and I have no reason to think that the

---

* *Phil. Mag.* Oct. 1879. [Art. LXII. p. 428.]

actual resolving-power fell far short. This is a considerable result to obtain with a photograph which may be taken in half an hour at a cost of two or three shillings.

The case is different, however, when we turn to the spectrum of the second order. Used in this way the original gives magnificent results; but they are not reproduced in the copies. Some parts of the photograph will sometimes show a faint spectrum of the second order; but it is usually traversed by one or more dark bands, whose nature will presently be examined more at length.

As a rule, glass (or at any rate transparent) originals only would be used for purposes of reproduction; but as a matter of curiosity I tried what could be done in copying an original ruled on speculum-metal. The specimen experimented upon was similar to my own, both as to the total number of lines and as to the degree of closeness; it belongs to Mr Spottiswoode, to whom I am indebted for the loan of it. In this case the light of the sun had to pass through the sensitive film before it could reach the speculum-metal; it was then reflected back, and in *returning* through the film impressed the ruled structure. No very brilliant result was to be expected; but I succeeded so far as to obtain a copy which gave very fair results when tested upon the sun.

In my former paper I mentioned that when a spectrum of high order is thrown upon the eye, there usually appear upon the grating a certain number of irregular dark bands. These are the places at which the copy fails to produce the spectrum in question. With lines not closer than 3000 or 6000 to the inch, and with reasonably flat glass as support to the photographic film, these bands rarely invade the first or second spectrum. When, however, we come to 17,000 lines to the inch, it requires pretty flat glass and some precautions in printing to keep even the first spectrum free from them.

It was obvious from the first that the formation of these bands was a question of the distance between the ruled surface of the original and the sensitive film; but it is only within the last year or so that I have submitted the point to special experiment. For this purpose I substitute for plane-parallel glass as a substratum for the sensitive film the convex surface of a lens of moderate curvature. As in the experiment of Newton's rings, we obtain in this way an interval gradually increasing from the point of contact outwards, and thus upon one plate secure a record of the effect upon the copy of varying degrees of closeness. When a spectrum of any order is thrown upon the eye, those places upon the grating where the spectrum in question fails appear as dark rings. My first experiment of this kind was made with the Rutherfurd grating, in order principally to find out how close a contact was really necessary for copying. From the diameter of the first dark ring, in conjunction with a rough estimate of the curvature of the lens, I concluded that the interval between the surfaces should nowhere much exceed $\frac{1}{10000}$ of an inch. It appeared at the same time that the chance was remote of obtaining a satisfactory performance in the spectrum of the second order. About this time the theoretical views occurred to me which will presently be explained, and I purposed to check them by more careful measurements than I had yet attempted. In the course of last summer, however, I found accidentally that Fox Talbot had made, many years ago[*], some kindred observations; and the perusal of his account of them induced me to alter somewhat my proposed line of attack. It will be convenient to quote here Fox Talbot's brief statement:—

"About ten or twenty feet from the radiant point, I placed in the path of the ray an equidistant grating[†] made by Fraunhofer, with its lines vertical. I then viewed the light which had passed through this grating with a lens of considerable magnifying-power. The appearance was very curious, being a regular alternation of numerous lines or bands of red and green colour, having their directions parallel to the lines of the grating. On removing the lens a little further from the grating, the bands gradually changed their colours, and became alternately blue and yellow. When the lens was a little more removed, the bands again became red and green. And this change continued to take place for an indefinite number of times, as the distance between the lens and grating increased. In all cases the bands exhibited two complementary colours.

"It was very curious to observe that, though the grating was greatly out of the focus of the lens, yet the appearance of the bands was perfectly distinct and well defined.

"This, however, only happens when the radiant point has a *very small* apparent diameter, in which case the distance of the lens may be increased, even up to one or two feet from the grating, without much impairing the beauty and distinctness of the coloured bands. So that if the source of light were a mere mathematical point, it appears possible that this distance might be increased without limit; or that the disturbance in the luminiferous undulations caused by the interposition of the grating continues indefinitely, and has no tendency to subside of itself."

It is scarcely necessary to point out that what was seen by the eye in this experiment in any position of the magnifying lens was the same as would have been depicted upon a photographic plate situated at its focus, at least if the same kind of rays had been operative in both cases. Talbot's

---

[*] *Phil. Mag.* Dec. 1836.

[†] A plate of glass covered with gold leaf, on which several hundred parallel lines are cut, in order to transmit the light at equal intervals.

observations are therefore to the point as determining the effect of varying intervals in photographic copying.

On the whole the above description agrees well with what I had expected from theory. It is indeed impossible to admit that the red and green coloration could disappear and revive an *indefinite* number of times. The appearance of colour at all shows that the phenomenon varies with the wave-length, and accordingly that it would (as in all such cases when white light is used) ultimately be lost. Besides the limit imposed by the apparent magnitude of the source of light, there must be another depending upon the variation of wave-length within the range concerned.

In trying to repeat Talbot's experiment I found that even the 3000-to-the-inch grating was too fine to be conveniently employed; and eventually I fell back upon a very coarse grating made some years ago by photographing (with the camera and lens) a piece of striped stuff. By comparison of coincidences with the divisions of fine ivory scale (vernier fashion), the period was determined as ·0104 inch. As a source of light I used a slit placed parallel to the lines of the grating and backed by a fish-tail gas-flame seen edgeways. In order to observe the appearances behind the grating, a lens of moderate magnifying-power was sufficient. This lens was moved gradually back until something distinctive was seen; the distance between the lens and the grating was then measured and recorded. In order to render the light more nearly monochromatic, pieces of red or green glass were usually held in front of the eye.

With red light the nearly equal bright and dark bars are seen in focus when the distance of the lens from the grating is $1\frac{3}{4}$ inch. As the distance is increased, the definition deteriorates, and is worst at a distance of $3\frac{5}{8}$. In this position the proper period (·0104 inch) is lost, but subordinate fluctuations of brightness in shorter periods prevent the formation of a thoroughly flat field of view. As the distance is further increased, the definition appears to improve, until at distances $5\frac{3}{4}$ and $6\frac{3}{4}$ it is nearly as good as at first. The definition in an intermediate position such as $6\frac{1}{4}$ is distinctly inferior, but is far from being lost as in position $3\frac{5}{8}$. From the theoretical point of view, to be presently explained, these two positions of extra good definition are not to be distinguished. They relate rather to the sharpness of the *edge* of the band, than to any special prominence of the proper period. At a distance of $7\frac{1}{2}$ we have again a place of worst definition, at $10\frac{1}{4}$ a revival, and so on. These alternations could be traced to a distance of *nine feet* behind the grating.

The accompanying table gives the positions of best and worst definition for red and green light respectively. Of these the places of *worst* definition could be observed with the greater accuracy; but none of the observations have any pretensions to precision. The star indicates the position for focus.

It is evident that the positions for red light gradually fall quite away from the corresponding positions for green light. At $19\frac{3}{4}$, for example, if we use a green glass, we lose sight of the proper period, and have before us an almost uniform field; but if without making any other change we substitute

Red light		Green light	
Best	Worst	Best	Worst
*$1\frac{3}{4}$	$3\frac{5}{8}$	*$1\frac{1}{4}$	$3\frac{3}{4}$
$5\frac{3}{4}$, $6\frac{3}{4}$	$7\frac{1}{2}$	$6\frac{1}{4}$	$8\frac{5}{8}$
$10\frac{1}{4}$	$12\frac{1}{2}$	11	$14\frac{1}{4}$
$15\frac{1}{8}$	17	$16\frac{1}{2}$	$19\frac{3}{4}$
$19\frac{3}{4}$	$21\frac{7}{8}$	$22\frac{1}{4}$	$25\frac{1}{4}$
$24\frac{3}{4}$	27	$27\frac{3}{4}$	$31\frac{1}{4}$
$29\frac{1}{4}$	$32\frac{1}{8}$	$33\frac{1}{4}$	37
$34\frac{1}{2}$	$37\frac{1}{2}$		

a red glass for the green one, we see the bands again with great distinctness. At about the greatest distance included in the table the positions of best definition are again in coincidence; but here there is an important remark to be made. If, using the green glass, we adjust a needle-point to the centre of a bright band, we find, on substituting the red glass, that the needle-point is now in the centre, not of a *bright*, but of a *dark* band. The fact is that at every revival of definition the image changes sign, in the photographic sense, from positive to negative, or from negative to positive—a clear proof that the appearance in question is not a mere *shadow* in any ordinary sense of the term.

With respect to the numerical values of the distances given in the table, theory indicates that the interval from worst to worst or from best to best definition should be a third proportional to the period of the grating d and the wave-length of the light λ, *i.e.* should be equal to d^2/λ. In the case of red light, the mean interval from worst to worst is 4·8 inches, and from best to best 4·7. The corresponding numbers for green light are 5·5 and 5·3. In the subsequent calculation, I have used the first stated intervals as probably the more correct.

For the grating employed the actual value of d was ·0104 inch; but a small correction is required for the want of parallelism of the light. The distance of the source was about 27 feet; so that, as the mean distance behind the grating at which the appearances were observed was $1\frac{1}{2}$ foot, the above value of d must be increased in the ratio of $28\frac{1}{2}$ to 27. Thus for the effective d in centimetres, we get

$$2 \cdot 54 \times \frac{57}{54} \times \cdot 0104.$$

Calculating from this and from the observed intervals a by means of the formula $\lambda = d^2/a$, we get in centimetres

$$\lambda_{\text{(red)}} = 6{\cdot}40 \times 10^{-5}, \qquad \lambda_{\text{(green)}} = 5{\cdot}59 \times 10^{-5}.$$

Direct determination of the mean wave-lengths of the lights transmitted by the red and green glasses respectively gave

$$\lambda_{\text{(red)}} = 6{\cdot}64 \times 10^{-5}, \qquad \lambda_{\text{(green)}} = 5{\cdot}76 \times 10^{-5}.$$

The true wave-lengths are certainly somewhat greater than those calculated from Talbot's phenomenon; but the difference is perhaps hardly outside the limits of experimental error. If the measurements were ever repeated, it would be advisable to use a collimating lens as well as a more accurate grating.

The problem of determining the illumination at various points behind a grating exposed to a parallel beam of homogeneous light, could probably be attacked with success by the usual methods of physical optics, if it were assumed that the grating presented uniform intervals alternately transparent and opaque. Actual gratings, however, do not answer to this description, and, indeed, vary greatly in character. I have therefore preferred to follow the comparatively simple method, explained in my book on *Sound*, §§ 268, 301, which is adequate to the determination of the leading features of the phenomenon.

Taking the axis of z normal to the grating, and parallel to the original direction of the light, and the axis of x perpendicular to the lines of the grating, we require a general expression for the vibration of given frequency which is periodic with respect to x in the distance d. Denoting the velocity of propagation of ordinary plane waves by a, and writing $k = 2\pi/\lambda$, we may take as this expression

$$A_0 \cos(kat - kz)$$

$$+ A_1 \cos\left(\frac{2\pi x}{d} + e_1\right)\cos(kat - \mu_1 z) + B_1 \cos\left(\frac{2\pi x}{d} + e_1{}'\right)\sin(kat - \mu_1 z)$$

$$+ A_2 \cos\left(\frac{4\pi x}{d} + e_2\right)\cos(kat - \mu_2 z) + B_2 \cos\left(\frac{4\pi x}{d} + e_2{}'\right)\sin(kat - \mu_2 z) + \dots,$$

where

$$\mu_1{}^2 = k^2 - \frac{4\pi^2}{d^2}, \qquad \mu_2{}^2 = k^2 - \frac{4 \cdot 4\pi^2}{d^2}, \qquad \mu_3{}^2 = k^2 - \frac{9 \cdot 4\pi^2}{d^2}, \qquad \&\text{c.}$$

[The terms represent the various spectra, and] the series is to be continued so long as μ^2 is positive, *i.e.* so long as the period of the component fluctuations parallel to x is greater than λ. Features in the wave-form whose period is less than λ cannot be propagated in this way, but are rapidly extinguished.

The intensity of vibration, measured by the square of the amplitude, is

$$\left[A_0 + A_1 \cos\left(\frac{2\pi x}{d} + e_1\right) \cos(kz - \mu_1 z) \right.$$

$$+ B_1 \cos\left(\frac{2\pi x}{d} + e_1{'}\right) \sin(kz - \mu_1 z) + A_2 \cos\left(\frac{4\pi x}{d} + e_2\right) \cos(kz - \mu_2 z) + \ldots \left. \right]^2$$

$$+ \left[- A_1 \cos\left(\frac{2\pi x}{d} + e_1\right) \sin(kz - \mu_1 z) \right.$$

$$+ B_1 \cos\left(\frac{2\pi x}{d} + e_1{'}\right) \cos(kz - \mu_1 z) - A_2 \cos\left(\frac{2\pi x}{d} + e_2\right) \sin(kz - \mu_2 z) + \ldots \left. \right]^2.$$

In order to apply this result to our present question, it is supposed as a rough approximation that the terms with suffixes higher than unity may be omitted. We thus obtain

$$A_0{}^2 + \tfrac{1}{2}A_1{}^2 + \tfrac{1}{2}B_1{}^2 + 2A_0 A_1 \cos\left(\frac{2\pi x}{d} + e_1\right) \cos(kz - \mu_1 z)$$

$$+ 2A_0 B_1 \cos\left(\frac{2\pi x}{d} + e_1{'}\right) \sin(kz - \mu_1 z)$$

$$+ \tfrac{1}{2}A_1{}^2 \cos\left(\frac{4\pi x}{d} + 2e_1\right) + \tfrac{1}{2}B_1{}^2 \cos\left(\frac{4\pi x}{d} + 2e_1{'}\right),$$

which as a function of z is periodic with a period determined by

$$kz - \mu_1 z = 2\pi,$$

or

$$z = \frac{\lambda}{1 - \sqrt{(1 - \lambda^2/d^2)}}.$$

In the cases with which we are concerned λ^2 is small in comparison with d^2, so that approximately $z = 2d^2/\lambda$. So far, then, as this theory extends, the phenomena behind the grating are reproduced with every retreat through a distance $2d^2/\lambda$; but, on account of the terms omitted, this conclusion does not apply to the subordinate periods (on which depends the performance of a copy in the spectra of higher order); nor does it apply rigorously even to the principal period itself.

Similar results to those given by direct inspection on the coarse grating have been obtained by photographic copying of finer ones, a lens (as already explained) being substituted for flat glass as a support for the sensitive film. When the copy is held so that the spectrum of the first order is formed upon the eye, several dark rings are visible, separated by intervals of brightness. With the 6000 Nobert the diameter of the first dark ring was ·54 inch, and at the centre round the point of contact there was a dark spot nearly as dark as the ring. In the second and third spectra the centres were also dusky, though not so black as in the first. The diameter of the first dark ring in the second spectrum was ·30 inch. [Inch = 2·54 cm.]

The occurrence of a dark centre is a point of interest, as showing that for purposes of reproduction it is possible for the contact to be too close, though I do not remember to have met with this in practice; and theoretically it is what would be expected when we consider that the original does not act by opacity. According to this view a different result should be obtained in copying an opaque grating; and such I have found to be the case. For this purpose I employed a copy of the same 6000 Nobert, taken some years ago on a tannin plate, and prepared the photographic film on the *same lens* as before. When the resulting photograph was examined, the spectra of the first three orders showed *bright centres*. The diameter of the first dark ring in the first spectrum was ·44 inch—smaller than before.

With the 3000 Nobert in place of the 6000, the ring-system is formed on a larger scale. The centres for the first four spectra are black, with the exception of the actual place of contact, where evidently the collodion film was impressed mechanically. The diameter of the first dark ring in the first spectrum is ·90 inch, not quite the double of ·54 inch, although the same lens as before was used. In the second spectrum the diameter of the first dark ring is ·56 inch, and in the third spectrum ·40 inch.

Interesting as these bands may be in theory, they are to be avoided as much as possible in the practical reproduction of gratings, not merely because a part of the area is lost, but also on account of the reversal which takes place at every revival of brightness. Without having examined the matter very closely, I had generally found the performance of gratings which showed these bands to be inferior; and now it would seem that the explanation is to be found in the above-mentioned reversals, which could not fail to interfere with the resolving-power.

During my early experiments it happened once that in the course of printing an accidental shifting took place, leading to the impression of a double image. A more perfect result was afterwards obtained by intentionally communicating to the plates a slight relative twist in the middle of the exposure. When a spectrum from such a grating is thrown upon the eye, parallel bars are seen perpendicular to the direction of the grooves; but the number and position of these bars depend upon the order of the spectrum. In one case twenty-five bars were counted in the first spectrum, and twice that number in the second. But it is unnecessary to dwell further upon these observations, as they correspond exactly with what the ordinary theory of gratings would lead us to expect.

73.

ON IMAGES FORMED WITHOUT REFLECTION OR REFRACTION.

[*Philosophical Magazine*, XI. pp. 214—218, 1881.]

THE function of a lens in forming an image is to compensate by its variable thickness the differences in phase which would otherwise exist between secondary waves arriving at the focal point from various parts of the aperture. If we suppose the diameter of the lens ($2r$) to be given, and its focal length f gradually to increase, these differences of phase at the image of an infinitely distant luminous point diminish without limit. When f attains a certain value, say f_1, the extreme error of phase to be compensated falls to $\frac{1}{4}\lambda$. Now, as I have shown on a previous occasion[*], an extreme error of phase amounting to $\frac{1}{4}\lambda$, or less, produces no appreciable deterioration in the definition; so that from this point onwards the lens is useless, as only improving an image already sensibly as perfect as the aperture admits of. Throughout the operation of increasing the focal length, the resolving-power of the instrument, which depends only upon the aperture, remains unchanged; and we thus arrive at the rather startling conclusion that a telescope of any degree of resolving-power might be constructed without an object-glass, if only there were no limit to the admissible focal length. This last proviso, however, as we shall see, takes away almost all practical importance from the proposition.

To get an idea of the magnitudes of the quantities involved, let us take the case of an aperture of $\frac{1}{5}$ inch [inch = 2·54 cm.], about that of the pupil of the eye. The distance f_1, which the actual focal length must exceed, is given by

$$\sqrt{\{f_1{}^2 + r^2\}} - f_1 = \tfrac{1}{4}\lambda;$$

---

[*] *Phil. Mag.* November 1879. [Art. LXII. § 4.]

so that
$$f_1 = 2r^2/\lambda.$$
Thus, if $\lambda = \frac{1}{40000}$, $r = \frac{1}{10}$, $f_1 = 800$.

The image of the sun thrown on a screen at a distance exceeding 66 feet, through a hole $\frac{1}{5}$ inch in diameter, is therefore at least as well defined as that seen direct.   In practice it would be better defined, as the direct image is far from perfect.   If the image on the screen be regarded from a distance f_1, it will appear of its natural angular magnitude.   Seen from a distance less than f_1, it will appear magnified.   Inasmuch as the arrangement affords a view of the sun with full definition and with an increased apparent magnitude, the name of a telescope can hardly be denied to it.

As the minimum focal length increases with the square of the aperture, a quite impracticable distance would be required to rival the resolving-power of a modern telescope.   Even for an aperture of four inches f_1 would be five miles.

A similar argument to that just employed to find at what point a lens begins to have an advantage over a simple aperture, may be applied to determine at what point an achromatic lens begins to assert a perceptible superiority over a single lens in forming a white image.   The question in any case is simply whether, when the adjustment is correct for the central rays of the spectrum, the error of phase for the most extreme rays (which it is necessary to consider) amounts to a quarter of a wave-length.   If not, the substitution of an achromatic lens will be of no advantage.

If μ be the refractive index for which the adjustment is perfect, then the error of phase for the ray of index $\mu + \delta\mu$ is $\delta\mu \cdot t$, where t is the "thickness" of the lens.   Now
$$(\mu - 1)t = r^2/2f;$$
so that, if the error of phase amount to $\frac{1}{4}\lambda$,
$$\frac{\delta\mu}{\mu - 1} = \frac{\lambda f_1}{2r^2}.$$

In order to apply this numerically, let us take the case of hard crown-glass, for which the indices are given by Hopkinson*.   The practical limits of the spectrum being taken at B and G, we have $\mu_B = 1\cdot5136$, $\mu_G = 1\cdot5284$, the difference of which is $\cdot0148$.   If the focus be correct for the mean value of μ, the extreme value of $\delta\mu$ is $\cdot0074$, and that of $\delta\mu/(\mu - 1)$ is $\cdot0074/\cdot521$, or $\cdot0142$.   In strictness we ought to take into account the variation of λ; but for such a purpose as the present we may put it at $\frac{1}{40000}$ inch; and then the fraction $\cdot0142$ expresses the admissible focus when a single lens is used as compared with the focus necessary when a lens is dispensed with altogether.

* *Proc. Roy. Soc.* 1877.

Thus, if the aperture be one-fifth of an inch, an achromatic lens has no advantage over a single one, if the focal length be greater than about 11 inches. If, on the other hand, we suppose the focal length to be 66 feet, a single lens is practically perfect up to an aperture of $1\cdot7$ inch. The effect of spherical aberration in disturbing definition was considered in my former paper. In such a case as that last specified it is altogether negligible. The advantage of a long focus was well understood by Huyghens and his contemporaries; but it may have been worth while to consider the matter for a moment from another point of view, from which it clearly appears that the substitution of an achromatic for a single lens serves no other purpose than to diminish the minimum admissible focal length.

Returning now to homogeneous light, let us consider the case of an *annular* aperture of radii r_1 and r_2. The extreme difference of phase at distance f is now $(r_2{}^2 - r_1{}^2) \div 2f$. If this be $\frac{1}{4}\lambda$, we get

$$f_1 = \frac{2\,(r_2{}^2 - r_1{}^2)}{\lambda} = \frac{2\,(r_2 + r_1)\,(r_2 - r_1)}{\lambda}$$

as the value of the minimum distance at which a lens can be dispensed with without loss. If $r_2 - r_1$ be small, f_1 is much smaller than for a full circle of radius r_2; and it might appear that a great advantage would be gained either in the diminution of f_1 or by an increase in r_2. The question, however, remains whether *with a lens* the definition due to an annular aperture of given outer radius r_2 is independent of the inner radius r_1.

The image of a mathematical point consists, it is known, of a central patch of brightness, surrounded by rings alternately dark and bright. If we conceive the radius of the central stop (i.e. r_1) gradually to increase from 0 to r_2, the diameter of the central luminous patch diminishes in the ratio $3\cdot83 : 2\cdot41$. From this it might be supposed that the definition due to the marginal rim acting alone would be superior to the definition due to the whole aperture*. It is true that there is at first some improvement in definition; but as r_1 approaches equality with r_2 a rapid deterioration sets in, notwithstanding the smallness of the central luminous patch. In order to understand this it is necessary to examine more minutely the distribution of light over the entire field.

If the point under consideration be distant ρ from the centre of the diffraction-pattern, the illumination for the full aperture is given by

$$I^2 = \frac{\pi^2 r^4}{\lambda^2 f^2} \left[\frac{2J_1\,(2\pi r\rho/\lambda f)}{2\pi r\rho/\lambda f} \right]^2 = \frac{\pi^2 r^4}{\lambda^2 f^2} \frac{4J_1{}^2\,(y)}{y^2},$$

if $y = 2\pi r\rho/\lambda f$, J_1 being the symbol of the Bessel's function of order unity. The dark rings correspond to the roots of J_1, and occur when $y = 3\cdot83,\ 7\cdot02,\ 10\cdot17,$ &c.

---

* See a paper on the Diffraction of Object-glasses, *Astr. Month. Notices*, 1872. [Art. XIX.]

The whole illumination within the area of the circle of radius ρ is given by

$$\int I^{2} 2\pi\rho \, d\rho = 2\pi r^{2} \int_{0}^{y} y^{-1} J_{1}^{2}(y) \, dy.$$

This integral may be transformed by known properties of Bessel's functions. Thus*,

$$\frac{J_{1}(y)}{y} = J_{0}(y) - \frac{dJ_{1}(y)}{dy} ;$$

so that

$$\frac{J_{1}^{2}(y)}{y} = J_{0}(y) . J_{1}(y) - J_{1}(y)\frac{dJ_{1}(y)}{dy} = -J_{0}(y)\frac{dJ_{0}(y)}{dy} - J_{1}(y)\frac{dJ_{1}(y)}{dy} .$$

We therefore obtain

$$2\int_{0}^{y} y^{-1} J_{1}^{2}(y) \, dy = 1 - J_{0}^{2}(y) - J_{1}^{2}(y).$$

If y be infinite, $J_{0}(y)$ and $J_{1}(y)$ vanish, and the whole illumination is expressed by πr^{2}, as is evident *à priori*. In general the proportion of the whole illumination to be found *outside* the circle of radius ρ is given by

$$J_{0}^{2}(y) + J_{1}^{2}(y).$$

For the dark rings $J_{1}(y) = 0$; so that the fraction of illumination outside any dark ring is simply $J_{0}^{2}(y)$. Thus, for the 1st, 2nd, 3rd, and 4th dark rings we get respectively ·161, ·090, ·062, and ·047, showing that more than $\frac{9}{10}$ of the whole light is concentrated within the area of the second dark ring.

The corresponding results for a narrow annular aperture would be very different, as we may easily convince ourselves. The illumination at any point of the central spot or of any of the bright rings is proportional to the *square* of the width of the annulus, while the whole quantity of light is proportional to the width itself. As, therefore, the annulus narrows, a less and less proportion of the whole light is contained in any finite number of luminous rings, and the definition of an image corresponding to an assemblage of luminous points is proportionally impaired.

The truth is that, so far as it is possible to lay down any general law at all, the definition depends rather upon the *area* than upon the *external diameter* of the aperture. If A be this area, the illumination at the focal point, where all the secondary waves concur in phase, is given by $I_{0}^{2} = A^{2}/\lambda^{2}f^{2}$, the primary illumination being taken as unity. The whole illumination passing the aperture is on the same scale represented by A. Hence if A' be the area over which an illumination I_{0}^{2} would give the actual total illumination,

---

* Todhunter's *Laplace's Functions*, p. 297.

$AA' = \lambda^2 f^2$; and A', being in some sense the area of the diffraction-pattern, may be taken as a criterion of the definition.

In the case of an annulus we saw that the minimum focal length allowing a lens to be dispensed with is also dependent upon the *area* of aperture— $\pi (r_2^2 - r_1^2)$; so that it would appear that if the object be to form at a given distance, and without a lens, as well-defined an image as may be, it is of comparatively little consequence whether or not an annular aperture be adopted. A moderate central stop would doubtless be attended with benefit; but it is probable that harm rather than good would result from any thing like extreme proportions.

P.S.—Reference should be made to a paper by Petzval on the Camera Obscura (*Phil. Mag.* Jan. 1859), in which the definition of images formed without lenses is considered. The point of view is different from that above adopted. [1899. The reader may also consult *Proc. Roy. Inst.* Feb. 1891, *Nature* XLIV. p. 249, 1891; *Phil. Mag.* XXXI. p. 87, 1891.]

74.

ON THE ELECTROMAGNETIC THEORY OF LIGHT.

[*Philosophical Magazine*, XII. pp. 81—101, 1881.]

THE claims of the theory propounded by Maxwell, according to which light consists of a disturbance in a medium susceptible of dielectric polarization, are so considerable that it is desirable to extend its application as far as possible to various optical phenomena. The question of the velocity of propagation in vacuum and in singly or doubly refracting transparent dielectrics was considered by Maxwell himself; and the agreement with experiment, though far from perfect, is sufficiently encouraging. More recently it has been shown by Helmholtz*, Lorentz†, Fitzgerald‡, and J. J. Thomson§, that the same theory leads to Fresnel's expressions for the intensity of light reflected and refracted at the surface of separation of transparent media, and that the auxiliary hypotheses necessary in this part of the subject agree with those required to explain the laws of double refraction. In this respect the electromagnetic theory has a marked advantage over the older view, which assimilated luminous vibrations to the ordinary transverse vibrations of elastic solids. According to the latter, Fresnel's laws of double refraction, fully confirmed by modern observation‖, require us to suppose that in a doubly-refracting crystal the rigidity of the medium varies with the direction of the strain; while, in order to explain the facts relating to the intensities of reflected light, we have to make the inconsistent assumption that the rigidity does not vary in passing from one medium to another. A further discussion of this subject will be found in papers published in the *Philosophical Magazine* during the year 1871. [Arts. VIII. IX. X. XI.]

---

* *Crelle*, Bd. LXXII. 1870.

‡ *Phil. Trans.* 1880.

‖ Glazebrook, *Phil. Trans.* 1879.

† Schlömilch, XXII. 1877.

§ *Phil. Mag.* April 1880.

If the dielectric medium be endowed with sensible conductivity, the electric vibrations will be damped; that is to say, the light will undergo absorption, with a rapidity which Maxwell has calculated. By supposing the conductivity to be so great that practically complete absorption takes place within a distance comparable with the wave-length, we may obtain a theory of metallic reflection which is not without interest, although the phenomena of abnormal dispersion show that it cannot be regarded as complete.

For an isotropic medium at rest we have the equations (Maxwell's *Electricity and Magnetism*, §§ 591, 598, 607, 610, 611)

$$u = p + df/dt, \text{&c.,} \quad\dots\dots\dots\dots\dots\dots\dots(1)$$

$$4\pi f = KP, \text{&c.,} \quad\dots\dots\dots\dots\dots\dots\dots(2)$$

$$p = CP, \text{&c.,} \quad\dots\dots\dots\dots\dots\dots\dots(3)$$

$$P = -dF/dt - d\Psi/dx, \text{&c.,} \quad\dots\dots\dots\dots(4)$$

$$a = dH/dy - dG/dz, \text{&c.,} \quad\dots\dots\dots\dots\dots(5)$$

$$a = \mu\alpha, \text{&c.,} \quad\dots\dots\dots\dots\dots\dots\dots(6)$$

$$4\pi u = d\gamma/dy - d\beta/dz, \text{&c. ;} \quad\dots\dots\dots\dots\dots(7)$$

in which f, g, h are the electric displacements, p, q, r the currents of conduction, u, v, w the total currents, P, Q, R the components of electromotive force, K the specific inductive capacity, C the conductivity, α, β, γ the components of magnetic force, a, b, c the components of magnetization, μ the magnetic capacity, F, G, H the components of electrokinetic momentum, and Ψ the electric potential.

From (2), (4), and (5) we get

$$4\pi \left(\frac{d}{dy} \frac{f}{K} - \frac{d}{dx} \frac{g}{K} \right) = -\frac{d}{dt} \left(\frac{dF}{dy} - \frac{dG}{dx} \right) = \frac{dc}{dt}, \text{&c}\dots\dots\dots(8)$$

In the case of K constant, equation (8) expresses that the electric displacement $\int (f\,dx + g\,dy)$ round a small circuit in the plane of xy corresponds to the electromotive force round the circuit, represented by dc/dt.

Again, from (1), (2), (3), (6), (7),

$$4\pi \left(\frac{df}{dt} + \frac{4\pi C}{K} f \right) = \frac{d}{dy} \frac{c}{\mu} - \frac{d}{dz} \frac{b}{\mu}, \text{&c}\dots\dots\dots\dots(9)$$

From equations (8) and (9)* the problem of reflection can be investigated. In order to limit ourselves to plane waves of simple type, we shall suppose

* [1899. These are the general circuital relations given by Maxwell, *Phil. Trans.* 1868; *Scientific Papers*, vol. II. p. 138.]

that K, μ, and C are independent of z, and that the electric and magnetic functions are independent of z and (as dependent upon the time) proportional to e^{int}. The two principal cases will be considered separately, (1) when the electric displacements are perpendicular to the plane of incidence, (2) when they are executed in that plane.

Case 1.   This is defined by the conditions

$$f = 0, \quad g = 0, \quad \text{and (accordingly) } c = 0.$$

Thus

$$ina = -4\pi \frac{d}{dy} \frac{h}{K}, \qquad inb = 4\pi \frac{d}{dx} \frac{h}{K}, \quad \ldots\ldots\ldots\ldots(10)$$

$$4\pi \left(in + \frac{4\pi C}{K}\right) h = \frac{d}{dx} \frac{b}{\mu} - \frac{d}{dy} \frac{a}{\mu}. \quad \ldots\ldots\ldots\ldots\ldots(11)$$

Eliminating a and b from (10) and (11), we get

$$\frac{d}{dx}\left(\frac{1}{\mu}\frac{d}{dx}\right)\frac{h}{K} + \frac{d}{dy}\left(\frac{1}{\mu}\frac{d}{dy}\right)\frac{h}{K} + K\left(n^2 - in\frac{4\pi C}{K}\right)\frac{h}{K} = 0. \quad \ldots\ldots(12)^*$$

Case 2.   Here the special conditions are

$$h = 0, \quad a = 0, \quad b = 0.$$

We have

$$4\pi \left(\frac{d}{dy}\frac{f}{K} - \frac{d}{dx}\frac{g}{K}\right) = inc, \quad \ldots\ldots\ldots\ldots\ldots(13)$$

$$4\pi \left(in + \frac{4\pi C}{K}\right)f = \frac{d}{dy}\left(\frac{c}{\mu}\right), \qquad 4\pi \left(in + \frac{4\pi C}{K}\right)g = -\frac{d}{dx}\left(\frac{c}{\mu}\right); \quad \ldots\ldots(14)$$

whence by elimination of f and g,

$$\frac{d}{dx}\left\{\frac{1}{K(n^2 - in.\,4\pi C/K)}\frac{d}{dx}\left(\frac{c}{\mu}\right)\right\}$$

$$+ \frac{d}{dy}\left\{\frac{1}{K(n^2 - in.\,4\pi C/K)}\frac{d}{dy}\left(\frac{c}{\mu}\right)\right\} + \mu\left(\frac{c}{\mu}\right) = 0. \quad \ldots(15)^*$$

Equations (12) and (15) simplify considerably in their application to a uniform medium, assuming the common form

$$d^2/dx^2 + d^2/dy^2 + \mu K\,(n^2 - in.\,4\pi\,C/K) = 0. \quad \ldots\ldots\ldots\ldots(16)^*$$

To express the boundary conditions let us suppose that $x = 0$ is the surface of transition between two uniform media.   From (12) we learn that the required conditions for case 1 are that

$$\frac{h}{K} \quad \text{and} \quad \frac{1}{\mu}\frac{d}{dx}\left(\frac{h}{K}\right)$$

must be continuous.

* [1899. A slip of the pen is here corrected.]

In like manner, for case 2 we see from (15) that

$$\frac{c}{\mu} \quad \text{and} \quad \frac{1}{K(n^2 - in \cdot 4\pi C/K)} \frac{d}{dx}\left(\frac{c}{\mu}\right)$$

must be continuous.

If the media are transparent, or but moderately opaque, we have to put $C = 0$. The differential equation is of the form

$$d^2/dx^2 + d^2/dy^2 + n^2\mu K = 0. \dots\dots\dots\dots\dots\dots(17)$$

In case 1 the boundary conditions are the continuity of the dependent variable and of $\mu^{-1}d/dx$, and in case 2 the continuity of the dependent variable and of $K^{-1}d/dx$. Analytically, the results are thus of the same form in both cases. If θ and θ_1 are respectively the angles of incidence and refraction, the ratio of the reflected to the incident vibration is in case 1

$$\frac{\tan\theta_1/\tan\theta - \mu/\mu_1}{\tan\theta_1/\tan\theta + \mu/\mu_1}, \dots\dots\dots\dots\dots\dots(18)$$

and in case 2

$$\frac{\tan\theta_1/\tan\theta - K/K_1}{\tan\theta_1/\tan\theta + K/K_1}, \dots\dots\dots\dots\dots(19)$$

in which K, μ relate to the first, and K_1, μ_1 to the second medium; while the relation between θ_1 and θ is

$$K_1\mu_1 : K\mu = \sin^2\theta : \sin^2\theta_1. \dots\dots\dots\dots\dots(20)$$

As Helmholtz has remarked, Fresnel's formulæ may be obtained on two distinct suppositions. If $\mu_1 = \mu$,

$$(18) = \frac{\sin(\theta_1 - \theta)}{\sin(\theta_1 + \theta)},$$

and

$$(19) = \frac{\tan(\theta_1 - \theta)}{\tan(\theta_1 + \theta)};$$

but if $K_1 = K$, then (19) identifies itself with the sine-formula, and (18) with the tangent-formula. Electrical phenomena, however, lead us to prefer the former alternative, and thus to the assumption that the *electric* displacements are perpendicular to the plane of polarization. The formulæ for the refracted waves, which follow from those of the reflected waves in virtue of the principle of energy alone, do not call for detailed consideration.

In the problem of perpendicular incidence, we have from (12), if μ be constant and C zero,

$$\frac{d^2}{dx^2}\frac{h}{K} + n^2\mu K\left(\frac{h}{K}\right) = 0. \dots\dots\dots\dots\dots(21)$$

For an application of this equation to determine the influence of defective suddenness in the transition between two uniform media, the reader is referred to a paper in the eleventh volume of the *Proceedings of the Mathematical Society*. [Art. LXIII. p. 462.]

In order to obtain a theory of metallic reflection, C must be considered to have a finite value in the second medium. The symbolical solution is not thereby altered from that applicable to transparent media, the effect of the finiteness of C being completely represented in both cases by the substitution of $K(1 - in^{-1} 4\pi C/K)$ for K. Thus, if μ be constant, the formula for the amplitude and phase of the reflected wave in case 1 is to be found by transformation of (18), in which the imaginary angle of refraction θ_1 is connected with θ by the relation

$$K_1(1 - in^{-1} 4\pi C/K_1) : K = \sin^2 \theta : \sin^2 \theta_1. \ldots\ldots\ldots\ldots(22)$$

In like manner the solution for case 2 is to be found by transformation of (19) under the same supposition.

With regard to the proposed transformations, the reader is referred to a paper by Eisenlohr [*] and to some remarks thereupon by myself[†]. The results are the formulæ published without proof by Cauchy. From the calculations of Eisenlohr it appears that Jamin's observations cannot be reconciled with the formulæ without supposing $K_1 : K$, *i.e.* the real part of the square of the complex refractive index, to be negative—a further proof that much remains to be done before the electrical theory of metallic reflection can be accepted as complete[‡].

The same fundamental equations (8) and (9) will now be applied to the problem of determining the effect on a train of plane waves of a small variation in the quantities K and μ which define the medium. A similar method will be adopted to that already used for light in a paper "On the Scattering of Light by small Particles"[§], and in my book *On the Theory of Sound*, § 296, the principle of which consists in an approximation depending upon the neglect of the higher powers of the small variations ΔK and $\Delta \mu$.

Let us suppose that a train of plane waves, in which the electric displacement is parallel to z, and magnetization parallel to y, propagates itself parallel to x undisturbed until it falls upon a region where the generally constant values of K and μ become $K + \Delta K$ and $\mu + \Delta \mu$. If ΔK and $\Delta \mu$

---

[*] *Pogg. Ann.* t. CIV. p. 368.

[†] *Phil. Mag.* May 1872. [Art. XVI. p. 146.]

[‡] July 15. I see that Lorentz, in a pamphlet *Over de Theorie der Terugkaatsing en Breking van het Licht* (Arnhem, 1875), has developed a theory of metallic reflection similar to that indicated in the text, and has noticed the same difficulty in the application to experiment.

[§] *Phil. Mag.* June 1871. [Art. IX.]

were zero, the wave would pass on as before; but under the circumstances secondary waves are generated, which diverge from the region of disturbance, and are ultimately, when ΔK and $\Delta \mu$ are small enough, proportional in magnitude to these quantities. As the expression of the primary waves we may take

$$h_0 = e^{int} \, e^{ikx}, \dots\dots\dots\dots\dots\dots\dots\dots\dots\dots(23)$$

and corresponding thereto, by (8),

$$b_0 = 4\pi k n^{-1} K^{-1} e^{int} \, e^{ikx}, \dots\dots\dots\dots\dots(24)$$

in which, if λ denote the wave-length, $k = 2\pi/\lambda$, and n/λ is the velocity of propagation $(K\mu)^{-\frac{1}{2}}$. The complete values of the functions being represented, as before, by f, g, h, a, b, c, we shall put

$$f = f_0 + f_1 + f_2 + \dots \&c., \qquad a = a_0 + a_1 + \dots \&c.,$$

$f_0 \dots a_0 \dots$ being independent of ΔK and $\Delta\mu$, $f_1 \dots a_1 \dots$ being of the first order, $f_2 \dots a_2 \dots$ of the second order, and so on, in these quantities. In the actual case f_0, g_0, a_0, c_0 vanish, and only h_0 and b_0 are finite.

From (8) and (9) with $C = 0$, we get

$$4\pi \left\{ \frac{df}{dy} - \frac{dg}{dx} + K^{-1} \frac{d}{dy} (f \Delta K^{-1}) - K^{-1} \frac{d}{dx} (g \Delta K^{-1}) \right\} = K \frac{dc}{dt},$$

$$4\pi \left\{ \frac{dg}{dz} - \frac{dh}{dy} + K^{-1} \frac{d}{dz} (g \Delta K^{-1}) - K^{-1} \frac{d}{dy} (h \Delta K^{-1}) \right\} = K \frac{da}{dt}, \quad \left.\right\} \dots(25)$$

$$4\pi \left\{ \frac{dh}{dx} - \frac{df}{dz} + K^{-1} \frac{d}{dx} (h \Delta K^{-1}) - K^{-1} \frac{d}{dz} (f \Delta K^{-1}) \right\} = K \frac{db}{dt}.$$

$$\frac{dc}{dy} - \frac{db}{dz} + \mu^{-1} \frac{d}{dy} (c \Delta \mu^{-1}) - \mu^{-1} \frac{d}{dz} (b \Delta \mu^{-1}) = 4\pi \mu \frac{df}{dt},$$

$$\frac{da}{dz} - \frac{dc}{dx} + \mu^{-1} \frac{d}{dz} (a \Delta \mu^{-1}) - \mu^{-1} \frac{d}{dx} (c \Delta \mu^{-1}) = 4\pi \mu \frac{dg}{dt}, \quad \left.\right\} \dots(26)$$

$$\frac{db}{dx} - \frac{da}{dy} + \mu^{-1} \frac{d}{dx} (b \Delta \mu^{-1}) - \mu^{-1} \frac{d}{dy} (a \Delta \mu^{-1}) = 4\pi \mu \frac{dh}{dt}.$$

By differentiation of the first equation of (26) and substitution from (25), we get, having regard to

$$df/dx + dg/dy + dh/dz = 0, \dots\dots\dots\dots\dots(27)$$

which is a consequence of (1), (3), (7),

$$\mu K \frac{d^2 f}{dt^2} = \frac{d^2 f}{dx^2} + \frac{d^2 f}{dy^2} + \frac{d^2 f}{dz^2} + K^{-1} \left(\frac{d^2}{dy^2} + \frac{d^2}{dz^2} \right) (f \Delta K^{-1})$$

$$- K^{-1} \frac{d^2}{dx\,dy} (g \Delta K^{-1}) - K^{-1} \frac{d^2}{dx\,dz} (h \Delta K^{-1})$$

$$+ \frac{\mu K}{4\pi} \frac{d^2}{dy\,dt} (c \Delta \mu^{-1}) - \frac{\mu K}{4\pi} \frac{d^2}{dz\,dt} (b \Delta \mu^{-1}),$$

or, remembering that the functions as dependent upon time vary as e^{int},

$$\nabla^2 f + k^2 f + \overset{-1}{K}\left(\frac{d^2}{dy^2} + \frac{d^2}{dz^2}\right)(f\Delta K^{+1})$$

$$- \overset{-1}{K}\frac{d^2}{dx\,dy}(g\Delta K^{-1}) - \overset{-1}{K}\frac{d^2}{dx\,dz}(h\Delta K^{-1})$$

$$+ \frac{in\overset{-1}{\mu}K}{4\pi}\frac{d}{dy}(c\Delta\mu^{-1}) - \frac{in\overset{-1}{\mu}K}{4\pi}\frac{d}{dz}(b\Delta\mu^{-1}) = 0, \quad\ldots\ldots\ldots\ldots(28)$$

with two similar equations in g and h.

Introducing now the expansion in powers of ΔK and $\Delta\mu$, we get as the first approximation

$$\nabla^2 f_1 + k^2 f_1 - \overset{-1}{K}\frac{d^2}{dx\,dz}(h_0\Delta K^{-1}) - \frac{in\overset{-1}{\mu}K}{4\pi}\frac{d}{dz}(b_0\Delta\mu^{-1}) = 0,$$

or, on substitution for b_0 in terms of h_0 from (23), (24),

$$\nabla^2 f_1 + k^2 f_1 - \overset{-1}{K}\frac{d^2}{dx\,dz}(h_0\Delta K^{-1}) - ik\overset{-1}{\mu}\frac{d}{dz}(h_0\Delta\mu^{-1}) = 0, \quad\ldots\ldots\ldots(29)$$

and

$$\nabla^2 g_1 + k^2 g_1 - \overset{-1}{K}\frac{d^2}{dy\,dz}(h_0\Delta K^{-1}) = 0, \quad\ldots\ldots\ldots\ldots\ldots\ldots\ldots\ldots(30)$$

$$\nabla^2 h_1 + k^2 h_1 + \overset{-1}{K}\left(\frac{d^2}{dx^2} + \frac{d^2}{dy^2}\right)(h_0\Delta K^{-1}) + ik\overset{-1}{\mu}\frac{d}{dx}(h_0\Delta\mu^{-1}) = 0. \quad\ldots(31)$$

The solution of (29) is

$$f_1 = -\frac{\overset{-1}{K}}{4\pi}\iiint \frac{e^{-ikr}}{r}\frac{d^2}{dx\,dz}(h_0\Delta K^{-1})\,dx\,dy\,dz$$

$$- \frac{ik\overset{-1}{\mu}}{4\pi}\iiint \frac{e^{-ikr}}{r}\frac{d}{dz}(h_0\Delta K^{-1})\,dx\,dy\,dz, \quad\ldots\ldots\ldots(32)$$

where r, equal to $\sqrt{\{(\alpha - x)^2 + (\beta - y)^2 + (\gamma - z)^2\}}$, is the distance of the element of volume $dx\,dy\,dz$ from the point α, β, γ at which f_1 is to be estimated.

In applying (32) to the calculation of a secondary wave at a distance from the region of disturbance, we may conveniently integrate it by parts. Thus,

$$f_1 = -\frac{K}{4\pi}\iiint h_0\Delta K^{+1}\frac{d^2}{dx\,dz}\left(\frac{e^{-ikr}}{r}\right)dx\,dy\,dz$$

$$+ \frac{ik\overset{-1}{\mu}}{4\pi}\iiint h_0\Delta K^{+1}\frac{d}{dz}\left(\frac{e^{-ikr}}{r}\right)dx\,dy\,dz.$$

From the general value of r,

$$\frac{d}{dz}\left(\frac{e^{-ikr}}{r}\right) = \frac{\gamma-z}{r}\frac{e^{-ikr}(1+ikr)}{r^2}, \quad \text{...........................(33)}$$

$$\frac{d^2}{dx\,dz}\left(\frac{e^{-ikr}}{r}\right) = \frac{\alpha-x}{r}\frac{\gamma-z}{r}\frac{e^{-ikr}(3+3ikr-k^2r^2)}{r^3} \quad \text{..........(34)}$$

If r be sufficiently great in comparison with λ, only the highest power of kr in the above expressions need be retained; and if r be also great in comparison with the dimensions of the region of disturbance, supposed to be situated about the origin of coordinates, $(\alpha-x)/r$ &c. may be replaced by α/r &c. Thus,

$$\frac{d}{dz}\left(\frac{e^{-ikr}}{r}\right) = \frac{\gamma}{r}\frac{ike^{-ikr}}{r}; \qquad \frac{d^2}{dz\,dx}\left(\frac{e^{-ikr}}{r}\right) = -\frac{\alpha\gamma}{r^2}\frac{k^2e^{-ikr}}{r};$$

and the expression for f_1 becomes

$$f = \frac{k^2}{4\pi r}\left[K\frac{\alpha\gamma}{r^2}\iiint h_0\Delta K^{-1}e^{-ikr}\,dx\,dy\,dz - \mu\frac{\gamma}{r}\iiint h_0\Delta\mu^{-1}e^{-ikr}\,dx\,dy\,dz\right].$$

For the sake of brevity we will write this

$$f_1 = \frac{k^2}{4\pi r}\left[KP\frac{\alpha\gamma}{r^2} - \mu Q\frac{\gamma}{r}\right], \quad \text{......................(35)}$$

where

$$P = \iiint h_0\Delta K^{-1}e^{-ikr}\,dx\,dy\,dz, \qquad Q = \iiint h_0\Delta\mu^{-1}e^{-ikr}\,dx\,dy\,dz. \text{ ... (36)}$$

In like manner from (30) and (31),

$$g_1 = \frac{k^2}{4\pi r}\left[KP\frac{\beta\gamma}{r^2}\right], \quad \text{................................(37)}$$

$$h_1 = \frac{k^2}{4\pi r}\left[-KP\frac{\alpha^2+\beta^2}{r^2} + \mu Q\frac{\alpha}{r}\right]. \quad \text{...............(38)}$$

Equations (35), (37), (38) express the electric displacement in the secondary waves. Since $\alpha f + \beta g + \gamma h = 0$, the displacement is perpendicular to the direction of the secondary ray. The general expression for the intensity is found by adding the squares of f, g, h; but it will be sufficient for our present purpose to limit ourselves to the case where the secondary ray is perpendicular to the primary ray, *i.e.* to the case $\alpha = 0$. Then

$$f^2 + g^2 + h^2 = \frac{k^4}{16\pi^2r^2}\left[K^2P^2\frac{\beta^2}{r^2} + \mu^2Q^2\frac{\gamma^2}{r^2}\right]. \quad \text{...............(39)}$$

If P and Q are both finite, there is no direction along which the secondary light vanishes. We find in experiment, however, that the light scattered by small particles on which polarized light impinges, does vanish in one direction perpendicular to the original ray; and thus either P or Q must vanish. Now, when the particles are very small, we have

$$P = h_0 \Delta K^{-1} e^{-ikr} \iiint dx\, dy\, dz, \qquad Q = h_0 \Delta \mu^{-1} e^{-ikr} \iiint dx\, dy\, dz; \quad \dots(40)$$

so that if P vanishes, $\Delta K = 0$; and if Q vanishes, $\Delta \mu = 0$. The optical evidence that either ΔK or $\Delta \mu$ vanishes is thus very strong; while electrical reasons lead us to conclude that it is $\Delta \mu$.

If we write T for the volume of the small particle, we get from (40), as the special forms of (35), (37), (38) applicable to this case,

$$f_1 = \frac{\pi T}{\lambda^2 r} e^{i(nt-kr)} \left[K\Delta K^{-1} \frac{\alpha\gamma}{r^2} - \mu\Delta\mu^{-1} \frac{\gamma}{r} \right], \quad \dots\dots\dots\dots(41)$$

$$g_1 = \frac{\pi T}{\lambda^2 r} e^{i(nt-kr)} \left[K\Delta K^{-1} \frac{\beta\gamma}{r^2} \right], \quad \dots\dots\dots\dots\dots\dots\dots\dots(42)$$

$$h_1 = \frac{\pi T}{\lambda^2 r} e^{i(nt-kr)} \left[-K\Delta K^{-1} \frac{\alpha^2+\beta^2}{r^2} + \mu\Delta\mu^{-1} \frac{\alpha}{r} \right]. \quad \dots\dots\dots(43)$$

If $\Delta \mu = 0$, as we shall henceforward suppose, $f : g = \alpha : \beta$, showing that the electrical displacement is in the plane containing the secondary ray and the direction of primary electrical displacement (z), and

$$f_1{}^2 + g_1{}^2 + h_1{}^2 \propto (\alpha^2 + \beta^2)/r^2;$$

so that the intensity is proportional to the square of the sine of the angle between the secondary ray and the direction of the primary electrical displacement. The blue colour of the light scattered from small particles is explained by the occurrence of λ^2 in the denominators of the expressions for f_1, g_1, h_1; but for further particulars on this subject the reader must be referred to my previous papers.

Equations (35), (36), &c, are rigorously applicable, however large the region of disturbance, if the square of ΔK may really be neglected. From them we see that, under the circumstances in question, each element of a homogeneous obstacle acts independently as a centre of disturbance, and that 'the aggregate effect in any direction depends upon the phases of the elementary secondary disturbances as affected by the situation of the element along the paths of the primary and of the secondary light. In fact,

$$P = \Delta K^{-1} e^{int} \iiint e^{ikx} e^{-ikr} dx\, dy\, dz.$$

If θ, ϕ be the angles defining (in the usual notation) the direction of the secondary ray, and r_0 correspond to the origin of coordinates, we have

$$P = \Delta K^{-1} e^{i(nt-kr_0)} \iiint e^{ik(x+x\sin\theta\cos\phi+y\sin\theta\sin\phi+z\cos\theta)} dx\, dy\, dz; \quad \dots(44)$$

and the question now before us for consideration is the value of the integral in (44) as dependent upon the size of the obstacle and the direction of the secondary ray. It is evident that the formulæ are applicable only when the whole retardation of the primary light in traversing the obstacle can be neglected in comparison with the wave-length; but if this condition be satisfied, there is no further limitation upon the size of the obstacle. In the case where the secondary ray forms the prolongation of the primary, or deviates sufficiently little from this direction, the exponential in (44) reduces to unity, signifying that every element of the obstacle acts alike, any retardation of phase at starting due to situation along the primary ray being balanced by an acceleration corresponding to a less distance to be travelled along the secondary ray. At a greater or less obliquity, according to the size of the obstacle, opposition of phase sets in; and at still greater obliquities the resultant can be found only by an exact integration. Its intensity is then less, and generally much less, than in the first case—a conclusion abundantly borne out by observation.

The simplest example of this kind is that afforded by an infinite cylinder (*e.g.* a fine spider-line), on which the light impinges perpendicularly to the axis, so that every thing takes places in two dimensions. This case is indeed not strictly covered by the preceding formulæ, on account of the infinite extension of the region of disturbance; but a moment's consideration will make it clear that each elementary column here acts according to the laws already described—that is to say, gives rise to a component disturbance whose phase is determined by the situation of the element along the primary and secondary rays. If the angle between the two rays be called χ, we have to consider the value of

$$\iint e^{ik\,(x + x\cos\chi + y\sin\chi)}\,dx\,dy.$$

Introducing polar coordinates r, θ, we find

$$x + x\cos\chi + y\sin\chi = 2r\cos\tfrac{1}{2}\chi\cos\left(\theta - \tfrac{1}{2}\chi\right);$$

so that the integral

$$= \iint e^{ikr\,.\,2\cos\frac{1}{2}\chi\,.\,\cos\theta}\,r\,dr\,d\theta$$

$$= \int_0^a \int_0^{2\pi} \{\cos\left(2kr\cos\tfrac{1}{2}\chi\cos\theta\right) + i\sin\left(2kr\cos\tfrac{1}{2}\chi\cos\theta\right)\}\,r\,dr\,d\theta$$

$$= 2\pi\int_0^a J_0\left(2kr\cos\tfrac{1}{2}\chi\right)r\,dr, \dots\dots\dots\dots\dots\dots(45)$$

J_0 denoting the Bessel's function of zero order.

The integration with respect to r indicated in (45) can be effected by known properties of Bessel's functions; and the result is expressible by a function of the first order. We get

$$\frac{\pi a}{k \cos \frac{1}{2}\chi} J_1(2ka \cos \frac{1}{2}\chi); \quad \dots\dots\dots\dots\dots\dots(46)$$

and J_1 is defined by

$$J_1(z) = \frac{z}{2}\left(1 - \frac{z^2}{2.4} + \frac{z^4}{2.4^2.6} - \frac{z^6}{2.4^2.6^2.8} + \dots \right). \quad \dots\dots\dots(47)$$

If $\cos \frac{1}{2}\chi = 0$ (*i.e.* in the direction of original propagation), (46) becomes πa^2, every element of the area acting alike. This is the maximum value. When χ is such that

$$2ka \cos \frac{1}{2}\chi = \pi \times 1\cdot2197,$$

the secondary light vanishes, at a greater angle revives, then vanishes again, and so on, the angles being of course functions of the wave-length. If we conceive the cylinder to increase in size gradually from zero, the scattered light vanishes first in the backward direction $\chi = 0$, in which direction evidently the greatest differences of phase occur. Every thing is determined by the course of the function J_1; and (46) within the limits of its application embodies the theory of Young's eriometer.

We will now consider the case of an obstacle in the form of a sphere. If z be a coordinate measured perpendicularly to the plane containing the primary and secondary rays, formula (46), multiplied by dz, will represent the effect of a slice of the sphere, whose radius is a and thickness dz, and what remains to be effected is merely the integration with respect to z. For this purpose we write $z = c \sin \phi$, $a = c \cos \phi$, where c is the radius of the sphere. The integral then takes the form

$$\frac{2\pi c^2}{k \cos \frac{1}{2}\chi}\int_0^{\frac{1}{2}\pi} J_1(2kc \cos \frac{1}{2}\chi \cos \phi) \cos^2\phi \, d\phi, \quad \dots\dots\dots\dots(48)$$

or, if we expand J_1 by (47), and integrate according to a known formula,

$$\frac{2\pi c^3}{3}\left\{2 - \frac{m^2}{5} + \frac{m^4}{7.5.4} - \frac{m^6}{9.7.5.4.6} + \frac{m^8}{11.9.7.5.4.6.8}\dots\right\}, \quad \dots(49)^*$$

in which m is written for $2kc \cos \frac{1}{2}\chi$. It will be understood that (49), after multiplication by $e^{int}\Delta K^{-1}$, gives merely the value of $\cdot P$ in (36), and that to find the complete expression for the secondary light in any direction other factors must be introduced in accordance with (35), (37), (38). The angle χ,

---

* July 15. I find for the first root of (49), $m = 4\cdot50$, giving as the smallest obliquity $(\pi - \chi)$ at which the secondary light vanishes,

$$\pi - \chi = 2 \sin^{-1}(4\cdot50/2kc).$$

being that included between the secondary ray and the axis of x, may be expressed by

$$\sin \chi = \sqrt{(\beta^2 + \gamma^2)} \div r. \quad \dots\dots\dots\dots\dots(50)$$

Our theory, as hitherto developed, shows that, whatever the shape and size of the particles, there is no scattered light in a direction parallel to the primary electric displacements, except such as may depend upon squares and higher powers of the difference of optical properties. In order to render an account of the "residual blue" observed by Tyndall when particles in their growth have reached a certain magnitude, it is necessary to pursue the approximation. By (28), with $\Delta\mu$ neglected, we have

$$\nabla^2 f_2 + k^2 f_2 + K \left(\frac{d^2}{dy^2} + \frac{d^2}{dz^2} \right) (f_1 \Delta K^{-1})$$

$$- K \frac{d^2}{dx\,dy} (g_1 \Delta K^{-1}) - K \frac{d^2}{dx\,dz} (h_1 \Delta K^{-1}) = 0, \quad \dots\dots\dots(51)$$

and two similar equations in g_2 and h_2. On the supposition that f_1, g_1, h_1 are known throughout the region of disturbance, these equations may be solved in the same way as (29), (30), and (31). For the sake of brevity we may confine ourselves to the particular direction for which the terms of the first order vanish. Thus at a sufficient distance r' along the axis of z,

$$f_2 = - \frac{k^2 K}{4\pi r'} \iiint f_1 \Delta K^{-1} e^{-ikr'} \, d\alpha\, d\beta\, d\gamma, \quad \dots\dots\dots\dots(52)$$

$$g_2 = - \frac{k^2 K}{4\pi r'} \iiint g_1 \Delta K^{-1} e^{-ikr'} \, d\alpha\, d\beta\, d\gamma, \dots\dots\dots\dots(53)$$

$$h_2 = 0. \quad \dots\dots\dots\dots\dots\dots\dots\dots\dots\dots\dots(54)$$

We have now to find the values of f_1 and g_1 within the region of disturbance, to which of course (35) &c. are not applicable. In the general solution (32), h_0 is a function of x only; so that the elements of the integral vanish in the *interior* of a homogeneous obstacle, and we have only to deal with the surface. Integrating by parts across this surface, we find

$$f_1 = \frac{K}{4\pi} \iiint \frac{d}{dz} (h_0 \Delta K^{-1}) \frac{d}{dx} \left(\frac{e^{-ikr}}{r} \right) dx\,dy\,dz$$

$$= - \frac{K}{4\pi} \frac{d}{d\alpha} \iiint \frac{d}{dz} (h_0 \Delta K^{-1}) . \frac{e^{-ikr}}{r} \, dx\,dy\,dz, \quad \dots\dots\dots(55)$$

r being a function of x and α only through $(\alpha - x)$. In like manner

$$g_1 = - \frac{K}{4\pi} \frac{d}{d\beta} \iiint \frac{d}{dz} (h_0 \Delta K^{-1}) . \frac{e^{-ikr}}{r} \, dx\,dy\,dz. \quad \dots\dots\dots(56)$$

In the case of a small homogeneous sphere, whose centre is taken as origin of coordinates, these formulæ lead to fairly simple results. The triple integral in (55), (56) may readily be exhibited in its real character of a surface-integral. Thus

$$\iiint \frac{d}{dz}(h_0 \Delta K^{-1}) \frac{e^{-ikr}}{r} dx\,dy\,dz = -\Delta K^{-1} \iint \frac{h_0 z}{c} \frac{e^{-ikr}}{r} dS, \quad \dots\dots(57)$$

where dS is an element of the surface whose radius is c. This applies to a sphere of any size; but we have now to introduce an approximation depending on the supposition that kc is small. As far as the first power of kc,

$$-\Delta K^{-1} \iint \frac{h_0 z}{c} \frac{e^{-ikr}}{r} dS = -\Delta K^{-1} \frac{e^{int}}{c} \iint \left(\frac{z + ikzx}{r} - ikz\right) dS$$

$$= -\Delta K^{-1} \frac{e^{int}}{c} \iint \frac{z + ikzx}{r} dS,$$

in which the double integral is the common potential of matter distributed over the spherical surface with density $(z + ikzx)$. Calling this for the moment V, we have (Thomson and Tait, *Nat. Phil.* § 536) at any internal point (α, β, γ),

$$V = 4\pi c\left(\tfrac{1}{3}\gamma + \tfrac{1}{5}ik\gamma\alpha\right)^*;$$

so that

$$\iiint \frac{d}{dz}(h_0 \Delta K^{-1}) \frac{e^{-ikr'}}{r} dx\,dy\,dz = -4\pi\Delta K^{-1} e^{int}\left(\tfrac{1}{3}\gamma + \tfrac{1}{5}ik\gamma\alpha\right). \quad \dots\dots(58)$$

Thus by (55), (56),

$$f_1 = \tfrac{1}{5} K\Delta K^{-1} ik\gamma e^{int}, \qquad g_1 = 0. \quad \dots\dots\dots(59)$$

We are now prepared to calculate f_2, g_2 from (52), (53). These formulæ apply to both directions along the axis of z; but in what follows it will be convenient to suppose that it is the positive direction which is under consideration. In this case, if ρ denote the distance from the centre of the sphere, $r' = \rho - \gamma$ and $e^{-ikr'} = e^{-ik\rho}(1 + ik\gamma)$ approximately; so that

$$f_2 = -\frac{k^2 (K\Delta K^{-1})^2 e^{i(nt-k\rho)}}{20\pi\rho} \iiint ik\gamma(1 + ik\gamma)\,d\alpha\,d\beta\,d\gamma$$

$$= \frac{k^4 (K\Delta K^{-1})^2}{20\pi} \frac{e^{i(nt-k\rho)}}{\rho} \iiint \gamma^2 \,d\alpha\,d\beta\,d\gamma;$$

or if, as before, T be the volume of the sphere,

$$f_2 = \frac{\pi T}{\lambda^2 \rho} e^{i(nt-k\rho)} (K\Delta K^{-1})^2 \frac{k^2 c^2}{25}, \qquad g_2 = 0. \quad \dots\dots\dots(60)$$

Comparing (60) and (41), we see that the amplitude of the light scattered along z is not only of higher order in ΔK, but is also of the order $k^2 c^2$ in

---

* [1899. A numerical error which occurred here and in the consequential equations (59), (60), (61), is now corrected. See Walker, *Quart. Journ. of Math.* vol. xxx. p. 204, 1898.]

comparison with that scattered in other directions. The incident light being white, the intensity of the component colours scattered along z varies as the inverse 8th power of the wave-length, so that the resultant light is a rich blue.

There is another point of importance to be noticed. Although, when the terms of the second order are included, the scattered light does not vanish along the axis of z, the peculiarity is not lost, but merely transferred to another direction. Putting together the terms of the first and second orders, we see that the scattered light will vanish in a direction in the plane of xz, inclined to z (towards x) at a small angle θ, such that

$$\theta = - K\Delta K^{-1} \frac{k^2 c^2}{25} = \frac{\Delta K}{K} \frac{k^2 c^2}{25} . \qquad \ldots\ldots\ldots\ldots\ldots\ldots(61)$$

In the usual case of particles optically denser than the surrounding medium, ΔK is positive, from which we gather that the direction in which the scattered light vanishes to the second order of approximation is inclined backwards, so that the angle through which the light may be supposed to be bent by the action of the particle is *obtuse*.

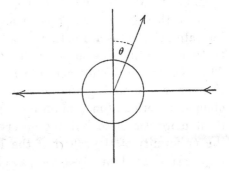

The fact that, when the primary light is polarized, there is in one perpendicular direction no light scattered by very small particles, was stated by Stokes[*]; but it is, I believe, to Tyndall that we owe the observation that with somewhat larger particles the direction of minimum illumination becomes oblique. I do not find, however, any record of the direction of the obliquity (that is, of the sign of the small angle θ), and have therefore made a few observations for my own satisfaction.

In a darkened room a beam of sunlight was concentrated by a large lens of 2 or 3 feet focus; and in the path of the light was placed a beaker glass, containing a dilute solution of hyposulphite of soda. On the addition of a small quantity of dilute sulphuric acid a precipitate of sulphur slowly forms, and during its growth manifests exceedingly well the phenomena under consideration. The more dilute the solutions, the slower is the progress of the precipitation. A strength such that there is a delay of four or five minutes before any effect is apparent, will be found suitable; but no great nicety of adjustment is necessary. By addition of ammonia in sufficient quantity to neutralize the acid, the precipitation may be arrested at any desired stage. More time is thus obtained to complete the examination;

_____

[*] *Phil. Trans.* 1852, § 183.

but the condition of things is not absolutely permanent, the already precipitated sulphur appearing to aggregate itself into larger masses.

In the optical examination we may, if we prefer it, polarize the primary
light; but it is usually more convenient to analyze the scattered light. In
the early stages of the precipitation the polarization is complete in a perpendicular direction, and incomplete in oblique directions. After an interval
the polarization begins to be incomplete in the perpendicular direction, the
light which reaches the eye when the nicol is in the position of minimum
transmission being of a beautiful blue, much richer than anything that can
be seen in the earlier stages. This is the moment to examine whether there
is a more complete polarization in a direction somewhat oblique; and it is
found that with θ positive there is in fact an oblique direction of more complete polarization, while with θ negative the polarization is more imperfect
than in the perpendicular direction itself.

The polarization in a distinctly oblique direction, however, is not perfect,
a feature for which more than one reason may be put forward. In the first
place, with a given size of particles, the direction of complete polarization
indicated by (61) is a function of the colour of the light, the value of θ being
three or four times as large for the violet as for the red end of the spectrum.
The experiment is, in fact, much improved by passing the primary light
through a coloured glass held in the window-shutter. Not only is the
oblique direction of maximum polarization more definite and the polarization
itself more complete, but the observation is easier than with white light, by
the uniformity of the colour of the light scattered in various directions. If
we begin with a blue glass, we may observe the gradually increasing obliquity
of the direction of maximum polarization; and then by exchanging the blue
glass for a red one, we may revert to the original condition of things, and
observe the transition from perpendicularity to obliquity over again. The
change in the wave-length of the light has the same effect as a change in
the size of the particles; and the comparison gives curious information as to
the rate of growth.

But even with homogeneous light it would be unreasonable to expect an
oblique direction of perfect polarization. So long as the particles are all
very small in comparison with the wave-length, there is complete polarization
in the perpendicular direction; but when the size is such that obliquity sets
in, the degree of obliquity will vary with the size of the particles, and the
polarization will be complete only on the very unlikely condition that the
size is the same for them all. It must not be forgotten, too, that a very
moderate increase in dimensions may carry the particles beyond the reach of
our approximations.

The fact that at this stage the polarization is a maximum when the
angle through which the light is turned *exceeds* a right angle is the more

worthy of note, as the opposite result would probably have been expected. By Brewster's law this angle in the case of [regular reflection from] a plate is *less* than a right angle; so that not only is the law of polarization for a very small particle different from that applicable to a plate, but the first effect of an increase of size is to augment the difference.

We must remember that our recent results are limited to particles of a spherical form. It is not difficult to see that, for elongated particles, the terms in $(\Delta K)^2$ may be of the same order with respect to kc as the principal term; so that if $(\Delta K)^2$ be sensible, mere smallness of the particles will not secure complete evanescence of scattered light along z. The general solution of the problem for an infinitesimal particle of arbitrary shape must raise the same difficulties as beset the general determination of the induced magnetism developed in a piece of soft iron when placed in a uniform field of force. In the case of an ellipsoidal particle the problem is soluble; but it is perhaps premature to enter upon it, until experiment has indicated the existence of phenomena likely to be explained thereby.

For an infinitesimal particle in the form of a sphere, we may readily obtain the complete solution without any approximation depending upon the smallness of ΔK. We know by the analogous theory of magnetism, that a dielectric sphere situated in a uniform field of electric force will undergo electric displacement of uniform amount, and in a direction parallel to that of the force. Thus the complete solution applicable to an infinitely small sphere is obtained from (29), (30), (31) by writing h for h_0; where by h is denoted the actual displacement (parallel to z) within the particle, and by h_0 the displacement in the enveloping medium under the same electric force. If K' be the specific inductive capacity for the particle, the ratio of $h : h_0$ is $3K' : K' + 2K$; and in this ratio the results expressed in (41), (42), (43) are to be increased. If we extract the factors $K\Delta K^{-1}$ which there occur, we get

$$\frac{3K'}{K'+2K} K\Delta K^{-1} = \frac{3K'K}{K'+2K}\left(\frac{1}{K'}-\frac{1}{K}\right) = -\frac{3(K'-K)}{K'+2K};$$

so that

$$f = -\frac{3(K'-K)}{K'+2K}\frac{\pi T}{\lambda^2 r}\frac{\alpha\gamma}{r^2}e^{i(nt-kr)}, \quad \&c. \dots\dots\dots\dots(62)$$

We learn from (62) that our former result as to the evanescence of the secondary light along z is true for an infinitely small spherical particle to *all* orders of ΔK *.

We will now return to the two-dimension problem with the view of determining the disturbance resulting from the impact of plane waves upon

* [1899. The completion of the solution for the sphere (with finite variation of μ as well as of K) and the extension to obstacles of ellipsoidal form is given in *Phil. Mag.* vol. XLIV. p. 48, 1897.]

a cylindrical obstacle whose axis is parallel to the plane of the waves. There are, as in the problem of reflection from plane surfaces, two principal cases— (1) when the electric displacements are parallel to the axis of the cylinder taken as axis of z, (2) when the electric displacements are perpendicular to this direction.

Case 1.   From (12), with $C = 0$, $\mu =$ constant,

$$\left(\frac{d^2}{dx^2} + \frac{d^2}{dy^2}\right)\frac{h}{K} + n^2\mu K \frac{h}{K} = 0 ;$$

or if, as before, $k = 2\pi/\lambda$,

$$\left(\frac{d^2}{dx^2} + \frac{d^2}{dy^2} + k^2\right)\frac{h}{K} = 0, \quad \dots\dots\dots\dots\dots\dots\dots\dots(63)$$

in which k is constant in each medium, but changes as we pass from one medium to another. From (63) we see that the problem now before us is analytically identical with that treated in my book on *Sound*, § 343, to which I must refer for more detailed explanations. The incident plane waves are represented by

$$e^{int}\,e^{ikx} = e^{int}\,e^{ikr\cos\theta}$$

$$= e^{int}\{J_0(kr) + 2iJ_1(kr)\cos\theta + \dots + 2i^mJ_m(kr)\cos m\theta + \dots\}; \quad \dots(64)$$

and we have to find for each value of m an internal motion finite at the centre, and an external motion representing a divergent wave, which shall in conjunction with (64) satisfy at the surface of the cylinder ($r = c$) the condition that the function and its differential coefficient with respect to r shall be continuous. The divergent wave is expressed by

$$B_0\psi_0 + B_1\psi_1\cos\theta + B_2\psi_2\cos 2\theta + \dots,$$

where ψ_0, ψ_1, &c. are the functions of kr defined in § 341. The coefficients B are determined in accordance with

$$B_m\left\{kc\,\frac{d\psi_m}{d\,.\,kc}\,J_m(k'c) - k'c\,\psi_m\,\frac{d}{d\,.\,k'c}\,J_m(k'c)\right\}$$

$$= 2\,i^m\{k'c\,J_m(kc)\,J_m'(k'c) - kc\,J_m(k'c)\,J_m'(kc)\},$$

except in the case of $m = 0$, when $2\,i^m$ on the right-hand side is to be replaced by i^m. In working out the result we suppose kc and $k'c$ to be small; and we find approximately for the secondary disturbance corresponding to (64)

$$\psi = \left(\frac{\pi}{2\,ikr}\right)^{\frac{1}{2}}e^{i\,(nt-kr)}\left[\frac{k'^2c^2 - k^2c^2}{2} - \frac{k^2c^2\,(k'^2c^2 - k^2c^2)}{8}\cos\theta\right]; \quad \dots\dots(65)$$

showing, as was to be expected, that the leading term is independent of θ.

For case 2, which is of greater interest, we have from (15),

$$\left(\frac{d}{dx}\frac{1}{k^2}\frac{d}{dx} + \frac{d}{dy}\frac{1}{k^2}\frac{d}{dy} + 1\right)c = 0. \quad\quad\ldots\ldots\ldots\ldots\ldots(66)^*$$

This is of the same form as (63) within a uniform medium, but gives a different boundary condition at a surface of transition. In both cases the function itself is to be continuous; but in that with which we are now concerned the second condition requires the continuity of the differential coefficient *after division by* k^2. The equation for B_m is therefore

$$B_m\left\{k'c\frac{d\psi_m}{d.kc}J_m(k'c) - kc\,\psi_m\frac{dJ_m(k'c)}{d.k'c}\right\}$$

$$= 2\,i^m\left\{kc\,J_m(kc)\,J_m{}'(k'c) - k'c\,J_m(k'c)\,J_m{}'(kc)\right\},$$

with the understanding that the 2 is to be omitted when $m = 0$. Corresponding to the primary wave $e^{i(nt+kx)}$, we find as the expression of the secondary wave at a great distance from the cylinder,

$$\psi = \left(\frac{\pi}{2ikr}\right)^{\frac{1}{2}}e^{i(nt-kr)}\left[-\frac{k^2c^2}{16}(k^2c^2 - k'^2c^2)\right.$$

$$\left. - k^2c^2\frac{k'^2 - k^2}{k'^2 + k^2}\cos\theta - \frac{1}{8}k^4c^4\frac{k^2 - k'^2}{k^2 + k'^2}\cos 2\theta\right]. \quad\ldots\ldots\ldots(67)$$

The term in $\cos\theta$ is now the leading term; so that the secondary disturbance approximately vanishes in the direction of the primary electrical displacements, agreeably with what has been proved before. It should be stated here that (67) is not complete to the order k^4c^4 in the term containing $\cos\theta$. The calculation of the part omitted is somewhat tedious in general; but if we introduce the supposition that the difference between k'^2 and k^2 is small, its effect is to bring in the factor $(1 - \frac{1}{4}k^2c^2)$.

Extracting the factor $(k'^2 - k^2)$, we may conveniently write (67)

$$\psi = -k^2c^2\frac{k'^2 - k^2}{k'^2 + k^2}\left(\frac{\pi}{2ikr}\right)^{\frac{1}{2}}e^{i(nt-kr)}\left[\cos\theta - \frac{k'^2c^2 + k^2c^2}{16} - \frac{k^2c^2}{8}\cos 2\theta\right], \quad\ldots(68)$$

in which

$$\cos\theta - \frac{k'^2c^2 + k^2c^2}{16} - \frac{k^2c^2}{8}\cos 2\theta = \cos\theta - \frac{k'^2c^2 - k^2c^2}{16} - \frac{k^2c^2}{4}\cos^2\theta. \quad\ldots\ldots(69)$$

In the directions $\cos\theta = 0$, the secondary light is thus not only of high order in kc, but is also of the second order in $(k' - k)$. For the direction in which the secondary light vanishes to the next approximation, we have

$$\tfrac{1}{2}\pi - \theta = \tfrac{1}{16}(k'^2c^2 - k^2c^2) = \frac{k^2c^2}{16}\frac{K' - K}{K}. \quad\ldots\ldots\ldots\ldots(70)$$

---

* In (66) c is the magnetic component, and not the radius of the cylinder. So many letters are employed in the electromagnetic theory, that it is difficult to hit upon a satisfactory notation.

This corresponds to (61) for the sphere; and is true if kc, $k'c$ be small enough, whatever may be the relation of k' and k. For the cylinder, as for the sphere, the direction is such that the primary light would be bent through an angle *greater* than a right angle.

If we neglect the square of $(k'^2 - k^2)$, the complete expression corresponding to (69) is

$$\cos\theta\,(1 - \tfrac{1}{4}k^2c^2) - \tfrac{1}{4}k^2c^2\cos^2\theta = \cos\theta\,[1 - \tfrac{1}{4}k^2c^2 - \tfrac{1}{4}k^2c^2\cos\theta].$$

This may be compared with the value obtained by the former method, viz. $\cos\theta\,J_1(2kc\cos\tfrac{1}{2}\theta) \div kc\cos\tfrac{1}{2}\theta$, and will be found to agree with it as far as the square of kc.

If we suppose the cylinder to be extremely small, we may confine ourselves to the leading terms in (65) and (67). Let us compare the intensities of the secondary lights emitted in the two cases along $\theta = 0$, *i.e.* directly backwards. From (65)

$$\psi \propto \tfrac{1}{2}(k'^2c^2 - k^2c^2),$$

while from (67)

$$\psi \propto -k^2c^2(k'^2 - k^2)/(k'^2 + k^2).$$

The opposition of sign is apparent only, and relates to the different methods of measurement adopted in the two cases. In (65) the primary and secondary disturbances are represented by h/K, but in (67) by the magnetic function c. If we express the solution in the second case in terms of the electric function g, we shall find (see 13) that the ratio of c to g changes sign when we pass from the primary light propagated along $-x$ to the secondary light propagated along $+x$. The actual ratio of amplitudes in the two cases is thus $(k'^2 + k^2)/2k^2$, or $(K' + K)/2K$. Unless the difference between K' and K be neglected, the two components of unpolarized light are scattered along this direction in different proportions, that component preponderating in which the electric displacement is parallel to the axis of the cylinder. The secondary light is therefore partially polarized in the plane perpendicular to the axis.

[1899. In a recent paper by Prof. Love (*Lond. Math. Soc. Proc.* vol. XXX. p. 316, 1899), results are arrived at (32, 33) which appear to contradict the doctrine (p. 529) that the "residual blue" is dependent upon the *square* of ΔK. The more complete solution, expressed in his equation (42), is however in harmony with the doctrine, and confirms the correctness of equations (60), (61) above.]

75.

ON THE VELOCITY OF LIGHT.

[*Nature*, XXIV. pp. 382, 383; XXV. p. 52, 1881.]

THE result announced by Young and Forbes (*Roy. Soc. Proc.* May 17, 1881) that blue light travels *in vacuo* about 1·8 per cent. faster than red light, raises an interesting question as to what it is that is really determined by observations of this character. If the crest of an ordinary water wave were observed to travel at the rate of a foot per second, we should feel no hesitation in asserting that this was the velocity of the wave; and I suppose that in the ordinary language of undulationists the velocity of light means in the same way the velocity with which an individual wave travels. It is evident however that in the case of light, or even of sound, we have no means of identifying a particular wave so as to determine its rate of progress. What we really do in most cases is to impress some peculiarity, it may be of intensity, or of wave-length, or of polarisation, upon a part of an otherwise continuous train of waves, and determine the velocity at which this *peculiarity* travels. Thus in the experiments of Fizeau and Cornu, as well as in those of Young and Forbes, the light is rendered intermittent by the action of a toothed wheel; and the result is the velocity of the group of waves, and not necessarily the velocity of an individual wave. In a paper on "Progressive Waves" (*Proc. Math. Soc.* vol. IX.), reprinted as an appendix to vol. II. of my book on the *Theory of Sound* [Art. XLVII.], I have investigated the general relation between the group-velocity U and the wave-velocity V. It appears that if k be inversely proportional to the wave-length,

$$U = \frac{d(kV)}{dk},$$

and is identical with V only when V is independent of k, as has hitherto been supposed to be the case for light in vacuum. If, however, as Young and Forbes believe, V varies with k, then U and V are different. The truth is however that these experiments tell us nothing in the first instance about the value of V. They relate to U; and if V is to be deduced from them it must be by the aid of the above given relation.

When we come to examine more closely the form of this relation, we see that a complete knowledge of V (as a function k) leads to a complete knowledge of U, but that a complete knowledge of U—all that experiments of this kind can ever give us—does not determine V, without the aid of some auxiliary assumption. The usual assumption is that V is independent of k, in which case U is also independent of k. If we have reason to conclude from observation that U is not independent of k, this assumption is disproved; but we can make no progress in determining V until we have introduced some other.

It is not easy to see how the missing link is to be supplied; but in order to have an idea of the probable magnitude of the difference in question, I have assumed the ordinary dispersion formula $V = A + Bk^2$ to be applicable. Taking the ratio of wave-lengths of the orange-red and green-blue lights employed as $6 : 5$, I find that for red light $V = U(1 - {\cdot}0273)$, so that the velocity of the wave would be nearly 3 per cent. less than that given by Young and Forbes as the result of the experiment.

Under these circumstances it becomes a matter of interest to examine the bearing of other evidence on the question of the velocity of light. Independently of the method of the toothed wheel, the velocity of light has been determined by Foucault and Michelson using the revolving mirror. It is not very obvious at first sight whether the value thus arrived at is the group-velocity or the wave-velocity, but on examination it will be found to be the former. The successive wave-fronts of the light after the first reflection are not parallel, with the consequence that (unless V be constant) an individual wave-front rotates in the air between the two reflections.

The evidence of the terrestrial methods relating exclusively to U, we turn to consider the astronomical methods. Of these there are two, depending respectively upon aberration and upon the eclipses of Jupiter's satellites. The latter evidently gives U. The former does not depend upon observing the propagation of a peculiarity impressed upon a train of waves, and therefore has no relation to U. If we accept the usual theory of aberration as satisfactory, the result of a comparison between the coefficient found by observation and the solar parallax is V—the wave-velocity.

The question now arises whether the velocity found from aberration agrees with the results of the other methods. A comparison of the two astronomical determinations should give the ratio $U : V$, independently of the solar parallax. The following data are taken from Mr Gill's " Determination of the Solar Parallax from observations of Mars made at the Island of Ascension in 1877."

The time T, required by light to travel a mean radius of the earth's orbit, has been determined by two astronomers from the eclipses of Jupiter's satellites. Delambre found, from observations made in the last century, $T = 493\cdot2$ s., but recently Glasenapp has obtained from modern observations the considerably higher value, $T = 500\cdot8$ s. $\pm 1\cdot02$. With regard to the constant of aberration, Bradley's value is $20''\cdot25$, and Struve's value is $20''\cdot445$. Mr Gill calculates as the mean of the best modern determinations (nine in number), $20''\cdot496$.

If we combine Glasenapp's value of T with Michelson's value of the velocity of light, we get for the solar parallax $8''\cdot76$. Struve's constant of aberration in conjunction with the same value of the velocity of light gives $8''\cdot81$. From these statements it follows that if we regard the solar parallax as known, we get almost the same velocity of light from the eclipses of Jupiter's satellites as from aberration, although the first result relates to the group-velocity, and the second to the wave-velocity. If instead of Struve's value of the constant of aberration we take the mean above spoken of, we get for the solar parallax $8''\cdot78$, allowing still less room for a difference between U and V.

Again, we may obtain a comparison without the aid of the eclipses of Jupiter's satellites by introducing, as otherwise known, the value of the solar parallax. Mr Gill's value from observations of Mars is $8''\cdot78$, agreeing exactly with Michelson's light-velocity and the mean constant of aberration. Some other astronomers favour a higher value of the solar parallax, such as $8''\cdot86$; but whichever value we adopt, and whether we prefer Cornu's or Michelson's determination of the light-velocity, the conclusion is that there can be no such difference between the group-velocity and the wave-velocity as 2 or 3 per cent., unless indeed the usual theory of aberration requires serious modification. These considerations appear to me to increase the already serious difficulties, which cause hesitation in accepting the views of Young and Forbes. The advent of further evidence will doubtless be watched with great interest by scientific men.

One other point I may refer to in conclusion. Speculations as to harmonic relations between various spectral rays emitted by a glowing gas proceed upon the assumption that the frequency of vibration is inversely proportional to the wave-length, or, in other words, that the velocity of propagation V is independent of the wave-length, the question now at issue. If the views of Young and Forbes are correct, calculations of this kind must be overhauled. On the other hand, the establishment of well-defined simple ratios between wave-lengths would tend to show that V does not vary.

In reply to Mr Macaulay (*Nature*, vol. XXIV. p. 556) I will endeavour to explain more clearly the statements made in my former communication on this subject (*Nature*, vol. XXIV. p. 382).

With reference to the group-velocity U, we know from Fourier's theorem that any disturbance travelling in one dimension can be regarded as resulting from the superposition of infinite trains of waves of the harmonic type, and of various amplitudes and wave-lengths. And we know that any one of these trains, of wave-length λ, is propagated unchanged with a velocity V, which we regard as a known function of λ, dependent upon the nature of the medium.

Unless we can deal with phases, a simple train of waves presents no mark by which its parts can be identified. The introduction of such a mark necessarily involves a departure from the original simplicity of a single train, and we have to consider how in accordance with Fourier's theorem the new state of things is to be represented. The only case in which we can expect a simple result is when the mark is of such a character that it leaves a considerable number of consecutive waves still sensibly of the given harmonic type, though the wave-length and amplitude may vary within moderate limits at points whose distance amounts to a very large multiple of λ. We will therefore suppose that the complete expression by Fourier's series involves only wave-lengths which differ but little from one another, and accordingly write it—

$$a_1 \cos\{(n + \delta n_1)\, t - (k + \delta k_1)\, x + \epsilon_1\} + a_2 \cos\{(n + \delta n_2)\, t - (k + \delta k_2)\, x + \epsilon_2\} + \ldots ,$$

or in the equivalent form—

$$\cos(nt - kx)\, \Sigma a_1 \cos(\delta n_1 t - \delta k_1 x + \epsilon_1) - \sin(nt - kx)\, \Sigma a_1 \sin(\delta n_1 t - \delta k_1 x + \epsilon_1),$$

where $k = 2\pi/\lambda$, and $n = kV$. From this we see that, as in accordance with the suppositions already made,

$$\delta n_1/\delta k_1 = \delta n_2/\delta k_2 = \ldots = dn/dk,$$

the deviations from the simple harmonic type travel with velocity dn/dk and not with velocity n/k, that is with velocity $d(kV)/dk$, and not with velocity V.

[1899. The latter part of this note was erroneous and is accordingly omitted. The result with reference to the revolving mirror was correctly stated in the first note (p. 538). The subject has been ably treated by Prof. W. Gibbs, and to his paper (*Nature*, XXXIII. p. 582, 1886) the reader who wishes to pursue the question further is referred.]

76.

ON A QUESTION IN THE THEORY OF LIGHTING.

[*Brit. Ass. Rep.* 1881, p. 526.]

IT is known that a large part of the radiation from terrestrial sources is non-luminous. Even in the case of the electric arc the obscure radiation amounts, according to Tyndall, to eight-ninths of the whole, and of the remainder probably no inconsiderable part is to be found in the extreme red rays of feeble luminosity. For practical purposes this obscure radiation is useless; and the question forces itself upon us, whether or no there is any necessity, absolutely inherent in the case, for so large a proportion of waste. The following arrangement, not of course proposed as practical, seems to prove that the question should be answered in the negative.

Conceive a small spherical body of infusible material, to which energy can be communicated by electricity or otherwise, to be surrounded by a concentric reflecting spherical shell. Under these circumstances no energy can escape; but if a small hole be pierced in the shell, radiation will pass through it. In virtue of the suppositions which we have made, the emergent beam will be of small angle, and may be completely dealt with at a moderate distance by a prism and lens. Let us suppose then that a spectrum of the hole is formed and is received upon a reflecting plate so held at the focus as to return the rays upon the lens and prism. These rays will re-enter the hole, and impinge upon the radiating body, which is thus again as completely isolated as if the shell were unperforated. We have now only to suppose a portion of the focal plate to be cut away in order to have an apparatus from which only one kind of radiation can escape. Whatever energy is communicated to the internal body must ultimately undergo transformation into radiation of the selected kind.

77.

EXPERIMENTS ON COLOUR*

[*Nature*, xxv. pp. 64—66, 1881.]

In a former paper with the above title (*Nature*, vol. iii. p. 234, Art. vii.) I described some combinations of absorbing media capable of transmitting the red and green, while stopping the other rays of the spectrum. In this way I obtained a *purely compound* yellow, made up of red and green, and free from homogeneous yellow light. In devising such combinations we have in the first place to seek an absorbing agent capable of removing the yellow of the spectrum, while allowing the red and green to pass. For this purpose I used an alkaline infusion of litmus, or solution of chloride of chromium, placed in a trough with parallel glass sides. In order to stop the blue rays we may avail ourselves of chromate of potassium. If a second trough be not objected to, it is best to use the bichromate, as exercising the most powerful absorption upon the upper end of the spectrum; but the bichromate cannot be mixed with litmus without destroying the desired action of the latter upon yellow. In this case we must content ourselves with the neutral chromate.

During the last year and a half I have resumed these experiments with the view, if possible, of finding solid media capable of the same effects, and so of dispensing with the somewhat troublesome troughs necessary for fluids. With this object we may employ films of gelatine or of collodion, spread upon glass and impregnated with various dyes, gelatine being chosen when the dye is soluble in water, and collodion when the dye is soluble in alcohol. Thus in the case of litmus a slightly warmed plate is coated with a hot and carefully filtered solution of gelatine, allowed to remain in a perfectly horizontal position until the gelatine is set, and then put aside to dry, by preference in a current of warm air. The films thus obtained are usually

* Read before Section A of the British Association, September 2, 1881.

somewhat rough upon the surface, so that I have preferred to use two pieces cemented together, coated sides inwards, with Canada balsam. In conjunction with the litmus we may employ a silver-stained orange glass, and so isolate the red and green rays. For the orange glass Mr C. Horner has substituted a film of collodion stained with aurine. Samples possibly vary; but that which I have used, though extremely opaque to the blue-green rays, and therefore so far very suitable for the purpose, allows a considerable quantity of the higher blue to pass. By spreading aurine upon a pale yellow glass, I obtained a very perfect absorption of the blue-green and higher rays. Plates prepared as above described answer the purpose very well; but I have found that in some cases the litmus in contact with the balsam becomes slowly reddened, the action creeping inwards from the edge. A dye, capable of replacing litmus, and free from this defect, is "soluble aniline blue," whose absorption, as I found rather unexpectedly, begins in the yellow and orange. Bichromate of potash and aniline blue may be mixed in the same solution, and there is no difficulty in so adjusting the proportions as to secure a good compound yellow. To obtain solid films gelatine must be used, as in the case of litmus, for the dye is not soluble in collodion. With aniline there is no difficulty from the Canada balsam, and two plates cemented together answer perfectly.

For systematic observations on compound colours nothing probably can be better than Maxwell's colour-box in its original form; but it seemed to me that for the examination of certain special questions a more portable arrangement would be convenient. In an instrument of this class a full degree of brightness requires that the width of the eye-slit, placed where the spectrum is formed, should not contract the aperture of the eye, i.e. should not be less than about one-fifth of an inch; and although the maximum of brightness is not necessary, considerations of this kind largely influence the design. If we regard the width of the eye-slit as given, a certain length of spectrum is necessary in order to attain the desired standard in respect of purity of colour; so that what we have to aim at is a sufficient linear extension of the spectrum. A suitable compromise can then be made between the claims of brightness and purity.

The necessary length of spectrum can be obtained by increasing either the angular dispersion of the prisms or the focal length of the lens by which the image is formed. If portability be no object, the latter is the preferable method, and the focal length may well be increased up to five or six feet; in this way we may obtain a field of view of given purity of colour and of maximum brightness, at the expense only of its angular extent. If, however, we desire an instrument which can be moved from one place to another without losing its adjustment, the focal length of the lenses must be kept down, and then a large prismatic dispersion is the only alternative.

Increased dispersion can of course be obtained by multiplication of prisms; but for the purpose in view, high resolving power is not wanted, and our object may be attained with a comparatively small total thickness of glass, either by the use of higher angles than usual, or by giving the light a more nearly grazing emergence. The latter was the course adopted in designing the first instrument of which I have to speak. A pair of prisms of 60°, cut from an ordinary single $1\frac{1}{4} \times 1\frac{1}{4}$-inch prism along a plane bisecting at right angles its refracting edge, were arranged in the corner of a shallow box, so as to form what Thollon calls a *couple*. Considered as a simple, rigidly connected refractor, the pair of prisms are placed so as to give minimum deviation, but the incident and emergent light makes smaller angles with the final surfaces, than if each prism were adjusted separately for minimum deviation. The collimating and focusing lenses are common spectacle-glasses of about 8″ focus. The box is $12″ \times 12″ \times 3″$. Light entering at a slit on one of the sides of the box would be turned by the prisms through an angle rather greater than a right angle, and throw a pure spectrum upon another side of the box. This side is cut away, and provided with movable screens of cardboard, so that any part may be open or closed as desired. When the eye is applied to the first slit, the prisms are seen uniformly illuminated with colours whose composition depends upon the situation and width of the slits between the cardboard screens through which light is allowed to enter. In this way we may obtain a uniform field of view lighted with any combination of spectral colours. My object, however, was to obtain an instrument for making comparisons between the simple and compound yellow, and for this purpose an addition was necessary. This consisted of a very acute-angled prism held close to the dispersing prisms in such a position that its refracting edge was horizontal, dividing the field of view into two equal parts. The action of this prism is most easily understood by again supposing the light to enter at the eye-slit. Half of the light proceeds as before, forming ultimately a pure spectrum upon the side of the box. The upper half of the beam, however, is deflected by the acute-angled prism, and the corresponding spectrum is thrown upwards, so as to lie somewhat higher upon the side of the box. This part is also cut away, and provided with movable screens. By the principle of reversibility the consequence is that an eye placed at the first slit sees *two* uniform patches of colour, the lower formed as before by light from the lower set of slits, the upper, covering the acute-angled prism, by light from the upper set of slits. These colours are in close juxtaposition, and may be compared with ease and accuracy.

The great difficulty in this class of instruments is to devise any efficient and reasonably simple method of controlling the position and widths of the slits. In the present case I contented myself with strips of blackened cardboard cemented to the side of the box with sealing-wax, or soft wax, according to the degree of permanence of adjustment aimed at. One part of the field was illuminated with homogeneous yellow (about the line D) from a single slit. The other half was lighted with a mixture of full red and full green, and the observation consisted in adjusting the widths of the slits through which the red and green were admitted, until the mixture was a match with the simple yellow.

The first trials of this instrument in the spring of last year revealed an interesting peculiarity of colour vision, quite distinct from colour blindness. The red and green mixture which to my eyes and to those of most people matches perfectly the homogeneous yellow of the line D, appeared to my three brothers-in-law hopelessly too red, "almost as red as red sealing-wax." In order to suit their eyes the proportion of red had to be greatly diminished, until to normal sight the colour was a fair green with scarcely any approach to yellow at all. So far as could be made out at the time, the three abnormal observers agreed well among themselves, a fact which subsequent measurements have confirmed. It appeared afterwards that a fourth brother was normal as well as the three sisters.

These peculiarities were quite unexpected. After the fact had been proved, I remembered a dispute some years before as to the colour of a dichromatic liquid, which appeared to me green, while one of my brothers-in-law maintained that it was red; but the observation was not followed up, as it ought to have been, each of us, I suppose, regarding the other as inaccurate. After the establishment of the difference I determined to carry out a plan, which I had tried with success some years before (October, 1877), for a colour-mixing arrangement depending on double refraction, by which I hoped to obtain an easily adjustable instrument suitable for testing the vision of a number of persons.

In my original experiments I used a 60° doubly refracting prism of quartz, which threw two spectra of the linear source upon the screen containing the eye-slit. These oppositely polarised spectra partially overlapped, and by suitable placing of the prism could be made to furnish red and green light to the eye. By the rotation of a small Nicol held immediately behind the eye-slit, the red or green could be isolated or mixed in any desired proportion. One advantage of this arrangement is that the two component lights come from the same slit, so that we are less dependent upon the uniformity of the light behind; but it is perhaps a greater merit that the adjustment of proportions is effected by a simple rotation at the eye-slit, allowing the observer to try the effect of small changes with ease and rapidity.

In the new instrument, which was completed during the autumn of last year, separate prisms were used to effect the dispersion and double refraction. For the sake of compactness, a direct vision prism by Browning, containing two flints and three crowns, was chosen, in conjunction with a small achromatic double image prism. At one end of a long narrow box, 24″ × 2″ × 2″, the light is admitted through a slit whose position and width can be adjusted by sliding its jaws along a divided scale. After travelling about 9¼″ it falls upon the double image prism mounted upon a small table so as to allow of rotation, and then after two more inches upon a collimating lens, by which the two beams are rendered parallel. Next comes the dispersing prism, and then the focusing lens, throwing pure spectra upon the other end of the box which carries the eye-slit. The distance between the two lenses is 3½″, and the entire length of the box is about 24″. The eye-slit is a fixture, and immediately behind it is the rotating Nicol, whose position is read by a pointer on a divided circle.

The parts of the spectrum from which the component lights are taken can be chosen over a sufficient range by use of the two adjustments already mentioned. By rotation of the table on which the double image prism is mounted, the separating power is altered, and one spectrum made to slide over the other, while by moving the entrance slit the spectra are shifted together without relative displacement.

It yet remains to describe the parts by which the comparison colour is exhibited. Between the double image prism and the collimating lens a small vertical reflector is mounted on a turn-table at an angle of about 45°. Its dimensions are such that it covers the lower half of the field of view only, leaving the upper half undisturbed, and its function is to reflect light coming from a lateral slit through the dispersing prism so as to throw a third spectrum upon the eye. The lateral slit is carried in a small draw-tube projecting about 2″ from the side of the box, and the light proceeding from it is rendered nearly parallel before reflection by a lens of short focus. No adjustment is provided for the position or width of the lateral slit; all that is necessary in this respect being attainable by rotating the mirror and by varying the brightness of the light behind. As sources of light I have found Argand gas flames, surrounded by opal globes, to be suitable. The gas tap supplying the lateral flame is within reach of the observer, who has thus the means of adjusting the match both with respect to colour and with respect to brightness, without losing sight of the subjects of comparison. The zero of the divided circle corresponds approximately to the complete exclusion of green, but readings were always taken on both sides of it so as to make the results independent of this adjustment. The circle is divided into 100 parts, green being excluded at 0 and 50, and red at 25 and 75. Tenths of a division could be estimated pretty correctly, an accuracy of reading fully sufficient for the purpose, as the observations of even practised observers would vary two or three tenths.

It is evident that the numbers obtained are dependent upon the quality of the light by which the principal slit is illuminated. In order to avoid errors in the comparison of different persons' vision arising from this source, it is advisable always to take simultaneous observations from some practised individual whose vision may be treated as a standard; but no evidence appeared of any variation in the quality of the gaslight. The special application of such instruments to the comparison of the qualities of various kinds of mixed light was alluded to at the end of my paper "On the Light from the Sky," &c. (*Phil. Mag.* April, 1871 [Art. VIII. p. 103]).

I have obtained matches between simple and compound yellow from twenty-three male observers, principally students in the laboratory. Of these sixteen agree with myself within the limits of the errors of observation. The remaining seven include my three brothers-in-law, and two others, Mr J. J. Thomson and Mr Threlfall, whose vision in this respect agrees very nearly with theirs. The vision of the other two observers differs from mine in the opposite direction. In one case the difference, though apparently real, is small, but in the other (Mr Hart), though there was some difficulty in getting a good observation, the difference is most decided. Among seven female observers whom I have tried, there is not one whose vision differs sensibly from my own.

Although the number examined is insufficient for statistical purposes, it is evident that the peculiarity is by no means rare, at least among men. As far as my experience has gone, it would seem too as if normal vision were not of the nature of an average, from which small deviations are more probable than larger ones; but this requires confirmation. In order to give a more precise idea of the amount of the difference in question, I have calculated from the laws of double refraction the relative quantities of red and green light required by Mr F. M. Balfour and myself to match the same yellow light. If we call R and G the maximum brightnesses of the red and green light (as they would reach the eye if the Nicol were removed), and r, g the actual brightnesses (as modified by the analyser) necessary for the match, then for Mr Balfour—

$$r/g = 1{\cdot}50 \, (R/G),$$

while for myself—

$$r/g = 3{\cdot}13 \, (R/G).$$

In other words, Mr Balfour requires only half as much red as myself, in order to turn a given amount of green into yellow. The corresponding numbers for the other four observers of this class would be substantially the same. On the other hand, Mr Hart requires much *more* red than I do in order to convert a given green into yellow—in the ratio of about $2{\cdot}6 : 1$.

Except in the case of Mr Hart, the colour vision of these observers is defective only in the sense that it differs from that of the majority. Their appreciation of small colour differences is as distinct as usual. In order to test this Mr G. W. Balfour made a complete series of colour matches with revolving disks in the manner described by Maxwell and in my former paper. Six matches, of which only two are really independent, were observed, the consistency of the set being a measure of the accuracy of observation. The average error proved to be only double of that which I have found in my own observations, and rather less than that usually met with in the case of observers whose vision is normal.

In connection with what has been described above with respect to trichromic vision, it is interesting to notice that corresponding and perhaps larger differences are to be found in the vision of the so-called colour-blind. The double-refraction apparatus may conveniently be used in this investigation. With the pointer adjusted to 0 or 25, we have in the upper half of the field pure red or pure green respectively, and in the lower half pure yellow as usual. By suitable adjustment of the gas taps two observers of this class, Mr T—— and Mr B——, are able to obtain perfect matches both between red and yellow, and between green and yellow, but the proportions necessary are very different for the two observers. In Mr T——'s red and yellow match, the red is to normal vision dazzlingly bright, and the yellow almost too dark to be recognised; while the green and yellow match, however extravagant as to hue, appears reasonable in respect of brightness. On the other hand, to Mr B——'s eyes, the red of the spectrum does not look nearly so dark, and the equivalent red and yellow appear to the normal eye to be much more nearly upon a level. Although these great differences exist, there is no doubt that the vision of both observers is strictly dichromic, and that, apart from brightness, all the rays of the spectrum, from red to green, have the same effect upon their eyes.

If we wish to go beyond the fact that this vision is dichromic, and inquire whether the case is one of red blindness or of green blindness, we must be careful to consider whether the question itself has a definite meaning. If trichromic vision were always the same, and if a particular case of colour-blind vision differed from it merely by the absence of the red sensation, that vision would intelligibly be characterised as red-blind. There is reason to believe that such cases exist. In all probability the suppression of my own red sensation would lead me to make matches very nearly the same as Mr T——'s; and in this sense he may fairly be called red-blind. But under the same circumstances my matches would be altogether rejected by Mr B——; and the question may be asked, whether his case, being certainly not one of simple red-blindness, can be brought under the head of green-blindness. To this the sufficient answer is, that if I became green-blind my matches would differ from those of Mr B—— far more than if I was red-blind. The test of green-blindness would be the possibility of matches between colours which to normal eyes appear green and purple or green and grey. Although a good deal has been said lately on this subject, I am not aware of a case in which accurate matches of this kind have been obtained from observers whose colour vision is in other respects acute. If such cases exist, inquiry should be instituted, in order to see how far the matches would correspond to green-blindness of an otherwise normal eye.

We see, then, that there is dichromic vision which cannot accurately be described as affected with red-blindness, and still less as affected with green-blindness. The difference from normal vision, being not simply one of defect, cannot be defined by any single phrase. To obtain a complete knowledge of it quantitative observations over the whole spectrum, such as those carried out by Maxwell, are necessary. It is fortunate that these observations are easier to arrange for dichromic than for trichromic vision.

That I might be able to form an opinion upon the general acuteness of his colour vision, Mr T—— was good enough to observe a series of five colour matches between red, white, blue, green, and yellow, one being left out each time. The results are given in the accompanying table; those

		Red	White	Blue	Green	Yellow	Dec. 2, 1880
1	{	76·2	23·8	− 23·3	− 76·7	0	Observed
	{	77·4	22·6	− 21	− 79	0	Calculated
2	{	56·6	43·4	− 52·3	0	− 47·7	Observed
	{	56·2	43·8	− 52·5	0	− 47·5	Calculated
3	{	68·2	5·5	0	− 100	26·3	Observed
	{	69·7	6·5	0	− 100	23·8	Calculated
4	{	60·3	0	8	− 100	31·7	Observed
	{	61·2	0	7·8	− 100	31	Calculated
5	{	0	32·5	− 43·5	67·5	− 56·5	Observed
	{	0	32·3	− 44·1	67·7	− 55·9	Calculated
A		522	424	− 511	35	− 470	—
B		641	405	− 470	− 199	− 377	—

marked "calculated" being a consistent set derived by elimination from the two marked A and B. The good general agreement of the two sets of numbers is a proof that within its restricted range Mr T——'s sense of colour is acute. The first observation in which a mixture of red and white is matched by a mixture of green and blue is the most characteristic.

In conclusion I will describe an apparatus by which it is possible to observe these colour-matches without rotating the disks. At the time of my first experiments, about ten years since, I was struck with the advantage which might ensue if it were possible to have the mixed colours in view during the time of actual adjustment, and I thought of a plan by which this object might be attained. The idea, which I carried out soon afterwards, was to spin an *image* of the disks instead of the disks themselves. An inverting prism was mounted in a tube which could be made to rotate. The axis of rotation is adjusted so as to point accurately to the centres of the disks mounted as usual. An eye applied to the prism sees the disks undisplaced as a whole, but inverted by reflection. As the tube rotates, the image of the disks rotates also, and with double angular velocity. When the speed is sufficient, the colours lying on any circle concentric with the disks are blended exactly as if the disks themselves revolved.

This apparatus is quite successful; but its real advantages of working at a smaller velocity, and of allowing adjustment while the rotation continues, are counterbalanced in practice by the inconvenience of having to look through a tube, and the uncertainty introduced by the possible disturbance of the match due to unequal illumination of the area occupied by the disks.

78.

ON THE INFINITESIMAL BENDING OF SURFACES OF REVOLUTION.

[Proceedings of the London Mathematical Society, XIII. pp. 4—16, 1881.]

WHEN a thin sheet of matter is subjected to stress, the force which it opposes to extension is great in comparison with that which it opposes to bending. Under ordinary circumstances, the deformation takes place approximately as if the sheet were inextensible as a whole, a condition which in a remarkable degree facilitates calculation, though (it need scarcely be said) even bending implies an extension of all but the central layers. The inextensibility postulated refers properly to the central layer, which is spoken of as a surface.

We will commence with the case of a surface naturally spherical, and investigate what kind of deformation it admits of, under the condition that no line traced upon the surface is altered in length. The radius of the sphere being a, let the point whose natural coordinates are a, θ, ϕ be displaced to the position $a + \delta r$, $\theta + \delta\theta$, $\phi + \delta\phi$, where δr, $\delta\theta$, $\delta\phi$ are to be treated as small. Since the element of arc ds is of the same length after displacement as before, we have

$$(ds)^2 = a^2 (d\theta)^2 + a^2 \sin^2 \theta \, (d\phi)^2$$
$$= (a + \delta r)^2 (d\theta + d\delta\theta)^2 + (d\delta r)^2 + (a + \delta r)^2 \sin^2 (\theta + \delta\theta)(d\phi + d\delta\phi)^2,$$

or, retaining only the first power of δr, $\delta\theta$, $\delta\phi$,

$$a\, d\theta\, d\delta\theta + \delta r\, (d\theta)^2 + \delta r \sin^2 \theta\, (d\phi)^2 + a \sin\theta \cos\theta\, \delta\theta\, (d\phi)^2 + a \sin^2\theta\, d\phi\, d\delta\phi = 0.$$

Now

$$d\delta\theta = \frac{d\delta\theta}{d\theta}\, d\theta + \frac{d\delta\theta}{d\phi}\, d\phi, \qquad d\delta\phi = \frac{d\delta\phi}{d\theta}\, d\theta + \frac{d\delta\phi}{d\phi}\, d\phi, \quad \ldots\ldots(1)$$

so that

$$(d\theta)^2 \left\{ \frac{d\delta\theta}{d\theta} + \frac{\delta r}{a} \right\} + d\theta\, d\phi \left\{ \frac{d\delta\theta}{d\phi} + \sin^2\theta \frac{d\delta\phi}{d\theta} \right\}$$

$$+ (d\phi)^2 \left\{ \frac{\delta r}{a} \sin^2\theta + \sin\theta\cos\theta\, \delta\theta + \sin^2\theta \frac{d\delta\phi}{d\phi} \right\} = 0. \quad\ldots\ldots(2)$$

In this equation $d\theta$ and $d\phi$ are arbitrary, so that the coefficients of $(d\theta)^2$, $d\theta\, d\phi$, and $(d\phi)^2$ must vanish separately. We thus obtain as the conditions of no extension,

$$\frac{d\delta\theta}{d\theta} + \frac{\delta r}{a} = 0, \qquad \frac{d\delta\theta}{d\phi} + \sin^2\theta \frac{d\delta\phi}{d\theta} = 0, \qquad \frac{\delta r}{a} + \cot\theta\, \delta\theta + \frac{d\delta\phi}{d\phi} = 0 \ldots(3, 4, 5)$$

From (3), (4), (5), by elimination of δr,

$$\frac{d}{d\phi} \left(\frac{\delta\theta}{\sin\theta} \right) + \sin\theta \frac{d\delta\phi}{d\theta} = 0, \qquad \frac{d\delta\phi}{d\phi} - \sin\theta \frac{d}{d\theta} \left(\frac{\delta\theta}{\sin\theta} \right) = 0 ; \quad \ldots(6, 7)$$

or, since

$$\sin\theta \frac{d}{d\theta} = \frac{d}{d\log\tan(\frac{1}{2}\theta)},$$

$$\frac{d}{d\phi} \left(\frac{\delta\theta}{\sin\theta} \right) + \frac{d\delta\phi}{d\log\tan(\frac{1}{2}\theta)} = 0, \qquad \frac{d\delta\phi}{d\phi} - \frac{d}{d\log\tan\frac{1}{2}\theta} \left(\frac{\delta\theta}{\sin\theta} \right) = 0. \quad (8, 9)$$

From (8) and (9) we see that both $\delta\phi$ and $(\sin\theta)^{-1}\delta\theta$ satisfy an equation of the second order of the same form, viz.,

$$\frac{d^2 u}{d(\log\tan\frac{1}{2}\theta)^2} + \frac{d^2 u}{d\phi^2} = 0. \quad\ldots\ldots\ldots\ldots\ldots\ldots(10)$$

From the nature of the case, u is a periodic function of ϕ, and can be expanded by Fourier's theorem in a series of sines and cosines of ϕ and its multiples. Moreover, each term of the series must satisfy the equations independently. Thus, if u varies as $\cos s\phi$ or as $\sin s\phi$, (10) becomes

$$\frac{d^2 u}{d(\log\tan\frac{1}{2}\theta)^2} - s^2 u = 0 ;$$

whence

$$u = A' \tan^s \tfrac{1}{2}\theta + B' \cot^s \tfrac{1}{2}\theta, \quad\ldots\ldots\ldots\ldots\ldots(11)$$

where A' and B' are independent of θ.

If we take

$$\delta\phi = (\cos s\phi, \sin s\phi) [A \tan^s \tfrac{1}{2}\theta + B \cot^s \tfrac{1}{2}\theta], \quad\ldots\ldots\ldots\ldots(12)$$

we get for the corresponding value of $\delta\theta$, from (8),

$$\delta\theta/\sin\theta = (-\sin s\phi, \cos s\phi) [A \tan^s \tfrac{1}{2}\theta - B \cot^s \tfrac{1}{2}\theta] ; \quad\ldots\ldots(13)$$

and thence, from (5),

$$\delta r/a = (\sin s\phi, -\cos s\phi)\left[A\,(s+\cos\theta)\tan^s \tfrac{1}{2}\theta + B\,(s-\cos\theta)\cot^s\tfrac{1}{2}\theta\right], \quad\dots(14)$$

in which the constants which accompany $\sin s\phi$ are independent of those which accompany $\cos s\phi$.

If we suppose $s = 1$, we get

$$\sin\theta\,\delta\phi = (\cos\phi, \sin\phi)\left[A + B - (A - B)\cos\theta\right],$$
$$\delta\theta = (-\sin\phi, \cos\phi)\left[A - B - (A + B)\cos\theta\right],$$
$$\delta r/a = (\sin\phi, -\cos\phi)\left[(A + B)\sin\theta\right].$$

The two displacements proportional to $A - B$ are rotations of the whole surface as a rigid body round the axes $\theta = \tfrac{1}{2}\pi$, $\phi = 0$, and $\theta = \tfrac{1}{2}\pi$, $\phi = \tfrac{1}{2}\pi$. Those proportional to $A + B$ are displacements parallel to the same axes.

The two other motions possible without bending are a rotation round the axis $\theta = 0$, represented by $\delta\theta = 0$, $\delta\phi = \text{const.}$, $\delta r = 0$, and a displacement parallel to the same axis represented by

$$\delta\phi = 0, \qquad \frac{d}{d\theta}\left(\frac{\delta\theta}{\sin\theta}\right) = 0, \qquad \frac{dr}{a} = -\cot\theta\,\delta\theta,$$

or

$$\delta\phi = 0, \qquad \delta\theta = \gamma\sin\theta, \qquad \delta r = -\gamma a\cos\theta.$$

These correspond to a zero value of s, and are readily obtained from the original equations (3), (4), (5).

If the sphere be complete, the displacements just considered, and corresponding to $s = 0$ and $s = 1$, are the only ones possible. For higher values of s, we see, from (14), that δr is infinite at one or other pole, unless A and B both vanish. In other words, the complete sphere is perfectly rigid, so far as concerns pure bending.

If neither pole be included in the actual surface, which for example we may suppose bounded by circles of latitude, finite values of both A and B are admissible, and therefore necessary for a complete solution of the problem. But if, as would more often happen, one of the poles, say $\theta = 0$, is included, the constants B must be considered to vanish. Under these circumstances, the solution is

$$\left.\begin{aligned}
\delta\phi &= A\tan^s\tfrac{1}{2}\theta\cos s\phi \\
\delta\theta &= -A\sin\theta\tan^s\tfrac{1}{2}\theta\sin s\phi \\
\delta r &= Aa\,(s+\cos\theta)\tan^s\tfrac{1}{2}\theta\sin s\phi
\end{aligned}\right\}, \quad\dots\dots\dots\dots(15)$$

to which is to be added that obtained by writing $s\phi - \tfrac{1}{2}\pi$ for $s\phi$, and changing the arbitrary constant.

From (15) we see that, along those meridians for which $\sin s\phi = 0$, the displacement is tangential and in longitude only, while along the intermediate meridians for which $\cos s\phi = 0$, there is no displacement in longitude, but one in latitude, and one normal to the surface of the sphere.

Along the equator $\theta = \frac{1}{2}\pi$, and then $\delta\phi = A \cos s\phi$, $\delta\theta = -A \sin s\phi$, $\delta r = Aas \sin s\phi$, so that the maximum displacements in latitude and longitude are equal.

The results embodied in (15) are applicable, even although neither the distribution of thickness nor the form of the boundary are symmetrical with respect to $\theta = 0$, but it is only under this condition that the displacements are *normal* in the mechanical sense. In what follows it will therefore be supposed that the boundary is a circle of latitude, and that the thickness, if variable at all, is a function of latitude only.

In order to find the potential energy of bending, the first step is the purely geometrical one of investigating the curvature at every point of the deformed surface. The principal curvatures differ from the original curvature of the sphere in opposite directions and to an equal amount, and the potential energy of bending corresponding to any element of the surface is proportional to the square of this excess or defect of curvature, without regard to the direction of the principal planes. The calculation of the curvature is somewhat tedious, and it may suffice here merely to give the result. I find that the curvature corresponding to (15) at any point θ, ϕ, and in a plane making with the meridian through the point an angle γ is expressed by

$$\frac{1}{\rho} = \frac{1}{a} - \frac{A}{a} \frac{(s^3 - s) \tan^s \frac{1}{2}\theta}{\sin^2 \theta} \sin(s\phi + 2\gamma). \quad \ldots\ldots\ldots\ldots(16)$$

The maximum value of $\delta(\rho^{-1})$ is therefore the same at every point along a circle of latitude, and is given by

$$\delta\left(\frac{1}{\rho}\right) = \frac{A(s^3 - s) \tan^s \frac{1}{2}\theta}{a \sin^2 \theta}, \quad \ldots\ldots\ldots\ldots\ldots(17)$$

vanishing, as was to be expected, in the cases $s = 0$, $s = 1$. With respect to the directions of the principal planes, we see, from (16), that along the meridians where $\delta\phi$ vanishes $(\cos s\phi = 0)$, the principal planes are the meridian and its perpendicular, while along the meridians where δr vanishes, the principal planes are inclined to the meridian at angles of 45°, as is indeed otherwise obvious.

The potential energy corresponding to the element of surface $a^2 \sin\theta\, d\theta\, d\phi$ may be denoted by $a^2 H \{\delta(\rho^{-1})\}^2 \sin\theta\, d\theta\, d\phi$, where H depends upon the material and the thickness. The whole potential energy V is to be found by integration with respect to ϕ between the limits 0 and 2π, and for θ between the limits 0 and θ.

We will now prove that the expression for V corresponding to any number of displacements of type (15) involves the squares only, and not the products, of the amplitudes A. In order to have full generality for each value of s, we write, in (15 and 16), $s\phi + \epsilon$ instead of $s\phi$.

We have

$$\frac{1}{\rho} - \frac{1}{a} = - \cos 2\gamma \, \Sigma \, \frac{(s^3 - s) \tan^s \frac{1}{2}\theta}{a \sin^2 \theta} A \sin (s\phi + \epsilon)$$

$$- \sin 2\gamma \, \Sigma \, \frac{(s^3 - s) \tan^s \frac{1}{2}\theta}{a \sin^2 \theta} A \cos (s\phi + \epsilon),$$

so that the excess of curvature in the principal plane is given by

$$\left(\delta\frac{1}{\rho}\right)^2 = \left[\Sigma \, \frac{(s^3 - s) \tan^s \frac{1}{2}\theta}{a \sin^2 \theta} A \sin (s\phi + \epsilon)\right]^2 + \left[\Sigma \, \frac{(s^3 - s) \tan^s \frac{1}{2}\theta}{a \sin^2 \theta} A \cos (s\phi + \epsilon)\right]^2.$$

It is now evident that, in the integration with respect to ϕ, the products $A_s A_{s'}$ will disappear in virtue of

$$\int_0^{2\pi} \sin (s\phi + \epsilon) \sin (s'\phi + \epsilon') \, d\phi = 0, \qquad \int_0^{2\pi} \cos (s\phi + \epsilon) \cos (s'\phi + \epsilon') \, d\phi = 0,$$

and the complete expression for V is simply

$$V = 2\pi \, \Sigma \, (s^3 - s)^2 \, A_s^2 \int_0^\theta H \sin^{-3} \theta \tan^{2s} \frac{1}{2}\theta \, d\theta. \quad \ldots\ldots\ldots(18)$$

We will now suppose that H is constant, in order to proceed with the integration. Writing for brevity t for $\tan \frac{1}{2}\theta$, we have

$$dt = \frac{d\theta}{2 \cos^2 \frac{1}{2}\theta}, \quad \text{so that} \quad d\theta = \frac{2\,dt}{1 + t^2}, \quad \text{and} \quad \sin \theta = \frac{2t}{1 + t^2}.$$

Thus

$$\int_0^\theta \sin^{-3} \theta \tan^{2s} \frac{1}{2}\theta \, d\theta = \frac{1}{8} \int (1 + t^2)^2 t^{2s-4} dt^2 = \frac{1}{8}\left(\frac{t^{2s-2}}{s-1} + \frac{2t^{2s}}{s} + \frac{t^{2s+2}}{s+1}\right). \quad \ldots(19)$$

In the case of a hemisphere $t = 1$, and (19) assumes the form

$$\frac{2s^2 - 1}{4 (s^3 - s)}. \quad \ldots\ldots\ldots\ldots\ldots\ldots\ldots(20)$$

Hence, for a hemisphere,

$$V = \tfrac{1}{2}\pi H \Sigma \, (s^3 - s)(2s^2 - 1) \, A_s^2. \quad \ldots\ldots\ldots\ldots(21)$$

If the extreme value of θ be $60°$ instead of $90°$, we get instead of (20)

$$\frac{8s^2 + 4s - 3}{4 \cdot 3^{s+1} (s^3 - s)}, \quad \ldots\ldots\ldots\ldots\ldots\ldots\ldots(22)$$

and

$$V = \tfrac{1}{2}\pi H \Sigma \, 3^{-(s+1)} \, (s^3 - s)(8s^2 + 4s - 3) \, A_s^2. \quad \ldots\ldots\ldots(23)$$

If we wish to take the terms in (15) involving $\cos s\phi$, $\sin s\phi$ separately, we have only to write, in (18), (21), (23), $A_s^2 + A_s'^2$ in place of A_s^2.

These expressions for V, in conjunction with (15), are sufficient for the solution of statical problems relative to the deformation of spherical shells under the action of given impressed forces. Suppose, for example, that a string of tension F connects the opposite points on the edge of a hemisphere represented by $\theta = \frac{1}{2}\pi$, $\phi = \frac{1}{2}\pi$, $\phi = \frac{3}{2}\pi$, and that it is required to find the deformation. From (15) we see that the work done by the impressed forces corresponding to the deformation δA_s is

$$- \delta A_s \, as \, (\sin \tfrac{1}{2} s\pi) + \sin (\tfrac{1}{2} s\pi + s\pi) \, F.$$

This vanishes if s be odd, and if s be even is equal to

$$- 2\delta A_s \, as \sin \tfrac{1}{2} s\pi . F.$$

Hence A_s vanishes if s be odd, and, by (21), if s be even,

$$dV/dA_s = \pi H (s^3 - s)(2s^2 - 1) A_s = - 2as \sin \tfrac{1}{2} s\pi . F,$$

whence

$$A_s = - \frac{2a F \sin \tfrac{1}{2} s\pi}{\pi H (s^2 - 1)(2s^2 - 1)}. \quad \dots\dots\dots\dots\dots(24)$$

Attaching numerical values to s, we find

$$A_1 = 0, \qquad\qquad A_3 = 0, \qquad\qquad A_5 = 0, \&c.$$

$$A_2 = \frac{2a F}{\pi H . 3 . 7}, \qquad A_4 = - \frac{2a F}{\pi H . 15 . 31}, \qquad A_6 = \frac{2a F}{\pi H . 35 . 71}, \&c.$$

By (24) and (15), the deformation is completely determined.

If, to take a case in which the force is tangential, we suppose that the hemisphere rests upon its pole with its edge horizontal, and that a rod of weight W is laid symmetrically along the diameter $\theta = \frac{1}{2}\pi$, we find in like manner

$$A_s = \frac{a W \sin \tfrac{1}{2} s\pi}{\pi H (s^3 - s)(2s^2 - 1)}, \quad \dots\dots\dots\dots\dots(25)$$

for even values of s, and zero for odd values of s. In this case the series is even more convergent than before.

Before we can find the frequencies of free vibration, we must investigate an expression for the kinetic energy. It may readily be proved that only the squares of A_s and A_s' are involved, so that these quantities are really the normal coordinates of the system. We have, from (15),

$$\left(\frac{d\delta r}{dt}\right)^2 + \left(\frac{ad\delta\theta}{dt}\right)^2 + \left(\frac{a \sin \theta \, d\delta\phi}{dt}\right)^2 = a^2 \left(\frac{dA_s}{dt}\right)^2 \tan^{2s} \tfrac{1}{2}\theta \, \{(s + \cos \theta)^2 \sin^2 s\phi + \sin^2 \theta\}.$$

When integrated with respect to ϕ, this gives

$$\pi a^2 \left(\frac{dA_s}{dt}\right)^2 \tan^{2s} \tfrac{1}{2}\theta \{(s + \cos\theta)^2 + 2\sin^2\theta\}.$$

We have now to multiply by $\sin\theta\, d\theta$, and integrate. Let

$$\int_0^\theta \tan^{2s} \tfrac{1}{2}\theta \{(s + \cos\theta)^2 + 2\sin^2\theta\} \sin\theta\, d\theta = f(s);$$

then, x being written for $1 + \cos\theta$,

$$f(s) = \int_x^2 \left(\frac{2-x}{x}\right) \{(s-1)^2 + 2x(s+1) - x^2\}\, dx. \quad \ldots\ldots\ldots(26)$$

For a hemisphere the lower limit of the integral is unity. If σ denote the surface density, the expression for the whole kinetic energy T is

$$T = \tfrac{1}{2}\pi a^4 \sigma \Sigma f(s) \left\{ \left(\frac{dA_s}{dt}\right)^2 + \left(\frac{dA_s'}{dt}\right)^2 \right\}. \quad \ldots\ldots\ldots\ldots(27)$$

From (27) and (21) the frequencies of free vibration for a hemisphere are immediately obtainable. The equation for A_s, or A_s', is

$$\sigma a^4 f(s) \frac{d^2 A_s}{dt^2} + H(s^3 - s)(2s^2 - 1) A_s = 0;$$

so that, if A_s varies as $\cos(p_s t + \epsilon)$,

$$p_s^2 = \frac{H(s^3 - s)(2s^2 - 1)}{\sigma a^4 f(s)}. \quad \ldots\ldots\ldots\ldots\ldots(28)$$

In like manner for the saucer of 120°, from (22)

$$p_s^2 = \frac{H(s^3 - s)(8s^2 + 4s - 3)}{\sigma a^4 f(s).3^{s+1}}. \quad \ldots\ldots\ldots\ldots(29)$$

The values of $f(s)$ can be calculated without difficulty in the various cases. Thus, for the hemisphere,

$$f(2) = \int_1^2 x^{-2}(4 - 4x + x^2)(1 + 6x - x^2)\, dx$$

$$= 20 \log 2 - 12\tfrac{1}{3} = 1\cdot52961,$$

$$f(3) = 57\tfrac{1}{3} - 80 \log 2 = 1\cdot88156,$$

$$f(4) = 200 \log 2 - 136\tfrac{1}{3} = 2\cdot29609, \ \&c.,$$

so that

$$p_2 = \frac{\sqrt{H}}{a^2\sqrt{\sigma}} \times 5\cdot2400, \qquad p_3 = \frac{\sqrt{H}}{a^2\sqrt{\sigma}} \times 14\cdot726, \qquad p_4 = \frac{\sqrt{H}}{a^2\sqrt{\sigma}} \times 28\cdot462.$$

In experiment, it is the *intervals* between the various tones with which we are most concerned. We find

$$p_3 : p_2 = 2\cdot8102, \qquad p_4 : p_2 = 5\cdot\overline{4}316.$$

In the case of glass bells, such as are used with air-pumps, the interval between the two gravest tones is usually somewhat smaller; the representative fraction being nearer to $2\cdot5$ than $2\cdot8$.

For the saucer of 120°, the lower limit of the integral in (26) is $\frac{3}{2}$, and we get on calculation

$$f(2) = \cdot12864, \qquad f(3) = \cdot054884,$$

giving

$$p_2 = \frac{\sqrt{H}}{a^2\sqrt{\sigma}} \times 7\cdot9947, \qquad p_3 = \frac{\sqrt{H}}{a^2\sqrt{\sigma}} \times 20\cdot911,$$

$$p_3 : p_2 = 2\cdot6157.$$

The pitch of the two gravest tones is thus decidedly higher than for the hemisphere, and the interval between them is less.

With reference to the theory of tuning bells, it may be worth while to consider the effect of a small change in the angle, for the case of a nearly hemispherical bell. In general

$$p_s^2 = \frac{4H(s^3 - s)^2 \int_0^\theta \sin^{-3}\theta \tan^{2s}\tfrac{1}{2}\theta\, d\theta}{a^4\sigma \int_0^\theta \tan^{2s}\tfrac{1}{2}\theta\,\{(s + \cos\theta)^2 + 2\sin^2\theta\}\sin\theta\, d\theta} \qquad \ldots\ldots\ldots(30)$$

If $\theta = \tfrac{1}{2}\pi + \delta\theta$, and P_s denote the value of p_s for the exact hemisphere, we get from previous results

$$p_s^2 = P_s^2\left[1 + \delta\theta\left\{\frac{4(s^3 - s)}{2s^2 - 1} - \frac{s^2 + 2}{f(s)}\right\}\right]. \qquad \ldots\ldots\ldots\ldots(31)$$

Thus

$$p_2^2 = P_2^2\left[1 + \delta\theta\left\{\frac{24}{7} - \frac{6}{1\cdot52961}\right\}\right] = P_2^2\,(1 - \cdot49\,\delta\theta)$$

$$p_3^2 = P_3^2\left[1 + \delta\theta\left\{\frac{96}{17} - \frac{11}{1\cdot88156}\right\}\right] = P_3^2\,(1 - \cdot20\,\delta\theta),$$

shewing that an increase in the angle depresses the pitch. As to the interval between the two gravest tones, we get

$$\left(\frac{p_3}{p_2}\right)^2 = \left(\frac{P_3}{P_2}\right)^2 \times (1 + \cdot29\,\delta\theta),$$

shewing that it increases with θ. This agrees with the results given above for $\theta = 60°$.

The fact that the form of the normal functions is independent of the distribution of density and thickness, provided that they vary only with latitude, allows us to calculate a great variety of cases, the difficulties being merely those of simple integration. If we suppose that only a narrow belt in co-latitude θ has sufficient thickness to contribute sensibly to the potential and kinetic energies, we have simply, instead of (30),

$$p_s^2 = \frac{4H\,(s^3 - s)^2\,\sin^{-4}\theta}{a^4\sigma\,\{(s + \cos\theta)^2 + 2\sin^2\theta\}}, \quad \dots\dots\dots\dots\dots(32)$$

whence

$$\frac{p_3}{p_2} = 4\sqrt{\left\{\frac{6 + 4\cos\theta - \cos^2\theta}{11 + 6\cos\theta - \cos^2\theta}\right\}}. \quad \dots\dots\dots\dots(33)$$

The ratio varies very slowly from 3, when $\theta = 0$, to $2\cdot954$, when $\theta = \frac{1}{2}\pi$.

If b denote the thickness at any co-latitude θ, $H \propto b^3$, $\sigma \propto b$. I have calculated the ratio of frequencies of the two gravest tones of a hemisphere on the suppositions (1) that $b \propto \cos\theta$, and (2) that $b \propto (1 + \cos\theta)$. The formula is that marked (30) with H and σ under the integral signs. In the first case, $p_3 : p_2 = 1\cdot7942$, differing greatly from the value for a uniform thickness. On the second more moderate supposition as to the law of thickness, $p_3 : p_2 = 2\cdot4591$, $p_4 : p_2 = 4\cdot4837$. It would appear that the smallness of the interval between the gravest tones of common glass bells is due in great measure to the thickness diminishing with increasing θ.

It is worthy of notice that the curvature $\delta\,(\rho^{-1})$, which by (17) varies as $\sin^{-2}\theta\,\tan^s\frac{1}{2}\theta$, vanishes at the pole for $s = 3$ and higher values, but is finite for $s = 2$.

We will now investigate the deformation of the general surface of revolution under the condition that no line traced upon it suffers extension, and for this purpose we will employ columnar coordinates z, r, ϕ.

We have

$$(ds + d\delta s)^2 = (dz + d\delta z)^2 + (r + \delta r)^2\,(d\phi + d\delta\phi)^2 + (dr + d\delta r)^2,$$

whence, if $d\delta s$ vanish,

$$dz\,d\delta z + r^2\,d\phi\,d\delta\phi + r\,\delta r\,(d\phi)^2 + dr\,d\delta r = 0.$$

Now

$$d\delta z = \frac{d\delta z}{dz}\,dz + \frac{d\delta z}{d\phi}\,d\phi, \qquad d\delta r = \frac{d\delta r}{dz}\,dz + \frac{d\delta r}{d\phi}\,d\phi,$$

and

$$dr = \frac{dr}{dz}\,dz + \frac{dr}{d\phi}\,d\phi,$$

which, by hypothesis, $dr/d\phi = 0$.

Thus

$$(dz)^2 \left\{ \frac{d\delta z}{dz} + \frac{dr}{dz}\frac{d\delta r}{dz} \right\} + (d\phi)^2 \left\{ r^2 \frac{d\delta \phi}{d\phi} + r\delta r \right\} + dz\,d\phi \left\{ \frac{d\delta z}{d\phi} + r^2 \frac{d\delta \phi}{dz} + \frac{dr}{dz}\frac{d\delta r}{d\phi} \right\} = 0.$$

In this equation the coefficients of $(dz)^2$, $dz\,d\phi$, $(d\phi)^2$ must vanish separately, if the surface is not extended, so that

$$\frac{d\delta z}{dz} + \frac{dr}{dz}\frac{d\delta r}{dz} = 0, \qquad r\frac{d\delta \phi}{d\phi} + \delta r = 0, \qquad \frac{d\delta z}{d\phi} + r^2 \frac{d\delta \phi}{dz} + \frac{dr}{dz}\frac{d\delta r}{d\phi} = 0. \dots(34, 35, 36)$$

From these, by elimination of δr,

$$\frac{d\delta z}{dz} - \frac{dr}{dz}\frac{d}{dz}\left(r\frac{d\delta \phi}{d\phi} \right) = 0, \qquad \frac{d\delta z}{d\phi} + r^2 \frac{d\delta \phi}{dz} - r\frac{dr}{dz}\frac{d^2 \delta \phi}{d\phi^2} = 0,$$

from which again, by elimination of δz,

$$\frac{d}{dz}\left(r^2 \frac{d\delta \phi}{dz} \right) - r\frac{d^2 r}{dz^2}\frac{d^2 \delta \phi}{d\phi^2} = 0. \dots\dots\dots(37)$$

As in the case of the sphere, if the distribution of thickness and the form of the boundary or boundaries be symmetrical with respect to the axis, the normal functions of the system are found by assuming $\delta\phi$ to be proportional to $\cos s\phi$ or $\sin s\phi$. Geometrically speaking, we are at liberty to make this resolution in any case. Thus the equation for $\delta\phi$ may be put into the form

$$r^2 \frac{d}{dz}\left(r^2 \frac{d\delta \phi}{dz} \right) + s^2 r^3 \frac{d^2 r}{dz^2}\,\delta\phi = 0. \dots\dots\dots(38)$$

The simplest application of these results is to the *cylinder* for which r is constant, equal say to a. Thus (34), (35), (36), (38) become simply

$$\frac{d\delta z}{dz} = 0, \qquad \delta r + a\frac{d\delta \phi}{d\phi} = 0, \qquad \frac{d\delta z}{d\phi} + a^2 \frac{d\delta \phi}{dz} = 0, \dots\dots(39)$$

$$\frac{d^2 \delta \phi}{dz^2} = 0. \dots\dots\dots(40)$$

By (40), if $\delta\phi \propto \cos s\phi$, we may take

$$a\delta\phi = (A_s a + B_s z)\cos s\phi, \dots\dots\dots\dots(41)$$

and then, by (39),

$$\delta r = s\,(A_s a + B_s z)\sin s\phi, \qquad \delta z = -s^{-1} B_s a \sin s\phi. \dots(42, 43)$$

Corresponding terms, with fresh arbitrary constants, obtained by writing $s\phi + \frac{1}{2}\pi$ for $s\phi$, may of course be added. If $B_s = 0$, the displacement is in two dimensions only. This kind of deformation is considered more at length in my book on the *Theory of Sound*, § 233.

If an inextensible disc be attached to the cylinder at $z = 0$, so as to form a kind of cup, the displacements δr and $\delta\phi$ must vanish for that value of z, exception being made of the case $s = 1$. Hence $A_s = 0$, and

$$a\delta\phi = B_s z \cos s\phi, \qquad \delta r = s B_s z \sin s\phi, \qquad \delta z = -s^{-1} B_s a \sin s\phi \ldots (44)$$

In the case of a *cone*, for which $r = \tan\gamma \cdot z$, the equations become

$$\left.\begin{array}{cc} \dfrac{d\delta z}{dz} + \tan\gamma \dfrac{d\delta r}{dz} = 0, & z\tan\gamma \dfrac{d\delta\phi}{d\phi} + \delta r = 0 \\[3mm] \dfrac{d\delta z}{d\phi} + z^2\tan^2\gamma \dfrac{d\delta\phi}{d\phi} + \tan\gamma \dfrac{d\delta r}{d\phi} = 0 \end{array}\right\} , \ldots\ldots(45)$$

$$\frac{d}{dz}\left(z^2 \frac{d\delta\phi}{dz}\right) = 0. \ldots\ldots\ldots\ldots\ldots(46)$$

If we take, as usual, $\delta\phi \propto \cos s\phi$, we get as the solution of (46)

$$\delta\phi = (A_s + B_s z^{-1})\cos s\phi, \ldots\ldots\ldots\ldots(47)$$

and corresponding thereto

$$\delta r = s\tan\gamma\,(A_s z + B_s)\sin s\phi, \ldots\ldots\ldots(48)$$

$$\delta z = \tan^2\gamma\,[s^{-1}B_s - s\,(A_s z + B_s)]\sin s\phi. \ldots\ldots(49)$$

If the cone be complete up to the vertex at $z = 0$, $B_s = 0$, so that

$$\delta\phi = A_s\cos s\phi, \qquad \delta r = sA_s r\sin s\phi, \qquad \delta z = -sA_s\tan\gamma r\sin s\phi. \ldots(50, 51, 52)$$

For the cone and the cylinder, the second term in the general equation (38) vanishes. We shall obtain a more extensive class of soluble cases by supposing that the surface is such that

$$r^3\,d^2r/dz^2 = \text{constant}, \ldots\ldots\ldots\ldots\ldots(53)$$

an equation which is satisfied by surfaces of the second degree in general. If

$$z^2/a^2 + r^2/b^2 = 1, \ldots\ldots\ldots\ldots\ldots(54)$$

we shall find

$$r^3\,d^2r/dz^2 = -b^4/a^2, \ldots\ldots\ldots\ldots(55)$$

and thus (38) takes the form

$$\frac{d^2\delta\phi}{d\alpha^2} - \frac{s^2 b^4}{a^2}\delta\phi = 0, \ldots\ldots\ldots\ldots(56)$$

if $\delta\phi \propto \cos s\phi$, and α is defined by

$$\alpha = \int r^{-2}\,dz, \ldots\ldots\ldots\ldots\ldots(57)$$

or in the present case

$$\alpha = \frac{a}{2b^2}\log\frac{a+z}{a-z}. \ldots\ldots\ldots\ldots(58)$$

36

The solution of (56) is

$$\delta\phi = \left[A \left(\frac{a+z}{a-z}\right)^{-\frac{1}{2}s} + B \left(\frac{a+z}{a-z}\right)^{+\frac{1}{2}s} \right] \cos s\phi. \quad \ldots\ldots\ldots(59)$$

The corresponding values of δr and δz are to be obtained from (35) and (36).

If the surface be complete through the vertex $z = a$, the term multiplied by B must disappear. Thus, omitting the constant multiplier, we may take

$$\delta\phi = \left(\frac{a-z}{a+z}\right)^{\frac{1}{2}s} \cos s\phi ; \quad \ldots\ldots\ldots\ldots\ldots\ldots(60)$$

whence, by (35), (36),

$$\delta r = n \frac{b}{a} \frac{(a-z)^{\frac{1}{2}s+\frac{1}{2}}}{(a+z)^{\frac{1}{2}s-\frac{1}{2}}} \sin s\phi, \qquad \delta z = (sz + a) \frac{b^2 (a-z)^{\frac{1}{2}s}}{a^2 (a+z)^{\frac{1}{2}s}} \sin s\phi. \quad \ldots(61, 62)$$

If we measure z' from the vertex, $z' = a - z$, and

$$\delta\phi = \left(\frac{z'}{r}\right)^s \cos s\phi, \quad \ldots\ldots\ldots\ldots\ldots\ldots\ldots\ldots\ldots\ldots\ldots(63)$$

$$\delta r = sr \left(\frac{z'}{r}\right)^s \sin s\phi, \quad \ldots\ldots\ldots\ldots\ldots\ldots\ldots\ldots\ldots\ldots\ldots(64)$$

$$\delta z = -\delta z' = \frac{b^2}{a^2} \left\{(s+1)a - sz'\right\} \left(\frac{z'}{r}\right)^s \sin s\phi. \quad \ldots\ldots\ldots\ldots(65)$$

For the parabola, a and b are infinite, while $b^2/a = 2a'$, and $r^2 = 4a'z'$. Thus

$$\delta\phi = r^s \cos s\phi, \qquad \delta r = sr^{s+1} \sin s\phi, \qquad \delta z = -2(s+1)a'r^s \sin s\phi. \quad \ldots(66)$$

END OF VOL. I.

Printed in the United States
By Bookmasters